RF and Microwave Transistor Oscillator Design

RF and Microwave Transistor Oscillator Design

Andrei Grebennikov
Infineon Technologies AG, Germany

John Wiley & Sons, Ltd

Telephone (+44) 1243 779777

Email (for orders and customer service enquiries): cs-books@wiley.co.uk
Visit our Home Page on www.wiley.com

Other Wiley Editorial Offices

John Wiley & Sons Inc., 111 River Street, Hoboken, NJ 07030, USA

Jossey-Bass, 989 Market Street, San Francisco, CA 94103-1741, USA

Wiley-VCH Verlag GmbH, Boschstr. 12, D-69469 Weinheim, Germany

John Wiley & Sons Australia Ltd, 42 McDougall Street, Milton, Queensland 4064, Australia

John Wiley & Sons (Asia) Pte Ltd, 2 Clementi Loop #02-01, Jin Xing Distripark, Singapore 129809

John Wiley & Sons Canada Ltd, 6045 Freemont Blvd, Mississauga, ONT, Canada L5R 4J3

Wiley also publishes its books in a variety of electronic formats. Some content that appears in print may not be available in electronic books.

Anniversary Logo Design: Richard J. Pacifico

British Library Cataloguing in Publication Data

A catalogue record for this book is available from the British Library

ISBN 978-0-470-02535-2 (HB)

Typeset in 10/12pt Times Roman by TechBooks, New Delhi, India.

This book is printed on acid-free paper responsibly manufactured from sustainable forestry in which at least two trees are planted for each one used for paper production.

Contents

About the Author

Dr Andrei Grebennikov, IEEE Senior Member, has obtained long-term academic and industrial experience. He worked with Moscow Technical University of Telecommunications and Informatics, Russia; Institute of Microelectronics, Singapore; M/A-COM, Ireland and Infineon Technologies, Germany, as an engineer, researcher, lecturer and educator.

Dr Grebennikov has lectured as a Guest Professor in University of Linz, Austria, and presented short courses as an Invited Speaker at the International Microwave Symposium, European Microwave Conference and Motorola Design Centre, Malaysia. He is also the author of more than 60 papers, 3 books and several US patents.

Preface

The main objective of this book is to present all relevant information necessary for RF and microwave transistor oscillator design including well-known and new theoretical approaches and practical circuit schematics and designs, as well as to suggest optimum design approaches, which combine effectively analytic calculations and computer-aided design. This book can be useful for lecturing to promote the analytical way of thinking and combine effectively theory and practice of RF and microwave engineering. As often happens, a new result is a long-forgotten old one. Therefore, not only new results based on new technologies or circuit schematics are given, but some old ideas, schematics or approaches are also introduced, that could be very useful in modern practice or could contribute to the development of new ideas or techniques.

As a result, this book is intended for and can be recommended to:

- *university-level professors and researchers*, as possible reference and well-founded material for creative research and teaching activity which will contribute to strong background for graduates and postgraduates students;
- *R&D staff*, to combine the theoretical analysis and practical aspects, including computer-aided design (CAD) and to provide a sufficient basis for new ideas in theory and practical circuit techniques;
- *practising RF designers and engineers*, as an anthology of many well-known and new practical transistor oscillator circuits with detailed descriptions of their operational principles and applications and clear practical demonstration of theoretical results.

Chapter 1 presents the most commonly used design techniques for analysing nonlinear circuits, in particular, transistor oscillators. There are several approaches to analyse and design nonlinear circuits, depending on their main specifications. That means an analysis both in the time domain to determine transient circuit behaviour and in the frequency domain to improve power and spectral performances when parasitic effects such as instability and spurious emission must be eliminated or minimized. Using the time-domain technique, it is relatively easy to describe a nonlinear circuit with differential equations, which can be solved analytically in explicit form for only some simple cases. Under the assumption of slowly varying amplitude and phase, it is possible to obtain the separate truncated first-order differential equations for the amplitude and phase of the oscillation process from the original second-order nonlinear differential equation. However, generally it is necessary to use numerical methods. The time-domain analysis is limited to its inability to operate with the circuit immittance (impedance or admittance) parameters as well as the fact that it can be practically applied only for circuits with lumped parameters or ideal transmission lines. The frequency-domain analysis is less ambiguous because a relatively complex circuit can often be reduced to one or more sets of immittances at each harmonic component. For example, using a quasilinear approach, the nonlinear circuit parameters averaged by the fundamental component allow one to apply a linear circuit analysis. Advanced modern CAD simulators incorporate both time-domain and frequency-domain methods as well as optimization techniques to provide all the necessary design cycles.

Chapter 2 introduces the principles of oscillator design, including start-up and steady-state operation conditions, basic oscillator configurations using lumped and transmission-line elements and simplified equation-based oscillator analysis and design techniques. An immittance design approach is introduced and applied to series and parallel feedback oscillators, including circuit design and simulation aspects. Numerous practical examples of RF and microwave oscillators using MOSFET, MESFET and bipolar devices, including the descriptions of their circuit realizations, are given.

Applying dc bias to the active device does not generally result in the negative resistance condition. This condition has to be induced in these devices and it is determined by the physical mechanism in the device and chosen circuit topology. The transistor in the oscillator circuits is mostly represented as the active two-port network, whose operation principle is reflected through its equivalent circuit. The influence of the circuit and transistor parameters can result in a hysteresis effect or oscillation instability in practical design. In high-frequency practical implementation, the presence of the parasitic device and circuit elements can contribute to the multi-resonant circuits. The possibility of an operation mode with different natural frequencies depends on the value of the coupling coefficient between resonant circuits. Therefore, the stability conditions for a steady-state single-frequency operation for a multi-resonant circuit, in general, and two coupled resonant circuits, in particular, are analytically derived. The several examples of stability criteria for different single-resonant and double-resonant oscillator circuits are described and analysed in Chapter 3. In addition, the phase plane method as a qualitative method of an analysis of the dynamics of the oscillation systems and a Nyquist stability criterion are shown and illustrated by several examples of the oscillator circuits described by second-order differential equations.

Generally, RF and microwave transistor oscillator design is a complex problem. Depending on the technical requirements, it is necessary to define the configuration of the oscillator circuit, choose a proper transistor type, evaluate and measure the parameters of the transistor nonlinear model under small- and large-signal conditions. Finally, an appropriate nonlinear simulator must be used to simulate the oscillator performance in time and frequency domains. An oscillator analysis can be based on the two-port network approach to describe the active device and feedback circuit. In this case, the basic parameters of the transistor equivalent circuit can be directly measured, or approximated on the basis of experimental data, with sufficient accuracy across a wide frequency range. However, the values of the external feedback circuit elements are initially unknown. The process of determining the optimum values of the feedback and load parameters can be time-consuming and, in a typical case, calls for much simulation. Consequently, it is convenient to use an analytic method of optimizing oscillator design. This method should incorporate the explicit expressions for feedback elements and load impedance in terms of the transistor equivalent circuit elements and its static volt–ampere and voltage–capacitance characteristics. Chapter 4 presents both the empirical and analytic optimum design approaches applied to series and parallel feedback oscillators, including circuit design and simulation aspects, and high-efficiency design techniques as well. Typical practical examples of RF and microwave oscillators using MOSFET, MESFET, HEMT, and bipolar devices, including the descriptions of their circuit configurations, are given.

Chapter 5 describes different oscillator noise models to express a clear relationship between the resonant circuit and active device noise model parameters. The simple Leeson linear model for a feedback oscillator, which was derived empirically, is based on the expectations that the contribution to the real oscillator output spectrum is provided by two basic processes. The first process is a result of the phase fluctuations due to the additive white noise at frequency offsets close to the carrier. The second process is a result of the low-frequency fluctuations or flicker noise up-converted to the carrier region because of the active device nonlinear effects. The nonlinear Kurokawa analysis based on the sinusoidal representation of the current in the negative-resistance oscillator extends the oscillator noise model by introducing relationships between the noise power, stability conditions and amplitude-to-phase conversion. However, such a noise generation mechanism does not consider the mixing effect from the inherent nonlinear behaviour of the active device when the current at the output of the active device must be represented by a Fourier series expansion. Thus, the phase noise generated around the fundamental frequency of the oscillation generally is an equal contribution of two simultaneous and correlated phenomena: additive phase noise due to phase modulation process and converted phase noise due to conversion from one sideband to another.

Voltage-controlled oscillators are key components in many applications, especially in wireless communication systems, measurement equipment, or military applications. A growing market of wireless applications requires highly integrated circuit solutions, where both high-performance transistors and passive elements with high quality factors can be used. Chapter 6 discusses the varactor modelling issues, varactor nonlinearity and its effect to frequency modulation, and resonant circuit techniques to improve VCO tuning linearity using lumped and transmission-line elements. Various practical examples of VCO implementation techniques based on using different types of active devices, circuit schematic approaches and hybrid or monolithic integrated circuit technologies are shown and described.

The rapid growth of new-generation wireless communication systems has created a strong demand for designing single-chip radio transceivers in a fully monolithic CMOS process with extremely small size due to better integration, low cost and low operating voltage. To increase the integration level, all passive components must be integrated monolithically into a single chip. In this case, the elements of a resonant *LC* circuit of the voltage-controlled oscillator as a core part of the synthesizers should feature high quality factors over frequency tuning range. Chapter 7 discusses the technological aspects to realize MOS varactors and spiral inductors, basic concepts of circuit design and implementation issues, oscillator phase noise and the effect of low-frequency flicker noise. Also included are various practical examples of differential, complementary and quadrature CMOS VCOs using different process technologies.

Wideband voltage-controlled oscillators are used in a variety of RF and microwave systems, including broadband measurement equipment, wireless and TV applications and military electronic countermeasure systems. Among wideband tunable signal sources such as YIG-tuned oscillators, wideband VCOs are preferable because of their small size, low weight, high settling time speed and capability of fully monolithic integration. Therefore, modern radar and communication applications demand VCOs that are capable of being swept across a wide range of potential threat frequencies with a speed and settling time far beyond that of the YIG-tuned oscillators. This chapter discusses the basic concepts of wideband VCO circuit design and gives specific circuit solutions using lumped elements and transmission lines to improve their frequency tuning characteristics. Various examples of the RF and microwave VCO circuit configurations using bipolar, MOSFET and MESFET devices are analysed, their circuit parameters are calculated or optimized to provide maximum tuning bandwidth or minimum tuning linearity. Also included are numerous practical examples of wideband VCOs for RF and microwave applications in radar or telecommunication systems.

Chapter 9 discusses phase noise reduction techniques and gives specific resonant circuit solutions using lumped and distributed parameters for frequency stabilization and phase noise reduction. Phase noise improvement can also be achieved by appropriate low-frequency loading and feedback circuitry optimization. The feedback system incorporated into the oscillator bias circuit can provide significant phase noise reduction over a wide frequency range from the high frequencies up to microwaves. Particular discrete implementations of a bipolar oscillator with collector and emitter noise feedback circuits are described. Also a filtering technique based on a passive *LC* filter to lower the phase noise in the differential oscillator is presented. Several topologies of fully integrated CMOS voltage-controlled oscillators using filtering techniques are shown and discussed. A novel noise-shifting differential VCO based on a single-ended classical three-point circuit configuration with common base can improve the phase noise performance by a proper circuit realization. An optimal design technique using an active element based on a tandem connection of a common source FET device and a common base bipolar transistor with optimum coupling of the active element to the resonant circuit is presented. The phase noise in microwave oscillators can also be reduced using negative resistance compensation increasing the loaded quality factor of the oscillator resonant circuit. Finally, a new approach utilizing a nonlinear feedback loop for phase noise suppression in microwave oscillators is discussed.

Acknowledgements

Dr Vladimir Nikiforov who was a patient tutor and wise teacher during our long-term joint research work at the Moscow Technical University of Communications and Informatics and whose invaluable human and scientific features have contributed to the author's research and publishing creative activity.

Dr Bill Chen, the first Director of the Institute of Microelectronics, Singapore, for making it possible to use the institute's excellent facilities and environment without which this book project could not be realized.

Alex Teo and his colleagues from Ansoft Corporation for their excellent professional software products and valuable technical assistance.

Ravinder Walia for helpful and useful discussions of CMOS oscillator design issues.

Prof Grigory Aristarkhov, Dr Vladimir Chernyshev, Dr Pavel Mikhnevich, Dr Vladimir Pashnin, Dr Nikolai Paushkin and Dr Elena Stroganova from the Moscow Technical University of Communications and Informatics, Russia; Dr Herbert Jaeger from the University of Linz, Austria; Dr Alberto Costantini from the University of Bologna, Italy; and Dr Rajinder Singh and Dr Lin Fujiang from the Institute of Microelectronics, Singapore, for encouragement and support.

The author especially wishes to thank his wife, Galina Grebennikova, for performing important numerical calculations and computer artwork design, as well as for her constant encouragement, inspiration, support and assistance.

Finally, I would like to express my sincere appreciation to all the staff at John Wiley & Sons involved in this publishing project for their cheerful professionalism and outstanding efforts.

1

Nonlinear circuit design methods

This chapter presents the most commonly used design techniques for analysing nonlinear circuits, in particular, transistor oscillators. There are several approaches to analyse and design nonlinear circuits, depending on their main specifications. This means an analysis both in the time domain to determine transient circuit behaviour and in the frequency domain to improve power and spectral performances when parasitic effects such as instability and spurious emission must be eliminated or minimized. Using the time-domain technique, it is relatively easy to describe a nonlinear circuit with differential equations, which can be solved analytically in explicit form for only a few simple cases. Under an assumption of slowly varying amplitude and phase, it is possible to obtain separate truncated first order differential equations for the amplitude and phase of the oscillation process from the original second-order nonlinear differential equation. However, generally it is required to use numerical methods. The time-domain analysis is limited to its inability to operate with the circuit immittance (impedance or admittance) parameters as well as the fact that it can be practically applied only for circuits with lumped parameters or ideal transmission lines. The frequency-domain analysis is less ambiguous because a relatively complex circuit can often be reduced to one or more sets of immittances at each harmonic component. For example, using a quasilinear approach, the nonlinear circuit parameters averaged by fundamental component allow one to apply a linear circuit analysis. Advanced modern CAD simulators incorporate both time-domain and frequency-domain methods as well as optimization techniques to provide all necessary design cycles.

This chapter also includes a brief introduction of simulator tools based on the Ansoft Serenade circuit simulator. In addition, some practical equations, such as the Taylor and Fourier series expansions, Bessel functions, trigonometric identities and the concept of the conduction angle, which simplify the circuit design procedure, are given.

1.1 SPECTRAL-DOMAIN ANALYSIS

The best way to understand the oscillator electrical behaviour and the fastest way to calculate its basic electrical characteristics such as output power, efficiency, phase noise, or harmonic suppression, is to use a spectral-domain analysis. Generally, such an analysis is based on the determination of the output response of the nonlinear active device when the multiharmonic

RF and Microwave Transistor Oscillator Design A. Grebennikov

signal is applied to its input port, which analytically can be written in the form

$$i(t) = f[v(t)] \tag{1.1}$$

where $i(t)$ is the output current, $v(t)$ is the input voltage and $f(v)$ is the nonlinear transfer function of the device. Unlike the spectral-domain analysis, time-domain analysis establishes the relationships between voltage and current in each circuit element in the time domain when a system of nonlinear integrodifferential equations is obtained applying Kirchhoff's law to the circuit to be analysed.

The voltage $v(t)$ in frequency domain generally represents the multiple frequency signal at the device input in the form

$$v(t) = V_0 + \sum_{k=1}^{N} V_k \cos(\omega_k t + \phi_k) \tag{1.2}$$

where V_0 is the constant voltage, V_k is the voltage amplitude and ϕ_k is the phase of the kth-order harmonic component ω_k, $k = 1, 2, \ldots, N$, and N is the number of harmonics.

The spectral domain analysis based on substituting Equation (1.2) in Equation (1.1) for a particular nonlinear transfer function of the active device determines an output spectrum as a sum of the fundamental-frequency and higher-order harmonic components, the amplitudes and phases of which will determine the output signal spectrum. Generally, this is a complicated procedure which requires a harmonic balance technique to numerically calculate an accurate nonlinear circuit response. However, the solution can be found analytically in a simple way when it is necessary to estimate only the basic performance of on oscillator in the form of the output power and efficiency. In this case, a technique based on a piecewise-linear approximation of the device transfer function can provide a clear insight into the basic oscillator behaviour and its operation modes. It can also serve as a good starting point for a final computer-aided design and optimization procedure.

The result of the spectral-domain analysis is shown as a summation of the harmonic components, the amplitudes and phases of which will determine the output signal spectrum. This problem can be solved analytically by using trigonometric identities, piecewise-linear approximation or Bessel functions.

1.1.1 Trigonometric identities

The use of trigonometric identities is very convenient when the transfer characteristic of the nonlinear element can be represented by the power series

$$i = a_0 + a_1 v + a_2 v^2 + \ldots + a_n v^n \tag{1.3}$$

If the effect of the input signal represents a single harmonic oscillation in the form

$$v = V \cos(\omega t + \phi) \tag{1.4}$$

then, by substituting Equation (1.4) into Equation (1.3), the power series can be written as

$$i = a_0 + a_1 V \cos(\omega t + \phi) + a_2 V^2 \cos^2(\omega t + \phi) + \ldots + a_n V^n \cos^n(\omega t + \phi) \tag{1.5}$$

To represent the right-hand side of Equation (1.5) as a sum of first-order cosine components, the following trigonometric identities, which replace the nth-order cosine components, can be used:

$$\cos^2\psi = \frac{1}{2}(1 + \cos 2\psi) \tag{1.6}$$

$$\cos^3\psi = \frac{1}{4}(3\cos\psi + \cos 3\psi) \tag{1.7}$$

$$\cos^4\psi = \frac{1}{8}(3 + 4\cos 2\psi + \cos 4\psi) \tag{1.8}$$

$$\cos^5\psi = \frac{1}{16}(10\cos\psi + 5\cos 3\psi + \cos 5\psi) \tag{1.9}$$

where $\psi = \omega t + \phi$.

By using the appropriate substitutions from Equations (1.6–1.9) and equating the signal frequency component terms, Equation (1.5) can be rewritten as

$$i = I_0 + I_1\cos(\omega t + \phi) + I_2\cos 2(\omega t + \phi) + I_3\cos 3(\omega t + \phi) + \ldots + I_n\cos n(\omega t + \phi) \tag{1.10}$$

where

$$I_0 = a_0 + \frac{1}{2}a_2V^2 + \frac{3}{8}a_4V^4 + \ldots$$

$$I_1 - a_1V + \frac{3}{4}a_3V^3 + \frac{5}{8}a_5V^5 + \ldots$$

$$I_2 = \frac{1}{2}a_2V^2 + \frac{1}{2}a_4V^4 + \ldots$$

$$I_3 = \frac{1}{4}a_3V^3 + \frac{5}{16}a_5V^5 + \ldots$$

Comparing Equations (1.3) and (1.10), we find:

- For nonlinear elements, the output spectrum contains frequency components which are multiples of the input signal frequency. The number of the highest-frequency component is equal to the maximum degree of the power series. Therefore, if it is necessary to know the amplitude of n-harmonic response, the volt–ampere characteristic of nonlinear element should be approximated by not less than an n-order power series.
- The output dc and even-order harmonic components are determined only by the even voltage degrees in the device transfer characteristic given by Equation (1.3). The odd-order harmonic components are defined only by the odd voltage degrees for the single harmonic input signal given by Equation (1.4).
- The current phase ψ_k of the kth-order harmonic component $\omega_k = k\omega$ is k times larger than the input signal current phase ψ:

$$\psi_k = \omega_k t + \phi_k = k(\omega t + \phi) \tag{1.11}$$

that is also applied to their initial phases defined as

$$\phi_k = k\phi \tag{1.12}$$

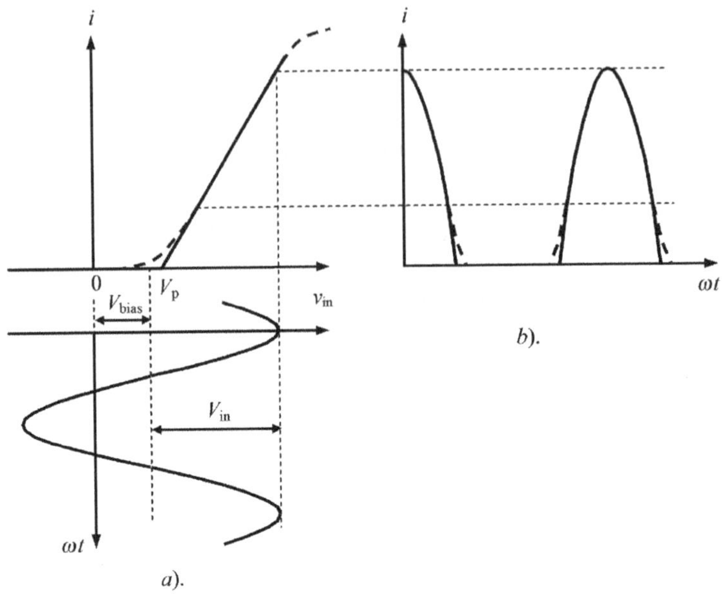

Figure 1.1 Piecewise-linear approximation technique

1.1.2 Piecewise-linear approximation

The piecewise-linear approximation of the active device current–voltage transfer characteristic is a result of replacing the actual nonlinear dependence $i = f(v_{in})$, where v_{in} the voltage applied to the device input, by an approximate one that consists of straight lines tangential to the actual dependence at the specified points. Such a piecewise-linear approximation for the case of two straight lines is shown in Figure 1.1a.

The output current waveforms for the actual current–voltage dependence (dashed curve) and its piecewise-linear approximation by two straight lines (solid curve) are plotted in Figure 1.1b. Under large-signal operation mode, the waveforms corresponding to these two dependencies are practically the same for the most part with negligible deviation for small values of the output current close to the pinch-off region of the device operation and significant deviation close to the saturation region of the device operation. However, the latter case results in a significant nonlinear distortion and is used only for high-efficiency operation modes when the active period of the device operation is minimized. Hence, at least two first output current components, dc and fundamental, can be calculated through a Fourier series expansion with a sufficient accuracy. Therefore, such a piecewise-linear approximation with two straight lines can be effective for a quick estimate of the oscillator output power and efficiency.

In this case, the piecewise-linear active device transfer current–voltage characteristic is defined by

$$i = \begin{cases} 0 & v_{in} \le V_p \\ g_m(v_{in} - V_p) & v_{in} \ge V_p \end{cases} \tag{1.13}$$

where g_m is the device transconductance, V_p is the pinch-off voltage.

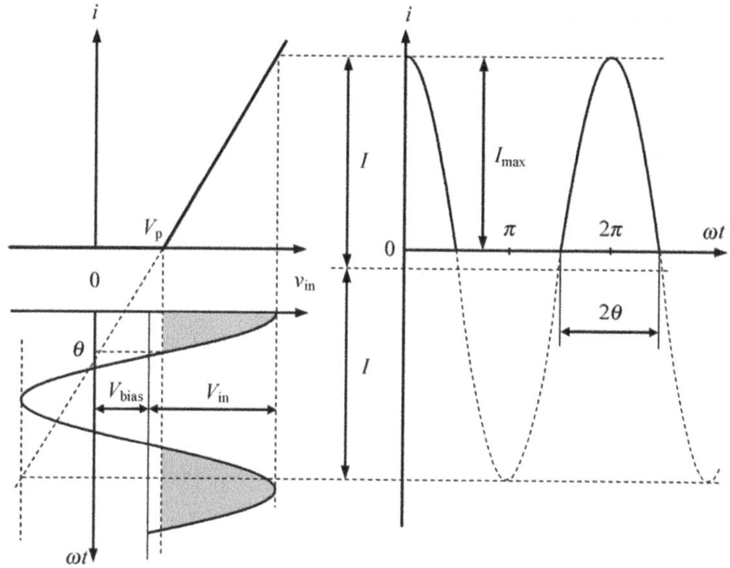

Figure 1.2 Schematic definition of conduction angle

Let us assume the input signal to be of cosinusoidal form

$$v_{in} = V_{bias} + V_{in} \cos \omega t \tag{1.14}$$

where V_{bias} is thc input dc bias voltagc.

At the point on the plot when voltage $v_{in}(\omega t)$ becomes equal to a pinch-off voltage V_p and where $\omega t = \theta$, the output current $i(\theta)$ has value zero. At this moment

$$V_p = V_{bias} + V_{in} \cos \theta \tag{1.15}$$

and θ can be calculated from

$$\cos \theta = -\frac{V_{bias} - V_p}{V_{in}} \tag{1.16}$$

As a result, the output current represents a periodic pulsed waveform described by the cosinusoidal pulses with the maximum amplitude I_{max} and width 2θ as

$$i = \begin{cases} I_q + I \cos \omega t & -\theta \le \omega t < \theta \\ 0 & \theta \le \omega t < 2\pi - \theta \end{cases} \tag{1.17}$$

where the conduction angle 2θ indicates the part of the RF current cycle during which device conduction occurs, as shown in Figure 1.2. When the output current $i(\omega t)$ has value zero, one can write

$$i = I_q + I \cos \theta = 0 \tag{1.18}$$

Taking into account that, for a piecewise-linear approximation, $I = g_m V_{in}$, Equation (1.17) can be rewritten as

$$i = g_m V_{in}(\cos \omega t - \cos \theta) \tag{1.19}$$

When $\omega t = 0$, then $i = I_{\text{max}}$ and

$$I_{\text{max}} = I(1 - \cos\theta) \tag{1.20}$$

The angle θ characterizes the class of the active device operation. If $\theta = \pi$ or 180°, the device operates in the active region during the entire period (class A operation). When $\theta = \pi/2$ or 90°, the device operates half a wave period in the active region and half a wave period in the pinch-off region (class B operation). The values of $\theta > 90°$ correspond to class AB operation with a certain value of the quiescent output current. Therefore, the double angle 2θ is called the conduction angle, the value of which directly indicates the class of the active device operation.

The Fourier series expansion of the even function when $i(t) = i(-t)$ contains only even component functions and can be written as

$$i(t) = I_0 + I_1 \cos\omega t + I_2 \cos 2\omega t + I_3 \cos 3\omega t + \ldots \tag{1.21}$$

where the dc, fundamental-frequency and nth-order harmonic components are calculated by

$$I_0 = \frac{1}{2\pi} \int_{-\theta}^{\theta} g_{\text{m}} V_{\text{in}} (\cos\omega t - \cos\theta)\, d(\omega t) = \gamma_0(\theta) I \tag{1.22}$$

$$I_1 = \frac{1}{\pi} \int_{-\theta}^{\theta} g_{\text{m}} V_{\text{in}} (\cos\omega t - \cos\theta) \cos\omega t\, d(\omega t) = \gamma_1(\theta) I \tag{1.23}$$

$$I_{\text{n}} = \frac{1}{\pi} \int_{-\theta}^{\theta} g_{\text{m}} V_{\text{in}} (\cos\omega t - \cos\theta) \cos(n\omega t)\, d(\omega t) = \gamma_{\text{n}}(\theta) I \tag{1.24}$$

where $\gamma_n(\theta)$ are called the coefficients of expansion of the output current cosinusoidal pulse or the current coefficients [1]. They can be analytically defined as

$$\gamma_0(\theta) = \frac{1}{\pi}(\sin\theta - \theta\cos\theta) \tag{1.25}$$

$$\gamma_1(\theta) = \frac{1}{\pi}\left(\theta - \frac{\sin 2\theta}{2}\right) \tag{1.26}$$

$$\gamma_n(\theta) = \frac{1}{\pi}\left[\frac{\sin(n-1)\theta}{n(n-1)} - \frac{\sin(n+1)\theta}{n(n+1)}\right] \tag{1.27}$$

where $n = 2, 3, \ldots$.

The dependencies of $\gamma_n(\theta)$ for the dc, fundamental-frequency, second- and higher-order current components are shown in Figure 1.3. The maximum value of $\gamma_n(\theta)$ is achieved when $\theta = 180°/n$. A special case is $\theta = 90°$, when odd current coefficients are equal to zero, i.e., $\gamma_3(\theta) = \gamma_5(\theta) = \ldots = 0$. The ratio between the fundamental-frequency and dc components $\gamma_1(\theta)/\gamma_0(\theta)$ varies from 1 to 2 for any values of the conduction angle, with a minimum value of 1 for $\theta = 180°$ and a maximum value of 2 for $\theta = 0°$. It is necessary to pay attention to the fact that, for example, the current coefficient $\gamma_3(\theta)$ becomes negative within the interval of $90° < \theta < 180°$. This implies appropriate phase changes of the third current harmonic component when its values are negative. Consequently, if the harmonic components for which

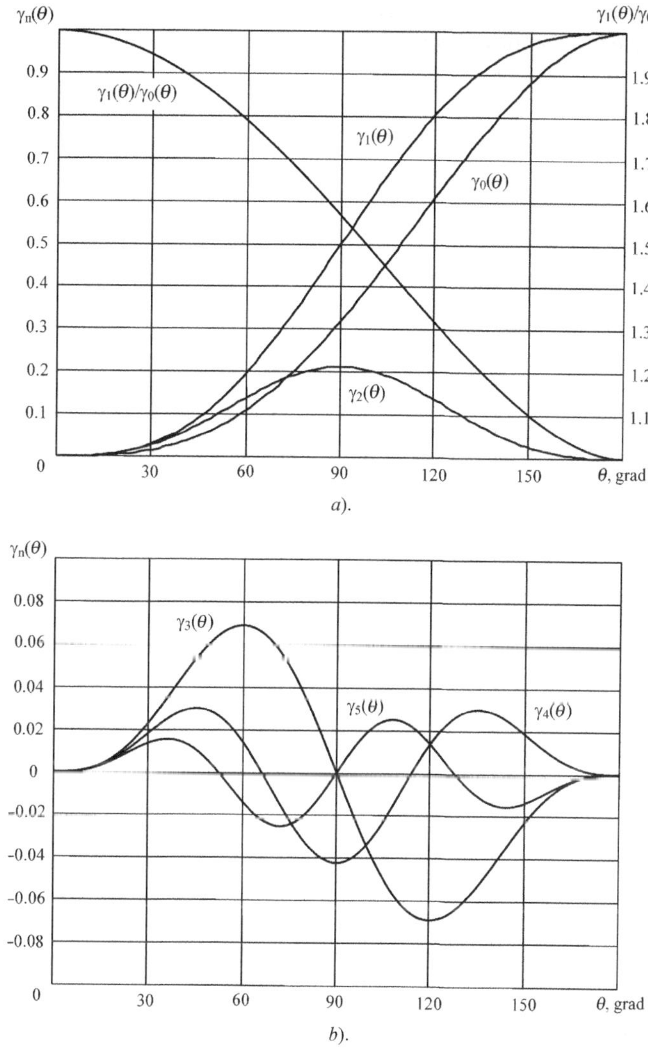

Figure 1.3 Dependencies of $\gamma_n(\theta)$ for dc, fundamental- and higher-order current components

$\gamma_n(\theta) > 0$ achieve positive maximum values at times corresponding to the midpoints of the current waveform, the harmonic components for which $\gamma_n(\theta) < 0$ can achieve negative maximum values at these times. As a result, combination of different harmonic components with proper loading will result in flattening of the current or voltage waveforms, thus improving efficiency of the oscillator. The amplitude of corresponding current harmonic component can be obtained as

$$I_n = \gamma_n(\theta) g_m V_{in} = \gamma_n(\theta) I \tag{1.28}$$

Sometimes it is necessary for an active device to provide a constant value of I_{max} at any value of θ. This requires an appropriate variation of the input voltage amplitude V_{in}. In this case, it is more convenient to use the other coefficients when the nth-order current harmonic amplitude I_n is related to the maximum current waveform amplitude $I_{\max}$, that is

$$\alpha_n = \frac{I_n}{I_{\max}} \tag{1.29}$$

From Equations (1.20), (1.28) and (1.29) it follows that

$$\alpha_n = \frac{\gamma_n(\theta)}{1 - \cos\theta} \tag{1.30}$$

and the maximum value of $\alpha_n(\theta)$ is achieved when $\theta = 120°/n$.

1.1.3 Bessel functions

The Bessel functions are used to analyse the oscillator operation mode when a nonlinear behaviour of the active device can be described by exponential functions. The transfer voltage–ampere characteristic of the bipolar transistor is approximated by the simplified exponential dependence neglecting reverse base–emitter current as

$$i(v_{\text{in}}) = I_{\text{sat}}\left[\exp\left(\frac{v_{\text{in}}}{V_T}\right) - 1\right] \tag{1.31}$$

where I_{sat} is the minority carrier saturation current and V_T is the temperature voltage. If the effect of the input signal given by Equation (1.14) is considered, then Equation (1.31) can be rewritten as

$$i(\omega t) = I_{\text{sat}}\left[\exp\left(\frac{V_{\text{bias}}}{V_T}\right)\exp\left(\frac{V_{\text{in}}\cos\omega t}{V_T}\right) - 1\right] \tag{1.32}$$

The current $i(\omega t)$ in Equation (1.32) is the even function of ωt and, consequently, it can be represented by the Fourier-series expansion given by Equation (1.21). To determine the Fourier components, the following expression is used:

$$\exp\left(\frac{V_{\text{in}}\cos\omega t}{V_T}\right) = I_0\left(\frac{V_{\text{in}}}{V_T}\right) + 2\sum_{k=1}^{\infty} I_k\left(\frac{V_{\text{in}}}{V_T}\right)\cos(k\omega t) \tag{1.33}$$

where $I_k(V_{\text{in}}/V_T)$ are the kth-order modified Bessel functions of the first kind for an argument of V_{in}/V_T, shown in Figure 1.4 for the zeroth- and first-order components. It should be noted that $I_0(0) = 1$ and $I_1(0) = I_2(0) = \ldots = 0$, and with an increase of the component number its amplitude appropriately decreases.

According to Equation (1.33), the current $i(\omega t)$ defined by Equation (1.31) can be rewritten as

$$i(\omega t) = I_{\text{sat}}\left[\exp\left(\frac{V_{\text{bias}}}{V_T}\right)I_0\left(\frac{V_{\text{in}}}{V_T}\right) - 1\right] + 2I_{\text{sat}}\exp\left(\frac{V_{\text{bias}}}{V_T}\right)I_1\left(\frac{V_{\text{in}}}{V_T}\right)\cos(\omega t)$$
$$+ 2I_{\text{sat}}\exp\left(\frac{V_{\text{bias}}}{V_T}\right)I_2\left(\frac{V_{\text{in}}}{V_T}\right)\cos(2\omega t) + 2I_{\text{sat}}\exp\left(\frac{V_{\text{bias}}}{V_T}\right)I_3\left(\frac{V_{\text{in}}}{V_T}\right)\cos(3\omega t) + \ldots \tag{1.34}$$

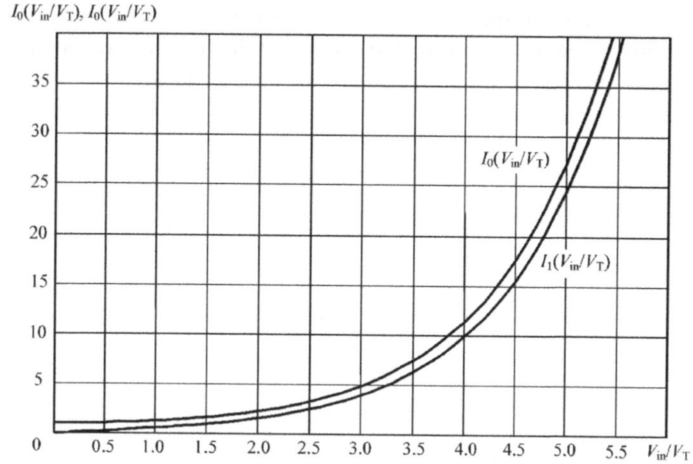

Figure 1.4 Zeroth- and first-order modified Bessel functions of the first kind

As a result, comparing Equations (1.34) and (1.21) allows the dc, fundamental-frequency and nth-order Fourier current components to be determined as

$$I_0 = I_{\text{sat}}\left[\exp\left(\frac{V_{\text{bias}}}{V_T}\right) I_0\left(\frac{V_{\text{in}}}{V_T}\right) - 1\right] \tag{1.35}$$

$$I_1 = 2I_{\text{sat}} \exp\left(\frac{V_{\text{bias}}}{V_T}\right) I_1\left(\frac{V_{\text{in}}}{V_T}\right) \tag{1.36}$$

$$I_n = 2I_{\text{sat}} \exp\left(\frac{V_{\text{bias}}}{V_T}\right) I_n\left(\frac{V_{\text{in}}}{V_T}\right) \tag{1.37}$$

where $n = 2, 3, \ldots$

When using the Bessel functions, the following relationships can be helpful:

$$2\frac{dI_n\left(V_{\text{in}}/V_T\right)}{d\left(V_{\text{in}}/V_T\right)} = I_{n+1}\left(\frac{V_{\text{in}}}{V_T}\right) + I_{n-1}\left(\frac{V_{\text{in}}}{V_T}\right) \tag{1.38}$$

$$\frac{dI_0\left(V_{\text{in}}/V_T\right)}{d\left(V_{\text{in}}/V_T\right)} = I_1\left(\frac{V_{\text{in}}}{V_T}\right) \tag{1.39}$$

$$\frac{2n}{\left(V_{\text{in}}/V_T\right)} I_n\left(\frac{V_{\text{in}}}{V_T}\right) = I_{n-1}\left(\frac{V_{\text{in}}}{V_T}\right) - I_{n+1}\left(\frac{V_{\text{in}}}{V_T}\right) \tag{1.40}$$

$$I_n\left(-\frac{V_{\text{in}}}{V_T}\right) = (-1)^n I_n\left(\frac{V_{\text{in}}}{V_T}\right) \tag{1.41}$$

1.2 TIME-DOMAIN ANALYSIS

A time-domain analysis establishes the relationships between voltage and current in each circuit element in the time domain when a system of equations is obtained, applying Kirchhoff's law to the circuit to be analysed. Normally, in a nonlinear circuit, such a system will be composed of nonlinear integrodifferential equations. The solution to this system can be found by applying numerical integration methods. Therefore, the choices of the time interval and the initial point are very important to provide a compromise between speed and accuracy of calculation; the smaller the interval, the smaller the error, but the number of points to be calculated for each period will be greater, which will make the calculation slower.

To analyse a nonlinear system in the time domain, it is necessary to know the voltage–current relationships for all circuit elements. For example, for linear resistance R, when the sinusoidal voltage applies and current are flowing through it, the voltage–current relationship in the time domain is given by

$$V = RI \tag{1.42}$$

where V is the voltage amplitude and I is the current amplitude.

For linear capacitance C

$$i(t) = \frac{dq(t)}{dt} = \frac{dq}{dv}\frac{dv}{dt} = C\frac{dv}{dt} \tag{1.43}$$

For linear inductance L

$$v(t) = \frac{d\varphi(t)}{dt} = \frac{d\varphi}{di}\frac{di}{dt} = L\frac{di}{dt} \tag{1.44}$$

where φ is the magnetic flux across the inductance.

Nonlinear dependencies, such as $q(v)$ or $\varphi(i)$, should each be expanded in a Taylor series by subtracting the dc components and substituting into Equations (1.43) and (1.44) to obtain the expressions for appropriate incremental capacitance and inductance. Then, for the quasilinear case, the capacitance and inductance can be defined by

$$C(V_0) = \left.\frac{dq(v)}{dv}\right|_{v=V_0} \tag{1.45}$$

and

$$L(I_0) = \left.\frac{d\varphi(i)}{di}\right|_{i=I_0} \tag{1.46}$$

where V_0 is the dc bias voltage across the capacitor and I_0 is the dc current flowing through the inductor.

Figure 1.5 shows the simplified (without bias circuits) electrical schematic of a transformer-coupled MOSFET oscillator with a parallel resonant circuit. To obtain the differential equations

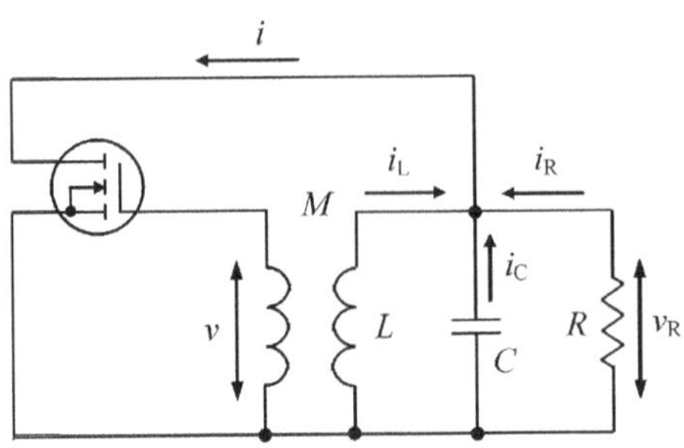

Figure 1.5 Schematic of a transformer-coupled MOSFET oscillator

for such an oscillator, the drain current i, the gate voltage v applied to the second winding of the transformer, and the load voltage v_R applied to the first winding of this transformer can be defined by

$$i = i_L + i_C + i_R \tag{1.47}$$

$$v_R = L\frac{di_L}{dt} = \frac{1}{C}\int i_C dt = i_R R \tag{1.48}$$

$$v = M\frac{di_L}{dt} = \frac{M}{L}v_R \tag{1.49}$$

where M is the transformer coupling factor.

To simplify the calculation, two preliminary assumptions can be used:

- the input current flowing to the gate terminal of the active device is negligible, enabling one to consider its input impedance as infinite;
- the effect of the output voltage v_R on the drain current i is ignored, i.e.,

$$i = f(v). \tag{1.50}$$

In this case, the derivative of current $i(v)$ with respect to time is written as

$$\frac{di}{dt} = \frac{di}{dv}\frac{dv}{dt} = g_m(v)\frac{dv}{dt} \tag{1.51}$$

where $g_m = di/dv$ is the small-signal transconductance of the device transfer characteristic given by Equation (1.50).

Substituting Equations (1.48) and (1.50) into Equation (1.47) gives

$$\frac{1}{L}\int v_R dt + C\frac{dv_R}{dt} + \frac{v_R}{R} = f(v) \tag{1.52}$$

Then, by differentiating Equation (1.52) and using Equations (1.49) and (1.51), we can write the second-order differential equation for the oscillator in the form

$$\frac{d^2v}{dt^2} + \frac{1}{C}\left[\frac{1}{R} - \frac{Mg_m(v)}{L}\right]\frac{dv}{dt} + \omega_0^2 v = 0 \tag{1.53}$$

where

$$\omega_0 = \frac{1}{\sqrt{LC}}$$

is the oscillator resonant frequency.

Equation (1.53) is a nonlinear equation because its second term depends on the unknown variable v. This nonlinearity is a result of the active device nonlinearity. From Equation (1.53), the start-up and steady-state oscillation conditions can be determined, as well as the particular features of the steady-state oscillations and oscillator transient response. To determine the start-up conditions, it is necessary to replace nonlinear Equation (1.53) by an appropriate linear one, with the linear small-signal transconductance g_m at the operating bias point. In this case, we are interested only in the result of the small deviation from an equilibrium point, whether the oscillations will grow or dissipate.

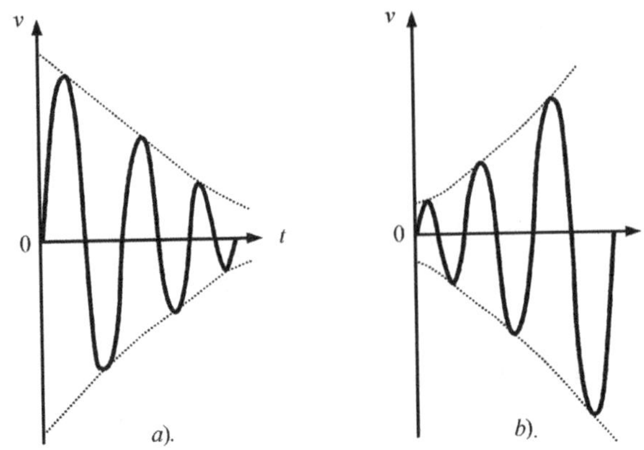

Figure 1.6 Oscillations with (a) low and (b) strong feedback factors

The solution of such a linear second-order differential equation is

$$v = V\exp(-\delta t)\sin(\omega_1 t + \phi) \tag{1.54}$$

where V and ϕ are the voltage amplitude and phase, respectively, depending on the initial conditions,

$$\delta = \frac{1}{2C}\left(\frac{1}{R} - \frac{Mg_{\mathrm{m}}}{L}\right) \tag{1.55}$$

is the dissipation factor, and

$$\omega_1 = \sqrt{\omega_0^2 - \delta^2} \tag{1.56}$$

is the free-running oscillation frequency.

From Equation (1.54) it follows that the voltage v at the device input provided by the feedback circuit creates current i at the device output, which delivers electrical energy to the oscillation system to compensate for the losses in it. At the same time, the required value of this energy is the result of the transformation of the energy of the dc current delivered from the dc current source to the energy of the ac current. If the feedback factor is sufficiently small when $\delta > 0$, the delivered energy compensates for the dissipated energy only partly. As a result, this leads to attenuation and dissipation of the oscillations, as shown in Figure 1.6a. For strong feedback factor when $\delta < 0$, the delivered energy exceeds the dissipated energy, and the oscillations increase with time, as shown in Figure 1.6b.

1.3 NEWTON–RAPHSON ALGORITHM

To describe circuit behaviour, it is necessary to solve the nonlinear algebraic equation, or system of equations, which do not generally admit a closed form solution analytically. One of the most common numerical methods to solve such equations is a method based on the Newton–Raphson algorithm [2]. The initial guess for this method is chosen using a Taylor series expansion of the nonlinear function. Consider a practical case when the device is represented by a two-port network where all nonlinear elements are functions of the two unknown voltages, input voltage

v_{in} and output voltage v_{out}. As a result, after combining linear and nonlinear circuit elements, a system of two equations can be written as

$$f_1(v_{in}, v_{out}) = 0 \tag{1.57}$$

$$f_2(v_{in}, v_{out}) = 0 \tag{1.58}$$

Assume that the variables v_{in0} and v_{out0} are the initial approximate solution of a system of Equations (1.57) and (1.58). Then, the variables can be written as $v_{in} = v_{in0} + \Delta v_{in}$ and $v_{out} = v_{out0} + \Delta v_{out}$, where Δv_{in} and Δv_{out} are the linear increments of the variables. Applying a Taylor series expansion to Equations (1.57) and (1.58) yields

$$f_1(v_{in0} + \Delta v_{in}, v_{out0} + \Delta v_{out}) = f_1(v_{in0}, v_{out0}) + \left.\frac{\partial f_1}{\partial v_{in}}\right|_{\substack{v_{in}=v_{in0}\\ v_{out}=v_{out0}}} \Delta v_{in} + \left.\frac{\partial f_1}{\partial v_{out}}\right|_{\substack{v_{in}=v_{in0}\\ v_{out}=v_{out0}}} \Delta v_{out} + o\left(\Delta v_{in}^2 + \Delta v_{out}^2 + \ldots\right) = 0 \tag{1.59}$$

$$f_2\left(v_{in0} + \Delta v_{in}, v_{out0} + \Delta v_{out}\right) = f_2\left(v_{in0}, v_{out0}\right) + \left.\frac{\partial f_2}{\partial v_{in}}\right|_{\substack{v_{in}=v_{in0}\\ v_{out}=v_{out0}}} \Delta v_{in} + \left.\frac{\partial f_2}{\partial v_{out}}\right|_{\substack{v_{in}=v_{in0}\\ v_{out}=v_{out0}}} \Delta v_{out} + o\left(\Delta v_{in}^2 + \Delta v_{out}^2 + \ldots\right) = 0 \tag{1.60}$$

where $o(\Delta v_{in}^2 + \Delta v_{out}^2 + \ldots)$ denotes the second- and higher-order components.

By neglecting the second- and higher-order components, Equations (1.59) and (1.60) can be rewritten in matrix form

$$-\begin{bmatrix} f_1 \\ f_2 \end{bmatrix} = \begin{bmatrix} \dfrac{\partial f_1}{\partial v_{in}} & \dfrac{\partial f_1}{\partial v_{out}} \\ \dfrac{\partial f_2}{\partial v_{in}} & \dfrac{\partial f_2}{\partial v_{out}} \end{bmatrix} \begin{bmatrix} \Delta v_{in} \\ \Delta v_{out} \end{bmatrix} \tag{1.61}$$

In the phasor form,

$$-\boldsymbol{F} = \boldsymbol{J}\Delta\boldsymbol{v} \tag{1.62}$$

where $\boldsymbol{J}$ is the Jacobian matrix of a system of Equations (1.57) and (1.58).

The solution of Equation (1.62) for a nonsingular matrix $\boldsymbol{J}$ can be obtained by

$$\Delta\boldsymbol{v} = -\boldsymbol{J}^{-1}\boldsymbol{F} \tag{1.63}$$

This means that if

$$\boldsymbol{v}_0 = \begin{bmatrix} v_{in0} \\ v_{out0} \end{bmatrix} \tag{1.64}$$

is the initial guess of this system of equation, then the next (more precise) solution can be written as

$$\boldsymbol{v}_1 = \boldsymbol{v}_0 - \boldsymbol{J}^{-1}\boldsymbol{F} \tag{1.65}$$

where

$$\boldsymbol{v}_1 = \begin{bmatrix} v_{in1} \\ v_{out1} \end{bmatrix} \tag{1.66}$$

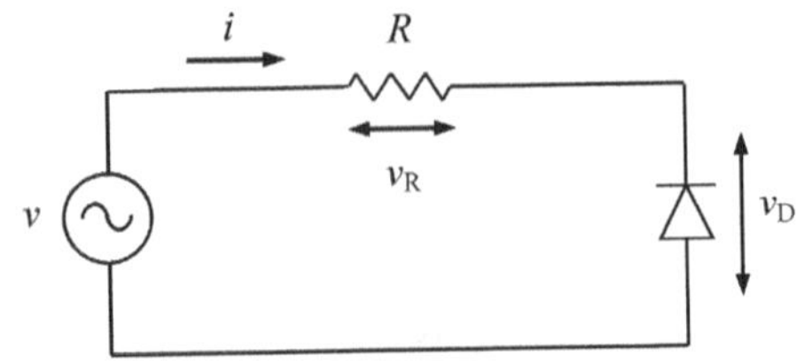

Figure 1.7 Circuit schematic with resistor, diode, and voltage source

Thus, starting with initial guess v_0, we compute v_1 at the first iteration. For the iteration $n+1$, we can write

$$v_{n+1} = v_n - J^{-1} F(v_n) \tag{1.67}$$

The iterative Equation (1.67) is given for a system of two equations; however it can be directly extended to a system of k nonlinear equations with k unknown parameters. This iterative procedure is terminated after $(n+1)$ iterations whenever

$$|x_{n+1} - x_n| = \sqrt{\sum_{k=1}^{K} \left(x_{n+1}^k - x_n^k\right)^2} < \varepsilon \tag{1.68}$$

where ε is a small positive number depending on the desired accuracy. For a practical purpose, it is desirable that the Newton–Raphson algorithm should converge in a few steps. Therefore, the choice of an appropriate initial guess is crucial to the success of the algorithm.

Consider the circuit shown in Figure 1.7. According to Kirchhoff's voltage law,

$$v = v_{\mathrm{R}} + v_{\mathrm{D}} \tag{1.69}$$

where $v_{\mathrm{R}} = iR$.

The electrical behaviour of the diode is described by

$$i(v_{\mathrm{D}}) = I_{\mathrm{sat}} \left[\exp\left(\frac{v_{\mathrm{D}}}{V_T}\right) - 1 \right] \tag{1.70}$$

Rearranging Equation (1.70) gives the equation for v_{D} in the form

$$v_{\mathrm{D}} = V_T \ln\left(\frac{i}{I_{\mathrm{sat}}} + 1\right) \tag{1.71}$$

Thus, from Equations (1.60) and (1.61) it follows that

$$v = iR + V_T \ln\left(\frac{i}{I_{\mathrm{sat}}} + 1\right) \tag{1.72}$$

This allows current i to be determined for a specified voltage v. However, because it is impossible to solve this equation analytically for current i in explicit form, the solution must be found numerically.

Consider a dc voltage source V with dc current I. For the sinusoidal voltage source, it is necessary to calculate the Bessel functions for dc, fundamental-frequency and higher-order

Table 1.1 Three-step iteration procedure

n	I_n, A	ε_n
0	0.05	0.899 371 786
1	0.878 469 005	0.070 902 781
2	0.948 955 229	0.000 416 557
3	0.949 371 786	

harmonic current components. It is convenient to rewrite Equation (1.72) as

$$f(I) = IR + V_T \ln\left(\frac{I}{I_{sat}} + 1\right) - V = 0 \tag{1.73}$$

from which

$$f'(I) = R + \frac{V_T}{I + I_{sat}} \tag{1.74}$$

Then, applying the iterative algorithm for a single variable, we can write

$$I_n = I_{n-1} - \frac{f\left(I_{n-1}\right)}{f'\left(I_{n-1}\right)}. \tag{1.75}$$

Using Equations (1.73) and (1.74) finally yields

$$I_n = I_{n-1} - \frac{I_{n-1}R + V_T \ln\left(\dfrac{I_{n-1}}{I_{sat}} + 1\right) - V}{R + V_T \dfrac{1}{I_{n-1} + I_{sat}}} \tag{1.76}$$

The results of the numerical calculation of the currents I_n for each iteration for $V_T = 25.9$ mA/V, $R = 5\,\Omega$, $V = 5$ V, $I_{sat} = 10$ μA and initial current $I_0 = 50$ mA are given in Table 1.1. The calculation error $\varepsilon_n = I_N - I_n$, where $n = 0, 1, \ldots, N$, illustrates the fast convergence to the solution for each iteration step. The error at each subsequent iteration step is approximately proportional to the square one of error at the previous step. If the required accuracy of $\varepsilon < 0.1\%$ is set in advance, the iteration procedure will be stopped at the third iteration step.

1.4 QUASILINEAR METHOD

To simplify the analysis and design procedure of the oscillator, in some cases it is enough to apply a quasilinear or Barkhausen–Moeller method based on the use of the ratios between the fundamental-frequency components of currents and voltages [3]. In this case, it is assumed that the self-oscillations must be close to sinusoidal. The derivation of equations for equivalent linear elements of the active device in terms of voltages and currents is based on its static voltage–ampere and voltage–capacitance characteristics.

For example, for a bipolar transistor, the simplified equivalent circuit of which is shown in Figure 1.8, all elements of its equivalent circuit are nonlinear, depending significantly on operation mode, especially the transconductance g_m and base-emitter capacitance C_π. The base-emitter capacitance C_π consists of the diffusion and junctions capacitances and, at

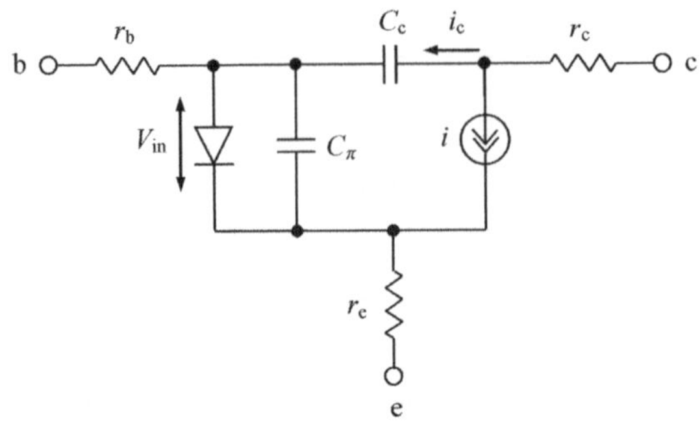

Figure 1.8 Bipolar transistor simplified equivalent circuit

high frequencies, its reactance is sufficiently high to shunt the base–emitter forward-biased diode. Taking into account that the transition frequency is obtained by $\omega_T = g_m/C_\pi$, it is sufficient to consider the only nonlinear elements g_m, ω_T and collector capacitance C_c, since the base resistance r_b poorly depends on a bias mode. The fundamentally averaged large-signal transconductance (or average transconductance) can be easily determined from Equation (1.36) by

$$g_{m1}(V_{in}) = \frac{I_1}{V_{in}} = \frac{2I_{sat}}{V_{in}} \exp\left(\frac{V_{bias}}{V_T}\right) I_1\left(\frac{V_{in}}{V_T}\right) \tag{1.77}$$

The collector capacitance C_c represents a junction capacitance and can be approximated by

$$C_c = \frac{C_{c0}}{\left(1 + \dfrac{v_c}{\varphi}\right)^\gamma} \tag{1.78}$$

where φ is the built-in junction potential, γ is the junction sensitivity and C_{c0} is the initial capacitance when $v_c = 0$.

If our interest is restricted to the fundamental frequency, and $v_c = V_{cc} + V_c \sin\omega t$, where V_{cc} is the collector dc supply voltage, then the following current flows through the collector capacitance which is defined for the quasilinear case as

$$i_c = C_c(v_c)\frac{dv_c}{dt} = \frac{\omega C_{c0} V_c \cos\omega t}{\left(1 + \dfrac{V_{cc}}{\varphi} + \dfrac{V_c}{\varphi}\sin\omega t\right)^\gamma}$$
$$= \frac{\omega C_c(V_{cc}) V_c \cos\omega t}{(1 + \xi\sin\omega t)^\gamma} \tag{1.79}$$

where $C_c(V_{cc})$ is the small-signal capacitance in the operating point and $\xi = V_c/(V_{cc} + \varphi)$.

As a result, the average large-signal collector capacitance C_{c1} can be calculated through the fundamental Fourier series component as

$$C_{c1}(V_c) = \frac{I_{c1}}{\omega V_c} = \frac{C_c(V_{cc})}{\pi}\int_0^{2\pi} \frac{\cos^2\omega t}{(1 + \xi\sin\omega t)^\gamma} d(\omega t) \tag{1.80}$$

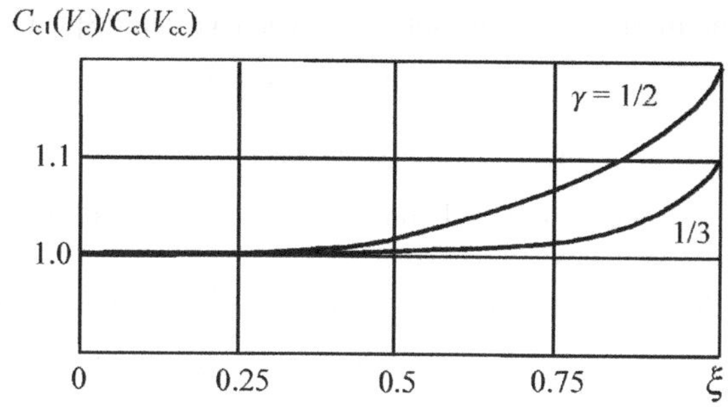

Figure 1.9 Large-signal behaviour of collector capacitance

Figure 1.9 shows the voltage dependencies of the average collector capacitance. Within a range of $\xi < 1$, the maximum large-signal value of $C_{c1}(V_c)$ deviates from the small-signal value of $C_c(V_{cc})$ by not more than 20% for an abrupt junction with $\gamma = 1/2$.

For a MESFET device with the simplified equivalent circuit shown in Figure 1.10, the drain current i_d is a function of the gate–source voltage v_{gs} and the drain–source voltage v_{ds}, which can be expanded in a two-dimensional Taylor series

$$i_d(v_{gs}, v_{ds}) = I_0 + \left.\frac{\partial f}{\partial v_{gs}}\right|_{\substack{v_{gs}=V_g \\ v_{ds}=V_{dd}}} (v_{gs} - V_g) + \left.\frac{\partial f}{\partial v_{ds}}\right|_{\substack{v_{gs}=V_g \\ v_{ds}=V_{dd}}} (v_{ds} - V_{dd})$$
$$+ \frac{1}{2}\left[\left.\frac{\partial^2 f}{\partial v_{gs}^2}\right|_{\substack{v_{gs}=V_g \\ v_{ds}=V_{dd}}} (v_{gs} - V_g)^2 + 2\left.\frac{\partial^2 f}{\partial v_{gs}\partial v_{ds}}\right|_{\substack{v_{gs}=V_g \\ v_{ds}=V_{dd}}} (v_{gs} - V_g)(v_{ds} - V_{dd})\right.$$
$$\left. + \left.\frac{\partial^2 f}{\partial v_{ds}^2}\right|_{\substack{v_{gs}=V_g \\ v_{ds}=V_{dd}}} (v_{ds} - V_{dd})^2 + \ldots\right] \tag{1.81}$$

where V_g is the gate dc bias voltage and V_{dd} is the drain dc supply voltage.

In the small-signal quasilinear case, the high-degree terms are neglected and

$$i_d(v_{gs}, v_{ds}) = I_0 + \left.\frac{\partial f}{\partial v_{gs}}\right|_{\substack{v_{gs}=V_g \\ v_{ds}=V_{dd}}} (v_{gs} - V_g) + \left.\frac{\partial f}{\partial v_{ds}}\right|_{\substack{v_{gs}=V_g \\ v_{ds}=V_{dd}}} (v_{ds} - V_{dd}) \tag{1.82}$$

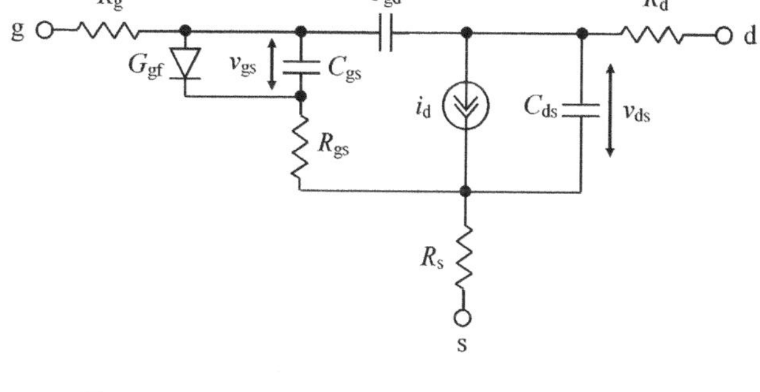

Figure 1.10 MESFET simplified equivalent circuit

The gate–source and drain–source instantaneous voltages can respectively be written as

$$v_{gs} = V_g + V_{gs}\cos(\omega t + \phi) \tag{1.83}$$

$$v_{ds} = V_{dd} + V_{ds}\cos\omega t \tag{1.84}$$

where V_{gs} and V_{ds} are the gate–source and drain–source voltage amplitudes and ϕ is the phase difference between these voltages.

Consequently, the instantaneous drain current given by Equation (1.81) can be rewritten as

$$i_d(\omega t) = I_0 + g_{m1}V_{gs}\cos(\omega t + \phi) + G_{ds1}V_{ds}\cos\omega t \tag{1.85}$$

where

$$g_{m1} = \left.\frac{I_d}{V_{gs}}\right|_{V_{ds}=0} \tag{1.86}$$

is the linearized large-signal transconductance,

$$G_{ds1} = \left.\frac{I_d}{V_{ds}}\right|_{V_{gs}=0} \tag{1.87}$$

is the differential output conductance, I_0 is the dc drain current, I_d is the fundamental drain current amplitude, and $G_{ds1} = 1/R_{ds1}$ [4].

Multiplying the right- and left-hand sides of Equation (1.85) by $\sin\omega t$ and integrating over the entire period of the oscillation result in the average transconductance g_{m1} obtained by

$$g_{m1} = -\frac{1}{\pi V_{gs}\sin\phi}\int_0^{2\pi} i_d(\omega t)\sin\omega t\, d(\omega t) \tag{1.88}$$

Similarly, multiplying by $\sin(\omega t + \phi)$ results in the average output conductance

$$G_{ds1} = \frac{1}{\pi V_{ds}\sin\phi}\int_0^{2\pi} i_d(\omega t)\sin(\omega t + \phi)\, d(\omega t) \tag{1.89}$$

The average large-signal gate–source capacitance C_{gs1} can be calculated similarly to that of for the abrupt collector capacitance C_{c1} of the bipolar transistor with $\gamma = 1/2$. The average large-signal gate forward conductance G_{gf1} is defined by

$$G_{gf1} = \frac{2I_{sat}}{V_{gs}} I_1\left(\frac{V_{gs}}{V_T}\right)\exp\left(\frac{V_g}{V_T}\right) \tag{1.90}$$

where I_{sat} is the saturation current of the Schottky barrier, $I_1(V_{gs}/V_T)$ is the first-order modified Bessel function of first kind.

The gate charging resistance R_{gs} varies with the gate–source capacitance C_{gs} in such a way that the charging time constant $\tau_g = R_{gs}C_{gs}$ varies insignificantly and it can be treated as a constant in a quasilinear approximation.

Now consider the transient response which can be obtained using the quasilinear method on the example of the MOSFET oscillator, the simplified schematic of which is shown in Figure 1.5. For a quasilinear approximation, the appropriate ratios can be obtained directly from the nonlinear differential Equation (1.53) by substituting voltage v and current i by their fundamental-frequency components. The average transconductance g_{m1} is considered as a function of the slowly varying fundamental voltage amplitude and, for a high quality factor of the oscillator resonant circuit, can be treated constant during the natural oscillation period. As a result, the nonlinear differential Equation (1.53) can be considered as linear written in the form of

$$\frac{d^2v}{dt^2} + \frac{1}{C}\left[\frac{1}{R} - \frac{Mg_{m1}(V)}{L}\right]\frac{dv}{dt} + \omega_0^2 v = 0 \quad (1.91)$$

where V is the fundamental-frequency voltage amplitude.

From Equation (1.54) it follows that the amplitude of the oscillations varies according to

$$\frac{dV}{dt} = -\delta(V)V \quad (1.92)$$

where

$$\delta(V) = \frac{1}{2C}\left(\frac{1}{R} - \frac{Mg_{m1}(V)}{L}\right)$$

Then, Equation (1.92) can be rewritten as

$$\frac{2}{\omega_0}\frac{dV}{dt} = \frac{1}{Q}\left(\frac{g_{m1}}{g_{m1}^0} - 1\right)V \quad (1.93)$$

where $Q = \omega_0 RC$ is the oscillator quality factor at the resonant frequency ω_0, and

$$g_{m1}^0 = \frac{L}{MR} \quad (1.94)$$

is the average transconductance in the steady-state oscillation mode.

The device voltage–ampere characteristic can be represented by a third-order power series given by Equation (1.3). Then, from Equation (1.10) it follows that

$$g_{m1}(V) = \frac{I_1}{V} = g_m - \frac{3}{4}g_{m3}V^2 \quad (1.95)$$

where $g_m = a_1$ is the small-signal transconductance at the operating bias point, $g_{m3} = -a_3$ and $a_1 > 0$, $a_3 < 0$ to provide soft start-up conditions.

Multiplying by V, separating variables and integrating the both parts of Equation (1.93) result in the amplitude transient response in the form

$$V = V_0 \Big/ \sqrt{1 + \left[\frac{V_0^2}{V^2(0)} - 1\right]\exp[-2|\delta(0)|t]} \quad (1.96)$$

where $V(0)$ is the amplitude V at $t = 0$, and

$$V_0 = \frac{2}{\sqrt{3g_{m3}}}\sqrt{g_m - g_{m1}^0} \quad (1.97)$$

is the voltage amplitude in the steady-state operation mode.

For the specified small-signal value of $\delta(0)$, the settling time of the oscillations will be defined by the ratio between the initial and steady state amplitudes. If $V(0) < V_0$, then the amplitude V increases monotonically, beginning with a small value $V(0)$ and nears the amplitude V_0, as shown in Figure 1.11. In this case, taking into account that at the beginning $V_0/V(0) \gg 1$ and neglecting the unit component in Equation (1.96), the amplitude increase

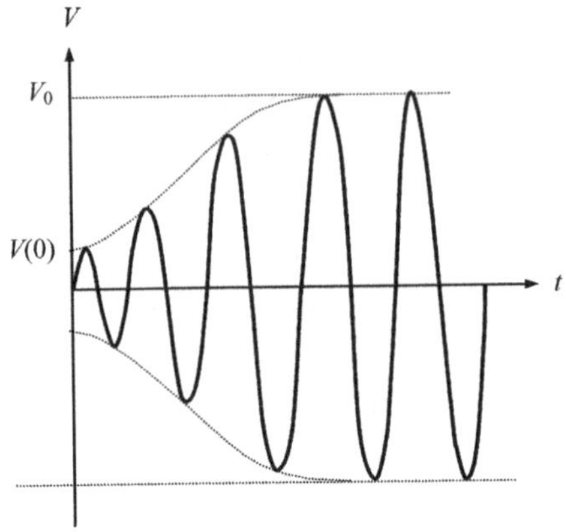

Figure 1.11 Transient response of transformer-coupled MOSFET oscillator

yields to an exponential law according to

$$V = V(0)\exp[|\delta(0)|t] \tag{1.98}$$

which gives, theoretically, an infinitely long settling time t_s.

Defining t_s as the time when the amplitude V increases up to $0.9V_0$, from Equation (1.96) we can obtain

$$t_s = \frac{1}{|\delta(0)|}\ln\left[\frac{2V_0}{V(0)}\right] \tag{1.99}$$

which means that an increase of the coupling factor M or small-signal transconductance g_m results in a shortening of the settling time. Hence, the settling time depends strongly on the initial amplitude $V(0)$ which is determined by the fluctuation process.

1.5 VAN DER POL METHOD

To illustrate a van der Pol method for analysing the behaviour of the oscillation systems described by the nonlinear second-order differential equations, let us consider once again the schematic of the transformer-coupled MOSFET oscillator shown in Figure 1.5. In this case, it is advisable to rewrite the nonlinear second-order differential equation given by Equation (1.53) in the form

$$\frac{d^2v}{dt^2} + 2\delta\frac{dv}{dt} + \omega_0^2 v = 2\delta R\frac{M}{L}\frac{di}{dt} \tag{1.100}$$

where the dissipation factor $\delta = 1/2RC$ includes only losses in the resonant circuit. Now we can use the method of slowly varying amplitudes (the van der Pol method) when Equation (1.100) is replaced by the corresponding truncated first-order differential equations for slowly varying amplitude and phase, respectively [3, 5, 6].

We shall seek a solution of Equation (1.100) in the form of the periodic oscillations of

$$v = V(t)\cos[\omega_0 t + \varphi(t)] \tag{1.101}$$

where $V(t)$ and $\varphi(t)$ are the slowly varying amplitude and phase, respectively. The term 'slowly varying' means that the relative variations of the amplitude and phase during the natural oscillation period are substantially smaller than unity. This means that the time derivatives of the amplitude and phase can be replaced by their average velocities during the oscillation period, i.e.,

$$\frac{dV}{dt} = \frac{\Delta V}{T} = \frac{\omega_0}{2\pi}\Delta V \tag{1.102}$$

$$\frac{d\varphi}{dt} = \frac{\Delta\varphi}{T} = \omega_0\frac{\Delta\varphi}{2\pi} \tag{1.103}$$

Then, under an assumption of slowly varying amplitude when $\Delta V/V \ll 1$, we can write

$$\frac{dV}{dt} \ll \omega_0 V \tag{1.104}$$

Accordingly, for higher-order derivatives of the voltage amplitudes, it is also assumed that

$$\frac{d^2V}{dt^2} \ll \omega_0\frac{dV}{dt} \qquad \frac{d^3V}{dt^3} \ll \omega_0\frac{d^2V}{dt^2} \qquad \ldots \tag{1.105}$$

Similarly, for the slowly varying phase when $\Delta\varphi/2\pi \ll 1$ and its higher-order derivatives, we can write

$$\frac{d\varphi}{dt} \ll \omega_0 \tag{1.106}$$

$$\frac{d^2\varphi}{dt^2} \ll \omega_0\frac{d\varphi}{dt} \qquad \frac{d^3\varphi}{dt^3} \ll \omega_0\frac{d^2\varphi}{dt^2} \qquad \ldots \tag{1.107}$$

The current $i(v)$ where v is defined by Equation (1.101) can be represented by a Fourier series

$$i(v) = I_0(V) + I_{1c}(V)\cos(\omega_0 t + \varphi) - I_{1s}(V)\sin(\omega_0 t + \varphi) + \ldots \tag{1.108}$$

where

$$I_{1c}(V) = \frac{1}{\pi}\int_{-\pi}^{\pi} i(V\cos\psi)\cos\psi\, d\psi \tag{1.109}$$

$$I_{1s}(V) = -\frac{1}{\pi}\int_{-\pi}^{\pi} i(V\cos\psi)\sin\psi\, d\psi \tag{1.110}$$

The first and second derivatives of Equation (1.101) can be calculated as

$$\frac{dv}{dt} = \frac{dV}{dt}\cos(\omega_0 t + \varphi) - V\left(\omega_0 + \frac{d\varphi}{dt}\right)\sin(\omega_0 t + \varphi) \tag{1.111}$$

$$\begin{aligned}\frac{d^2v}{dt^2} &= \frac{d^2V}{dt^2}\cos(\omega_0 t + \varphi) - 2\frac{dV}{dt}\left(\omega_0 + \frac{d\varphi}{dt}\right)\sin(\omega_0 t + \varphi)\\ &\quad + V\frac{d^2\varphi}{dt^2}\sin(\omega_0 t + \varphi) - V\left[\omega_0^2 + 2\omega_0\frac{d\varphi}{dt} + \left(\frac{d\varphi}{dt}\right)^2\right]\cos(\omega_0 t + \varphi)\end{aligned} \tag{1.112}$$

Equation (1.112) can be simplified by neglecting the terms with second derivatives and square of the first derivative to

$$\frac{d^2v}{dt^2} = -2\omega_0\frac{dV}{dt}\sin(\omega_0 t + \varphi) - \left(\omega_0^2 + 2\omega_0\frac{d\varphi}{dt}\right)V\cos(\omega_0 t + \varphi) \tag{1.113}$$

Similarly, by taking into account only linear terms in Equation (1.108), the current derivative $di(v)/dt$ can be obtained as

$$\begin{aligned}\frac{di(v)}{dt} &= -I_{1c}(V)\left(\omega_0 + \frac{d\varphi}{dt}\right)\sin(\omega_0 t + \varphi) - \frac{dI_{1c}(V)}{dt}\cos(\omega_0 t + \varphi)\\ &\quad - I_{1s}(V)\left(\omega_0 + \frac{d\varphi}{dt}\right)\cos(\omega_0 t + \varphi) - \frac{dI_{1s}(V)}{dt}\sin(\omega_0 t + \varphi)\end{aligned} \tag{1.114}$$

As a result, for small values of the dissipation factor and derivatives of the slowly varying functions, Equation (1.100) can be rewritten using Equations (1.111), (1.113) and (1.114) as

$$\begin{aligned}&-2\omega_0\frac{dV}{dt}\sin(\omega_0 t + \varphi) - 2\delta\omega_0 V\sin(\omega_0 t + \varphi) - 2\omega_0 V\frac{d\varphi}{dt}\cos(\omega_0 t + \varphi)\\ &= -2\delta\omega_0\frac{M}{L}RI_{1c}(V)\sin(\omega_0 t + \varphi) - 2\delta\omega_0\frac{M}{L}RI_{1s}(V)\cos(\omega_0 t + \varphi)\end{aligned} \tag{1.115}$$

Finally, equating the terms with sinusoidal and cosinusoidal components in Equation (1.115) results in separate equations for the time-varying amplitude $V(t)$ and phase $\varphi(t)$ in the form

$$\frac{1}{\delta}\frac{dV}{dt} + V = \frac{M}{L}RI_{1c}(V) \tag{1.116}$$

$$\frac{1}{\delta}V\frac{d\varphi}{dt} = \frac{M}{L}RI_{1s}(V) \tag{1.117}$$

For the algebraic transfer function $i(v)$ and cosinusoidal input voltage v in Equation (1.101), the integral for $I_{1s}(V)$ given by Equation (1.110) is equal to zero. Physically this means that the active device has no reactive elements and the oscillator resonant frequency is fully defined by the resonant frequency ω_0 of the parallel LC circuit. Therefore, as follows from Equation (1.117), the phase $\varphi(t)$ of the oscillations given by Equation (1.101) is constant and no longer a function of time. At the same time, the amplitude of the oscillations, which behaviour is described by Equation (1.116), varies exponentially, depending on the dissipation factor δ. The similar result obtained by a quasilinear method is presented by Equation (1.92).

Generally, a procedure of the derivation of the truncated lower-order differential equations from the original nonlinear differential equations is very complicated and time-consuming, even for the simple cases. However, using a symbolic representation of the nonlinear oscillation behaviour and following a Evtyanov approach make it possible to speed up the procedure of obtaining the truncated equations [6, 7].

According to a Evtyanov approach, the oscillator can generally be represented by the ideal current source $i(v)$ and linear two-port network $Z(p)$, where $p \equiv d/dt$ is the differential operator, as shown in Figure 1.12. A symbolic equation to describe the behaviour of the oscillator is written as

$$v = Z(p)i(v) \tag{1.118}$$

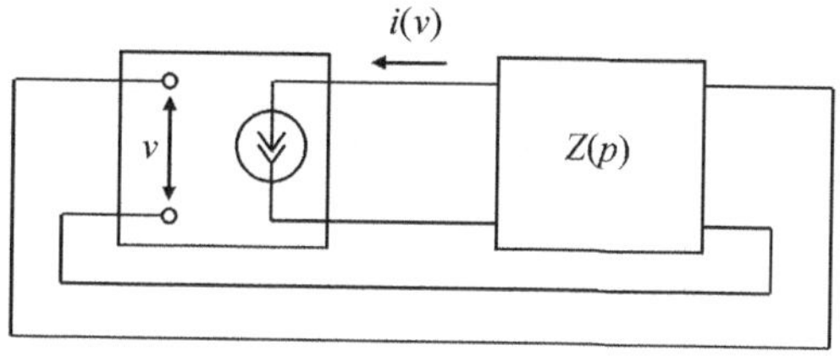

Figure 1.12 General oscillator model

where

$$Z(p) = \frac{\delta P(p,\delta)}{Q(p,\delta)} \tag{1.119}$$

$P(p, \delta)$ and $Q(p, \delta)$ are the polynomial functions of p and δ is the small parameter. For the oscillation systems with high quality factors, the small parameter δ usually represents a dissipation factor for one of its natural frequencies. The differential equation corresponding to a symbolic Equation (1.118) can be obtained by

$$Q(p,\delta)\, v = \delta P(p,\delta)\, i(v) \tag{1.120}$$

A solution of the differential Equation (1.120) is seeking a sum of the oscillations with the slowly varying amplitudes and phases according to

$$v = V_0(t) + \sum_{k=1}^{K} V_k(t)\cos[\omega_{0k}t + \varphi_k(t)] \tag{1.121}$$

where $k = 1, 2, \ldots, K$.

The complex voltage and current amplitudes can be written as

$$\boldsymbol{V}_k = V_k \exp(j\varphi_k) \tag{1.122}$$

$$\boldsymbol{I}_k = I_k \exp(j\varphi_k) \tag{1.123}$$

Then, using a two-dimensional Maclaurin series expansion about $p = 0, \delta = 0$ and neglecting the terms of order δ^2, the following system of truncated differential equations in a complex form can be obtained:

$$\left\{\left(\frac{\partial Q(p,\delta)}{\partial p}\right]_k p + \left(\frac{\partial Q(p,\delta)}{\partial \delta}\right]_k \delta\right\} \boldsymbol{V}_k = \delta(P(p,\delta)]_k \boldsymbol{I}_k \tag{1.124}$$

Dividing both sides of Equation (1.124) by

$$\left(\frac{\partial Q(p,\delta)}{\partial \delta}\right]_k \delta$$

allows us to rewrite Equation (1.124) in the final form

$$(T_k p + 1)\, \boldsymbol{V}_k = Z_k \boldsymbol{I}_k \tag{1.125}$$

where

$$T_k = \left(\frac{\partial Q(p,\delta)}{\partial p}\right]_k \bigg/ \delta \left(\frac{\partial Q(p,\delta)}{\partial \delta}\right]_k \tag{1.126}$$

$$Z_k = (P(p,\delta)]_k \bigg/ \left(\frac{\partial Q(p,\delta)}{\partial \delta}\right]_k \tag{1.127}$$

Substituting the complex voltage and current amplitudes from Equations (1.122) and (1.123) into Equation (1.125) and equating the real and imaginary parts after its differentiating result in two separate differential equations for amplitude and phase, respectively:

$$(T_k p + 1)V_k = R_k I_k(V_0, V_1, \ldots, V_K) \tag{1.128}$$

$$T_k V_k p\varphi_k = X_k I_k(V_0, V_1, \ldots, V_K) \tag{1.129}$$

where $R_k = \text{Re}Z_k$, $X_k = \text{Im}Z_k$, $k = 1, 2, \ldots, K$.

Comparison of a system of Equations (1.128) and (1.129) with a system of Equations (1.116) and (1.117) shows that these truncated differential equations for the slowly varying amplitude and phase are identical when $k = 1$, $V_0 = 0$ and $T_1 = 1/\delta$.

1.6 COMPUTER-AIDED ANALYSIS AND DESIGN

To analyse the nonlinear oscillator circuit, it is necessary to provide its frequency-domain and time-domain simulations giving the device, time, spectral or sweep presentations of the electrical characteristics and optimization of the circuit parameters to realize the optimum solution depending on customer requirements. The algorithm for nonlinear oscillator analysis used in Microwave Harmonica, which is a part of the Ansoft circuit simulator Serenade, is a modified harmonic balance method and can be divided into two steps: search and analysis [8, 9]. In the search mode, an external test source is added to an oscillator circuit to inject ac power, and forces the system away from the degenerate solutions when all ac currents are equal to zero is also a solution of Kirchhoff's current law. This approach includes the Kurokawa oscillation condition, which ensures that the degenerate solution is not obtained. The final steady-state solution of the search mode is treated as the initial estimate for the analysis mode. In the analysis mode, the system equation is solved using the modified harmonic balance method, in which oscillating frequency is used as an independent variable.

The search method includes the injection of an external ac source into the oscillator circuit and finding the steady-state operation mode using harmonic balance conditions and the Newton iteration scheme to solve a set of the system equations expressed in the frequency domain. By sweeping the injected frequency and power and examining both the magnitude and phase of the injected current, the oscillating condition of the circuit can be determined. The free-running oscillator is treated as a one-port network, as shown in Figure 1.13, where Y is the input admittance of overall one-port network including the load admittance Y_L, Y_{in} is the equivalent input admittance of the oscillator resonant circuit with the active device.

The steady-state condition for a single-oscillation frequency ω can be written as

$$Y = Y_{in} + Y_L = 0 \tag{1.130}$$

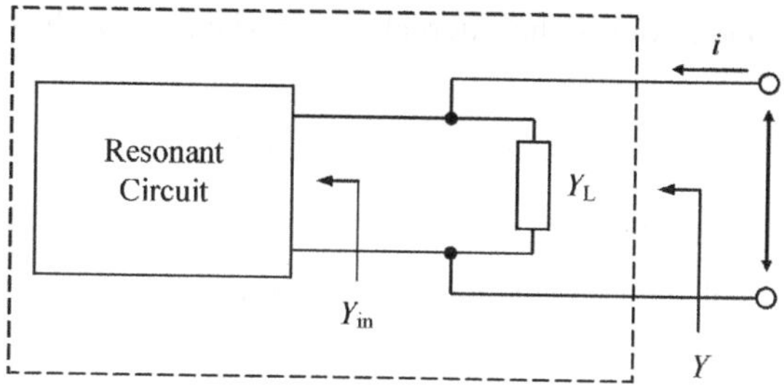

Figure 1.13 Application of the modified harmonic balance method to a free-running oscillator

which can be applied to each frequency component. Therefore, if K different frequency components ω_k are present, the oscillation conditions are determined by

$$Y_k = Y_{\text{in},k} + Y_{\text{L},k} = \text{Re}Y_k + j\text{Im}Y_k = 0 \tag{1.131}$$

where $k = 1, 2, \ldots, K$ indicates that the admittance Y_k is evaluated at each frequency ω_k.

As a result, if current $i(t)$ and voltage $v(t)$ are represented in the time domain by

$$i(t) = \sum_{k=0}^{K} i_k(t) = \sum_{k=0}^{K} I_k \cos(\omega_k t + \phi_k) \tag{1.132}$$

$$v(t) = \sum_{k=0}^{K} v_k(t) - \sum_{k=0}^{K} V_k \cos(\omega_k t + \theta_k) \tag{1.133}$$

where I_k and ϕ_k are the current amplitude and phase, V_k and θ_k are the voltage amplitude and phase, then the input admittance Y_k can be determined by phasor voltage $\boldsymbol{V}_k$ and phasor current $\boldsymbol{I}_k$ as

$$Y_k = \frac{\boldsymbol{I}_k}{\boldsymbol{V}_k} \tag{1.134}$$

By separating the real and imaginary parts of Y_k in Equation (1.131), we can obtain

$$\text{Re}Y_k = \text{Re}(\boldsymbol{I}_k/\boldsymbol{V}_k) = 0 \tag{1.135}$$

and

$$\text{Im}Y_k = \text{Im}(\boldsymbol{I}_k/\boldsymbol{V}_k) = 0 \tag{1.136}$$

which implies

$$\text{Re}I_k = \text{Im}I_k = 0 \tag{1.137}$$

and

$$V_k \neq 0 \tag{1.138}$$

The results obtained by Equations (1.135–1.138) show that, if the voltage amplitude of the injected test source in Figure 1.13 is large enough and the current amplitude of the test source under the harmonic balance condition is zero, the circuit is in oscillation condition. After changing the frequency and power of the external source and monitoring the injected current value, the oscillation condition of the test circuit can be determined. For an efficient search of the frequency and power of the external source when $k = 1$, the following algorithm has been used:

- the external power is set to be constant and at a low level;
- the frequency of the external source is swept until $\mathrm{Im}I_1$ is close to zero and $\mathrm{Re}I_1$ is negative;
- the power of the external source is increased stepwise and tracks the frequency until both $\mathrm{Re}I_1$ and $\mathrm{Im}I_1$ are close to zero.

When both $\mathrm{Re}I_1$ and $\mathrm{Im}I_1$ are reduced to very small values, highly accurate analysis results may be obtained. Furthermore, in order to avoid a large number of computations, a near target solution of about 0.1% error provides a good initial estimate for the analysis mode.

In the analysis mode, the external injected source is excluded and a modified harmonic balance technique is used to obtain a true and rigorous oscillator circuit analysis result. For a nonautonomous circuit analysis, the state variable vector $\boldsymbol{X}$ is composed of the state variables, including device port voltages, and error vector $\boldsymbol{E}$ is composed of the elements of the system errors, including corresponding port current errors. However, for the oscillator analysis, the structure of vector $\boldsymbol{X}$ is modified with the oscillating frequency f_1 as an additional variable, so the phases of the harmonic state variable voltages are referred to the phase of the voltage of the first state variable. In the same manner, to eliminate the degenerate solution, the error function vector $\boldsymbol{E}$ is reconstructed by replacing the error function elements at the fundamental frequency by a function designed to avoid the degenerate solution.

One of the main concerns in the oscillator circuit analysis is to improve the convergence property of the oscillator circuit simulation, especially when it is designed with high-Q resonant tank circuit. In this case, the system error near the oscillation frequency can be very large. Therefore, two techniques, (1) initial frequency setting and (2) fundamental frequency searching, have been used in the circuit simulator [8]:

1. All the node voltages and edge currents can be randomly initialized within a certain range that can reflect the practical level of the oscillator output power P_{out}. The oscillating frequency f_1 should be initialized as sufficiently close to the actual value of the fundamental frequency f, for example, within $0.1f \leq f_1 \leq 10f$.
2. Another way to improve the convergence ability is to first decrease the number of harmonics in order to simplify the error surface and gradually restore it to the desired value. This means a consecutive consideration of the analysis at the fundamental frequency, and then repeating it with the second-harmonic signal present. After obtaining the convergence with the third-harmonic signal present, the number of harmonics is then increased to the maximum number specified by user. This is very important to avoid aliasing when a small number of frequency components are taken into account.

The optimization procedure, which is crucial to provide fast and accurate circuit design by adjusting the values of certain circuit parameters, is based on an iterative process, in which the circuit is simulated to ascertain its electrical responses as compared with the optimization goals. Circuit parameters are adjusted to produce improved circuit responses. The optimization process continues until the selected number of iterations is completed or the optimization goal is achieved. Each goal gives rise to an error value that represents the discrepancy between the simulated circuit response and the appropriate goal limit. If the response satisfies the limits, then the error value is zero. Otherwise, the error value depends on the magnitude of the difference between the simulated response and the appropriate goal limit. In this case, the error function serves as a figure of merit during optimization procedure to select the best optimizable values. This error function value is a sum of the individual goal errors, which are weighted measures of the difference between the simulated circuit response and the desired response, as specified in the goal values. Weights are associated with each goal in order to allow the emphasis of certain goals over others. In a common case, the error function EF is defined by

$$\mathrm{EF} = \sum_{\text{phrases}} \sum_{\text{groups}} \sum_{f} \left[\sum_{i} (w_i e_i) / N_f \right] \tag{1.139}$$

where e_i is the error function contribution from the ith goal at one frequency, w_i is the weighting factor associated with the ith goal, N_f is the number of frequencies for the goal group containing e_i, $\sum_i$ means the summation over all optimization goals in a group, $\sum_f$ means the summation over all frequencies for which a group of goals is specified, $\sum_{\text{groups}}$ means the summation over all groups and $\sum_{\text{phrases}}$ means the summation over all optimization phrases [9]. Each line of goals is considered as a group and may contain only one frequency range, at which all the goals in the group are evaluated, whereas a number of goal groups may be defined within each optimization phrase.

The optimization procedure may include different optimization methods, for example, several such methods as random search, gradient search and minimax search are used in Ansoft circuit simulator Serenade. The *gradient search* is based on a quasi-Newton algorithm, which uses the exact gradient and approximate inverse of the Hessian matrix of the error function to find a direction of improvement for each optimization value. The first search direction is in the direction of the gradient vector along a line in n-dimensional space where n is the number of optimized values. Once a minimum is found in the first direction, a second search along another line in the same n-dimensional space is performed. In the second and subsequent iterations, the direction of search depends on the gradient vector, which is not the same as the overall gradient. The direction of search is modified to accelerate convergence as a minimum is approached. However, the gradient search is susceptible to local minimum points when, once a local minimum region of the error function is reached, the gradient search method may have difficulty in selecting optimizable values outside that minimal region.

The *random search* selects new optimizable values following a Monte Carlo approach. Starting from an initial set of optimizable values, for which the error function value is known, a new set of values is obtained using a random-number generator within the applicable optimizable value ranges. The error function is re-evaluated and these optimizable values are retained if a decrease in its value is identified. This is a trial-and-error process, in which random search, step-by-step, finds at least minimum of the error function. The random search repeats this procedure for as many times as the number of iterations specified before the optimization is started in order to approximate to the global minimum of the error function as close as possible. Initially, new optimizable values are drawn according to a uniform Gaussian distribution for each optimizable value and these optimizable variables are treated as independent Gaussian variables. After each iteration, whether it is successive or not, the distribution is modified and becomes non-Gaussian to its skewing towards lower error function values and away from higher ones. The random search tends to proceed in the direction of error function reduction, but it is not restricted to such areas completely, allowing improvement of the search efficiency without the risk of trapping the search in local minimum.

The *minimax method* provides the minimization of the largest weighted goal errors, i.e., the minimization of the maximum contributions to the error function value. The minimax error function always represents only the worst case violation of the optimization goals, where the desired circuit response specifications are either most severely violated when EF > 0, or satisfied with the smallest margin when EF < 0. In this case, the error function to be minimized may be defined in general as

$$\text{EF} = \text{MAX}_{\text{phrases}}\,\text{MAX}_{\text{groups}}\,\text{MAX}_f\,\text{MAX}_{\text{goals}}(w_i e_i) \tag{1.140}$$

where e_i is the discrete error function associated with a phrase at one frequency, w_i is the weighting factor associated with e_i, $\text{MAX}_{\text{goals}}$ means maximum value in the set over all goals of a phrase, MAX_f means maximum value in the set over all frequencies of a group, $\text{MAX}_{\text{groups}}$ means maximum value over all goal groups and $\text{MAX}_{\text{phrases}}$ means maximum value over all optimization phrases. A minimax solution means that the goal specifications are met in an optimal, typically equal-ripple manner. The sophisticated minimax search method proceeds in two stages. In the first stage of the search, the minimax problem is solved using a linear programming technique and, in the second stage, the search employs a quasi-Newton algorithm with second-order derivatives. A minimax iteration requires one evaluation of the objective function and its gradient and therefore is less time-consuming than an iteration of the gradient search.

REFERENCES

1. A. I. Berg, *Theory and Design of Tube Generators* (in Russian), Moskva: GEI (1932).
2. J. K Fidler and C. Nightingale, *Computer Aided Circuit Analysis*, Middlesex: Thomas Nelson & Sons (1978).
3. A. A. Andronov, A. A. Vitt and S. E. Khaikin, *Theory of Oscillations*, New York: Dover Publications (1987).
4. Y. Tajima, B. Wrona and K. Mishima, 'GaAs FET Large-Signal Model and its Application to Circuit Designs', *IEEE Trans. Electron Devices*, **ED-28**, 171–175 (1981).
5. B. van der Pol, 'The Nonlinear Theory of Electric Oscillations', *Proc. IRE*, **22**, 1051–1086 (1934).
6. M. V. Kapranov, V. N. Kuleshov and G. M. Utkin, *Theory of Oscillations in Radio Engineering* (in Russian), Moskva: Nauka (1984).
7. S. I. Evtyanov, *Electron Tube Oscillators* (in Russian), Moskva: Svyaz (1967).
8. C. R. Chang, M. B. Steer, S. Martin and E. Reese, 'Computer-Aided Analysis of Free-Running Microwave Oscillators', *IEEE Trans. Microwave Theory Tech.*, **MTT-39**, 1735–1745 (1991).
9. Microwave Harmonica®, *Reference Volume, Version 7.5 P.C.*, Ansoft Corporation (1998).

2

Oscillator operation and design principles

This chapter presents the principles of oscillator design, including start-up and steady-state operation conditions, basic oscillator configurations using lumped and transmission-line elements and simplified equation-based oscillator analysis and design techniques. An immittance design approach is introduced and applied to series and parallel feedback oscillators, including circuit design and simulation aspects. Numerous practical examples of RF and microwave oscillators using MOSFET, MESFET and bipolar devices, including the descriptions of their circuit realizations, are given.

2.1 STEADY-STATE OPERATION MODE

A simple feedback oscillator model is shown in Figure 2.1a where an oscillator circuit is decomposed into a forward nonlinear network and a feedback linear network, both of which are two-port networks. Figure 2.1b shows an example of a transformer-coupled MOSFET oscillator without bias circuit to illustrate common features of the feedback oscillator. Because an oscillator is an autonomous circuit, electronic noise in the active device or power supply turn-on transient leads to the self-excitation of the oscillations. This provides the initial oscillation build-up. As the oscillation amplitude grows, the active device displays larger nonlinearity and then limits the amplitude increase.

In a steady-state operation mode, the following complex equation, also known as the *Barkhausen criterion*, can be written for the parallel feedback oscillator:

$$\boldsymbol{T}(V_{\text{in}}, \omega) = \boldsymbol{A}(\boldsymbol{V}_{\text{in}}, j\omega)\boldsymbol{\beta}(j\omega) = 1 \tag{2.1}$$

where $\boldsymbol{A} = \boldsymbol{I}_{\text{out}}/\boldsymbol{V}_{\text{in}}$ is the forward transfer function and $\boldsymbol{\beta} = \boldsymbol{V}_{\text{in}}/\boldsymbol{I}_{\text{out}}$ is the feedback transfer function. This equation means that the oscillator complex loop gain is equal to unity [1, 2]. The feedback transfer function can be represented as

$$\boldsymbol{\beta}(j\omega) = \boldsymbol{K}(j\omega)\boldsymbol{Z}(j\omega) \tag{2.2}$$

where $\boldsymbol{K} = \boldsymbol{V}_{\text{in}}/\boldsymbol{V}_{\text{out}}$ is the voltage feedback coefficient and $\boldsymbol{Z} = \boldsymbol{V}_{\text{out}}/\boldsymbol{I}_{\text{out}}$ is the oscillator resonant circuit impedance. Presenting each of these complex quantities in the form of $\boldsymbol{A} = A\exp(j\phi_{\text{A}})$, $\boldsymbol{K} = K\exp(j\phi_{\text{K}})$, $\boldsymbol{Z} = Z\exp(j\phi_{\text{Z}})$, the following equations for magnitudes and

RF and Microwave Transistor Oscillator Design A. Grebennikov

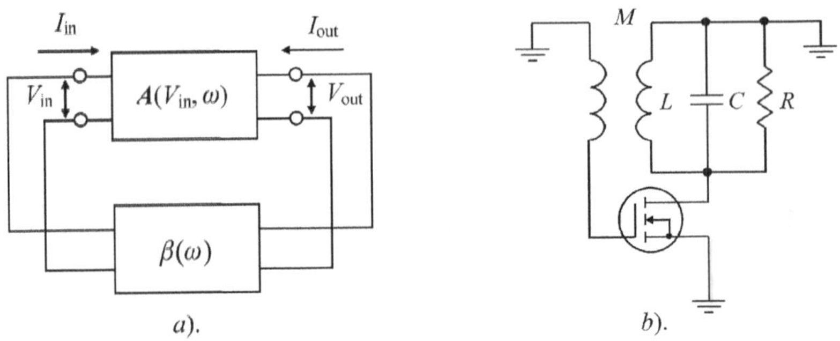

Figure 2.1 Schematics of (a) parallel feedback oscillator and (b) transformer-coupled MOSFET oscillator

phases directly follow from Equation (2.1):

$$A(V_{in}, \omega)K(\omega)Z(\omega) = 1 \tag{2.3}$$

$$\phi_A + \phi_K + \phi_Z = 0, 2\pi, \ldots \tag{2.4}$$

Equation (2.3), which is called the *amplitude balance condition*, means that the oscillator loop gain is equal to unity in the steady-state stationary operation mode. In this equation it is assumed that two quantities, K and Z, depend on frequency only. Consequently, the amplitude balance condition is carried out only under the appropriate value of input voltage amplitude V_{in}. To define the value of this amplitude, let us rewrite Equation (2.3) in the form

$$A(V_{in}, \omega) = 1/K(\omega)Z(\omega) \tag{2.5}$$

In Figure 2.2, the amplitude dependence $A(V_{in})$ and feedback straight line $1/KZ$ are plotted. The intersection point of these dependencies is determined the steady-state oscillation amplitude V_{in}^{o}.

Equation (2.4), which is called the *phase balance condition*, means that the sum of all oscillator loop phase shifts must be equal to zero or $2\pi n$, where $n = 1, 2 \ldots$. This equation defines the value of the oscillation frequency f_{osc}. In the simple case when $\phi_K = 0$ (transistor

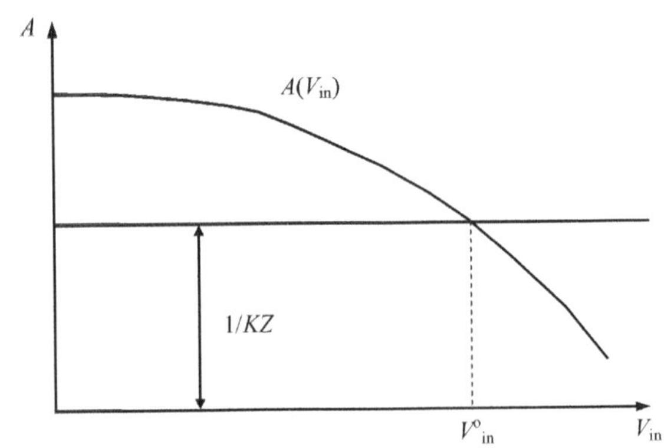

Figure 2.2 Graphic balance amplitude condition

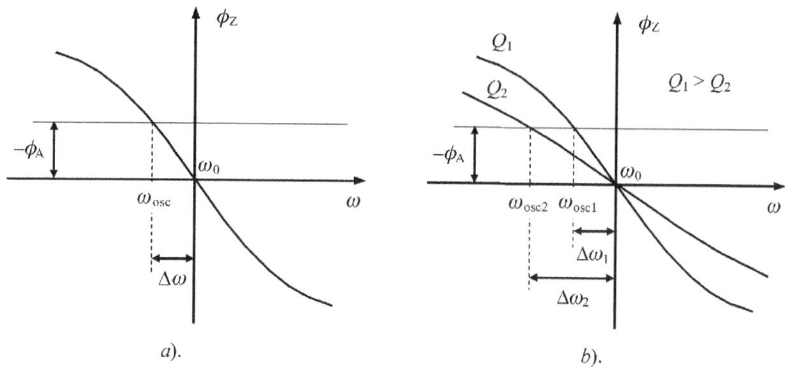

Figure 2.3 Deviation of oscillation frequency f_{osc} from resonant frequency f_0

input admittance is equal to zero and feedback magnitude K depends only on a mutual coupling coefficient between primary and secondary inductances M) and the active device doesn't produce the phase shift, that is $\phi_A = 0$, then $\phi_Z = 0$ and the oscillation frequency f_{osc} is equal to parallel resonant circuit frequency $f_0 = \omega_0/2\pi = 1/2\pi\sqrt{LC}$. If $\phi_A \neq 0$, the oscillation frequency f_{osc} will differ from f_0 in order to fully compensate the available phase shift ϕ_Z according to

$$\phi_Z = -\tan^{-1}\left(2Q\frac{\Delta\omega}{\omega_0}\right) = -\phi_A \tag{2.6}$$

where $Q = 1/\omega CR$ is a quality factor of the oscillator resonant circuit, $\Delta\omega = 2\pi(f_{osc} - f_0)$ [2].

Equation (2.6) can be rewritten in the form

$$\frac{\Delta\omega}{\omega_0} = \frac{\tan\phi_A}{2Q} \tag{2.7}$$

which determines the deviation of the oscillation frequency f_{osc} from resonant frequency f_0 as a function of the phase of the forward transfer function of the active device and oscillator quality factor.

From graphical representation of Equation (2.6) shown in Figure 2.3a it follows that the oscillation frequency f_{osc} is smaller than resonant frequency f_0. Furthermore, the lower the quality factor Q the smaller the oscillation frequency f_{osc}, as shown in Figure 2.3b, where the deviation $\Delta\omega_2$ from resonant frequency ω_0 becomes greater for the case of $Q_2 < Q_1$.

2.2 START-UP CONDITIONS

The start-up conditions can be illustrated on the plane of the transfer function $i_{out} = f(v_{in})$, as shown in Figure 2.4. Let us choose the bias voltage V'_g corresponding to maximum small-signal transfer function when $\partial i_{out}/\partial v_{in} = 0$. Such a condition is adequate to the maximum forward transfer function $A(V_{in} = 0)$ shown in Figure 2.2. According to Figure 2.2, the amplitude V_{in} of the oscillations grows monotonically whereas the transfer function A decreases. The amplitude of the output current i_{out} is trying to reach its maximum value corresponding to the stable steady-state operation mode when $V_{in} = V_{in}^{o}$. Such a behaviour of the transfer function

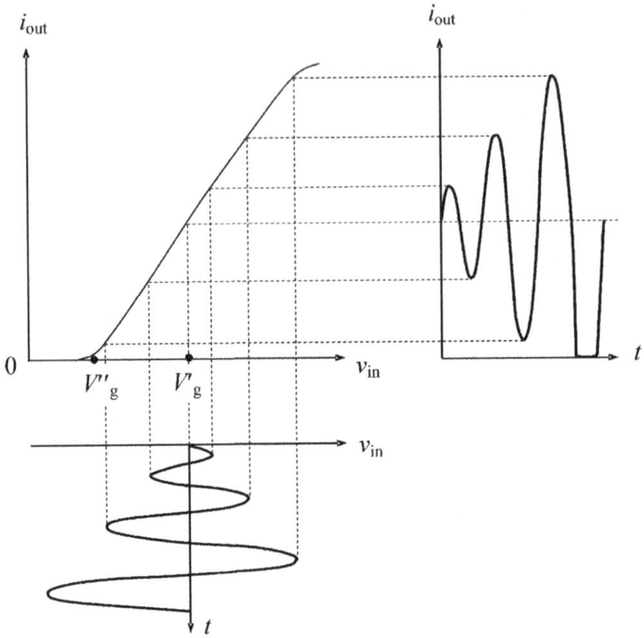

Figure 2.4 Start-up of oscillations

A means that the start-up oscillation conditions can be defined as

$$A(\omega)K(\omega)Z(\omega) > 1 \tag{2.8}$$

$$\phi_A + \phi_K + \phi_Z = 0, 2\pi, \ldots \tag{2.9}$$

Let the bias voltage V_g'' be chosen close to the device threshold voltage as shown in Figure 2.4. Such a bias condition corresponds to the small initial value of transfer function $A(V_{in} = 0)$, which is not enough to establish the stable steady-state stationary oscillations because $A < 1/KZ$. This is demonstrated by curve *II* in Figure 2.5a. Therefore, the oscillation system with such a bias condition requires an impulse to initiate self-oscillations, as it cannot oscillate by

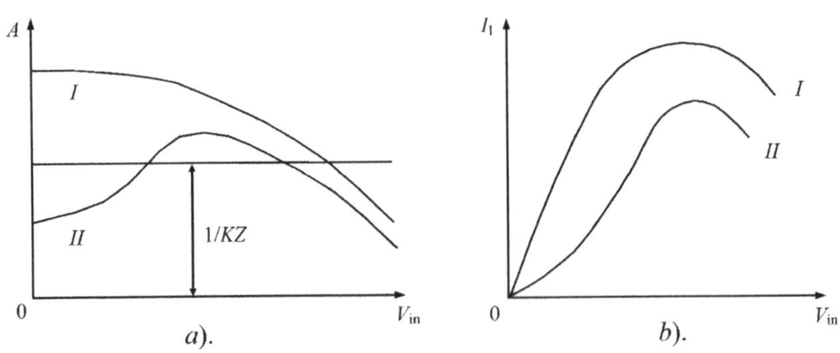

Figure 2.5 Balanced amplitude conditions for different biasing points

itself. The result of this impulse should be an output current amplitude resulting in a condition of $A > 1/KZ$. Hence, the system has hard operating conditions and the process of the oscillation establishment is called a *hard build-up* of self-oscillations. This means that oscillations of finite amplitude are established suddenly under some external influence.

Figure 2.4 shows the process of establishment of the stable self-oscillations as a result of the nonlinearity of the transfer function of the active device operating in pinch-off and active regions. This means that the start-up oscillation conditions correspond to a Class A operation mode of the active device having maximum or close to maximum value of the small-signal transconductance. The steady-state oscillation conditions are established when the active device operates in Class AB mode characterized by the conduction angle which particular value depends on the initial bias conditions. Let us assume a high value of the resonant circuit quality factor when the input cosinusoidal voltage

$$v_{in} = V_{in} \cos \omega t \tag{2.10}$$

is applied to the active device, representing an idealized nonlinear voltage-controlled current source. Then, the output current i_{out} contains the harmonic components and, being an even function, can be written as

$$i_{out} = I_0 + I_1 \cos \omega t + I_2 \cos 2\omega t + I_3 \cos 3\omega t + \ldots \tag{2.11}$$

For the transfer function $i_{out} = f(v_{in})$, the fundamental component I_1 can be obtained from Equations (2.10) and (2.11) using a Fourier formula according to

$$I_1 = \frac{1}{\pi} \int_{-\pi}^{\pi} f(V_{in} \cos \omega t) \cos \omega t \, d(\omega t) \tag{2.12}$$

Consequently, dividing the output fundamental amplitude I_1 by the input voltage amplitude V_{in} provides an analytical expression to calculate the fundamentally averaged value of the forward transfer function $A(V_{in})$ in a quasilinear approximation. The dependence of the output fundamental current amplitude I_1 on the input voltage amplitude V_{in}, expressed generally as

$$I_1 = f_1(V_{in}) \tag{2.13}$$

can be called the *amplitude characteristic* of the oscillator. Figure 2.5 shows the dependencies of (a) the amplitude characteristic and (b) the averaged transfer function as functions of the input voltage amplitude V_{in} for different operation modes where curves *I* and curves *II* correspond to the *soft* and *hard* start-up conditions for self-oscillation establishment, respectively.

Consider the influence of the harmonic components on the process of the establishment of the self-oscillations for finite value of the quality factor Q. Then, the input voltage according to Figure 2.1b can be written as

$$v_{in} = V_{in1} \cos \omega t + V_{in2} \cos\left(2\omega t - \frac{\pi}{2}\right) + V_{in3} \cos\left(3\omega t - \frac{\pi}{2}\right) + \ldots \tag{2.14}$$

where the resonant circuit is tuned to the fundamental and has the capacitive reactance for the second and higher-order harmonics. If we provide a second-order polynomial approximation of the nonlinear transfer function of the active device in the form

$$i_{out} = a_0 + a_1 v_{in} + a_2 v_{in}^2 \tag{2.15}$$

and confine our attention to the first two components in Equation (2.14), the fundamental component of the output current can be obtained by

$$i_1' = I_1 \cos \omega t + a_2 V_{\text{in}1} V_{\text{in}2} \cos\left(\omega t - \frac{\pi}{2}\right) \tag{2.16}$$

where $I_1 = a_1 V_{\text{in}1}$.

Equation (2.16) can be rewritten in the form

$$i_1' = I_1' \cos(\omega t + \phi_{\text{A}}) \tag{2.17}$$

where

$$I_1' = I_1 \sqrt{1 + (a_2/a_1)^2 \, V_{\text{in}2}^2} \tag{2.18}$$

and

$$\tan \phi_{\text{A}} = -a_2 V_{\text{in}2}/a_1 \tag{2.19}$$

Thus, the second component in Equation (2.16) results in the variation of the output current fundamental amplitude when $I_1' \neq I_1$ and the appearance of a negative value of the phase ϕ_{A} of the device transfer function, i.e., the phase shift between the fundamental voltage and fundamental current. This means that an increase in the harmonic content of the output voltage spectrum (smaller Q-factor) causes a decrease in the frequency of self-oscillations, as follows from Figure 2.3b.

Physically, the influence of the harmonic content on the frequency variation can be explained as follows: if the oscillations are purely sinusoidal, the energy distribution in both arms of the resonant LC circuit is equal; when the harmonics appear, the currents corresponding to them flow mainly through the capacitive arm, and therefore they increase the electrostatic energy of this arm in comparison with the inductive arm; in order to keep the energy equal in both arms, the fundamental frequency must diminish slightly with respect to the frequency given by the tank circuit only [3].

In Figure 2.6a, for the example of the transformer-coupled MOSFET oscillator shown in Figure 2.1b, the dependencies $I_1(V_{\text{in}})$ for a soft start-up condition with different values of

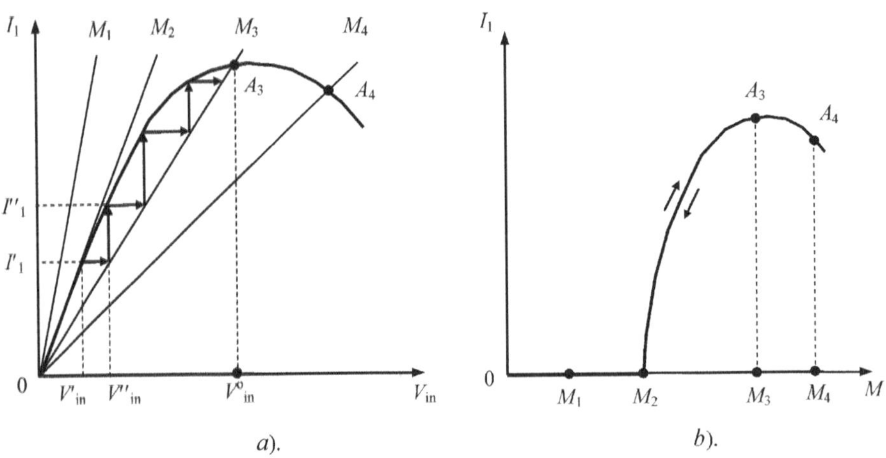

Figure 2.6 Start-up conditions with different values of the mutual coupling coefficient M

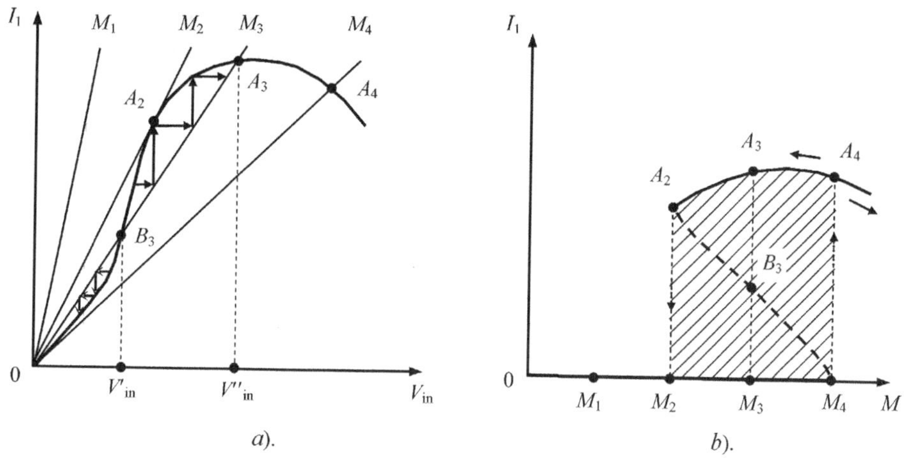

Figure 2.7 Hard start-up condition versus mutual coupling coefficient M

mutual coupling coefficient $M(M_1 < M_2 < M_3 < M_4)$ are presented [1, 2]. As M increases from zero to M_2, the only stable equilibrium corresponds to the static operation mode at the point $V_{in} = 0$. When $M > M_2$, from two potentially existing equilibrium conditions the dynamic conditions at the points A_3 and A_4 are stable. To check the stable equilibrium at the point A_3 with amplitude $V_{in} = V_{in}^o$ and unstable at the $V_{in} = 0$, assume an oscillation rise of the small input amplitude V'_{in} due to some internal or external effect. This would cause an appearance of the output current with the amplitude I'_1, which is determined according to the oscillator amplitude characteristic $I_1(V_{in})$. At that time this current makes conditional the input voltage with amplitude V''_{in}. As a result, the oscillation amplitude that has arisen accidentally increases up to the equilibrium value of V_{in}^o at the point A_3. With the growth of M, the output current amplitude I_1 changes smoothly, as shown in Figure 2.6b. When M decreases, the amplitude I_1 changes in accordance with the $I_1(M)$ curve and under $M = M_2$ the oscillations disappear.

In Figure 2.7a, the dependencies $I_1(V_{in})$ for hard start-up condition with different values of mutual-coupling coefficient $M(M_1 < M_2 < M_3 < M_4)$ are shown [1, 2]. At the point $M = M_3$ both curves intersect at three points, corresponding to three stationary modes: 0 is an equilibrium condition, B_3 and A_3 are the dynamic modes with amplitudes V'_{in} and V''_{in}, respectively. From the process of changing of the output fundamental current amplitude I_1 with initial voltage deviations from V'_{in} and V''_{in}, it follows that the stable conditions correspond to the points 0 and A_3, whereas the condition in the point B_3 is unstable.

Let us define the dependence of output fundamental current amplitude I_1 versus feedback coefficient M. As M increases from zero up to M_4 when the straight line corresponding to M_4 and the curve $I_1(V_{in})$ are tangents in the origin, the only stable equilibrium corresponds to the static operation mode at the point $V_{in} = 0$ when the small fluctuations can not produce an oscillation arise. At the point $M = M_4$, the above-mentioned operation mode becomes unstable and small oscillations grow up to a large amplitude value corresponding to the point A_4. The subsequent increase of M leads to the amplitude change along the curve $I_1(M)$, as shown in Figure 2.7b. If, then, to decrease M down to M_2, the collapse of the oscillation can occur only at the point A_2 because the dynamic operations modes corresponding to the points A_3 and A_4 are stable. This fact results from the qualitative process examination of Figure 2.7a where at the same time the points A_2 and B_3 are unstable. So, between the points A_2 and

A_4 in the limits of $M_2 < M < M_4$, the *hysteresis region* occurs. Thus, the hard build-up of the self-oscillation is characterized by spasmodic development of the oscillation with large amplitude under smooth increase of the feedback coefficient M and spasmodic collapse of the oscillation under smooth decrease of M.

2.3 OSCILLATOR CONFIGURATIONS AND HISTORICAL ASPECTS

The first oscillator configurations using vacuum tubes were based on the electromagnetic coupling between the output and input circuits. By providing a close enough coupling with the output, sufficient energy is supplied to the input circuit to keep the continuous self-sustaining oscillations. In this case, the oscillation frequency is approximately equal to the resonant frequency of the parallel LC circuit if the coupling and active device parasitics are small enough.

There are two basic configurations of a transformer-coupled oscillator. The oscillator configuration with a parallel resonant circuit at the input electromagnetically coupled with the output, as shown in Figure 2.8a, is called the *Armstrong oscillator* in honour of the outstanding American engineer E. H. Armstrong who first developed and described the condition of obtaining the self-sustaining oscillations using such a configuration [4]. Approximately at the same time, A. Meissner in Germany described the transformer-coupled oscillator with a parallel resonant circuit at the output electromagnetically coupled with the input, as shown in Figure 2.8b, which is called the *Meissner oscillator* [5]. High-purity stable oscillations can be obtained by a fully balanced version of Meissner oscillator with a three-turn link providing an excellent isolation of the resonant circuit when very little energy is taken from the resonator to provide the voltages to drive the input and sustain the oscillations [6].

To realize a phase balance condition in the transformer-coupled oscillator, the transformer should provide a phase shift of 180°. If the primary and secondary windings of the transformer have the same direction of wind, it is necessary to connect the secondary winding in the opposite direction relative to the primary one, that is, to connect the end of the secondary winding where the voltage is in-phase with the drain (collector) voltage to the ground. The coupling coefficient of the transformer is chosen to provide a soft start-up oscillation mode.

The schematic of the *Hartley oscillator* is very close to the schematic of the Meissner oscillator [7]. The difference is that the inductor L having an additional output replaces the transformer, as shown in Figure 2.9a. The inductance ratio determines the feedback coefficient. The Hartley oscillator can be represented by a well-known inductive three-point configuration

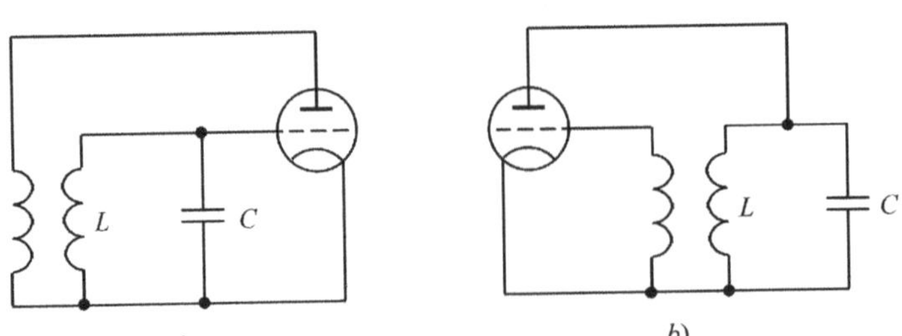

Figure 2.8 Schematics of (a) Armstrong and (b) Meissner oscillators

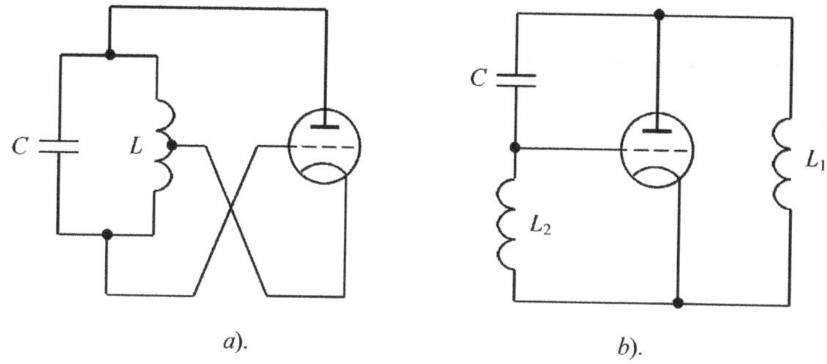

Figure 2.9 Schematics of Hartley oscillators

shown in Figure 2.9b, where the feedback element from the output to the input is the capacitor C. The inductors L_1 and L_2 represent the output and input circuits, respectively.

Unlike the Hartley oscillator with electromagnetic coupling, the schematic of a *Colpitts oscillator* is based on electrostatic coupling using the capacitive divider, the capacitance ratio of which determines the feedback coefficient, as shown in Figure 2.10a [8]. The Colpitts oscillator can be represented by a well-known capacitive three-point configuration shown in Figure 2.10b, where the feedback element from the output to the input is the inductor L. The capacitors C_1 and C_2 represent the output and input circuits, respectively.

Depending on which electrode of the active device is grounded, we can distinguish three basic configurations representing a Colpitts family of oscillators shown in Figure 2.11 for the example of the MOSFET device. The Colpitts oscillator with grounded source is shown in Figure 2.11a, the Colpitts oscillator with grounded drain anode is shown in Figure 2.11b, and the Colpitts oscillators with grounded gate is shown in Figure 2.11c, d. The connection of the appropriate electrode of the tube (or transistor) to the ground have important effects upon the following:

- the manner in which the dc power is fed to the oscillator and the corresponding loading effects on the resonant circuit;
- the manner in which the output power is fed to the external load;
- the distribution of the stray elements to the ground plane which has important effects, particularly for microwaves.

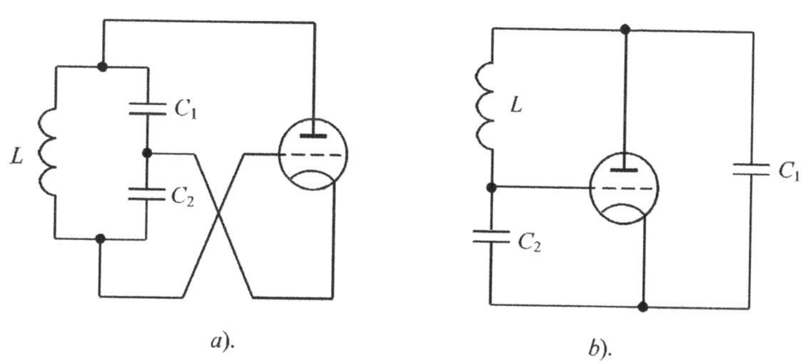

Figure 2.10 Schematics of Colpitts oscillators

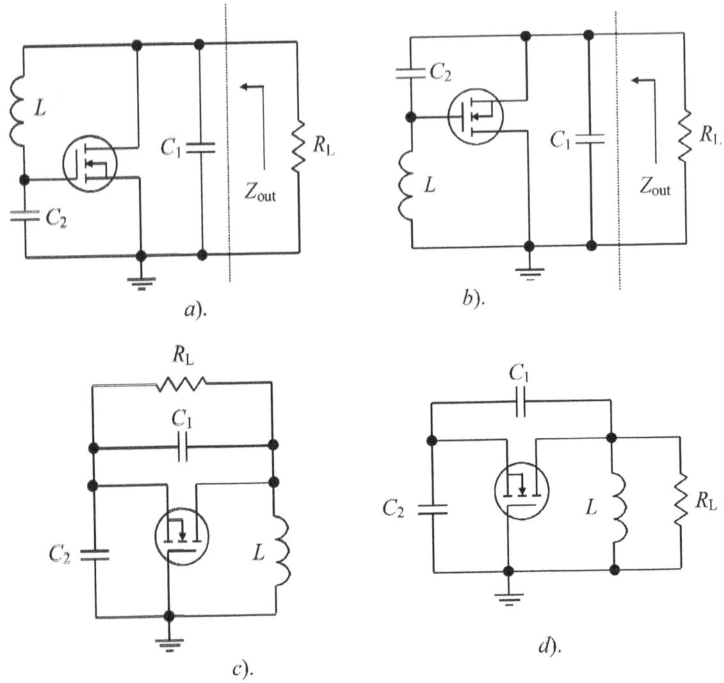

Figure 2.11 Colpitts family of oscillators

To understand whether the electrical performance of the same oscillator can change depending on the grounded electrode or terminal of the active device, let us consider two schematics of the Colpitts oscillator, with grounded source (emitter) shown in Figure 2.11a and grounded drain (collector) shown in Figure 2.11b. In both schematics, each of which represents the three-port network in a common case, the load resistance R_L is connected in parallel to the capacitance C_1, i.e., between drain and source terminals, respectively.

The indefinite admittance Y-matrix of a three-port network can be written as

$$[Y] = \begin{bmatrix} Y_{11} & Y_{12} & -(Y_{11}+Y_{12}) \\ Y_{21} & Y_{22} & -(Y_{21}+Y_{22}) \\ -(Y_{11}+Y_{21}) & -(Y_{12}+Y_{22}) & Y_{11}+Y_{12}+Y_{21}+Y_{22} \end{bmatrix} \tag{2.20}$$

By choosing successively terminal 1, 2 and 3 as the datum terminal of the active device shown in Figure 2.12, we can obtain the appropriate three definite two-port admittance matrices of the initial three-port network. Consider finding these two-port matrices in the form of an admittance matrix of (a) the common source (emitter), (b) the common gate (base) and (c) the common drain (collector) circuit connection of the bipolar transistor, as shown in Figure 2.12. We will use the term 'common terminal' for the grounded terminal, similarly to that of using at the power amplifier design. If the common source device is treated as a two-port circuit and characterized by four Y-parameters (Y_{11}, Y_{12}, Y_{21} and Y_{22}), the two-port matrix of the common drain circuit with grounded-drain terminal is simply obtained by deleting the second row and the second column in the matrix (Equation 2.20). For the common gate circuit with

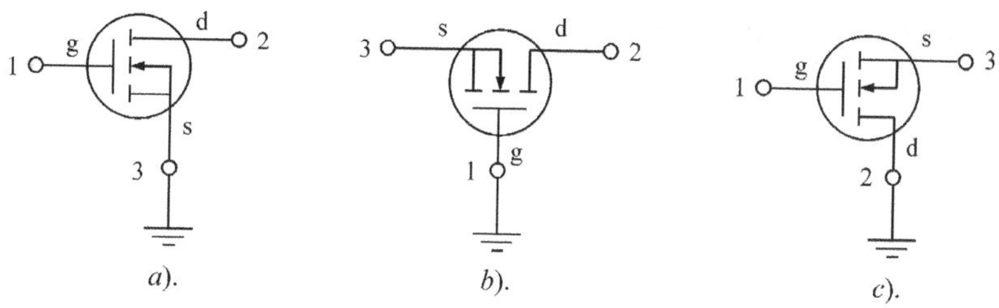

Figure 2.12 MOSFET devices with different common terminals

grounded-gate terminal, the first row and the first column should be deleted because the source terminal is considered the input terminal.

A similar approach can be applied to the indefinite three-port impedance network. This allows the impedance Z-parameters of the common gate (base) and common drain (collector) circuits to be determined through the initially known impedance Z-parameters of the common source (emitter) circuit. Parameters of the three-port network with different common terminals for bipolar and FET devices are given in Table 2.1.

For the schematic with common source shown in Figure 2.11a, we can write

$$Y_{out} = Y_{22} + j\omega C_1 + \frac{1}{j\omega L} - \frac{\left(Y_{12} - \dfrac{1}{j\omega L}\right)\left(Y_{21} - \dfrac{1}{j\omega L}\right)}{Y_{11} + j\omega C_2 + \dfrac{1}{j\omega L}} \tag{2.21}$$

where $Y_{out} = 1/Z_{out}$ is the equivalent output admittance.

For the oscillator schematic with common drain shown in Figure 2.11b, by using the relationships from Table 2.1, we can write

$$Y_{out} = Y_{11} + Y_{12} + Y_{21} + Y_{22} + j\omega C_1 + j\omega C_2 - \frac{(-Y_{11} - Y_{12} - j\omega C_2)(-Y_{11} - Y_{21} - j\omega C_2)}{Y_{11} + j\omega C_2 + \dfrac{1}{j\omega L}} \tag{2.22}$$

Table 2.1 Active device Y- and Z-parameters with different common terminals

	Y-parameters		Z-parameters	
Common source (emitter)	Y_{11}	Y_{12}	Z_{11}	Z_{12}
	Y_{21}	Y_{22}	Z_{21}	Z_{22}
Common gate (base)	$Y_{11} + Y_{12} + Y_{21} + Y_{22}$	$-(Y_{12} + Y_{22})$	$Z_{11} + Z_{12} + Z_{21} + Z_{22}$	$-(Z_{12} + Z_{22})$
	$-(Y_{21} + Y_{22})$	Y_{22}	$-(Z_{21} + Z_{22})$	Z_{22}
Common drain (collector)	Y_{11}	$-(Y_{11} + Y_{12})$	Z_{11}	$-(Z_{11} + Z_{12})$
	$-(Y_{11} + Y_{21})$	$Y_{11} + Y_{12} + Y_{21} + Y_{22}$	$-(Z_{11} + Z_{21})$	$Z_{11} + Z_{12} + Z_{21} + Z_{22}$

Equation (2.22) can be rewritten as

$$Y_{\text{out}} = Y_{22} + j\omega C_1 + \frac{1_1}{j\omega L} + \frac{\left(Y_{11} + Y_{12} + Y_{21} + j\omega C_2 - \dfrac{1}{j\omega L}\right)\left(Y_{11} + j\omega C_2 + \dfrac{1}{j\omega L}\right)}{Y_{11} + j\omega C_2 + \dfrac{1}{j\omega L}}$$

$$- \frac{(Y_{11} + Y_{12} + j\omega C_2)(Y_{11} + Y_{21} + j\omega C_2)}{Y_{11} + j\omega C_2 + \dfrac{1}{j\omega L}}$$

$$= Y_{22} + j\omega C_1 + \frac{1}{j\omega L} - \frac{\left(Y_{12} - \dfrac{1}{j\omega L}\right)\left(Y_{21} - \dfrac{1}{j\omega L}\right)}{Y_{11} + j\omega C_2 + \dfrac{1}{j\omega L}} \tag{2.23}$$

which is identical to Equation (2.21) corresponding to the output admittance Y_{out} of the oscillator schematic with common source. Consequently, grounding of any terminal of the oscillator circuit does not change its electrical performance, provided there are no changes in the connection of the feedback elements and load to the active device.

The schematic with a common gate shown in Figure 2.11c is characterized by the same electrical behaviour as the oscillator circuits with common source and common drain. However, in such a configuration there is no connection of any load terminal to the ground, which makes its practical implementation more complicated. Figure 2.11d shows the oscillator schematic with a common gate with the load connection in parallel to the inductance L, that is between the drain (collector) and gate (base) terminals. However, another load connection with one grounded port results in different electrical properties of the oscillator compared with common source or common drain configurations.

The Gouriet–Clapp oscillator is a variation of the Colpitts oscillator with a tank inductor replaced by the series combination of the inductor L and variable capacitor C_{v}. In this case, frequency stability is improved because the reactance of such a series circuit varies more rapidly with frequency than that of a single inductor. However, the possibility of improving the oscillator stability by connecting a capacitor in series to one or each circuit inductor was first found, based on a Hartley type of oscillator [9]. In modern practical oscillator design, the Gouriet–Clapp oscillator configuration is called the *Clapp oscillator*, although G. Gouriet and J. Clapp had developed it independently [10]. A parallel counterpart of the Clapp oscillator shown in Figure 2.13b was described by E. Seiler and, therefore, is called the *Seiler*

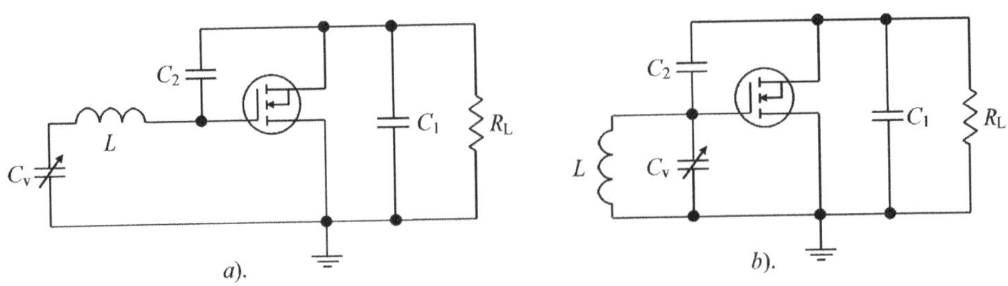

Figure 2.13 Schematics of (a) Gouriet–Clapp and (b) Seiler oscillators

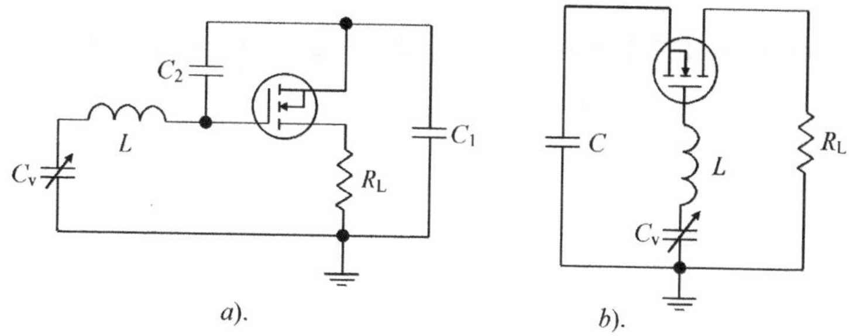

Figure 2.14 Schematics of modified Clapp oscillator with common gate

oscillator [10]. Such an oscillator configuration is useful for wideband frequency tuning when the capacitance C_v in the parallel circuit is variable.

Figure 2.13 shows the schematics of the Clapp and Seiler oscillators with common drain. However, to realize their common gate configurations, it is necessary to change the location of the load resistance to provide its grounding. For example, the load resistance R_L can be connected between the drain terminal and the ground, as shown in Figure 2.14a for the Clapp oscillator. Such a modified Clapp configuration, which is shown in Figure 2.14b, represents the common gate (base) oscillator with a series circuit in the gate (base). Usually, to provide a stable soft start-up oscillation condition, the reactance in the source (emitter) circuit should be capacitive, being inductive in the gate (base) circuit.

The *Vackar oscillator* is the modified Clapp oscillator with additional variable capacitor C_v, which simplified equivalent circuit representation is shown in Figure 2.15a [10, 11]. It combines the features of the circuits with the series and parallel arrangements and is useful for very wideband frequency tuning. When $C_v = 0$, the schematic of the Vackar oscillator becomes similar to the schematic of the Clapp oscillator shown in Figure 2.13a. The configuration of the Vackar oscillator shown in Figure 2.15b demonstrates the main difference compared with the Clapp schematic. Here, the capacitance C_{v2} represents the phase-varying capacitance providing an additional phase shift. With this circuit it is possible to utilize the maximum value of the oscillator quality factor over the complete tuning range, and the circuit has a substantially constant amplitude characteristic. Both oscillator schematics are equivalent, since the π-representation of the capacitances C_v, C_x and C_2 are replaced by the T-representation of the capacitances C_{v1}, C_{v2} and C_{v3} in accordance with equivalent π- to T-circuit transformation.

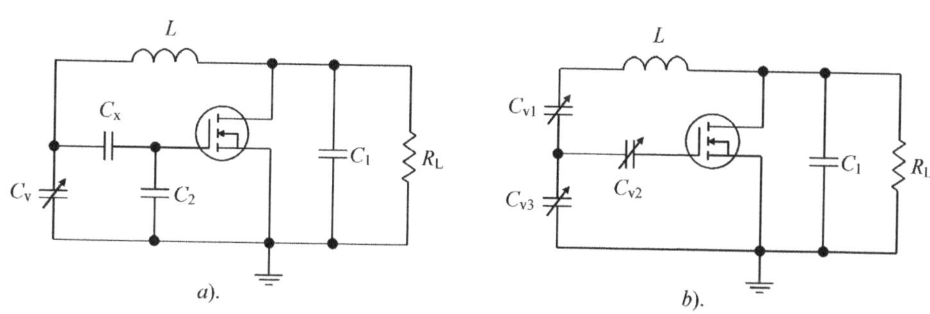

Figure 2.15 Schematics of Vackar oscillator

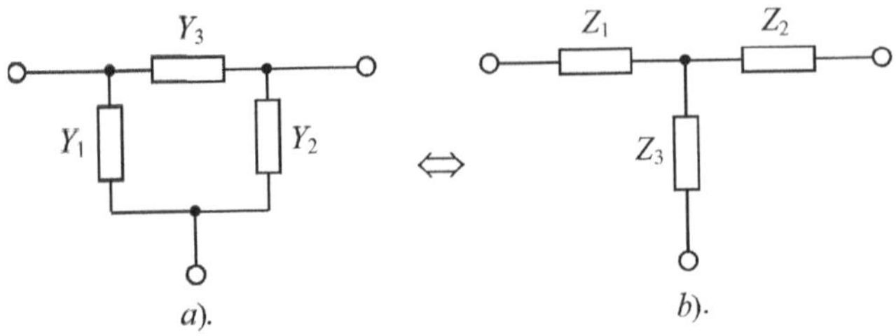

Figure 2.16 Equivalence of π- and T-circuits

For the appropriate relationships between the impedances of a T-circuit and admittances of a π-circuit, these two circuits shown in Figure 2.16 become equivalent with respect to the effect on any other two-port network. The resulting ratios between admittances for a π-circuit and impedances for a T-circuit are given in Table 2.2.

By substituting the corresponding capacitive admittances for the oscillator schematic shown in Figure 2.15a into equations for π- to T-circuit transformation, the corresponding capacitances connected in a T-configuration for the oscillator schematic shown in Figure 2.15b can be calculated by

$$C_{v1} = C_v + C_x + \frac{C_v C_x}{C_1} \tag{2.24}$$

$$C_{v2} = C_1 + C_x + \frac{C_1 C_x}{C_v} \tag{2.25}$$

$$C_{v3} = C_v + C_1 + \frac{C_v C_1}{C_x} \tag{2.26}$$

The transformer-coupled oscillator based on a differential amplifier is shown in Figure 2.17a, where the coupling coefficient is chosen to provide a stable soft start-up condition. The simple oscillator configuration using a differential transistor pair is shown in Figure 2.17b. Since the voltage at the gate of the transistor connected to the parallel resonant circuit is in-phase to the voltage at the drain of the transistor with grounded gate, the feedback in such a differential-pair oscillator is positive.

The push–pull connection of the transistors in power amplifiers is usually used to increase the resulting output power, simplify the output matching with load and improve spectral performance by suppressing even-order harmonics. The same concept can also be applied to the

Table 2.2 Relationships between π- and T-circuit parameters.

T- to π-transformation	π- to T-transformation
$Y_1 = \dfrac{Z_2}{Z_1Z_2 + Z_2Z_3 + Z_1Z_3}$	$Z_1 = \dfrac{Y_2}{Y_1Y_2 + Y_2Y_3 + Y_1Y_3}$
$Y_2 = \dfrac{Z_1}{Z_1Z_2 + Z_2Z_3 + Z_1Z_3}$	$Z_2 = \dfrac{Y_1}{Y_1Y_2 + Y_2Y_3 + Y_1Y_3}$
$Y_3 = \dfrac{Z_3}{Z_1Z_2 + Z_2Z_3 + Z_1Z_3}$	$Z_3 = \dfrac{Y_3}{Y_1Y_2 + Y_2Y_3 + Y_1Y_3}$

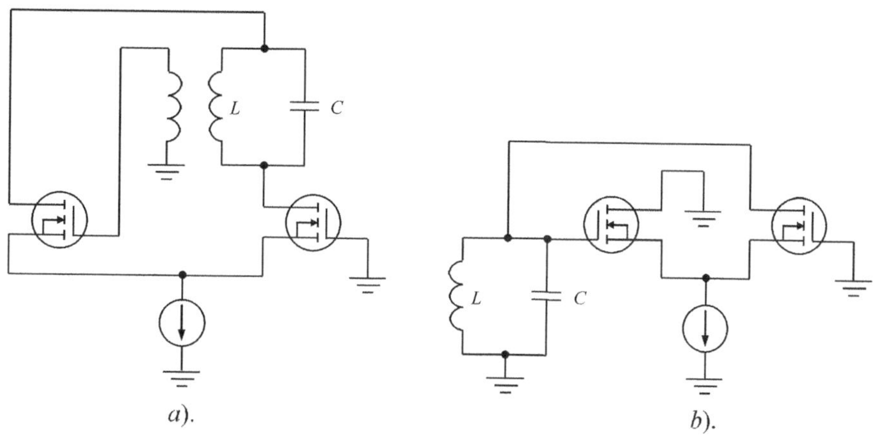

Figure 2.17 Schematics of differential-pair oscillators

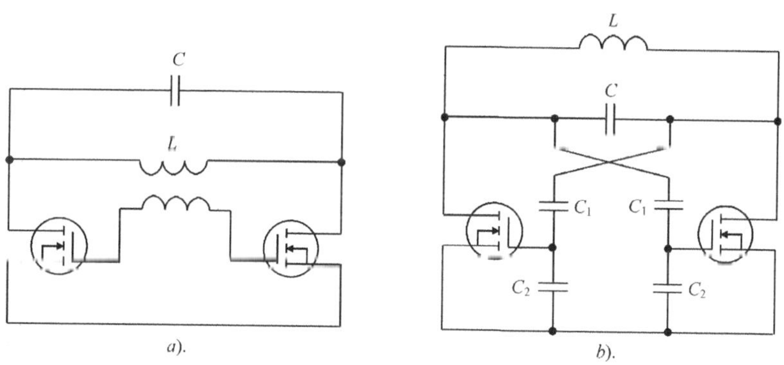

Figure 2.18 Schematics of push–pull oscillators

oscillator design. The push–pull oscillator circuit shown in Figure 2.18a is based on the two single-ended Meissner oscillators where the transistors are turned on and off alternately. Since the voltage at the gate of the one transistor is in-phase to the voltage at the drain of the other transistor, there is no need to invert phase using the secondary winding. Another push–pull oscillator configuration based on two single-ended Seiler oscillators is shown in Figure 2.18b, where the positive feedback is formed by the capacitive divider based on the capacitors C_1 and C_2.

2.4 SELF-BIAS CONDITION

To realize more effective operation mode of the oscillator, a self-bias resistor is usually included in the oscillator circuit. Figure 2.19 illustrates the circuit principle of self-biasing by two examples. In a common source transformer-coupled oscillator shown in Figure 2.19a, the self-bias resistor R_s is shunted by the capacitor C_s to minimize the RF signal losses. In a common gate oscillator shown in Figure 2.19b, the RF choke is connected in series with the self-bias resistor R_s for the same purpose.

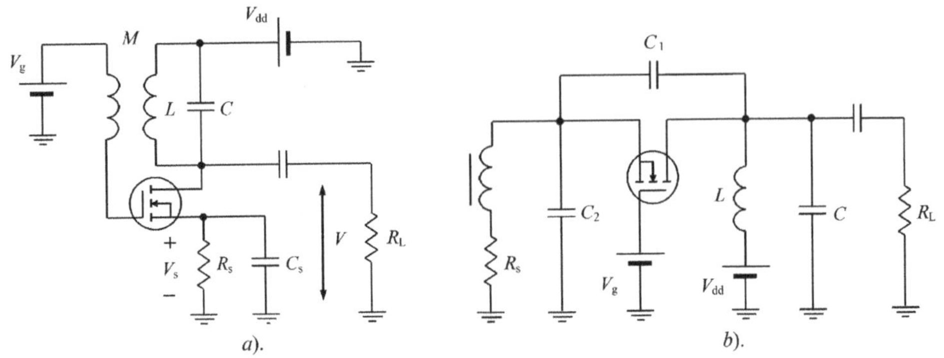

Figure 2.19 MOSFET oscillators with self-bias resistors

Let us consider the principle of the self-bias operation. The start-up conditions are satisfied under the initial gate–source bias corresponding to the large value of the small-signal transconductance when $V_s = 0$. As the oscillation amplitude grows, the dc bias gate–source voltage $V_{gs} = V_g - V_s$ decreases due to dc voltage drop across the self-bias resistor with the dc collector current increase ($V_s > 0$), as shown in Figure 2.20a. The decrease of the gate–source bias voltage leads to an appropriate decrease of the large-signal transconductance and to the high-efficiency steady-state operation mode with the gate–source fundamental voltage amplitude V_{in}. As a result, the self-bias condition combines the soft self-excitation of the oscillation with high efficiency of the Class AB, Class B or Class C operation mode under hard self-excitation of the oscillation with a certain value of the conduction angle.

Using a piecewise-linear approximation of the device transfer characteristic, the dc drain current I_0 as a function of the input gate voltage amplitude V_{in} can be determined by means of the conduction angle from Equation (1.22) as

$$I_0 = g_m V_{in} \gamma_0 (\theta) \tag{2.27}$$

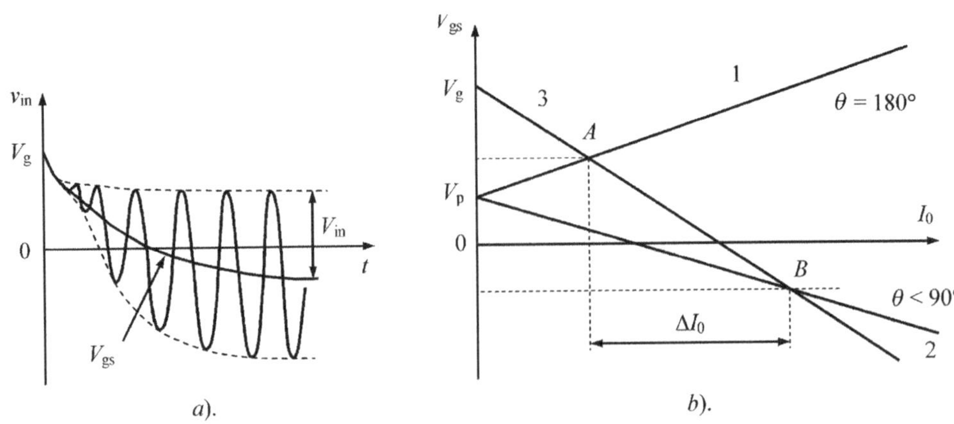

Figure 2.20 Self-bias operation

where g_m is the device small-signal transconductance, and

$$\cos\theta = -\frac{V_{gs} - V_p}{V_{in}} \tag{2.28}$$

where 2θ is the conduction angle and V_p is the device pinch-off voltage.

Substituting Equation (2.28) into Equation (2.27) allows us to obtain the relationship between gate–source bias voltage V_{gs} and dc drain current I_0 in the form

$$V_{gs} = V_p - \frac{I_0}{g_m}\frac{\cos\theta}{\gamma_0(\theta)} \tag{2.29}$$

On the other hand, a similar relationship can be written as

$$V_{gs} = V_g - I_0 R_s \tag{2.30}$$

Figure 2.20b shows the graphical solution of a system of these two equations where Equation (2.29) with the conduction angle of $2\theta = 360^\circ$ is plotted by curve 1, Equation (2.29) with the conduction angle of $2\theta < 180^\circ$ is plotted by curve 2 and Equation (2.30) is plotted by curve 3. The intersection of curve 3 with curve 1 at the point A corresponds to the start-up oscillation mode, whereas the point B corresponds to the steady-state oscillation mode with constant amplitude. During the oscillation build-up, the dc drain current increases by the value of ΔI_0.

Consequently, during the process of self-oscillations build-up, the two separate processes occur simultaneously: the increase of the oscillation amplitude across the resonant circuit and the decrease of the bias gate–source voltage due to the presence of the self-bias resistor R_s. Generally, both these processes are interdependent. The velocity of the first process is defined by the time constant of the resonant circuit given by

$$T = Q/\omega_0 \tag{2.31}$$

where $Q = \omega_0 C R_L$ is the oscillator quality factor at the resonant frequency ω_0. The velocity of the second process is defined by the time constant of the self-bias circuit as

$$T_s = R_s C_s \tag{2.32}$$

If $T_s \gg T$, the self-oscillations grow very rapidly, corresponding to the soft start-up condition for the bias voltage of $V_g - V_s = V_g(V_s = 0)$. The intersection of the oscillator amplitude characteristic $I_1(V_{in})$ with $V_s = 0$ and feedback line provides the voltage amplitude across the resonant circuit corresponding to point 1 shown in Figure 2.21a. The dc drain current grows with the same velocity, resulting in the slow growth of the self-bias voltage V_s due to the slow transient response of the self-bias circuit. This process contributes to the gradual transition to the points 2 and then 3, characterized by different types of the oscillator amplitude characteristic and smaller voltage amplitudes, with further collapse of the self-oscillations. The dc drain current becomes zero, resulting in the onset of the discharging process of the self-bias circuit. When the discharging process is finished, the process of the self-oscillations build-up will start once again. Thus, such a self-oscillation process is accompanied by *self-pulse modulation*, as shown in Figure 2.21b. For a smaller difference between T_s and T, the process of the oscillation amplitude decrease may not reach the collapse point when the oscillation amplitude starts to grow. As a result, the self-oscillations have an amplitude modulation in the steady-state operation mode. This process is called *self-modulation*. To eliminate the self-modulation effect, it is necessary to choose the condition when $T_s < T$.

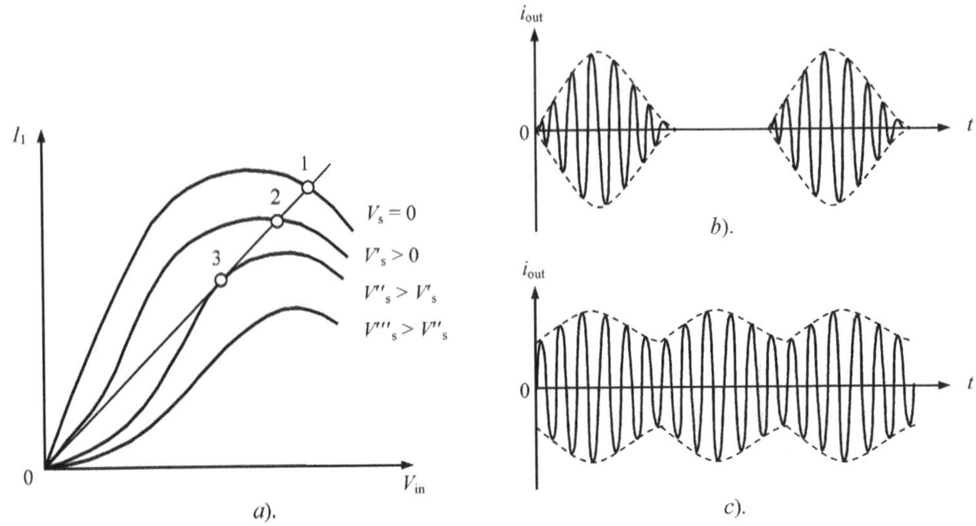

Figure 2.21 Self-pulse modulation and self-modulation phenomena

To establish an analytical relationship determining the amplitude stability, consider the truncated first-order equations for both the slowly varying fundamental amplitude V across the load resistor R_L and gate–source bias voltage V_{gs}, which can be derived in similar way to Equation (1.116) given in Chapter 1 in the form

$$T\frac{dV}{dt} + V = R_L I_1(V_{gs}, V) \tag{2.33}$$

$$T_s\frac{dV_{gs}}{dt} + V_{gs} = -R_s I_0(V_{gs}, V) \tag{2.34}$$

where the drain fundamental current I_1 and dc current I_0 are functions of two variables, V_{gs} and V [12]. It should be noted that the truncated equation for the gate–source bias voltage is similar to the truncated equation for drain voltage amplitude. The difference is that Equation (2.34) contains the slowly varying dc component and its right-hand side has opposite sign compared with Equation (2.33).

To investigate the circuit stability, we will use the method of small perturbations in the vicinity of the steady-state mode when

$$V = V^0 + \Delta V \tag{2.35}$$

$$V_{gs} = V_{gs}^0 + \Delta V_{gs} \tag{2.36}$$

where V^0 and V_{gs}^0 are the steady-state voltage values, ΔV and ΔV_{gs} are small voltage perturbations. Expanding the drain fundamental current I_1 and dc current I_0 into a Taylor series and limiting to the linear terms only yields

$$I_1 = I_1^0 + \frac{\partial I_1}{\partial V}\Delta V + \frac{\partial I_1}{\partial V_{gs}}\Delta V_{gs} \tag{2.37}$$

$$I_0 = I_0^0 + \frac{\partial I_0}{\partial V}\Delta V + \frac{\partial I_0}{\partial V_{gs}}\Delta V_{gs} \tag{2.38}$$

As a result, the following system of two first-order differential equations describing the behaviour of the oscillator with self-bias circuit near the steady-state mode can be written:

$$\left(T'\frac{d}{dt}+1\right)\Delta V-\xi\Delta V_{gs}=0 \tag{2.39}$$

$$-\xi_s\Delta V+\left(T'_s\frac{d}{dt}+1\right)\Delta V_{gs}=0 \tag{2.40}$$

where

$$\xi=\frac{\dfrac{\partial I_1}{\partial V_{gs}}}{\dfrac{1}{R_L}-\dfrac{\partial I_1}{\partial V}}\qquad \xi_s=\frac{-\dfrac{\partial I_0}{\partial V}}{\dfrac{1}{R_s}+\dfrac{\partial I_0}{\partial V_{gs}}}$$

$$T'=\frac{C}{\dfrac{1}{R_L}-\dfrac{\partial I_1}{\partial V}}\qquad T'_s=\frac{C_s}{\dfrac{\partial I_0}{\partial V_{gs}}+\dfrac{1}{R_s}}$$

The conditions for amplitude stability can be found by solving the corresponding second-order characteristic equation and represented in the form of the two following inequalities:

$$\frac{1}{T'}+\frac{1}{T'_s}>0 \tag{2.41}$$

$$\frac{1-\xi\xi_s}{T'T'_s}>0 \tag{2.42}$$

The border of the stability condition is determined by equating Equation (2.41) to unity, resulting in

$$\frac{1}{T'}=-\frac{1}{T'_s} \tag{2.43}$$

or

$$\frac{T_s}{T}=\frac{\dfrac{\partial I_0}{\partial V_{gs}}R_s+1}{\dfrac{\partial I_1}{\partial V}R_L-1} \tag{2.44}$$

For a piecewise-linear approximation of the device transfer characteristic, the fundamental drain current I_1 as a function of the input gate voltage amplitude V_{in} can be determined by means of the conduction angle from Equation (1.23) given in Chapter 1 as

$$I_1=g_m V_{in}\gamma_1(\theta) \tag{2.45}$$

Then, Equation (2.44) can be rewritten using Equations (2.29) and (2.45) in the form

$$\frac{T_s}{T}=\frac{1+g_m R_s(\theta/\pi)}{g_m\gamma_1(\theta)KR_L-1} \tag{2.46}$$

where $K=V_{in}/V$ is the magnitude of the voltage feedback coefficient.

For amplitude stability, it is necessary to create a condition of minimal value of the right-hand side of Equation (2.46) when the effect of the parasitic self-modulation is impossible. From

Equation (2.46) it follows that its numerator has minimal value for $\theta = 0^\circ$, corresponding to the pinch-off region, whereas its denominator has a minimal value for $\theta = 180^\circ$, corresponding to the purely active region. As a result, the sufficient condition for the amplitude stability can be obtained from

$$\frac{T_s}{T} = \frac{1}{g_m K R_L - 1} \tag{2.47}$$

which usually gives a condition for amplitude stability with sufficient margin.

In this case, it is interesting to estimate the filtering properties of such a bias circuit. Taking into account that $T_s = R_s C_s$ and $T = 2/\omega_0 \delta$, Equation (2.47) can be rewritten as

$$\frac{1}{\omega_0 C_s R_s} = \frac{\delta}{2} (g_m K R_L - 1) \tag{2.48}$$

which means that, for regeneration factor $g_m K R_L = 5$ and dissipation factor $\delta = 0.01$, the capacitive reactance $1/\omega_0 C_s$ is 50 times less than the bias resistance R_s.

Let us illustrate the amplitude stability situation graphically based on two diagrams, the dependence $V_{in}(V_{gs})$ as a function of $\gamma_1(\theta)$ and the dependence $V_{gs}(V_{in})$ as a function of $\gamma_0(\theta)$. In a steady-state mode, from Equation (2.45) it follows that

$$\gamma_1(\theta) = \frac{1}{g_m K R_L} \tag{2.49}$$

which defines the relationship between the conduction angle 2θ and regeneration factor $g_m K R_L$. Equation (2.28) can be rewritten as

$$V_{in} = -\frac{V_{gs} - V_p}{\cos\theta} \tag{2.50}$$

which demonstrates the linear dependence of the input amplitude V_{in} on gate–source bias voltage V_{gs} for constant conduction angle 2θ.

Figure 2.22a shows a family of amplitude curves corresponding to Equation (2.50) for different values of regeneration factor, beginning from the pinch-off voltage V_p and having the slope coefficient $-1/\cos\theta$. The limit points correspond to $g_m K R_L = 1$ when $\theta = 180^\circ$ and $g_m K R \to \infty$ when $\theta = 0^\circ$, whereas the vertical line is realized when $g_m K R_L = 2$ for $\theta = 90^\circ$. Such a linear behaviour of these curves is valid for the active operation region (below

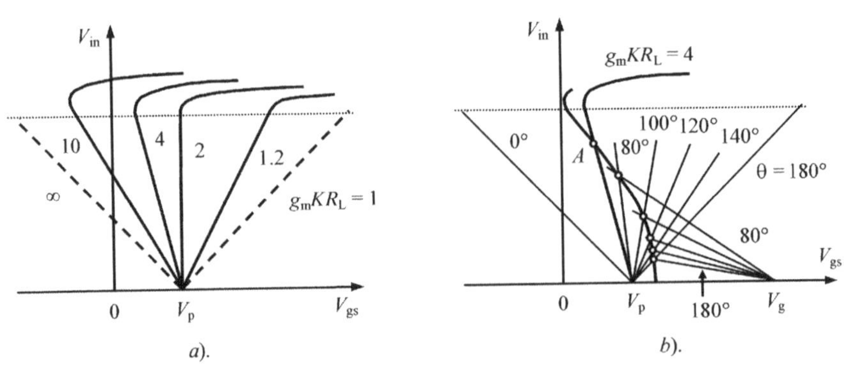

Figure 2.22 Input amplitude V_{in} versus bias voltage V_{gs}

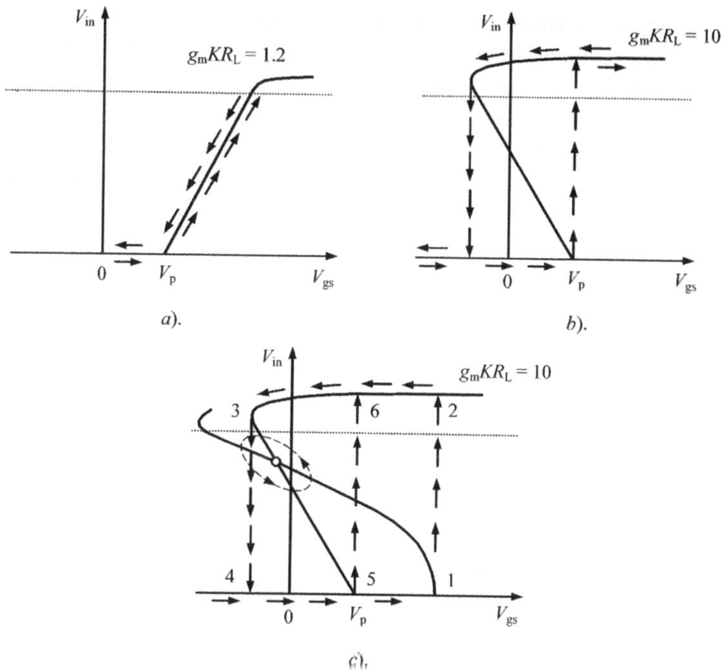

Figure 2.23 Amplitude stability for different biasing

the dotted line) when the output fundamental voltage amplitude V is less than supply voltage V_{dd}. In the saturation region (above the dotted line), the input voltage amplitude V_{in} is almost independent of V_{gs}, and its behaviour can be approximated by the horizontal curves.

Substituting Equation (2.27) into Equation (2.30) enables us to express the dc bias voltage V_{gs} as a function of the input amplitude V_{in} by

$$V_{gs} = V_g + g_m \gamma_0 (\theta) R_s V_{in} \tag{2.51}$$

To determine analytical relationship $V_{gs}(V_{in})$, it is necessary to exclude the parameter θ from Equations (2.28) and (2.51). Since both equations contain the transcendental functions of the parameter θ, it is best to use graphical plots. For each value θ, there is a linear relationship between V_{gs} and V_{in}, which can be plotted in the form of a family of linear curves, beginning from the point V_g for Equation (2.51) and from the point V_p for Equation (2.28). The intercept points corresponding to the same θ will give the required points of the bias curve $V_{gs}(V_{in})$. Figure 2.22b shows the bias curve which intersects the amplitude curve plotted for $g_m K R_L = 4$ at the point A corresponding to the values of the bias voltage V_{gs} and oscillation amplitude V_{in}, which are established in a steady-state oscillation mode.

Based on the behaviour of the amplitude curves, we can make some predictions regarding oscillator stability conditions. For example, for a simple case of external biasing only when $R_s = 0$ and $V_{gs} = V_g$, the positive slope of amplitude curve shown in Figure 2.23a when the regeneration factor is less than 2 and $\theta > 90^\circ$ implies the stable condition for initial point $V_{in} = 0$ when $V_g < V_p$ and its unstable condition when $V_g > V_p$ resulting in soft oscillation build-up. In this case, the oscillation amplitude V_{in} increases smoothly with the increase of the bias voltage V_g and decreases with its decrease. However, the negative slope of the amplitude curve when the regeneration factor is greater than 2 and $\theta < 90^\circ$ results in a hysteresis region characterized by unstable oscillation conditions shown in Figure 2.23b. In this case, the hard start-up self-oscillation conditions with certain amplitude V_{in} can be realized only when $V_g > V_p$.

Similar unstable behaviour corresponding to hard start-up conditions is observed for the oscillator using self-bias circuit when $T_s \gg T$. In this situation, the transient response of the bias circuit is much slower than that of the resonant circuit and the self-bias conditions cannot affect the stability conditions. Consequently, if the steady-state mode was unstable under external biasing only, it will remain unstable under self-bias conditions as well, accompanied by the self-pulse modulation or self-modulation phenomena. The oscillation process in this case can be developed as shown in Figure 2.23c. When the voltage supply is turned on, the amplitude V_{in} begins growing from point 1 to point 2, corresponding to the saturation region for the same bias voltage V_{gs}. Then, the bias voltage changes very slowly from point 2 to point 3 with almost constant amplitude V_{in}. Further amplitude reduction results in the collapse of the oscillation corresponding to point 4, and bias voltage V_{gs} begins to increase. When it reaches point 5, the oscillations with amplitude corresponding to point 6 are established once again due to the forward-biased transistor. Such a repeatable cyclic process results in a self-modulation operation mode of the oscillator. For lower values of T_s with faster charging and discharging processes of the shunt capacitor C_s, the time variations of the oscillation amplitude V_{in} and bias voltage V_{gs} become comparable and cyclic process changes its form from self-pulse modulation to self-modulated continuous oscillations, as shown in Figure 2.23c by the elliptical curve. Further reduction in T_s when $T_s < T$ leads to disappearance of the self-modulation effect. In this case, the bias voltage V_{gs} follows instantaneously the variations of the oscillation amplitude V_{in} according to Equation (2.50). The steady-state mode which is defined by the intercept point of the bias and amplitude curves is stable, even if this point is located at the amplitude curve with negative slope. This is because any deviations of the oscillation amplitude are followed by the appropriate fast variations of the bias voltage restoring the steady-state conditions at the intercept point.

2.5 OSCILLATOR ANALYSIS USING MATRIX TECHNIQUES

The determination of the start-up and steady-state oscillation conditions is very often based upon a loop or nodal analysis of the circuit. However, the oscillator analysis using matrix techniques brings out the similarities between several types of the oscillators and results in one group of equations, which can be used to analyse different oscillator configurations [13]. In this case, a two-port network can represent both the active device and feedback element. Depending on the oscillator configuration with parallel or series feedback, use of Y- or Z-parameters can be respectively applied.

2.5.1 Parallel feedback oscillator

A generalized equivalent circuit of the parallel feedback oscillator is shown in Figure 2.24a, where Y_1, Y_2 and Y_3 are the feedback elements. To calculate the steady-state stationary operation mode, it is convenient to add the matrix of the parallel feedback element Y_2 and the matrix of

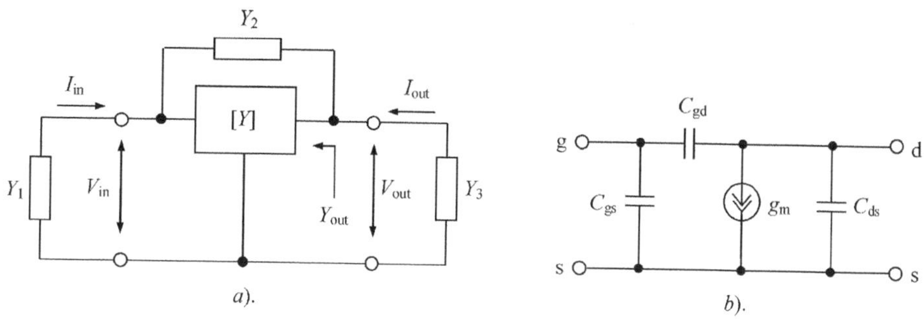

Figure 2.24 Equivalent circuit of the parallel feedback oscillator

the active device Y-parameters according to

$$[Y]+[Y_2]=\begin{bmatrix} Y_{11}+Y_2 & Y_{12}-Y_2 \\ Y_{21}-Y_2 & Y_{22}+Y_2 \end{bmatrix} \tag{2.52}$$

In this case, a system of two equations for the two-port network input current I_{in}, output current I_{out}, input voltage V_{in} and output voltage V_{out} can be written as

$$I_{in}=(Y_{12}+Y_2)\,V_{in}+(Y_{12}-Y_2)\,V_{out} \tag{2.53}$$

$$I_{out}=(Y_{21}-Y_2)\,V_{in}+(Y_{22}+Y_2)\,V_{out} \tag{2.54}$$

Since, for the oscillator shown in Figure 2.24a, the boundary conditions are obtained in the form

$$I_{in}=-Y_1V_{in} \tag{2.55}$$

$$I_{out}=-Y_3V_{out} \tag{2.56}$$

the following matrix equation can be written as

$$\begin{bmatrix} Y_{11}+Y_1+Y_2 & Y_{12}-Y_2 \\ Y_{21}-Y_2 & Y_{22}+Y_2+Y_3 \end{bmatrix}\begin{bmatrix} V_{in} \\ V_{out} \end{bmatrix}=0 \tag{2.57}$$

Thus, the steady-state stationary condition can be expressed by

$$\begin{vmatrix} Y_{11}+Y_1+Y_2 & Y_{12}-Y_2 \\ Y_{21}-Y_2 & Y_{22}+Y_2+Y_3 \end{vmatrix}=0 \tag{2.58}$$

After some simplification, Equation (2.58) can be rewritten as

$$Y_{22}+Y_2+Y_3-\frac{(Y_{12}-Y_2)(Y_{21}-Y_2)}{Y_{11}+Y_1+Y_2}=0 \tag{2.59}$$

As a result, the steady-state oscillation condition for the parallel feedback oscillator represented as a one-port negative conductance oscillator can be written as

$$Y_{out}+Y_L=0 \tag{2.60}$$

where $Y_3=Y_L$ and

$$Y_{out}=Y_{22}+Y_2-\frac{(Y_{12}-Y_2)(Y_{21}-Y_2)}{Y_{11}+Y_1+Y_2} \tag{2.61}$$

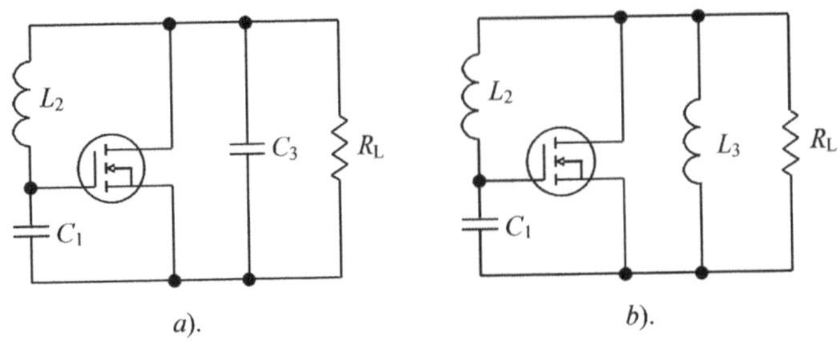

Figure 2.25 Electrical circuits of parallel feedback oscillators

Consequently, the separate equations for real and imaginary parts of the admittances of a negative conductance oscillator, which are similar to Equations (2.3) and (2.4) for magnitude and phase of the loop gain of a parallel feedback oscillator corresponding to the steady-state oscillation process, can be obtained as

$$\mathrm{Re}Y_{\mathrm{out}} + \mathrm{Re}Y_{\mathrm{L}} = 0 \tag{2.62}$$

$$\mathrm{Im}Y_{\mathrm{out}} + \mathrm{Im}Y_{\mathrm{L}} = 0 \tag{2.63}$$

Similarly, the start-up conditions for the parallel feedback oscillator given by Equations (2.8) and (2.9) can be rewritten as

$$\mathrm{Re}Y_{\mathrm{out}} + \mathrm{Re}Y_{\mathrm{L}} < 0 \tag{2.64}$$

$$\mathrm{Im}Y_{\mathrm{out}} + \mathrm{Im}Y_{\mathrm{L}} = 0 \tag{2.65}$$

To obtain the analytical relationships between the active device and resonant circuit parameters, let us consider the simplified intrinsic MOSFET high-frequency small-signal equivalent circuit shown in Figure 2.24b. The Y-parameters of the equivalent circuit are

$$\begin{aligned} Y_{11} &= j\omega(C_{\mathrm{gs}} + C_{\mathrm{gd}}) \qquad & Y_{12} &= -j\omega C_{\mathrm{gd}} \\ Y_{21} &= g_{\mathrm{m}} - j\omega C_{\mathrm{gd}} \qquad & Y_{22} &= j\omega(C_{\mathrm{ds}} + C_{\mathrm{gd}}) \end{aligned} \tag{2.66}$$

An assumption of the lossless feedback elements allows us to provide a simple qualitative evaluation of the oscillator start-up conditions. For a Colpitts oscillator, the basic of which circuit schematic is shown in Figure 2.25a, the feedback and load admittances are $Y_1 = j\omega C_1$, $Y_2 = 1/j\omega L_2$, and $Y_{\mathrm{L}} = 1/R_{\mathrm{L}} + j\omega C_3$.

Then, by substituting all admittances in Equation (2.64), the small-signal device transconductance g_{m} required to excite the self-oscillations will be determined by

$$g_{\mathrm{m}} > \frac{1}{R_{\mathrm{L}}} \frac{C_{\mathrm{gs}} + C_1}{C_{\mathrm{ds}} + C_3} \tag{2.67}$$

The self-oscillation frequency that depends on both transistor equivalent circuit parameters and feedback elements can be defined from

$$\omega^2 = \frac{1}{L_2} \frac{C_{\mathrm{gs}} + C_{\mathrm{ds}} + C_1 + C_3}{(C_{\mathrm{gs}} + C_{\mathrm{gd}} + C_1)(C_{\mathrm{ds}} + C_3) + (C_{\mathrm{gs}} + C_1)C_{\mathrm{gd}}} \tag{2.68}$$

If the value of the MOSFET intrinsic feedback capacitance C_{gd} is negligible, the expression for the oscillation is simplified to

$$\omega = \sqrt{\frac{1}{L_2}\left(\frac{1}{C_{gs}+C_1}+\frac{1}{C_{ds}+C_3}\right)} \tag{2.69}$$

From Equation (2.67) it follows that, at lower frequencies compared with the device transition frequency f_T when the effect of the elements of the device equivalent circuit is not significant, in order to provide the soft build-up of the oscillation, it is necessary to choose the feedback elements, load resistance and dc bias point when

$$g_m > \frac{1}{R_L}\frac{C_1}{C_3} \tag{2.70}$$

Because it was assumed that the value of the feedback susceptance $B_3 = \text{Im}Y_3$ is positive, consequently, to satisfy the start-up and steady-state oscillation conditions, it is necessary to use the capacitance C_3. By rewriting Equation (2.69) in the form

$$B_3 = \omega\left(\frac{C_{gs}+C_1}{\omega^2 L_2(C_{gs}+C_1)-1} - C_{ds}\right) \tag{2.71}$$

it is easy to show that when

$$L_2 < \frac{1}{\omega^2(C_{gs}+C_1)} \tag{2.72}$$

the value of B_3 becomes negative. This means that such an oscillator feedback element must be inductive with the value of $L_3 = -1/\omega B_3$. The circuit configuration for this oscillator is shown in Figure 2.25b.

2.5.2 Series feedback oscillator

A generalized equivalent circuit of the series feedback oscillator is shown in Figure 2.26a, where Z_1, Z_2 and Z_3 are the feedback elements. To calculate the steady-state operation mode, it is convenient to add the matrix of the series feedback element Z_2 and the matrix of the active device Z-parameters according to

$$[Z] + [Z_2] = \begin{bmatrix} Z_{11}+Z_2 & Z_{12}+Z_2 \\ Z_{21}+Z_2 & Z_{22}+Z_2 \end{bmatrix} \tag{2.73}$$

In this case, a system of two equations for the two-port network input current I_{in}, output current I_{out}, input voltage V_{in} and output voltage V_{out} can be written as

$$V_{in} = (Z_{12}+Z_2)\,I_{in} + (Z_{12}+Z_2)\,I_{out} \tag{2.74}$$

$$V_{out} = (Z_{21}+Z_2)\,I_{in} + (Z_{22}+Z_2)\,I_{out} \tag{2.75}$$

Since, for the oscillator shown in Figure 2.26a, the boundary conditions are obtained in the form

$$V_{in} = -Z_1 I_{in} \tag{2.76}$$

$$V_{out} = -Z_3 I_{out} \tag{2.77}$$

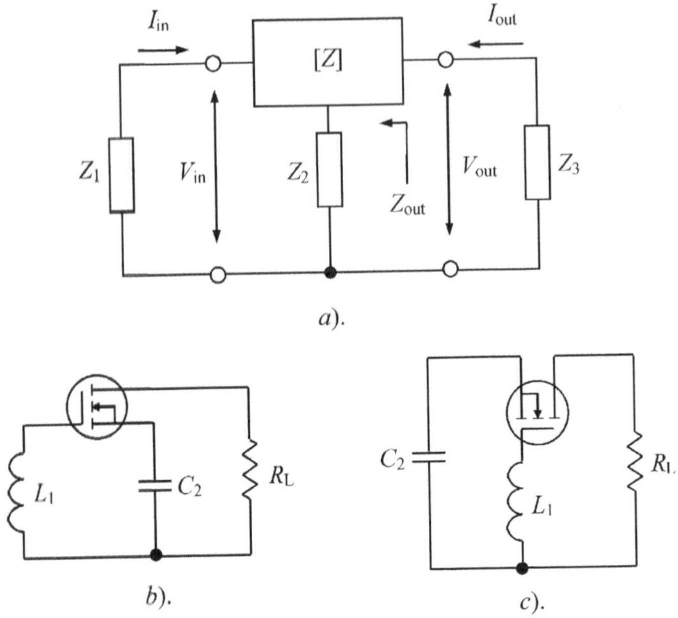

Figure 2.26 Equivalent circuits of the series feedback oscillator

the following matrix equation can be written as

$$\begin{bmatrix} Z_{11}+Z_1+Z_2 & Z_{12}+Z_2 \\ Z_{21}+Z_2 & Z_{22}+Z_2+Z_3 \end{bmatrix} \begin{bmatrix} I_{in} \\ I_{out} \end{bmatrix} = 0 \tag{2.78}$$

Thus, the steady-state oscillation condition can be expressed by

$$\begin{vmatrix} Z_{11}+Z_1+Z_2 & Z_{12}+Z_2 \\ Z_{21}+Z_2 & Z_{22}+Z_2+Z_3 \end{vmatrix} = 0 \tag{2.79}$$

After some simplifications, Equation (2.79) can be rewritten as

$$Z_{22}+Z_2+Z_3-\frac{(Z_{12}+Z_2)(Z_{21}+Z_2)}{Z_{11}+Z_1+Z_2}=0 \tag{2.80}$$

As a result, the steady-state oscillation condition for the series feedback oscillator represented as a one-port negative resistance oscillator can be written as

$$Z_{out}+Z_L=0 \tag{2.81}$$

where $Z_3 = Z_L$ and

$$Z_{out}=Z_{22}+Z_2-\frac{(Z_{12}+Z_2)(Z_{21}+Z_2)}{Z_{11}+Z_1+Z_2}$$

Consequently, the separate equations for real and imaginary parts of the negative resistance oscillator, which are similar to Equations (2.62) and (2.63) for the parallel feedback oscillator corresponding to the steady-state oscillation process, can be obtained as

$$\mathrm{Re}Z_{\mathrm{out}} + \mathrm{Re}Z_{\mathrm{L}} = 0 \tag{2.82}$$

$$\mathrm{Im}Z_{\mathrm{out}} + \mathrm{Im}Z_{\mathrm{L}} = 0 \tag{2.83}$$

Similarly, the start-up conditions for the series feedback oscillator can be rewritten as

$$\mathrm{Re}Z_{\mathrm{out}} + \mathrm{Re}Z_{\mathrm{L}} < 0 \tag{2.84}$$

$$\mathrm{Im}Z_{\mathrm{out}} + \mathrm{Im}Z_{\mathrm{L}} = 0 \tag{2.85}$$

Let us obtain the analytical relationships between the active device and resonant circuit parameters of the oscillator based on the MOSFET device, the admittance Y-parameters of the equivalent circuit of which are given by Equation (2.66). The ratios between Y-parameters and Z-parameters can be expressed in the form

$$Z_{11} = \frac{Y_{22}}{\Delta Y} \quad Z_{12} = -\frac{Y_{12}}{\Delta Y} \quad Z_{21} = -\frac{Y_{21}}{\Delta Y} \quad Z_{22} = \frac{Y_{11}}{\Delta Y} \tag{2.86}$$

Then, the steady-state condition for the series feedback oscillator can be written as

$$\frac{1 + Z_2 (Y_{11} + Y_{12} + Y_{21} + Y_{22}) + Z_1 (Y_{11} + Z_2 \Delta Y)}{Y_{22} + (Z_1 + Z_2)\,\Delta Y} + Z_3 = 0 \tag{2.87}$$

where $\Delta Y = 1/\Delta Z$ and $\Delta Y = Y_{11}Y_{22} - Y_{12}Y_{21}$. For a simple oscillator circuit configuration with external gate inductance shown in Figure 2.26b, the feedback elements including load are defined by $Z_1 = j\omega L_1$, $Z_2 = 1/j\omega C_2$ and $Z_3 = R_{\mathrm{L}}$.

Usually, in a wide frequency range up to $0.3f_{\mathrm{T}}$, it is possible to neglect the intrinsic gate–drain capacitance C_{gd} without the substantial loss of accuracy. Then, the small-signal transconductance g_{m} corresponding to the soft start-up condition and the frequency of the self-oscillations can be evaluated, respectively, as

$$g_{\mathrm{m}} > C_{\mathrm{gs}} R_{\mathrm{L}} \frac{(\omega C_{\mathrm{ds}})^2}{C_{\mathrm{ds}} + C_2} \tag{2.88}$$

$$\omega = \sqrt{\frac{1}{L_1}\left(\frac{1}{C_{\mathrm{gs}}} + \frac{1}{C_{\mathrm{ds}} + C_2}\right)} \tag{2.89}$$

From Equation (2.88) it follows that the oscillation build-up can be easily provided by choosing the MOSFET device with the smaller drain–source capacitance C_{ds} and using a sufficiently large value of the feedback capacitance C_2.

2.6 DUAL TRANSISTOR OSCILLATORS

As an alternative to the basic oscillator configurations using a single active device element, there is an approach where an active element can be composed by the two active devices connected in cascade. For example, the Franklin oscillator uses two transistor stages having the same common terminal (emitter for bipolar device or source for FET device) when the greater power gain and better isolation from the resonant circuit is possible compared with the case of a single-stage configuration [14]. Figure 2.27 shows the oscillator configurations with (a) parallel and (b) series resonant circuits. In the case of a parallel resonant circuit configuration, the resonant LC circuit is isolated from the input of the first stage and the output

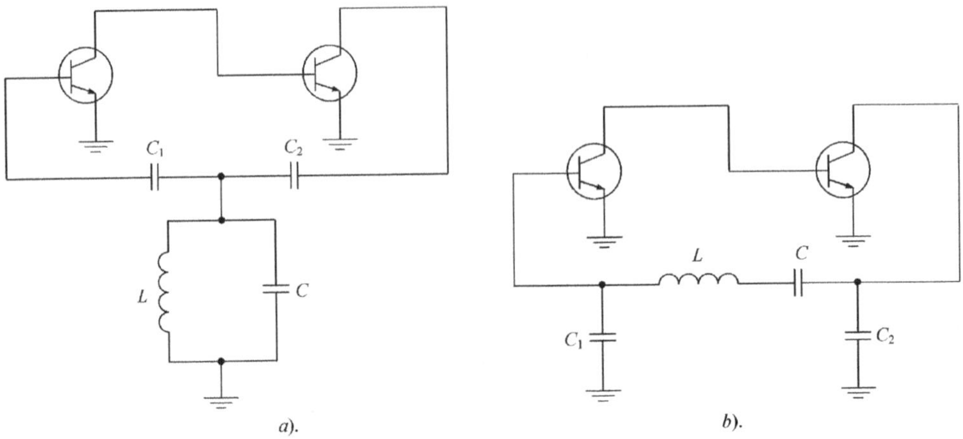

Figure 2.27 Schematics of bipolar Franklin oscillator

of the second stage by means of small shunt capacitances C_1 and C_2 having high reactances at the resonant frequency. In the case of a series resonant circuit configuration, using two large shunt capacitances C_1 and C_2 (having small reactances at the resonant frequency) provides the isolation of the resonant LC circuit from the active device path.

Figure 2.28a shows the practical schematic of a Franklin oscillator with series resonant circuit based on the bipolar devices 2N2369A [14]. In this case, the series resonant circuit is isolated by means of the shunt large capacitance $C_1 = 3.9$ nF and small inductance $L_2 = 0.2$ μH at the input and output of the amplifying circuit, respectively. As a general rule, these reactances should not exceed 20 Ω or 3% of the resonant circuit reactances. Small reactance values will enhance stability of the oscillations; however, too small values will prevent the oscillation build-up. By proper choice of emitter resistances, the Franklin oscillator can be made to exhibit a U-shaped frequency variation versus supply voltage, as shown in Figure 2.28b. Being adjusted for optimum supply voltage with zero slope of the curve where frequency is nearly independent of the supply voltage, the frequency stability with respect to the supply voltage variations can be maximized and become even comparable to many crystal oscillators.

The oscillator circuit can also be realized using a two-stage power amplifier, in which the first stage is assembled in a common drain configuration and the second stage is assembled in a common gate configuration, as shown in Figure 2.29a [15]. The first transistor composing the source follower circuit has high input impedance, minimizing its effect on the parallel resonant LC-circuit. At the same time, due to high current amplification, the output impedance of the first stage is low. Therefore, the second amplifying stage is based on the transistor in a common gate configuration, which provides low input impedance and high output impedance. As a result, the influence of the transistor impedances on the resonant circuit is minimal, resulting in a higher value of its loaded quality factor. By properly connecting the input of the first stage (port 1) with the output of the second stage (port 2), the loop is closed, and the circuit operates as an oscillator with the oscillation frequency determined mainly by the shunt capacitance C and parallel inductance L. Such a source-coupled oscillator is similar to the differential-pair oscillator shown in Figure 2.17b.

Figure 2.29b shows the circuit schematic of the source-coupled differential oscillator. By using two junction field-effect transistors (JFET) as active devices, such an oscillator is best

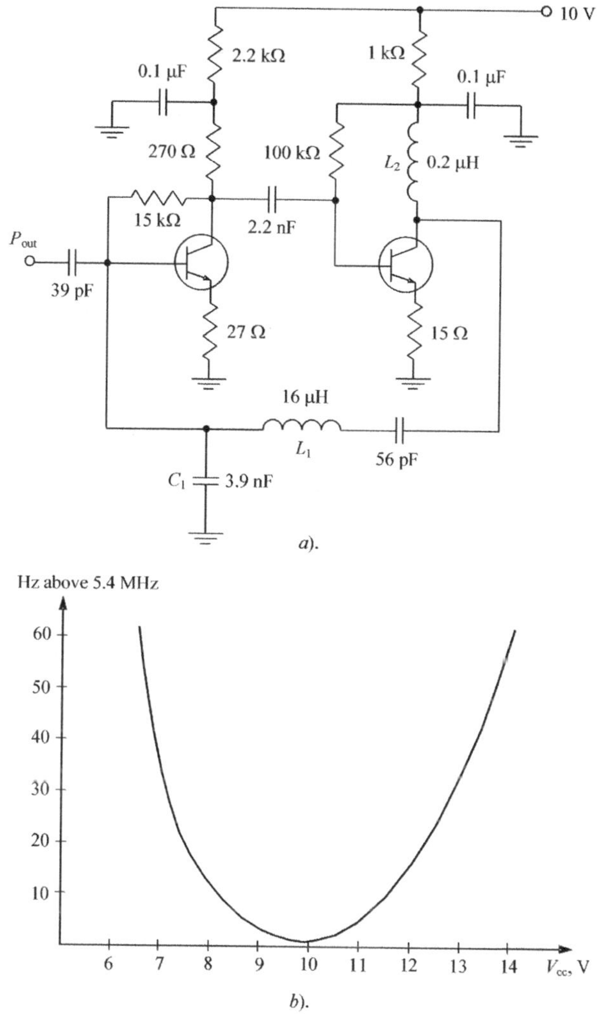

Figure 2.28 Franklin oscillator with series resonant circuit [14]

suited to operate in a frequency range from 100 MHz to 3 GHz [15]. The source resistor R_s results in a galvanic coupling of both transistors, and the gate of the second device gets its negative voltage biasing due to the voltage drop across R_s. Because of the high amplification factor, a very small amount of the output current is required to the input through the feedback capacitance C_f to provide an oscillation build-up. Since the output impedance of the first transistor in a common source configuration is relatively low, the standard 50-Ω load can be directly connected in parallel to the source resistor R_s through the dc blocking capacitance.

Another version of a dual JFET oscillator, the circuit schematic of which is shown in Figure 2.30, is also suitable for use at operating frequencies up to 3 GHz, having sufficient gain and providing good noise performance [16]. In the oscillator circuit, the gate of the first active device is connected through the blocking capacitance to a tank LC circuit. Series capacitances C_1 and C_2, representing a voltage divider, provide a partial connection of the second active device to the resonant tank circuit. The second JFET is connected in a common

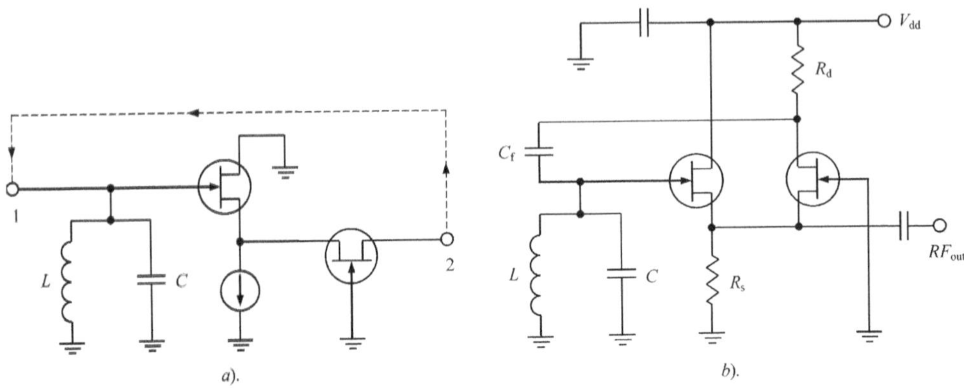

Figure 2.29 Schematics of the source-coupled JFET oscillator

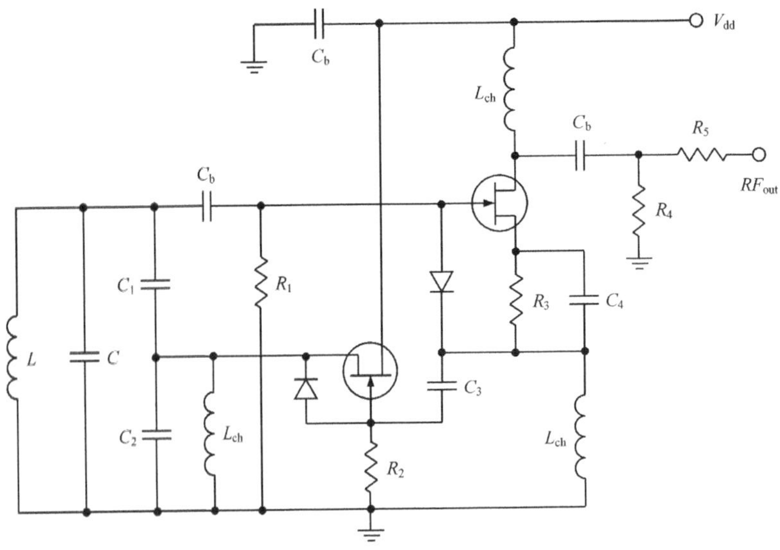

Figure 2.30 Circuit schematic of the dual JFET oscillator [16]

drain configuration and serves as a source follower to increase the loop gain of the circuit. Its gate terminal is coupled to the source terminal of the first JFET by capacitor C_3 and parallel connection of a resistor R_3 and a capacitor C_4. The two Schottky diodes are necessary to provide the limiting mechanism for positive peak signals, preventing forward conduction of the gate–source junctions of both JFETs and resulting in an improvement of the oscillator noise performance.

The cascode configuration of the oscillator gives a tremendous possibility in providing very stable high-frequency oscillations using its convenient injection locking opportunity by locking it to a more stable lower-frequency source. Moreover, a cascode connection of the active devices is very useful when it is necessary to provide the functions of both local oscillator and mixer simultaneously when one device operates as a stable oscillator and the input signal is applied to the second device. Figure 2.31a shows the bipolar cascode oscillator schematic (without

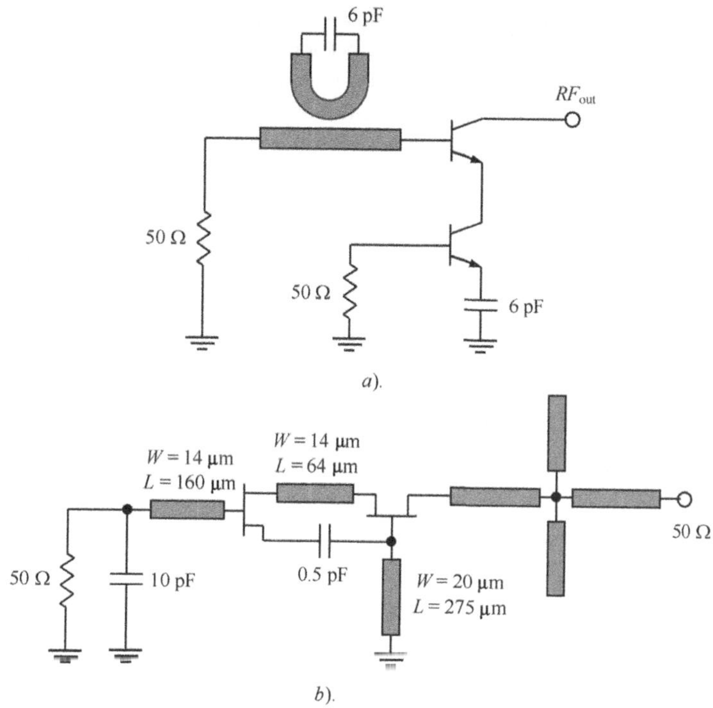

Figure 2.31 Microstrip cascode oscillator circuit schematics (© 1994 IEEE)

biasing circuits and blocking capacitors), in which the resonator is connected to the base of the top transistor in the common base configuration whereas the base of the second device is loaded by 50-Ω resistor [17]. For such a cascode circuit, the oscillator locking can be easily realized by connection of a lower frequency stable source to the base of the second device. To provide a negative resistance condition, the series capacitor is connected between the emitter of the transistor in the common emitter configuration and the ground. The resonator representing a parallel connection of the microstrip line and capacitor is coupled to the base circuit so that the value of the loaded quality factor is very close to its unloaded value, thus resulting in the best possible phase noise performance for such a resonator. By using the MRF559 bipolar transistors, the output power of 20.3 dBm with phase noise of −86.3 dBc/Hz at 10 kHz offset was achieved at an oscillation frequency of 847.6 MHz.

For millimetre-wave oscillators, it is not so easy to provide a negative resistance at their output terminal due to limited power gain of the transistor at these frequencies. Therefore, a cascode oscillator schematic is a promising alternative to the single-device oscillators with optimized feedback to increase a negative resistance at the desired frequency by increasing the gain, by using a dual transistor configuration. Figure 2.31b shows the oscillator circuit schematic with two cascode-connected HEMT devices, the first one in a common source configuration followed by the second common gate device [18]. The cascode-connected HEMT devices provide about 8 dB more power gain than individual HEMT device at 100 GHz when they are operating as a power amplifier. As a result, oscillators built with them show twice as much negative resistance. The single HEMT device had two 45-μm-wide gate fingers, providing the small-signal transconductance of 0.8 S/mm. The common source configuration of the first device is implemented with a series capacitive source feedback realized by a narrow microstrip line with electrical length greater than a quarter-wavelength. The two HEMT devices are dc decoupled to provide independent bias control for each transistor. Three bias circuits are necessary to connect to both drain terminals and one source terminal consisting of a quarter-wave high-impedance microstrip line and radial stub each. To suppress undesired oscillations at lower frequencies and, in particular, the parametric oscillations at half the oscillation frequency, a 50-Ω resistor is used in parallel to a bypass capacitor. An output microstrip impedance transformer provides the optimum load at the drain terminal for maximum oscillation power of 2 dBm achieved at the oscillation frequency of 102.5 GHz at a drain bias voltage of 0.9 V.

2.7 TRANSMISSION-LINE OSCILLATOR

Different sets of two-port network parameters can be used for oscillator design, including impedance Z-parameters, admittance Y-parameters or scattering S-parameters. The choice of any type of these parameters depends on the operating frequency, required design accuracy and implementation technique, availability of the small- or large-signal active device parameters or the possibility of using proper measurement and simulation tools. For example, in terms of one-port negative resistance approach, the conditions for self-oscillations expressed through the small-signal S-parameters can be written as

$$K = \frac{1 + |\Delta|^2 - |S_{11}|^2 - |S_{22}|^2}{2\,|S_{12}S_{21}|} < 1 \tag{2.92}$$

$$\Gamma_{\text{in}}\Gamma_{\text{S}} = 1 \tag{2.93}$$

$$\Gamma_{\text{out}}\Gamma_{\text{L}} = 1 \tag{2.94}$$

where K is the stability factor, $\Delta = S_{11}S_{22} - S_{12}S_{21}$, Γ_{in} and Γ_{S} are the input and source reflection coefficients, Γ_{out} and Γ_{L} are the output and load reflection coefficients [19]. The stability factor should be less than unity for any possibility of self-oscillations. The passive terminations Γ_{S} and Γ_{L} must be added to resonate input and output ports at the oscillation frequency. It should be noted that the conditions described by Equations (2.93) and (2.94) are interrelated, and if one condition is satisfied, then the other condition should be satisfied as well. The large-signal S-parameter of the transistor optimized for maximum output power can be measured by varying the input drive level and load impedance or injecting signal into the output port [20]. By using the measured S-parameters, the required ratios of terminal voltages and currents can be calculated.

By converting S-parameters to Z-parameters, the active device impedance parameters can be obtained as

$$\begin{bmatrix} Z_{11} & Z_{12} \\ Z_{21} & Z_{22} \end{bmatrix} = \frac{Z_0}{(1 - S_{11})(1 - S_{22}) - S_{12}S_{21}} \times \begin{bmatrix} (1 + S_{11})(1 - S_{22}) + S_{12}S_{21} & 2S_{12} \\ 2S_{21} & (1 - S_{11})(1 + S_{22}) + S_{12}S_{21} \end{bmatrix} \tag{2.95}$$

which allows us to calculate the input or output impedance of the loaded active device in the parallel feedback two-port network or negative resistance series feedback one-port network oscillator configurations.

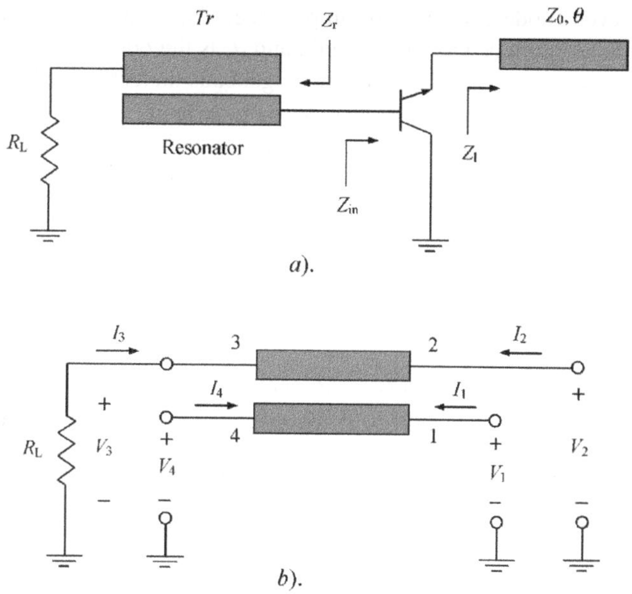

Figure 2.32 Block diagram of the bipolar negative resistance microstrip oscillator

Figure 2.32a shows the equivalent circuit of the negative resistance microwave bipolar oscillator with a common collector transistor loaded by a transmission-line open-circuited stub. The oscillation conditions will be satisfied if the real part of the input impedance Z_{in} is negative, exceeding in magnitude the real part of the resonant circuit impedance Z_r and its imaginary part is inductive or capacitive. Thus, it is very important issue to choose a proper resonant circuit providing the required reactance at the operating frequency and connection to the standard load. As an example, let us consider the two coupled open-circuited transmission lines representing the resonant circuit, where the active device is connected to one resonator whereas the load is connected to the opposite side of another resonator [21].

The impedance Z-matrix of a system of two identical lossless transmission lines in an inhomogeneous dielectric medium shown in Figure 2.32b can be written as

$$\begin{bmatrix} V_1 \\ V_2 \\ V_3 \\ V_4 \end{bmatrix} = \begin{bmatrix} Z_{11} & Z_{12} & Z_{13} & Z_{14} \\ Z_{21} & Z_{22} & Z_{23} & Z_{24} \\ Z_{31} & Z_{32} & Z_{33} & Z_{341} \\ Z_{41} & Z_{42} & Z_{43} & Z_{44} \end{bmatrix} \begin{bmatrix} I_1 \\ I_2 \\ I_3 \\ I_4 \end{bmatrix} \tag{2.96}$$

where

$$Z_{11} = Z_{22} = Z_{33} = Z_{44} = -\frac{j}{2}(Z_{0e}\cot\theta_e + Z_{0o}\cot\theta_o)$$

$$Z_{12} = Z_{21} = Z_{34} = Z_{43} = -\frac{j}{2}(Z_{0e}\cot\theta_e - Z_{0o}\cot\theta_o)$$

$$Z_{13} = Z_{31} = Z_{24} = Z_{42} = -\frac{j}{2}(Z_{0e}\csc\theta_e - Z_{0o}\csc\theta_o)$$

$$Z_{14} = Z_{41} = Z_{23} = Z_{32} = -\frac{j}{2}(Z_{0e}\csc\theta_e + Z_{0o}\csc\theta_o)$$

where Z_{0e} is the even mode characteristic impedance, Z_{0o} is the odd mode characteristic impedance, θ_e is the even mode electrical length, and θ_o is the odd mode electrical length of the transmission lines, respectively [22]. In a homogeneous dielectric medium, the electrical lengths for even and odd modes are equal, that is $\theta = \theta_e = \theta_o$.

For a system of two coupled lines with open-circuited ports 2 and 4 with boundary conditions of $I_2 = I_4 = 0$, the remaining two-port network can be described by the following matrix equation:

$$\begin{bmatrix} V_1 \\ V_3 \end{bmatrix} = \begin{bmatrix} Z_{11} & Z_{13} \\ Z_{31} & Z_{33} \end{bmatrix} \begin{bmatrix} I_1 \\ I_3 \end{bmatrix} \tag{2.97}$$

where

$$\begin{bmatrix} Z_{11} & Z_{13} \\ Z_{31} & Z_{33} \end{bmatrix} = -\frac{j}{2} \begin{bmatrix} Z_{0e} \cot\theta_e + Z_{0o} \cot\theta_o & Z_{0e} \csc\theta_e - Z_{0o} \csc\theta_o \\ Z_{0e} \csc\theta_e - Z_{0o} \csc\theta_o & Z_{0e} \cot\theta_e + Z_{0o} \cot\theta_o \end{bmatrix}$$

Taking into account that $V_3 = -R_L I_3$, the input impedance of a loaded two-port network can be written as

$$Z_r = Z_{11} - \frac{Z_{13} Z_{31}}{Z_{33} + R_L} \tag{2.98}$$

For a particular case of coupled quarter-wave transmission lines in a homogeneous medium when $\theta = 90^\circ$, such a system can provide an impedance transformation from the standard load $R_L = 50\,\Omega$ to a purely resistive impedance obtained by

$$Z_r = \frac{(Z_{0e} - Z_{0o})^2}{200} \tag{2.99}$$

In a common case, to define the characteristic impedances Z_{0e} and Z_{0o} and the appropriate effective dielectric constants under even and odd excitation modes, it is necessary to calculate the total capacitances of the coupled transmission lines for a given geometric structure [23]. However, the characteristic impedances Z_{0e} and Z_{0o} can be expressed through the characteristic impedance Z_0 of a single transmission line and midband (quarter-wavelength) voltage coupling coefficient C by [24]

$$Z_0 = \sqrt{Z_{0e} Z_{0o}} \qquad C = \frac{Z_{0e} - Z_{0o}}{Z_{0e} + Z_{0o}} \tag{2.100}$$

As a result, the even mode characteristic impedance Z_{0e} and odd mode characteristic impedance Z_{0o} can be obtained as

$$Z_{0e} = Z_0 \sqrt{\frac{1 + C}{1 - C}} \tag{2.101}$$

$$Z_{0o} = Z_0 \sqrt{\frac{1 - C}{1 + C}} \tag{2.102}$$

The frequency dependencies of the real and imaginary parts of the resonant circuit impedance Z_r for different coupling coefficients C are shown in Figure 2.33. As is seen from Figure 2.33a, the real part of Z_r depends insignificantly on the coupling coefficient C, especially at its lower values. The imaginary part of Z_r behaves almost as a linear function of frequency, starting from high capacitive reactances and having increasing inductive reactance

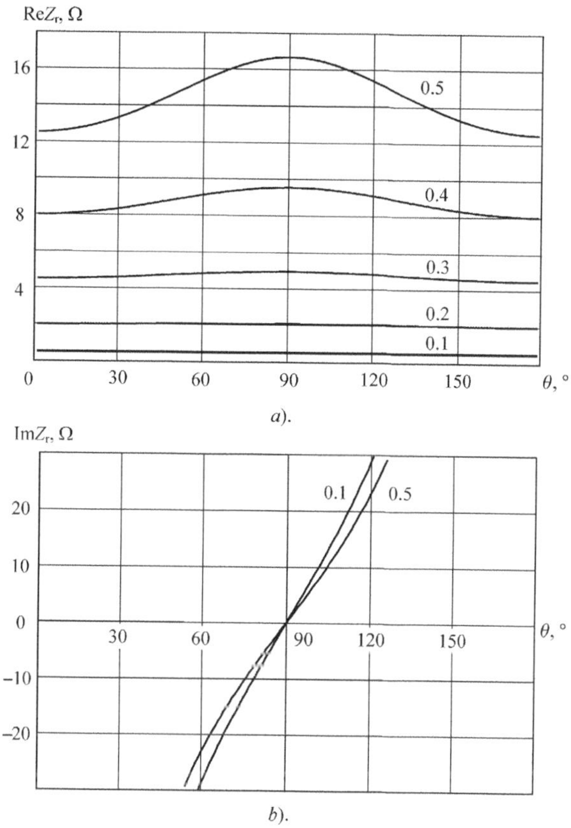

Figure 2.33 Frequency dependencies of resonant circuit impedance Z_r

for electrical length of the transmission lines more than a quarter-wavelength with very small variations over different values of the coupling coefficient C.

Figure 2.34a shows the equivalent Gummel–Poon SPICE model of the bipolar transistor, which can be used to simulate both its dc and high-frequency behaviour close to the transition frequency $f_T = g_m/2\pi C_\pi$ [25]. At sufficiently low frequencies, such an equivalent representation of the bipolar transistor can be simplified to the circuit shown in Figure 2.34b, where C_c is the collector capacitance and r_b is the base resistance. To represent the analytical results in an explicit form, it is advisable to ignore the effect of a base-width modulation (Early effect), so that the value of the resistance r_{ce} can be assumed as infinite. In addition, it is convenient to consider the base resistance r_b as an element of the external circuit. Then, the transistor common emitter Y-parameters can be obtained as

$$\begin{bmatrix} Y_{11} & Y_{12} \\ Y_{21} & Y_{22} \end{bmatrix} = \begin{bmatrix} j\dfrac{\omega}{\omega_T}(g_m + \omega_T C_c) & -j\omega C_c \\ g_m - j\omega C_c & j\omega C_c \end{bmatrix} \tag{2.103}$$

where g_m is the device transconductance and $\omega_T = 2\pi f_T$.

By using the ratios between common emitter and common collector Y-parameters from Table 2.1, the input impedance Z_{in} seen from the base terminal of the loaded common collector

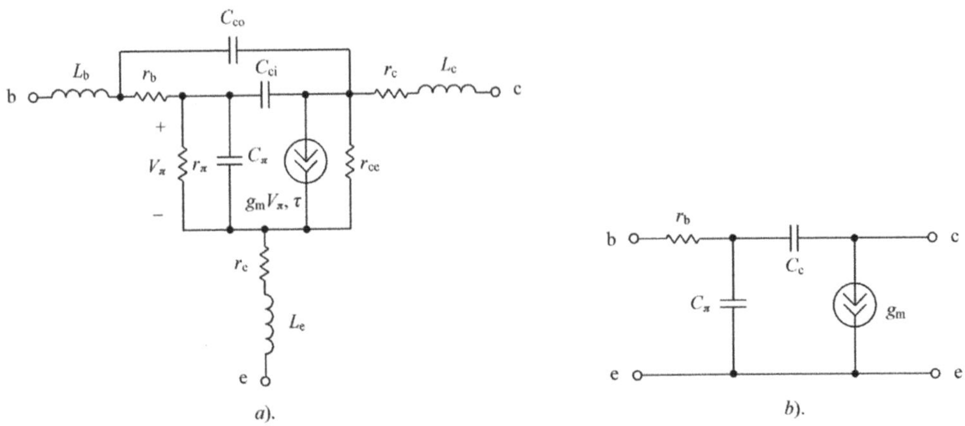

Figure 2.34 Simplified bipolar transistor equivalent circuits

transistor, being expressed through the common emitter Y-parameters, can be obtained as

$$Z_{in} = \frac{1 + (Y_{11} + Y_{12} + Y_{21} + Y_{22})\, Z_l}{Y_{11} + \Delta Y Z_l} \tag{2.104}$$

where Z_l is the input impedance of the open-circuited transmission line, as shown in Figure 2.32a.

Generally, the input impedance of the transmission line Z_l as a function of a transmission-line electrical length θ_l with arbitrary load impedance Z_R can be obtained as

$$Z_l = Z_0 \frac{Z_R + jZ_0 \tan\theta_l}{Z_0 + jZ_R \tan\theta_l} \tag{2.105}$$

where Z_0 is the characteristic impedance of the transmission line.

When $Z_R = \infty$,

$$Z_l = -jZ_0 \cot\theta_l \tag{2.106}$$

which corresponds to the capacitive input reactance for $\theta_l < 90^\circ$ and inductive input reactance for $\theta_l > 90^\circ$.

Substituting the load and device equivalent circuit parameters into Equation (2.104) yields

$$Z_{in} = \frac{1}{g_m} \frac{b}{\vartheta} \frac{1 + a\vartheta - ja}{a + j(b + a\vartheta + 1)} \tag{2.107}$$

where $a = g_m Z_0 \cot\theta_l$, $b = g_m/(\omega_T C_c)$, and $\vartheta = \omega/\omega_T$.

The frequency dependencies of the normalized real and imaginary parts of the input impedance Z_{in} for different values of parameter a and fixed value $b = 20$ are shown in Figure 2.35, where $R_{in} = \mathrm{Re}Z_{in}$ and $X_{in} = \mathrm{Im}Z_{in}$. From Figure 2.35a it follows that, for a certain value of parameter b, the maximum negative resistance is realized at the optimum value of parameter a (near $a = 10$ for $b = 20$). For the known values of the device transconductance g_m and characteristic impedance of the open-circuited stub Z_0, its optimum electrical length θ_l can be calculated from

$$\theta_l = \tan^{-1}\left(\frac{g_m Z_0}{a}\right) \tag{2.108}$$

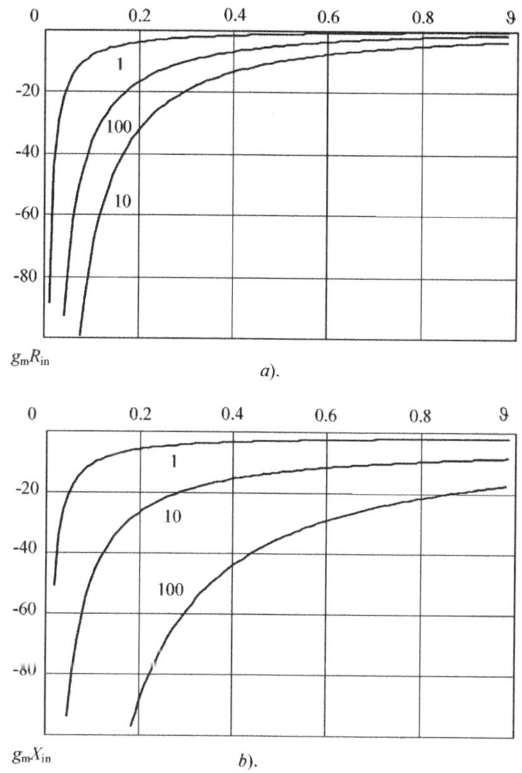

Figure 2.35 Frequency dependencies of normalized input impedance Z_{in}

Figure 2.35b shows that, for any value of parameter a, the input reactance ImZ_{in} is capacitive, becoming greater for the greater values of a. This means that the electrical length of the resonator should be more than a quarter-wavelength, as shown in Figure 2.33b, providing the inductive reactance to fulfil the phase balance oscillator condition. The choice of the proper coupling coefficient C between two coupled transmission lines depends on the output power requirement and substrate parameters. It should be noted that the bipolar transistors are potentially stable at low frequency, and an increase in the negative resistance when $\vartheta \to 0$ can be explained by simplifying the device equivalent circuit. In practice, at low frequencies, it is necessary to take into account the dynamic base-emitter resistance r_π and Early collector-emitter resistance r_{ce}, the presence of which substantially increases the value of the device stability factor. At high frequencies close to f_T, the effect of the parasitic series inductances can lead to the inductive input reactance and disappearance of the negative resistance as well. However, in general, the bipolar oscillator circuit configuration with common collector can provide negative resistance conditions over a very wide frequency range.

2.8 PUSH–PUSH OSCILLATOR

Using a common collector configuration of the transistors and a series resonant circuit located between the devices in the push–pull oscillators simplifies the load connection and allows

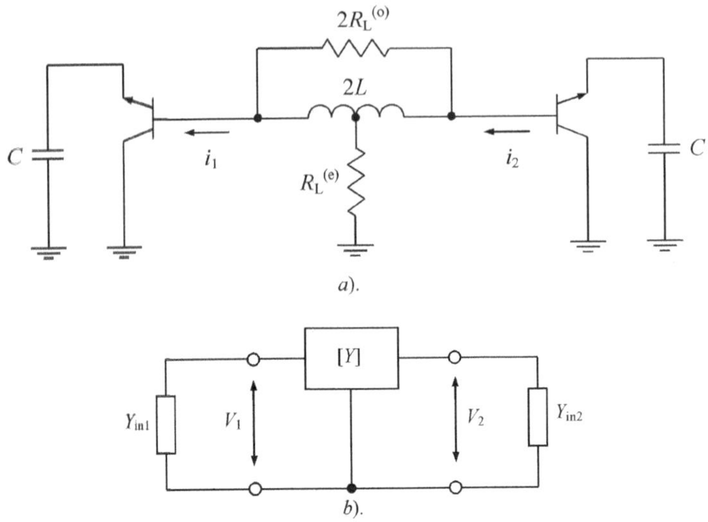

Figure 2.36 Simplified circuit schematics of bipolar push–pull oscillator

operation at twice the operating frequency. Figure 2.36a shows the general simplified equivalent circuit of the balanced oscillator having two load resistors connected in parallel to the series resonant circuit inductor and into its centre point, respectively. The series inductor located between the two active devices is common to both oscillators.

For a simplified circuit analysis, let us represent the oscillator schematic in the form of a general negative conductance oscillator with two active devices connected to a common two-port network, as shown in Figure 2.36b. The two-port network is characterized by the admittance Y-parameters and terminal voltages represented by voltage phasors given as

$$\mathbf{V}_1 = V_1 \exp(j\phi_1) \qquad \mathbf{V}_2 = V_2 \exp(j\phi_2) \tag{2.109}$$

where V_1, V_2, ϕ_1, and ϕ_2 are the magnitudes and phases of two voltage phasors, respectively.

The relationships between the circuit currents and voltages in a steady-state operation mode through the admittance parameters in a matrix form can be written as

$$\begin{bmatrix} -Y_{in1}\mathbf{V}_1 \\ -Y_{in2}\mathbf{V}_2 \end{bmatrix} = \begin{bmatrix} Y_{11} & Y_{12} \\ Y_{21} & Y_{22} \end{bmatrix} \begin{bmatrix} \mathbf{V}_1 \\ \mathbf{V}_2 \end{bmatrix} \tag{2.110}$$

where Y_{in1} and Y_{in2} represent the input admittances of the negative conductance devices.

Since the oscillator consists of two identical nonlinear active halves when $V_1 = V_2 = V$ and $V_{in1} = V_{in2} = V_{in}$ and symmetrical passive linear two-port network when $Y_{11} = Y_{22}$ and $Y_{12} = Y_{21}$, the matrix of Equation (2.110) can be rewritten in the form of two equations as

$$-Y_{in} = Y_{11} + Y_{12}\exp(j\phi) \tag{2.111}$$

$$-Y_{in} = Y_{12}\exp(-j\phi) + Y_{11} \tag{2.112}$$

where $\phi = \phi_2 - \phi_1$ is the phase difference between voltage phasors.

Simplified analysis of the steady-state operation modes shows that, in such a balanced oscillator, there may exist two modes with equal amplitudes [26, 27]:

- odd mode with

$$\phi = (2k + 1)\pi \qquad \text{where } k = 0, 1, 2, \ldots$$

- even mode with

$$\phi = 2k\pi \qquad \text{where } k = 0, 1, 2, \ldots$$

The frequency and amplitude of each of the two oscillation modes can be determined by solving the following equation:

$$Y_{in}(V, \omega) + Y_{11}(\omega) \pm Y_{12}(\omega) = 0 \tag{2.113}$$

In a time domain, assuming symmetrical current waveforms flowing into the base of both transistors, their Fourier series expansions in a common case can be written as

$$i_1 = I_1^{(1)} \cos \omega t + I_2^{(1)} \cos 2\omega t + I_3^{(1)} \cos 3\omega t + \ldots + I_n^{(1)} \cos n\omega t \tag{2.114}$$

$$i_2 = I_1^{(2)} \cos(\omega t + \phi) + I_2^{(2)} \cos 2(\omega t + \phi) + I_3^{(2)} \cos 3(\omega t + \phi) \ldots + I_n^{(2)} \cos n(\omega t + \phi) \tag{2.115}$$

where n is the harmonic number.

Consequently, in the odd mode, the currents are flowing in opposite directions, providing a push–pull operation of both transistors having out-of-phase base-emitter voltages. In this case, the circuit becomes antisymmetric and, at its centre point, a virtual ground will be formed at the fundamental frequency f_0 with zero fundamental voltage at this point. The output current flowing into the load $2R_L^{(o)}$ is a result of the out-of-phase summation of currents given by Equations (2.114) and (2.115) as

$$i^{(o)} = 2I_1 \cos \omega t + 2I_3 \cos 3\omega t + \ldots + 2I_{2k+1} \cos(2k + 1)\omega t \tag{2.116}$$

which contains only odd current components, and there are no odd components flowing into the load $R_L^{(e)}$. At the same time, the output current flowing into the load $R_L^{(e)}$ is a result of the in-phase summation of currents as

$$i^{(e)} = 2I_2 \cos 2\omega t + 2I_4 \cos 4\omega t + \ldots + 2I_{2k} \cos 2k\omega t \tag{2.117}$$

which contains only even current components.

In the even mode, the currents are flowing in the same direction, providing a push–push operation of both transistors having in-phase base-emitter voltages. In this case, the circuit becomes symmetric with load resistor $R_L^{(e)}$ at the centre of symmetry. For such an operation, the odd current components obtained by Equation (2.116) will flow to this load, whereas the even components obtained by Equation (2.117) will be dissipated at the load $2R_L^{(o)}$.

The equivalent circuits of the oscillator under odd and even operation modes are shown in Figure 2.37. For oscillations to occur in the odd mode, the negative conductance generated by the oscillator, the schematic of which is shown in Figure 2.37a, should exceed the resistive losses in the resonant circuit, that is

$$\text{Re}Y_{in} + \text{Re}Y_L^{(o)} < 0 \tag{2.118}$$

For oscillations to occur in the even mode, the negative resistance generated by the oscillator, shown in Figure 2.37b, should exceed the resistive losses in such a resonant circuit according

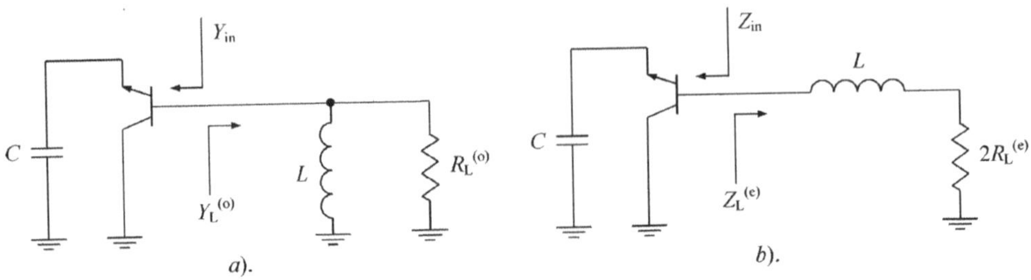

Figure 2.37 Equivalent oscillator circuits with (a) odd and (b) even modes

to

$$\mathrm{Re}Z_{in} + \mathrm{Re}Z_L^{(e)} < 0 \tag{2.119}$$

Generally, both odd and even operation modes can exist in the oscillator simultaneously, depending on the values of the load resistances $R_L^{(o)}$ and $R_L^{(e)}$. Under a start-up condition given by Equation (2.118), the even mode will be stable for the case of $R_L^{(o)} = \infty$ corresponding to the situation of the summation of the output powers of individual oscillators with resistive coupling [26, 28]. A simplified criterion for the odd mode stable operation for the oscillator circuit shown in Figure 2.36a can be obtained as

$$R_L^{(o)} < \frac{2R_L^{(e)}}{1 + \left(\dfrac{\omega L}{2R_L^{(e)}}\right)^2} \tag{2.120}$$

when the odd components will flow into the load $2R_L^{(o)}$, while the even components can be singled out at the load $R_L^{(e)}$. To improve stability of the odd operation mode, it is best to inhibit the oscillation condition given by Equation (2.119).

Such a balanced oscillator configuration with a series resonant circuit creates a simple opportunity to double the oscillation frequency when each half-circuit oscillates at the resonant frequency f_0, while the output signal at the load oscillates at the frequency $2f_0$. In this case, it is necessary to provide stable operation in the odd mode and to inhibit the oscillations in the even mode by removing the load resistor $R_L^{(o)}$. This results in the oscillator circuit configurations shown in Figure 2.38, where the variable capacitors C_v can be used for frequency tuning. For a lossless tank inductor L, the start-up amplitude oscillation conditions can be rewritten as

$$\mathrm{Re}Z_{in} < 0 \tag{2.121}$$

$$\mathrm{Re}Z_{in} + 2R_L > 0 \tag{2.122}$$

when the even harmonic components dissipate at the load R_L. A significant margin in negative resistance can be achieved compared with an equivalent single-ended oscillator, even at very high frequencies [29]. The outputs for the out-of-phase odd components can be connected to the device emitters. Because, for the even output current components, the transistors are operated in phase, it is called a *push–push operation mode*. Push–push operation is the inverse of push–pull operation, in that the load is either conductively or capacitively coupled to the centre point of the inductance or transmission line located between the bases (or gates) of the active devices [30, 31]. In this case, the currents flow into the load in the same direction

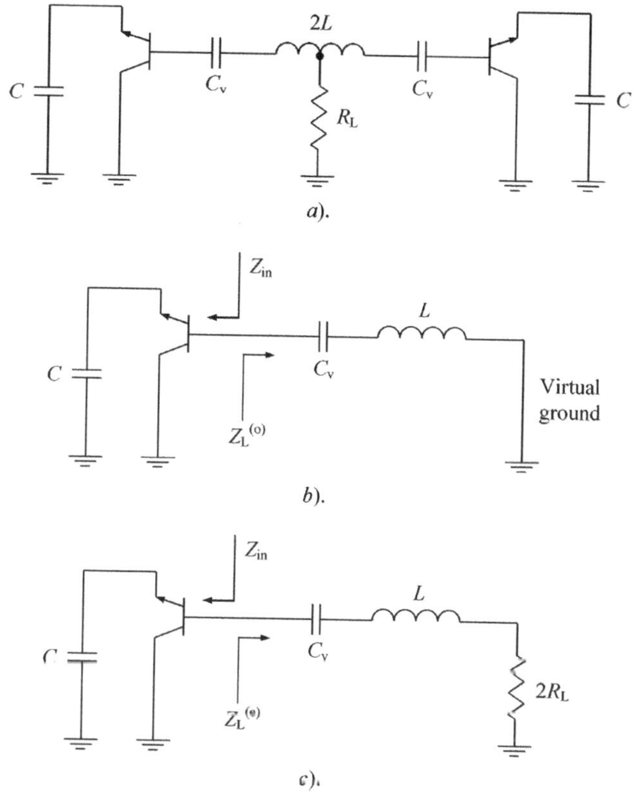

Figure 2.38 Simplified circuit schematics of bipolar push–push oscillator

during both half periods or 180° phases of the active device. Since there are two 180° phases during each cycle, the circuit acts as a frequency doubler. Such a push–push oscillator is a result of its odd operation mode with virtual ground; therefore, ideally this design is independent of the output load pulling compared with a fundamental oscillator approach.

For microwaves, the resonant circuits usually incorporate the transmission lines as inductive elements or resonators. Figure 2.39a shows the push–push oscillator using a half-wavelength or $\lambda/2$ microstrip resonator [32, 33]. For the fundamental frequency f_0, this resonator has a null point at the centre of microstrip line, being a point of the oscillator symmetry which is considered as a virtual ground or short-circuited point. In this case, the resonance voltage has maximum values at both ends of the resonator with a phase difference of 180°, and the resonance voltage is zero at the centre of the resonator. For the desired second harmonic frequency $2f_0$, such a point could be regarded as an open-circuited point. The null point is an ideal output port to combine the second harmonic signals from both active devices effectively. To achieve maximum second harmonic output signal and provide significant harmonic suppression, the series capacitors of 0.5 pF were chosen. The operating bias points with zero gate voltages and drain voltages of 4 V provide nonlinear operation of the used HEMT devices generating a high-power second harmonic signal. The impedance of half-wavelength microstrip open-circuited resonator was optimized to make the impedance of the output port close to 50 Ω. By using a

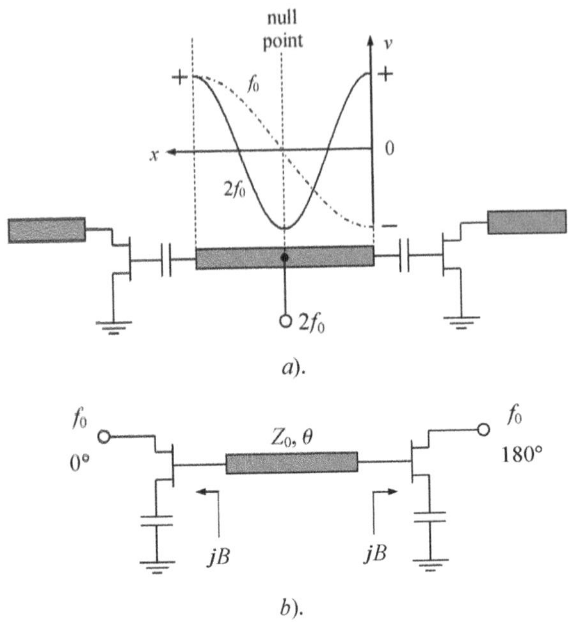

Figure 2.39 Push–push and balanced microstrip oscillator configurations

Fujitsu FHX35LG device, a maximum power of 8.4 dBm at the second harmonic frequency of 21.68 GHz with the fundamental frequency suppression of −26 dBc was achieved. The measured phase noise was −100.5 dBc/Hz at 1 MHz offset from the carrier frequency.

The inherent physical symmetry of the balanced oscillator circuits makes them very attractive in applications where two balanced out-of-phase outputs are required. Due to the excellent amplitude and phase balance as well as better noise performance compared with a single-ended configuration, they are widely used in balanced mixers, phase lock loops or synthesizers where out-of-phase signals from local oscillators are needed. Figure 2.39b shows the circuit schematic of such a balanced microstrip oscillator where the two out-of-phase fundamental signals are taken from the drain terminals of each transistor [34]. To provide a negative conductance operation at the oscillation frequency, the common source capacitive feedback is applied to both active devices. In this circuit, the transmission line located between the device gates serves as a part of the resonant circuit, being shorter than a quarter-wavelength, and its electrical length is calculated from

$$\theta = \tan^{-1}\frac{2BZ_0}{(BZ_0)^2 - 1} \tag{2.123}$$

where B is the negative conductance susceptance seen from each device input and Z_0 is the characteristic impedance of the transmission line. By using two 4×75 μm MESFET devices with gate length of 0.5 μm in a monolithic implementation, the stable oscillations at 19.3 GHz with output power of 5 dBm at supply voltage of 5 V were achieved at each output. An exact out-of-phase relationship is maintained within injection locking bandwidth of 10 MHz with 20 dB locking gain. This demonstrates the possibility of using a balanced oscillator as an active balun where the input signal is used for injection locking the balanced outputs. The increased

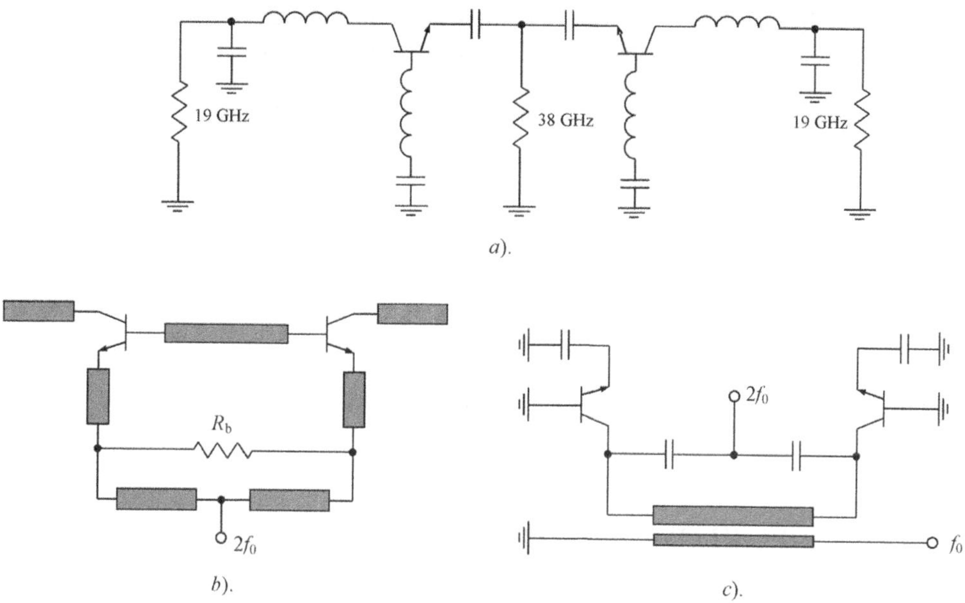

Figure 2.40 Bipolar push–push oscillator configurations

oscillation frequency of 39.7 GHz was achieved at supply voltage of 3 V using two $2 \times 60\mu m$ HEMT devices with gate length of 0.25 μm.

The push–push oscillator can be designed to operate at both fundamental and second harmonic frequencies simultaneously. Figure 2.40a shows the push–push microwave oscillator schematic where the centre point between emitter terminals is an output port to combine the second harmonic signals from both active devices, whereas fundamental out-of-phase signals are taken from both collectors separately [35]. To make transistors unstable, a lumped series resonant circuit at the base of each device has an inductive reactance at the resonant frequency. Using a series resonant circuit instead of a single inductance provides a steeper slope of the oscillator phase characteristic, improving the quality factor of its resonant circuit. At the same time, the two 50-Ω outputs are connected to the collectors through the L-type matching circuits each. At the emitters, the symmetrically located capacitors of equal values fulfil the oscillator phase balance condition providing the virtual ground point for the second harmonic output. The monolithic coplanar implementation of the microwave oscillator was realized by using InGaP/GaAs HBT devices with $f_T = 38$ GHz, resulting in phase noise levels of −96 dBc/Hz and −89 dBc/Hz at 100 kHz offset frequency and output powers of −5 and −25 dBm at the fundamental frequency of 19 GHz and second harmonic of 38 GHz, respectively.

The second harmonic signal from both devices can be collected using an in-phase Wilkinson power combiner, as shown in Figure 2.40b [36]. In this case, the out-of-phase fundamental signals will flow into the ballast resistor R_b. By proper choice of the characteristic impedances of the transmission lines, the required impedance transformation can be made to connect a standard 50-Ω load. For the designed hybrid 38 GHz push–push oscillator using commercial bipolar BFP405 devices with $f_T = 25$ GHz, the output power of −11.5 dBm with phase noise of −80 dBc/Hz at 100 kHz offset frequency was achieved. Suppression of the fundamental frequency signal was about 11 dB only, which is the result of some circuit asymmetry due to fabrication tolerances of bond wires and transistor chip parameters. If it is necessary to combine the second harmonic signal only, the output port can be directly connected to the centre point of the microstrip line located between two transistor emitters, with electrical length chosen to provide the oscillation conditions [37]. Figure 2.40c shows a simplified circuit schematic of the push–push bipolar oscillator where the balun consisting of asymmetric coupled lines is used to combine outputs for the fundamental frequency [38]. The combined output for the second harmonics can be obtained from the middle point of the circuit consisting of two capacitances connected in series between the transistor emitters. High suppression of undesired harmonic components can be achieved using double-sided integrated circuit technology. The push–push oscillator using the two HEMT FHX35LG devices and incorporating a microstrip line on the top of the Teflon dielectric substrate and a slot line on its reverse side could provide output power of 4.17 dBm at the oscillation frequency of 21.25 GHz with phase noise of −99.68 dBc/Hz at 1 MHz offset frequency and harmonic suppression of −30 dBc [39].

2.9 TRIPLE-PUSH OSCILLATOR

The possibility of developing sinewave low-frequency oscillators in three phases employing three identical sinewave fundamental oscillators was first described by Kaplan and Bachar [40]. At higher frequencies, the operation condition with the oscillation frequency three times greater than the fundamental frequency of each basic oscillator can be provided by using a parallel connection of the transistors and series transmission lines to the load. Such an oscillator is called a *triple-push oscillator* Figure 2.41a shows the general simplified equivalent circuit of a high-frequency triple-push oscillator having three identical common collector transistors connected to a common load R_L through their resonant circuit series inductances [41].

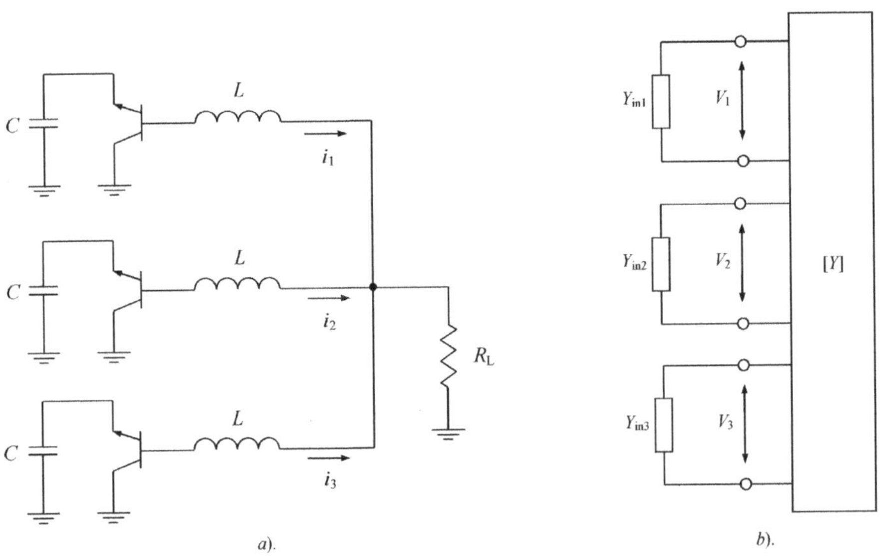

Figure 2.41 Simplified circuit schematics of the bipolar triple-push oscillator

For a simplified circuit analysis, let us represent the oscillator schematic in the form of a general negative conductance oscillator with three active devices connected to a common three-port network, as shown in Figure 2.41b. The three-port network is characterized by the admittance Y-parameters and terminal voltages represented by voltage phasors given by

$$\boldsymbol{V}_1 = V_1 \exp(j\phi_1) \quad \boldsymbol{V}_2 = V_2 \exp(j\phi_2) \quad \boldsymbol{V}_3 = V_3 \exp(j\phi_3) \tag{2.124}$$

where V_1, V_2, V_3, ϕ_1, ϕ_2 and ϕ_3 are the magnitudes and phases of three voltage phasors, respectively.

The relationships between the circuit currents and voltages in a steady-state operation mode through the admittance parameters in a matrix form can be written as

$$\begin{bmatrix} -Y_{\text{in1}}\boldsymbol{V}_1 \\ -Y_{\text{in2}}\boldsymbol{V}_2 \\ -Y_{\text{in3}}\boldsymbol{V}_3 \end{bmatrix} = \begin{bmatrix} Y_{11} & Y_{12} & Y_{13} \\ Y_{21} & Y_{22} & Y_{23} \\ Y_{31} & Y_{32} & Y_{33} \end{bmatrix} \begin{bmatrix} \boldsymbol{V}_1 \\ \boldsymbol{V}_2 \\ \boldsymbol{V}_3 \end{bmatrix} \tag{2.125}$$

where Y_{in1}, Y_{in2} and Y_{in3} represent the input admittances of the negative conductance devices.

Since the oscillator consists of three identical nonlinear active circuits when $V_1 = V_2 = V_3 = V$ and $V_{\text{in1}} = V_{\text{in2}} = V_{\text{in3}} = V_{\text{in}}$ and a symmetric passive linear two-port network when $Y_{11} = Y_{22} = Y_{33}$ and $Y_{12} = Y_{21} = Y_{13} = Y_{31} = Y_{23} = Y_{32}$, the matrix Equation (2.125) can be rewritten in the form of three equations as

$$-Y_{\text{in}} = Y_{11} + Y_{12}\exp(j\phi) + Y_{12}\exp[j(\phi + \psi)] \tag{2.126}$$

$$-Y_{\text{in}} = Y_{12}\exp(-j\phi) + Y_{11} + Y_{12}\exp(j\psi) \tag{2.127}$$

$$-Y_{\text{in}} = Y_{12}\exp[-j(\phi + \psi)] + Y_{12}\exp(-j\psi) + Y_{11} \tag{2.128}$$

where $\phi = \phi_2 - \phi_1$ and $\psi = \phi_3 - \phi_2$ are the phase difference between voltage phasors.

Simplified analysis of the steady-state operation modes shows that there may exist three modes, one even and two odd, with equal amplitudes [42, 43]:

- even mode with

$$\phi = \psi = 2k\pi \qquad \text{where } k = 0, 1, 2, \ldots$$

- odd modes with

$$\phi = \psi = \left(2k \pm \frac{2}{3}\right)\pi \qquad \text{where } k = 0, 1, 2, \ldots$$

The frequency and amplitude of each oscillation mode can be determined by solving the following equations:

- for even mode

$$Y_{\text{in}}(V,\omega) + Y_{11}(\omega) + 2Y_{12}(\omega) = 0 \tag{2.129}$$

- for odd modes

$$Y_{\text{in}}(V,\omega) + Y_{11}(\omega) - Y_{12}(\omega) = 0 \tag{2.130}$$

Hence, in the even mode, the in-phase currents are flowing in the same direction, providing a push–push operation of all three transistors. In this case, the circuit becomes symmetric with load resistor R_{L} at the centre of symmetry. In the odd mode, each current must have 120° phase shift relative to one another. In this case, a virtual ground will be formed at the

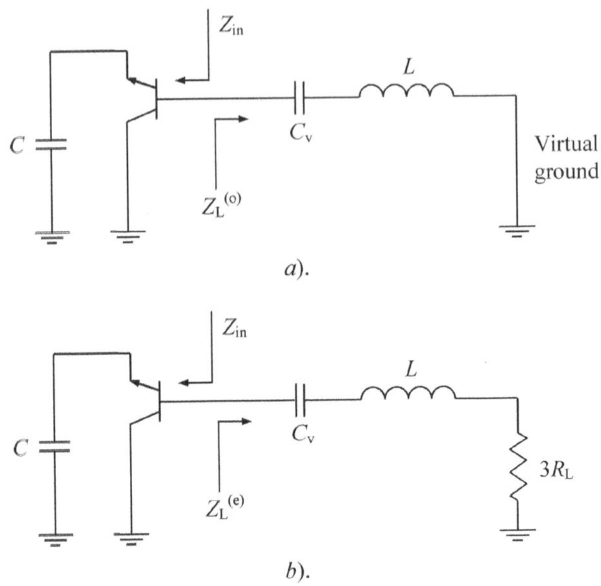

Figure 2.42 Equivalent oscillator circuits with (a) odd and (b) even modes

fundamental frequency f_0 with zero fundamental voltage at centre point. However, the third harmonic components, having the same 360° phase shift, will combine in phase at the load resistor R_L. In this case, it is necessary to provide a stable operation in the odd mode and to inhibit the oscillations in the even mode. This results in the oscillator circuit configurations shown in Figure 2.42, where the variable capacitors C_v can be used for frequency tuning. For a lossless tank inductor L, the start-up amplitude oscillation conditions can be written as

$$\mathrm{Re}Z_{in} < 0 \quad (2.131)$$

$$\mathrm{Re}Z_{in} + 3R_L > 0 \quad (2.132)$$

when the third harmonic components flow into the load R_L.

Using commercially available bipolar BFG540 transistors, the 4.85 GHz triple-push oscillator was fabricated on FR-4 substrate, based on three identical fundamental oscillators, each having the circuit schematic shown in Figure 2.43a [41, 43, 44]. The base capacitance of 4 pF is used to generate negative resistance, while the feedback capacitances of 0.5 pF are used to determine the oscillation frequency. The emitter resistance of 51 Ω is necessary to improve circuit performance versus temperature variations by stabilizing the bias point. At the supply voltage of 3.1 V, the oscillator exhibits the output power of 1 dBm with a phase noise of −110 dBc/Hz at 1 MHz offset frequency, consumes dc current of 17 mA and provides suppressions of 12 and 15.5 dB for the fundamental and second harmonics, respectively.

The 28.4 GHz triple-push oscillator chip was fabricated using InGaP/GaAs HBT technology with a common collector configuration of the HBT devices having $f_T = 32$ GHz and emitter size of 1.4 × 9 μm each [43]. Figure 2.43b shows the circuit schematic of a single fundamental oscillator where the emitter capacitor and collector inductor values were optimized

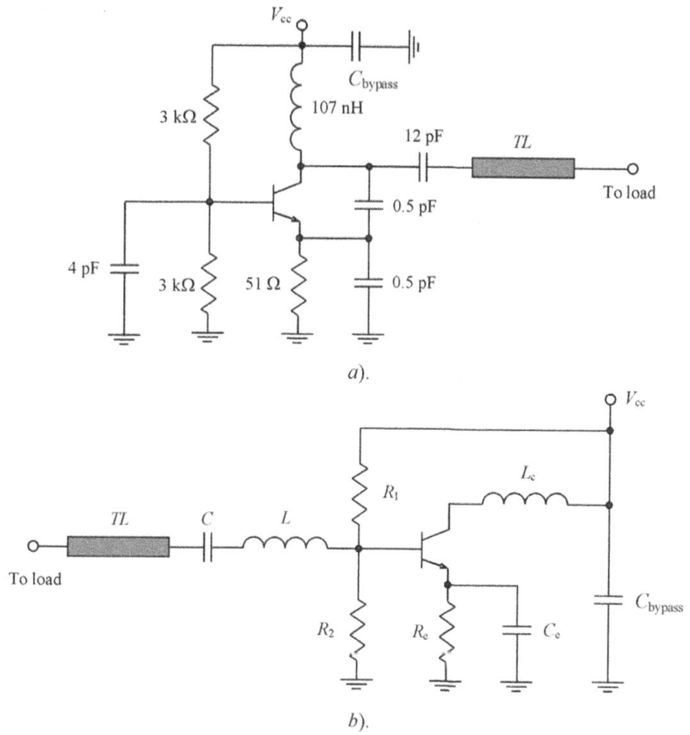

Figure 2.43 Circuit schematics of bipolar microwave triple-push oscillators (© 2001 IEEE)

to maximize the negative resistance looking into the base terminal. A resonator connected to the base is formed using a series connection of the inductor, capacitor and transmission line for determining the oscillation frequency. At a supply voltage of 8.8 V, such a triple-push oscillator exhibits output power of −15.4 dBm, while the output powers for the fundamental and second harmonics were suppressed to −21 and −34 dBm, respectively. By optimizing the circuit symmetry and bias conditions, the output powers at the fundamental and second harmonics were minimized to −37 dBm [45]. By using a cascode configuration of each single fundamental oscillator composing a triple-push oscillator, the output power of −13.5 dBm at the oscillation frequency of 39 GHz was achieved, with fundamental and second harmonic suppression more than 33 dB [46]. Varying the capacitance in the series resonant circuit in each fundamental oscillator could provide the frequency tuning from 39 to 46 GHz with better than 8 dB fundamental and second harmonic suppression over the tuning bandwidth.

2.10 OSCILLATOR WITH DELAY LINE

In real oscillator circuits, their state at any particular moment generally depends on the entire history of the system rather than on its instantaneous state. Especially it is a concern at very high frequencies when there is a certain time delay between the input signal and device output response due to the distributed structure of its equivalent circuit. Schematically, a delay line connected to the input terminal of the idealized active device can represent the effect of the

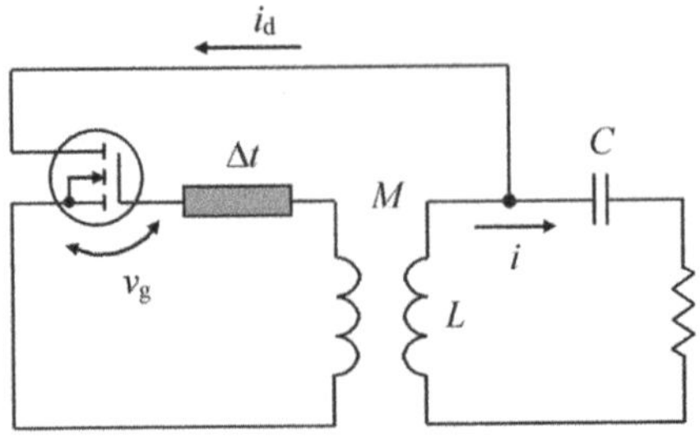

Figure 2.44 Schematic of transformer-coupled MOSFET oscillator with delay line

time delay. Figure 2.44 shows the electrical schematic of the transformer-coupled MOSFET oscillator with the series resonant circuit having a delay line with delay time Δt in the feedback circuit. The electrical behaviour of such an oscillator circuit can be described by the following integrodifferential equation:

$$L\frac{\mathrm{d}i}{\mathrm{d}t} + Ri + \frac{1}{C}\int (i - i_\mathrm{d})\,\mathrm{d}t = 0 \tag{2.133}$$

where i is the current flowing in the series resonant circuit and i_d is the drain current.

Differentiating Equation (2.133) with respect to time yields

$$\frac{\mathrm{d}^2 i}{\mathrm{d}t^2} + 2\delta\frac{\mathrm{d}i}{\mathrm{d}t} + \omega_0^2 i = \omega_0^2 i_\mathrm{d} \tag{2.134}$$

where $2\delta = R/L$ and $\omega_0^2 = 1/LC$. For simplicity of calculation, let us represent the drain current by a step function when $i_\mathrm{d} = 0$ for $v_\mathrm{g} < 0$ and $i_\mathrm{d} = I_0$ for $v_\mathrm{g} \geq 0$.

Assuming a high quality factor of the resonant circuit when the gate voltage can be considered as sinusoidal with delay $\theta = \omega\Delta t$, i.e., $v_\mathrm{g} = v_\mathrm{g}\ \sin(\omega t - \theta)$, we will seek the solution for the resonant circuit current in the form of $i = -I\ \cos\omega t$. As a result, expanding the right-hand side of Equation (2.134) into a Fourier series, substituting i into the left-hand side of Equation (2.134) and equating the corresponding terms with sines and cosines yield [47]

$$I\left(\omega^2 - \omega_0^2\right) = -\frac{2I_0}{\pi}\omega_0^2 \sin\theta \tag{2.135}$$

$$2\delta\omega I = \frac{2I_0}{\pi}\omega_0^2 \cos\theta \tag{2.136}$$

The oscillation amplitude I can be readily obtained from Equation (2.136) as

$$I = \frac{I_0}{\omega\pi\delta}\omega_0^2 \cos\theta \tag{2.137}$$

Substituting Equation (2.137) into Equation (2.135) allows us to calculate the oscillation frequency ω as

$$\omega = \omega_0\sqrt{1 + \left(\frac{\delta\ \tan\theta}{\omega_0}\right)^2} - \delta\ \tan\theta \tag{2.138}$$

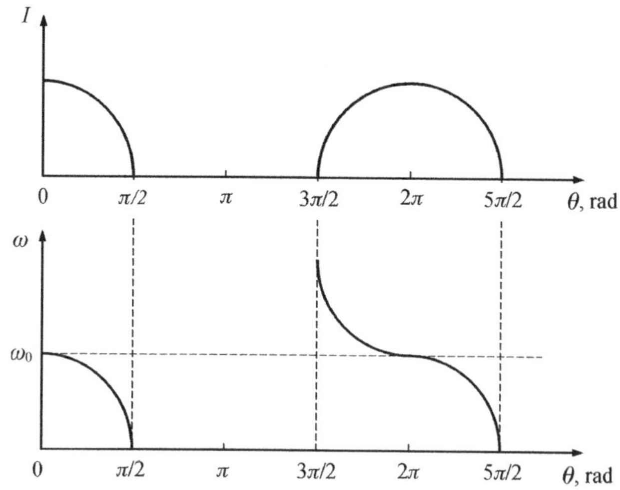

Figure 2.45 Oscillation amplitude and frequency versus time delay

For the oscillation systems with sufficiently high quality factor of their resonant circuits when $\delta \tan\theta/\omega_0 < 0.3$, Equations (2.137) and (2.138) can be simplified respectively to

$$I \cong \frac{I_0}{\pi\delta}\omega_0 \cos\theta \tag{2.139}$$

$$\omega \cong \omega_0 - \delta \tan\theta \tag{2.140}$$

Figure 2.45 shows the dependencies of the oscillation amplitude and frequency on the phase shift in the feedback circuit. It follows from Figure 2.45a that the oscillation amplitude I reduces from its maximum values at $\theta = 2n\pi$ to zero values at $\theta = (2n + 1)\pi/2$, where $n = 0, 1, 2, \ldots$. At the same time, the oscillation frequency ω is a function of the time delay as well, and deviates from the resonant frequency ω_0 with phase delay θ, as shown in Figure 2.45b. One of the specific features of such an oscillation circuit is that, for sufficiently large values of time delay Δt or phase shift θ, even a slight change in the oscillation frequency may result in a change of the excitation region. This is due to the distributed nature of the delay element (for example, transmission line) that introduces the possibility of multi-frequency oscillations within the oscillator resonant circuit frequency bandwidth, for which the self-oscillation conditions are fulfilled. However, which oscillations frequency will be established depends on the initial conditions, and if these conditions are predominant for one frequency, then the oscillations at this frequency will reach their steady-state amplitude, whereas the possible oscillations at other frequencies will be suppressed.

Now we can use the method of slowly varying amplitudes (the van der Pol method) to analyse the stability conditions for the oscillator with a delay line. The desired solution is in the form

$$i = I_c(t) \cos \omega t + I_s(t) \sin \omega t \tag{2.141}$$

where $I_c(t)$ and $I_s(t)$ are slowly varying current amplitudes. The solution given by Equation (2.141) can also be rewritten using the slowly varying amplitude $I(t)$ and phase $\varphi(t)$ in the form

$$i = I(t)\cos[\omega_0 t + \phi(t)] \tag{2.142}$$

where $I_c = I\cos\phi$, $I_s = -I\sin\phi$ and $I^2 = I_c^2 + I_s^2$.

To analyse the start-up conditions, it is sufficient to confine attention to the linear term of the device transfer characteristic $i_d = I_0 + g_m v_g$, where g_m is the device small-signal transconductance. As a result, the truncated or shortened first-order equations for the slowly varying amplitudes can be obtained as [47]

$$\frac{dI_c(t)}{dt} + \varphi(V_c, V_s) = 0 \tag{2.143}$$

$$\frac{dI_s(t)}{dt} + \psi(V_c, V_s) = 0 \tag{2.144}$$

where

$$\varphi(V_c, V_s) = -\frac{1}{2}\left(\frac{Mg_m\omega_0^2}{\omega}\cos\theta - \frac{2\delta}{\omega}\right)V_c + \frac{1}{2}\left(\frac{Mg_m\omega_0^2}{\omega}\sin\theta + \xi\right)V_s$$

$$\psi(V_c, V_s) = -\frac{1}{2}\left(\frac{Mg_m\omega_0^2}{\omega}\sin\theta + \xi\right)V_c - \frac{1}{2}\left(\frac{Mg_m\omega_0^2}{\omega}\cos\theta - \frac{2\delta}{\omega}\right)V_s$$

$$\xi = 1 - \left(\frac{\omega_0}{\omega}\right)^2$$

Equations (2.143) and (2.144) allow us to determine the conditions of stability and instability of the oscillation system. In this case, the determinant of a system of two equations must be equal to zero, that is

$$\begin{vmatrix} \dfrac{\partial\varphi}{\partial V_c} + \lambda & \dfrac{\partial\varphi}{\partial V_c} \\ \dfrac{\partial\psi}{\partial V_s} & \dfrac{\partial\psi}{\partial V_s} + \lambda \end{vmatrix} = 0 \tag{2.145}$$

solving of which gives the eigenvector λ in the form

$$\lambda_{1,2} = \frac{1}{2}\left(\frac{2\delta}{\omega} - \frac{Mg_m\omega_0^2}{\omega}\cos\theta\right) \pm j\frac{1}{2}\left(\frac{Mg_m\omega_0^2}{\omega}\sin\theta + \xi\right) \tag{2.146}$$

The oscillation system is unstable at the origin, resulting in the start-up of the oscillations if and only if the real part of eigenvector λ has negative real part, i.e., when

$$Mg_m\omega_0^2\cos\theta > 2\delta \tag{2.147}$$

where $g_m\cos\theta$ is the reduced device transconductance due to time delay and 2δ is the losses in the oscillation system. It follows from Equation (2.147) that the presence of the delay line reduces the oscillator regeneration factor. For the fixed values of the phase shift θ and δ, build-up of the self-sustained oscillations is possible when

$$\theta < \cos^{-1}\left(\frac{2\delta}{Mg_m\omega_0^2}\right) \tag{2.148}$$

Thus, due to the presence of the element providing the time delay (or phase shift) in the feedback circuit, the amplitude of the oscillations reduces and oscillation frequency deviates from the resonant frequency ω_0. Physically the worsening of the start-up conditions and oscillation amplitude degradation can be explained by the fact that the build-up of the oscillations occurs at a frequency different from the resonant circuit frequency ω_0 when the resonant circuit has some reactive impedance for the drain current. This results in lower power (proportional to the factor $\cos\theta$) flowing from the active device to compensate the losses in the resonant circuit.

It should be noted that using a delay line as an additional element in a feedback circuit can improve the oscillator performance. For example, by properly choosing the parameters of the voltage-controlled oscillator with time-delay feedback using a delay line, both noise improvement from degeneration and frequency agility from the periodic response can be obtained [48]. In this case, the open-loop gain should be chosen as large as possible for better noise degeneration, and the frequency bandwidth in the feedback circuit should be adjusted to meet the stability conditions. The delay line can also be used to implement a self-injection lock technique when a part of the oscillator output signal is used to injection lock the oscillator itself, resulting in phase noise reduction [49, 50].

REFERENCES

1. A. A. Andronov, A. A. Vitt and S. E. Khaikin, *Theory of Oscillations*, New York: Dover Publications (1987).
2. V. S. Andreev, *Theory of Nonlinear Electrical Circuits* (in Russian), Moskva: Radio i Svyaz (1982).
3. J. Groszkowski, 'The Interdependence of Frequency Variation and Harmonic Content, and the Problem of Constant-Frequency Oscillators', *Proc. IRE*, **21**, 958–9813 (1933).
4. E. H. Armstrong, 'Some Recent Developments in the Audion Receiver', *Proc. IRE*, **3**, 215–247 (1915); reprinted in *Proc. IEEE*, **85**, 685–697 (1997).
5. G. Kurz, *Oszillatoren*, Berlin: VEB Verlag Technik (1988).
6. N. Demma, 'Balanced Meissner Oscillator Circuits', *RF Design*, 72–74 (December 1993).
7. R. V. L. Hartley, Oscillation Generator, U.S. Patent 1,356,763 (October 1920).
8. E. H. Colpitts, Oscillation Generator, U.S. Patent 1,624,537 (April 1927).
9. F. B. Llewellyn, Constant frequency Oscillator, U.S. Patent 1,896,781 (Feburary 1933).
10. J. K. Clapp, 'Frequency Stable *LC* Oscillators', *Proc. IRE*, **42**, 1295–1300 (August 1954).
11. J. Vackar, Stabilization of Resonant Circuits, U.S. Patent 2,706,249 (April 1955).
12. S. I. Evtyanov, *Electron Tube Oscillators* (in Russian), Moskva: Svyaz (1967).
13. A. J. Cote, 'Matrix Analysis of Oscillators and Transistor Applications', *IRE Trans. Circuit Theory*, **CT-5**, 181–188 (1958).
14. F. Brown, 'Stable *LC* Oscillators', *RF Design* (March 1987); reprinted in G. A. Breed (Ed.), *Oscillator Design Handbook*, Englewood: Cardiff Publishing (1991), pp. 76–80.
15. B. Koster, P. Waldow and I. Wolff, 'A Unique, Low-Voltage, Source-Coupled J-FET VCO', *RF Design*, 58–68 (Apr. 2001).
16. R. S. Kaltenecker, R. E. Stengel, R. T. Enderby and J. S. Irwin, Dual FET Oscillator, U.S. Patent 4,785,264 (November 1988).
17. W. S. Chan, K. F. Tsang and G. B. Morgan, 'The Design of Oscillators Using the Cascode Circuit', *1994 IEEE Int. Symp. Circuits and Systems Dig.*, **5**, 689–692 (1994).
18. Y. Kwon, D. Pavlidis, P. Marsh, T. Brock and D. C. Streit, 'A 100-GHz Monolithic Cascode InAlAs/InGaAs HEMT Oscillator', *IEEE Microwave and Guided Wave Lett.*, **4**, 135–137 (1994).
19. G. D. Vendelin, A. M. Pavio and U. L. Rohde, *Microwave Circuit Design Using Linear and Nonlinear Techniques*, New York: John Wiley & Sons (1990).
20. K. L. Kotzebue, 'A Technique for the Design of Microwave Transistor Oscillators', *IEEE Trans. Microwave Theory Tech.*, **MTT-32**, 719–721 (1984).

21. D. Cohn and R. Mitchel, 'Design of Microstrip Transistor Oscillators', *Int. J. Electronics*, **35**, 385–395 (1973).
22. G. I. Zysman and A. K. Johnson, 'Coupled Transmission Line Networks in an Inhomogeneous Dielectric Medium', *IEEE Trans. Microwave Theory Tech.*, **17**, 753–759 (1969).
23. K. C. Gupta, R. Garg and R. Chadha, *Computer-Aided Design of Microwave Circuits*, Dedham: Artech House (1981).
24. H. Howe, *Stripline Circuit Design*, Dedham: Artech House (1974).
25. P. Antognetti and G. Massobrio, *Semiconductor Device Modeling with SPICE*, New York: McGraw-Hill (1993).
26. A. A. Dvornikov and G. M. Utkin, *Phased Oscillators in Radio Transmitters* (in Russian), Moskva: Energiya (1980).
27. Z. Ding and K. Chang, 'Modes and Their Stability of a Symmetric Two-Element Coupled Negative Conductance Oscillator Driven Spatial Power Combining Array', *IEEE Trans. Microwave Theory Tech.*, **MTT-44**, 1628–1636 (1996).
28. K. D. Stephan, 'Inter-Injection-Locked Oscillators for Power Combining and Phased Arrays', *IEEE Trans. Microwave Theory Tech.*, **MTT-34**, 1017–1025 (1986).
29. K. W. Kobayashi, A. K. Oki, L. T. Tran, J. C. Cowles, A. Gutierrez-Aitken, F. Yamada, T. R. Block and D. C. Streit, 'A 108-GHz InP-HBT Monolithic Push–Push VCO with Low Phase Noise and Wide Tuning Bandwidth', *IEEE J. Solid-State Circuits*, **SC-34**, 1225–1231 (1999).
30. A. Shipow, 'Linearity in Solid State Microwave Voltage Tuned Oscillators', *Microwave J.*, **26**, 130–137 (1983).
31. J. R. Bender and C. Wong, 'Push–Push Design Extends Bipolar Frequency Range', *Microwaves & RF*, **22**, 91–98 (1983).
32. H. Xiao, T. Tanaka and M. Aikawa, 'Push–Push Oscillator with Simplified Circuit Structure', *Electronics Lett.*, **38**, 1545–1547 (2002).
33. H. Xiao, T. Tanaka and M. Aikawa, 'Push–Push Oscillator Using $\lambda_g/2$ Microstrip Resonator', *IEICE Trans. Electron.*, **E86-C**, 1438–1443 (2003).
34. K. S. Ang, M. J. Underhill and I. D. Robertson, 'Balanced Monolithic Oscillators at *K*- and *Ka*-Band', *IEEE Trans. Microwave Theory Tech.*, **MTT-48**, 187–193 (2000).
35. M. Schott, H. Kuhnert, F. Lenk, J. Hilsenbeck, J. Wurt and W. Heinrich, '38 GHz push–push GaAs HBT MMIC Oscillator', *2002 IEEE MTT-S Int. Microwave Symp. Dig.*, 839–842 (2002).
36. F. X. Sinnesbichler, H. Geltinger and G. R. Olbrich, 'A 38-GHz Push–Push Oscillator Based on 25-GHz f_T BJT's', *IEEE Microwave and Guided Wave Lett.*, **9**, 151–153 (1999).
37. R. T. Gorczak, 'Voltage-Controlled Push–Push Oscillator', U.S. Patent 5,402,087 (March 1995).
38. H. Yabuki, M. Sagawa and M. Makimoto, 'New Type of Push–Push Oscipliers for the Frequency Synthesizer', *1992 IEEE MTT-S Int. Microwave Symp. Dig.*, 1085–1088 (1992).
39. K. Kawahata, T. Tanaka and M. Aikawa, 'A K-Band Push–Push Oscillator with High Suppression of Undesired Harmonic Signals', *IEICE Trans. Electron.*, **E86-C**, 1433–1437 (2003).
40. B. Z. Kaplan and S. T. Bachar, 'A Simple Oscillator Based on Conservative Models for Generating Three Phase Waveforms', *Proc. IEEE*, **67**, 1446–1448 (1979).
41. Y.-L. Tang and H. Wang, 'A Novel Triple-Push Oscillator Approach', *2000 IEEE MTT-S Int. Microwave Symp. Dig.*, **2**, 1201–1204 (2000).
42. K. Kurokawa, 'An Analysis of Rucker's Multidevice Symmetrical Oscillator', *IEEE Trans. Microwave Theory Tech.*, **MTT-18**, 967–969 (1970).
43. Y.-L. Tang and H. Wang, 'Triple-Push Oscillator Approach: Theory and Experiment', *IEEE J. Solid-State Circuits*, **SC-36**, 1472–1479 (2001).
44. H. Wang and Y.-L. Tang, Multiple-Push Oscillator, U.S. Patent 6,310,522 (October 2001).
45. Y.-L. Tang and H. Wang, 'A 24.6-GHz MMIC HBT Triple-Push Oscillator', *Proc. 31st European Microwave Conf.*, **1**, 113–116 (2001).

46. P.-Y. Chen, Y.-L. Tang, H. Wang, Y.-C. Wang, P.-C. Chao and C.-H. Chen, 'A 39-46 GHz MMIC HBT Triple-Push VCO Using Cascode Configuration', *Proc. 2002 Asia-Pacific Conf. ASIC*, 61–64 (2002).
47. V. V. Migulin, V. I. Medvedev, E. R. Mustel and V. N. Parygin, *Basic Theory of Oscillations*, Moscow: Mir Publishers (1983).
48. J. B. L. Rao and W. M. Waters, 'Voltage-Controlled Oscillator with Time-Delay Feedback', *Proc. 32nd Annual Frequency Control Symp.*, 378–384 (1978).
49. T. Ohta and K. Murakami, 'Reducing Negative Resistance Oscillator Noise by Self-Injection', *Electron. Commun. Japan*, **51-B**, 80–82 (1968).
50. H.-C. Chang, 'Phase Noise in Self-Injection-Locked Oscillators — Theory and Experiment', *IEEE Trans. Microwave Theory Tech.*, **MTT-51**, 1994–1999 (2003).

3

Stability of self-oscillations

Applying dc bias to the active device does not generally result in the negative resistance condition. This condition has to be induced in these devices and it is determined by the physical mechanism in the device and chosen circuit topology. The transistor in the oscillator circuits is mostly represented as the active two-port network, whose operation principle is reflected through its equivalent circuit. The influence of the circuit and transistor parameters can result in the hysteresis effect or oscillation instability in practical design. In high-frequency practical implementation, the presence of the parasitic device and circuit elements can contribute to the multi-resonant circuits. The possibility of an operation mode with different natural frequencies depends on a value of the coupling coefficient between resonant circuits. Therefore, the stability conditions for a steady-state single-frequency operation for a multi-resonant circuit, in general, and two coupled resonant circuits, in particular, are analytically derived. The several examples of stability criteria for different single-resonant and double-resonant oscillator circuits are described and analysed [1, 2]. In addition, the phase plane method, as a qualitative method of analysis of the dynamics of the oscillation systems, and a Nyquist stability criterion are shown and illustrated by several examples of the oscillator circuits described by second-order differential equations.

3.1 NEGATIVE-RESISTANCE OSCILLATOR CIRCUITS

The circuit state equations in a general form for (a) series and (b) parallel single-resonant oscillator circuits shown in Figure 3.1 are

for a series circuit

$$C\frac{dv_C}{dt} - i = 0 \tag{3.1}$$

$$L\frac{di}{dt} + R_L i + v + v_C = 0 \tag{3.2}$$

for a parallel circuit

$$L\frac{di_L}{dt} - v = 0 \tag{3.3}$$

$$C\frac{dv}{dt} + G_L v + i + i_L = 0 \tag{3.4}$$

RF and Microwave Transistor Oscillator Design A. Grebennikov

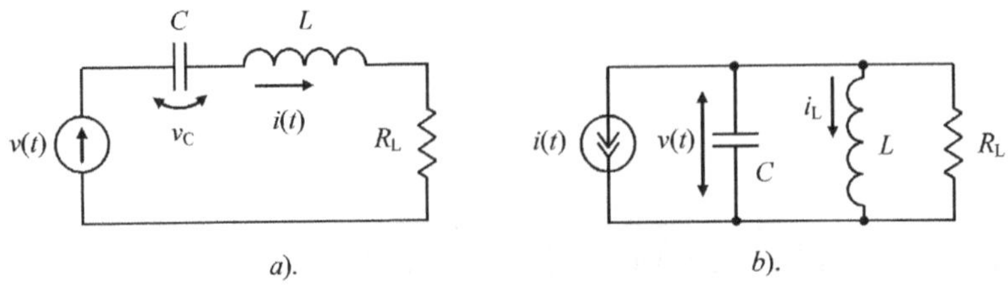

Figure 3.1 Single-resonant (a) series and (b) parallel oscillator circuits

where $G_L = 1/R_L$ is the load conductance, v_C is the instantaneous voltage on the capacitor C, and i_L is the instantaneous current in the inductor L.

Because of the sufficiently high value of the Q-factor of the resonant circuits the effects of higher harmonics can be neglected. As a result, the voltage v for parallel oscillator circuit and the current i for series oscillator circuit can be considered as sinusoidal. Taking into consideration the fundamental dependencies $v_1 = R_{out}(I)i_1$ and $i_1 = G_{out}(V)v_1$, where $R_{out}(I)$ and $G_{out}(V)$ are the fundamentally averaged large-signal active device equivalent output resistance and conductance, respectively, (which can be treated as slowly varying functions of the fundamental amplitudes I and V in comparison with the natural period during the frequency-varying process), the second-order differential equations can be found from Equations (3.1–3.4) as

$$LC\frac{d^2 i_1}{dt^2} + [R_{out}(I) + R_L]C\frac{di_1}{dt} + i_1 = 0 \tag{3.5}$$

$$LC\frac{d^2 v_1}{dt^2} + [G_{out}(V) + G_L]L\frac{dv_1}{dt} + v_1 = 0 \tag{3.6}$$

The characteristic equation corresponding to both Equations (3.5) and (3.6) can be written as

$$p^2 + 2\delta p + \omega_0^2 = 0 \tag{3.7}$$

where

$$\delta = \frac{R_{out}(I) + R_L}{2L} = \frac{G_{out}(V) + G_L}{2C} \tag{3.8}$$

$$\omega_0 = \frac{1}{\sqrt{LC}} \tag{3.9}$$

In a common case, Equation (3.7) has two possible roots

$$p_{1,2} = -\delta \pm \sqrt{\delta^2 - \omega_0^2} \tag{3.10}$$

If these roots are different, then a common solution of Equation (3.7) is

$$a(t) = A_1 \exp(p_1 t) + A_2 \exp(p_2 t) \tag{3.11}$$

where $a(t)$ corresponds to the instantaneous voltage $v_1(t)$ or current $i_1(t)$, A_1 and A_2 are assumed to be slowly varying functions of time.

For a practical case when a condition $0 < \delta < \omega_0$ is valid, a solution of Equation (3.7) is a pair of complex conjugate numbers with negative active parts shown in Figure 3.2a in the

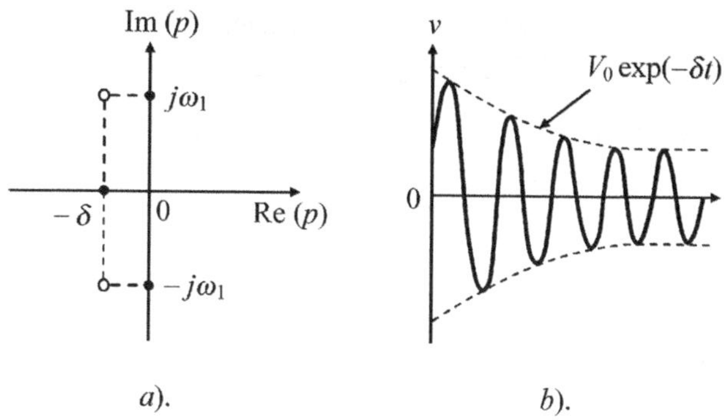

Figure 3.2 Resonant circuit transient performance

form

$$p_{1,2} = -\delta \pm j\sqrt{\omega_0^2 - \delta^2} = -\delta \pm j\omega_1 \tag{3.12}$$

By substituting Equation (3.12) in Equation (3.11), a common solution of the characteristic Equation (3.7) can be written as

$$a(t) = A_0 \exp(-\delta t) \cos(\omega_1 t + \varphi_0) \tag{3.13}$$

where the amplitude A_0 and the phase φ_0 are slowly varying functions of time. An example of the transient characteristic corresponding to Equation (3.12) is shown in Figure 3.2b.

A steady-state free-running oscillation build-up is provided with the velocity due to dissipation factor δ, the value of which is reduced with the decrease of negative output resistance $R_{\rm out}(I)$ or conductance $G_{\rm out}(V)$. It becomes zero in a steady-state oscillation mode when $R_{\rm out}(I) + R_{\rm L} = 0$ or $G_{\rm out}(V) + G_{\rm L} = 0$, or in a general immitance form

$$\text{Re}W_{\rm out}(A_0) + \text{Re}W_{\rm L} = 0 \tag{3.14}$$

A stability criterion for such simple oscillator circuits, with only an amplitude dependence of output resistance or conductance, can be easily calculated by the perturbation method. Let us consider small perturbation $\Delta A > 0$ in the operating point $\text{Re}W(A_0)$ when $A = A_0 + \Delta A$. Using the linear Taylor series expansion results in

$$\text{Re}W(A_0 + \Delta A) \cong \text{Re}W(A_0) + \frac{\partial \text{Re}W(A_0)}{\partial A}\Delta A \tag{3.15}$$

Hence, a steady-state oscillation mode will be stable if the amplitude of the oscillation dissipates according to exp($-\delta$t). This takes place when $\text{Re}W(A_0 + \Delta A) > \text{Re}W(A_0)$. Therefore, a stability criterion is

$$\frac{\partial \text{Re}W(A_0)}{\partial A} > 0 \tag{3.16}$$

which means that, as the active device negative resistance or conductance reduces with increase of the oscillation amplitude, stable oscillations are established in the oscillator.

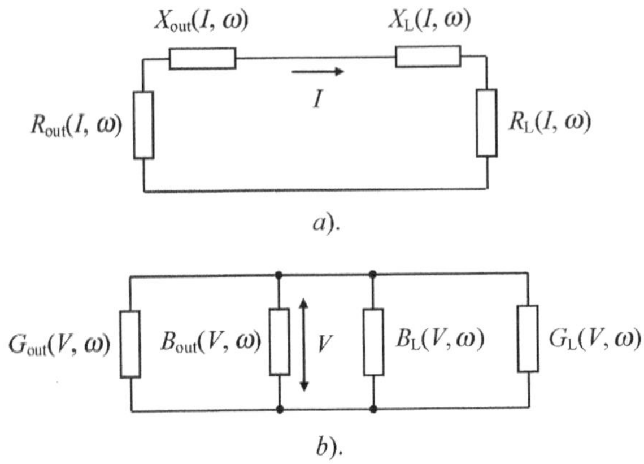

Figure 3.3 One-port (a) negative resistance and (b) negative conductance oscillator circuits

3.2 GENERAL SINGLE-FREQUENCY STABILITY CONDITION

In a common case of (a) negative resistance or (b) negative conductance one-port transistor oscillator circuit model shown in Figure 3.3, a complex equation consisting of the nonlinear immitances in the steady-state stationary operation mode can be expressed as

$$W(A_0, \omega_0) = \mathrm{Re}W(A_0, \omega_0) + j\mathrm{Im}W(A_0, \omega_0) = 0 \tag{3.17}$$

where

$$\mathrm{Re}W(A_0, \omega_0) = \mathrm{Re}W_{out}(A_0, \omega_0) + \mathrm{Re}W_L(A_0, \omega_0) \tag{3.18}$$

$$\mathrm{Im}W(A_0, \omega_0) = \mathrm{Im}W_{out}(A_0, \omega_0) + \mathrm{Im}W_L(A_0, \omega_0) \tag{3.19}$$

Assuming a solution of Equation (3.17) as $a(t) = A_0 \cos\omega_0 t$ or $a(t) = A_0 \,\mathrm{Re}[\exp(j\omega_0 t)]$, one can find the amplitude A_0 and frequency ω_0 of free-running oscillations. Then, consider the small perturbation from the steady-state stationary mode. The amplitude and frequency of the oscillation turned out to be $A = A_0 + \Delta A$ and $\omega = \omega_0 + \Delta\omega$ respectively, where $\Delta A \ll A_0$, $\Delta\omega \ll \omega_0$. Under conditions of small perturbations from the steady-state mode a nonlinear system behaves as linear, and amplitude of the oscillation dissipates exponentially. Therefore, the expected solution can be written as

$$a(t) = (A_0 + \Delta A)\exp(-\Delta\delta t)\cos(\omega_0 + \Delta\omega)t \tag{3.20}$$

or

$$a(t) = (A_0 + \Delta A)\mathrm{Re}\left[\exp j\left(\omega_0 + \Delta\omega + j\Delta\delta\right)t\right] \tag{3.21}$$

According to Equation (3.21), the steady-state oscillation condition in the form of Equation (3.17) can be rewritten as follows:

$$\mathrm{Re}W(A_0 + \Delta A, \omega_0 + \Delta\omega + j\Delta\delta) + j\mathrm{Im}W(A_0 + \Delta A, \omega_0 + \Delta\omega + j\Delta\delta) = 0 \tag{3.22}$$

By applying the linear Taylor series expansion of Equation (3.22) by the degrees of the small parameters ΔA, $\Delta\omega$ and $\Delta\delta$ for each components and replacing one complex equation with two equations for real and imaginary parts, in consideration of Equation (3.17) the following system of two equations can be written:

$$\frac{\partial \mathrm{Re}W}{\partial A}\Delta A + \frac{\partial \mathrm{Re}W}{\partial \omega}\Delta\omega + \frac{\partial \mathrm{Re}W}{\partial \delta}\Delta\delta = 0 \tag{3.23}$$

$$\frac{\partial \mathrm{Im}W}{\partial A}\Delta A + \frac{\partial \mathrm{Im}W}{\partial \omega}\Delta\omega + \frac{\partial \mathrm{Im}W}{\partial \delta}\Delta\delta = 0 \tag{3.24}$$

It is known that for analytic function $W(\omega + j\Delta\delta) = \mathrm{Re}W(\omega, \Delta\delta) + j\mathrm{Im}W(\omega, \Delta\delta)$, the Cauchy–Riemann equations can be written as

$$\frac{\partial \mathrm{Re}W}{\partial \omega} = \frac{\partial \mathrm{Im}W}{\partial \delta} \tag{3.25}$$

$$\frac{\partial \mathrm{Re}W}{\partial \delta} = -\frac{\partial \mathrm{Im}W}{\partial \omega} \tag{3.26}$$

As a result, by excluding $\Delta\omega$ from Equations (3.23) and (3.24) and using Equations (3.25) and (3.26), the following equation can be written:

$$\frac{\Delta\delta}{\Delta A} = \frac{\dfrac{\partial \mathrm{Re}W}{\partial A}\dfrac{\partial \mathrm{Im}W}{\partial \omega} - \dfrac{\partial \mathrm{Im}W}{\partial A}\dfrac{\partial \mathrm{Re}W}{\partial \omega}}{\left(\dfrac{\partial \mathrm{Re}W}{\partial \omega}\right)^2 + \left(\dfrac{\partial \mathrm{Im}W}{\partial \omega}\right)^2} \tag{3.27}$$

The steady-state stationary mode will be stable if a small perturbation of the oscillation amplitude ΔA dissipates in time and the values ΔA and $\Delta\delta$ have the same signs. Thus, stable oscillations are established in the oscillator if

$$\frac{\partial \mathrm{Re}W}{\partial A}\frac{\partial \mathrm{Im}W}{\partial \omega} - \frac{\partial \mathrm{Im}W}{\partial A}\frac{\partial \mathrm{Re}W}{\partial \omega} > 0 \tag{3.28}$$

which corresponds to a general form in terms of immitance parameters of a stability condition obtained by Kurokawa [3].

3.3 SINGLE-RESONANT CIRCUIT OSCILLATORS

3.3.1 Series resonant circuit oscillator with constant load

Figure 3.4 shows (a) the high-frequency (without bias circuits) electrical and (b) the equivalent circuits of the series resonant circuit negative-resistance oscillator with a common collector. In this case, a transistor has been configured so that the output resistance is negative. The capacitors C_1 and C_2 are used to increase the regeneration factor and to provide steady-state oscillation conditions of the amplitude and phase.

In this series circuit configuration with the negative imaginary part of the output transistor impedance, the load is constant and the negative output transistor resistance decreases with increase of the oscillation frequency. As a result,

$$\frac{\partial X}{\partial \omega} = \frac{\partial}{\partial \omega}\left(\omega L - \frac{1}{\omega C_{\mathrm{out}}}\right) = L + \frac{1}{\omega^2 C_{\mathrm{out}}} > 0 \tag{3.29}$$

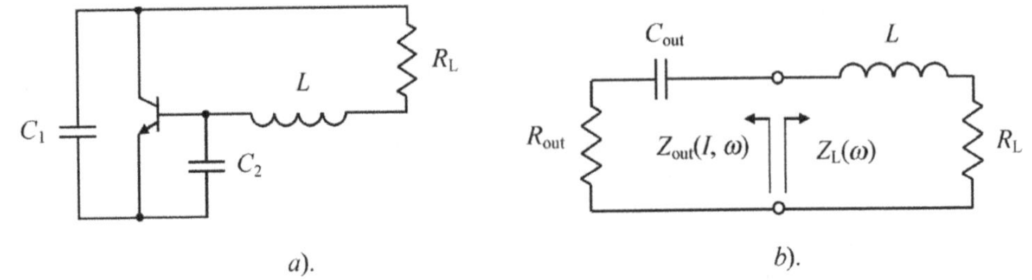

Figure 3.4 Parallel feedback (a) electrical and (b) equivalent oscillator circuits

and

$$\frac{\partial R}{\partial \omega} = \frac{\partial}{\partial \omega}(R_{out} + R_L) = \frac{\partial R_{out}}{\partial \omega} > 0 \tag{3.30}$$

Consequently, obtaining stable oscillations in a steady-state operation mode requires satisfying the following sufficient conditions for the series resonant circuit negative-resistance oscillator with constant load

$$\frac{\partial R_{out}}{\partial I} > 0 \qquad \frac{\partial X_{out}}{\partial I} < 0 \tag{3.31}$$

3.3.2 Parallel resonant circuit oscillator with nonlinear load

In some practical cases, the nonlinear active devices can be used in the capacity of a load or together with it; for example, for amplitude restriction when a separate parallel diode is used, as shown in Figure 3.5a, or for amplification of the output oscillator signal when the oscillator is loaded on the input buffer amplifier transistor circuit, as shown in Figure 3.5b, the nonlinear properties of which can be described by the diode too.

To approximate the diode volt–ampere characteristic, a well-known exponential model can be applied:

$$i_D(t) = I_{Dsat}\left(\exp\frac{v}{V_T} - 1\right) \tag{3.32}$$

where $v = V_{bias} + V\cos\omega t$, V_{bias} is the diode bias voltage, I_{Dsat} is the diode reverse-bias saturation current, V_T is the temperature voltage. Since the instantaneous current $i_D(t)$ in

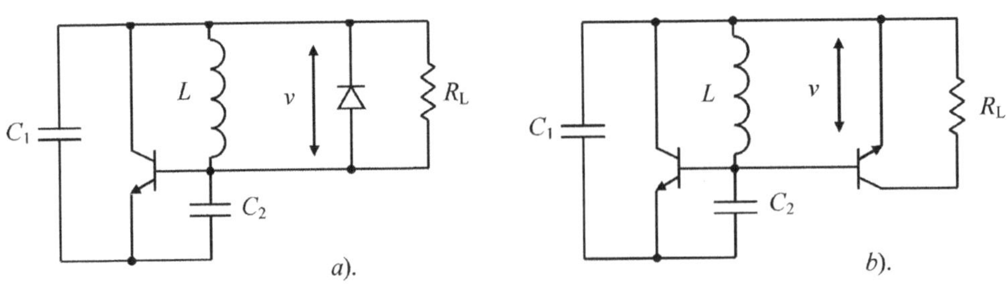

Figure 3.5 Loaded oscillator circuits

Equation (3.32) can be expressed through the Fourier series, it is convenient to use the Bessel functions for the Fourier components.

Then, as far as our interest has been restricted to the fundamental frequency, the value of the large-signal diode conductance G_{D1} is defined as

$$G_{\mathrm{D1}} = \frac{I_{\mathrm{D1}}}{V} = \frac{2I_{\mathrm{Dsat}}}{V}\exp\left(\frac{V_{\mathrm{bias}}}{V_{\mathrm{T}}}\right) I_1\left(\frac{V}{V_{\mathrm{T}}}\right) \tag{3.33}$$

where $I_1(V/V_{\mathrm{T}})$ is the modified first-order Bessel function of first kind given by

$$I_1(z) = \frac{z}{2}\sum_{r=0}^{\infty}\frac{1}{r!\Gamma(r+2)}\left(\frac{z}{2}\right)^{2r} \tag{3.34}$$

where $\Gamma(r+2)$ is a gamma-function.

By substituting Equation (3.34) into Equation (3.33), the large-signal diode conductance G_{D1} will be expressed as

$$G_{\mathrm{D1}} = \frac{I_{\mathrm{Dsat}}}{V_{\mathrm{T}}}\exp\left(\frac{V_{\mathrm{bias}}}{V_{\mathrm{T}}}\right)\sum_{r=0}^{\infty}\frac{1}{r!\Gamma(r+2)}\left(\frac{V}{2V_{\mathrm{T}}}\right)^{2r} \tag{3.35}$$

It follows from Equation (3.35) that the large-signal diode conductance G_{D1} is a monotonically and infinitely increasing function of the voltage amplitude V. Hence,

$$\frac{\partial G_{\mathrm{D1}}}{\partial V} > 0 \tag{3.36}$$

Thus, according to the stability criterion given by Equation (3.28), using a nonlinear device (diode or transistor) in a load network contributes to the stable steady-state operation mode of parallel negative conductance oscillator, provided the influence of a diode reactance is negligible.

3.4 DOUBLE-RESONANT CIRCUIT OSCILLATOR

Previous stability analysis has been done for the single-resonant circuit transistor oscillators. But there are many cases when, for theoretical or practical research, it is required to use a double-resonant circuit, for instance, when it is necessary to take into consideration the reactances of an active device package or input buffer amplifier transistor circuit or to increase the oscillation frequency sensitivity to the capacitance tuning in the case of the voltage-controlled oscillators. So, for an oscillator with the series single-resonant circuit shown in Figure 3.4b, the relative frequency deviation is given by

$$\frac{\partial X}{\partial \omega} = L + \frac{1}{\omega^2 C_{\mathrm{out}}} = 2\frac{Q_{\mathrm{L}}R_{\mathrm{L}}}{\omega} \Rightarrow \frac{\partial \omega}{\omega} = \frac{\partial X}{2Q_{\mathrm{L}}R_{\mathrm{L}}} \tag{3.37}$$

where $Q_{\mathrm{L}} = \omega L/R_{\mathrm{L}} = 1/\omega C_{\mathrm{out}}|R_{\mathrm{out}}| = Q_{\mathrm{T}}$. In order to increase the magnitude of the frequency deviation, the loaded resonant circuit quality factor Q_{L} must be reduced. However, it cannot be made smaller than the transistor quality factor Q_{T}, the value of which is limited to the finite value of output transistor reactance. Therefore, it is advisable to consider the possibility of adding another resonant circuit, as shown in Figure 3.6.

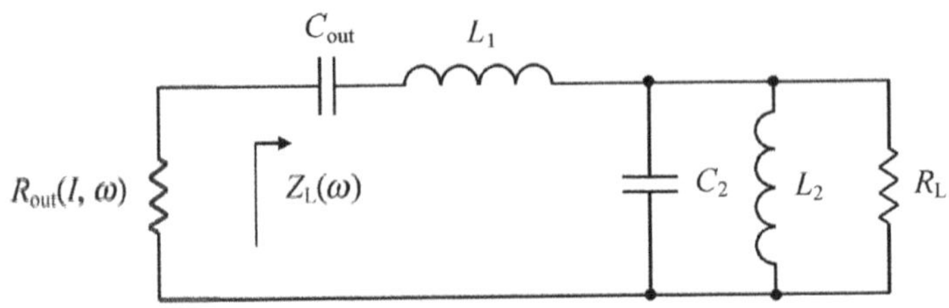

Figure 3.6 Double-resonant oscillator circuit

The double-resonant oscillator circuit impedance Z_L can be expressed as

$$Z_L = jQ_1R_L\left(\frac{\omega}{\omega_0} - \frac{\omega_0}{\omega}\right) + \frac{R_L}{1 + jQ_2\left(\frac{\omega}{\omega_0} - \frac{\omega_0}{\omega}\right)} \tag{3.38}$$

where the loaded quality factors of the series and parallel resonant circuits, Q_1 and Q_2, and resonant angular frequency $\omega_0 = 2\pi f_0$ are defined by

$$Q_1 = \frac{\omega_0 L_1}{R_L} \qquad Q_2 = \frac{\omega_0 L_2}{R_L}$$

$$\omega_0 = \frac{1}{\sqrt{L_1 C_{out}}} = \frac{1}{\sqrt{L_2 C_2}}$$

In order to provide the stable steady-state operation mode according to Equation (3.28), consider the slopes of the active and reactive parts of the double-resonant circuit impedance Z_L. As a result, the value of the derivative $\partial \mathrm{Re} Z_L/\partial\omega$ at the resonant frequency ω_0 is equal to

$$\frac{\partial \mathrm{Re} Z_L}{\partial \omega} = -\frac{2Q_2^2 R_L}{\omega} \left. \frac{\left(\frac{\omega}{\omega_0}\right)^2 - \left(\frac{\omega_0}{\omega}\right)^2}{\left[1 + Q_2^2\left(\frac{\omega}{\omega_0} - \frac{\omega_0}{\omega}\right)^2\right]^2} \right|_{\omega=\omega_0} = 0 \tag{3.39}$$

Consequently, to determine the stability condition, it is sufficient to know the value of the output transistor resistance slope $\partial R_{out}/\partial\omega$. The value of the reactance slope is given by

$$\frac{\partial \mathrm{Im} Z_L}{\partial \omega} = \frac{R_L}{\omega}\left(\frac{\omega}{\omega_0} + \frac{\omega_0}{\omega}\right) \left. \left\{ Q_1 - Q_2 \frac{1 - Q_2^2\left(\frac{\omega}{\omega_0} - \frac{\omega_0}{\omega}\right)^2}{\left[1 + Q_2^2\left(\frac{\omega}{\omega_0} - \frac{\omega_0}{\omega}\right)^2\right]^2} \right\} \right|_{\omega=\omega_0}$$

$$= \frac{2R_L}{\omega_0}\left(Q_1 - Q_2\right) \tag{3.40}$$

It follows from Equation (3.40) that the positive reactance slope is realized for $Q_1 > Q_2$, and there are no problems associated with a monotonic frequency tuning. Besides, as Q_2 approaches Q_1, the relative frequency deviation $\partial\omega/\omega_0$ increases for a given small variation

∂X_L, where $X_L = \mathrm{Im} Z_L$, according to

$$\frac{\partial\omega}{\omega_0} = \frac{1}{2(Q_1 - Q_2)}\frac{\partial X}{R_L} \tag{3.41}$$

However, if Q_2 is close enough to Q_1, a dramatic increase of the oscillator circuit sensitivity to the reactance variation will be accompanied by very noisy operation, high nonlinearity of the frequency tuning characteristic and high bandwidth ripple of output power [3]. According to the stability condition in the form of Equation (3.28), such a situation with approximately equal values of the loaded quality factors for series and parallel resonant circuits ($Q_1 \geq Q_2$) produces substantially worse a steady-state single-frequency stability condition for the double-resonant circuit oscillator where the possibility of simultaneous self-oscillations for two independent frequencies can occur.

3.5 STABILITY OF MULTI-RESONANT CIRCUITS

Generally, in a system of the two coupled resonant circuits two possible oscillations with different natural frequencies can exist simultaneously. The possibility of such an operation mode depends on the value of the coupling coefficient between two coupled resonant circuits. A high probability of satisfying the start-up conditions for both frequencies simultaneously exists for the case of strong coupling between resonant circuits near the hysteresis region when their natural frequencies are approximately equal. However, it is very important to provide stable steady-state single-frequency operation conditions without a jumping effect between two possible self-oscillation frequencies. It should be noted that, in real RF and microwave practical implementation, the presence of the parasitic device and circuit elements can contribute to the multi-resonant circuits. Depending on the coupling between these resonant circuits and start-up conditions, the potential probability of multi-frequency operation is sufficiently high. Therefore, let us derive first a general stability criterion for a multi-resonant circuit oscillator. Using a stability analysis method based on the perturbation of the harmonic balance system equation leads to eigenvalue loci, which demonstrate a very complicated behaviour and have to be interpreted in order to determine stability [4]. To analyse the negative-resistance one-port oscillator, it is advisable to apply a convenient quasilinear method, based on the replacement of nonlinear elements by their fundamentally averaged linear equivalents. In this case, the number of general eigenvalue equations will be determined by the proper number of coupled resonant circuits provided the periodic steady-state oscillation conditions are fulfilled for appropriate resonant frequencies.

3.5.1 General multi-frequency stability criterion

Basically, such an approach is an extension of the single-frequency one-port oscillator shown in Figure 3.3 to the oscillator with N independent frequencies and time-varying amplitude $A(t)$, voltage $V(t)$ or current $I(t)$, across the output transistor immitance written as

$$A(t) = \sum_{n=1}^{N} A_n \cos(\omega_n t + \varphi_n) \tag{3.42}$$

where the amplitudes A_n and phase φ_n are slowly varying functions of time.

Application of the steady-state operation conditions in the form of Equation (3.17) for each resonant frequency results in

$$\begin{aligned}&\mathrm{Re}W_k(A_1,\ldots,A_n,\omega_1,\ldots,\omega_n,\varphi_1,\ldots,\varphi_\mathrm{n})\\&\quad+j\mathrm{Im}W_k(A_1,\ldots,A_n,\omega_1,\ldots,\omega_n,\varphi_1,\ldots,\varphi_\mathrm{n})=0\end{aligned} \tag{3.43}$$

where $k = 1, 2, \ldots, N$.

For stability analysis, consider small perturbations ΔA_k, $\Delta\omega_k$ and $\Delta\varphi_k$ from the steady-state mode in the operating point A_{k0}, ω_{k0} and $\Delta\varphi_{k0}$ such that

$$A_k = A_{k0} + \Delta A_k \qquad \omega_k = \omega_{k0} + \Delta\omega_k \qquad \varphi_k = \varphi_{k0} + \Delta\varphi_k \tag{3.44}$$

Applying a Taylor series expansion about the operating point to Equation (3.43), where $\mathrm{Re}W_k = \mathrm{Re}W_{k0}$ and $\mathrm{Im}W_k = \mathrm{Im}W_{k0}$ at the operating point under steady-state conditions, and restricting consideration to the linear terms results in a matrix equation

$$D\Delta a + B\Delta\omega + C\Delta\varphi = 0 \tag{3.45}$$

where Δa, $\Delta\omega$ and $\Delta\varphi$ are the vector quantities defined by

$$\Delta a = \left[\frac{\Delta A_1}{A_{10}}, \frac{\Delta A_2}{A_{20}}, \ldots, \frac{\Delta A_\mathrm{N}}{A_\mathrm{N0}}\right]^\mathrm{T}$$

$$\Delta\omega = [\Delta\omega_1, \Delta\omega_2, \ldots, \Delta\omega_\mathrm{N}]^\mathrm{T}$$

$$\Delta\varphi = [\Delta\varphi_1, \Delta\varphi_2, \ldots, \Delta\varphi_\mathrm{N}]^\mathrm{T}$$

where D, B and C are the complex square matrices whose elements can be written as

$$D_\mathrm{kl} = A_{l0}\left(\frac{\partial \mathrm{Re}W_\mathrm{k0}}{\partial A_l} + j\frac{\partial \mathrm{Im}W_\mathrm{k0}}{\partial A_l}\right)$$

$$B_\mathrm{kl} = \frac{\partial \mathrm{Re}W_\mathrm{k0}}{\partial \omega_l} + j\frac{\partial \mathrm{Im}W_\mathrm{k0}}{\partial \omega_l}$$

$$C_\mathrm{kl} = \frac{\partial \mathrm{Re}W_\mathrm{k0}}{\partial \varphi_l} + j\frac{\partial \mathrm{Im}W_\mathrm{k0}}{\partial \varphi_l}$$

and index T denotes the transposed vector and $l = 1, 2, \ldots, N$.

According to Kurokawa's fundamental theory [3], the small perturbations ΔA_k, $\Delta\omega_k$ and $\Delta\varphi_k$ as slowly time varying functions are connected by

$$\Delta\omega_k = \frac{\mathrm{d}(\Delta\varphi_\mathrm{k})}{\mathrm{d}t} - \frac{j}{A_{k0}}\frac{\mathrm{d}(\Delta A_\mathrm{k})}{\mathrm{d}t} \tag{3.46}$$

Then, Equation (3.45) can be rewritten in the form of differential complex equation defining a multi-frequency stability criterion of negative-resistance one-port oscillator with a multiple-resonant circuit as

$$D\Delta a + B\frac{\mathrm{d}(\Delta\varphi)}{\mathrm{d}t} - jB\frac{\mathrm{d}(\Delta a)}{\mathrm{d}t} + C\Delta\varphi = 0 \tag{3.47}$$

A similar approach to derive a general multi-frequency oscillator stability criterion in terms of admittances was used with regard to diode harmonic oscillators [5]. If our interest is restricted to the independent frequencies, then C must be a zero matrix because an arbitrary phase reference associated with each independent frequency can be chosen. By equating the real and imaginary parts to zero, Equation (3.47) can be written in the form of a system of $2N$ equations by

$$\mathrm{Re}(D)\Delta a + \mathrm{Re}(B)\frac{\mathrm{d}(\Delta\varphi)}{\mathrm{d}t} + \mathrm{Im}(B)\frac{\mathrm{d}(\Delta a)}{\mathrm{d}t} = 0 \tag{3.48}$$

$$\mathrm{Im}(D)\Delta a + \mathrm{Im}(B)\frac{\mathrm{d}(\Delta\varphi)}{\mathrm{d}t} - \mathrm{Re}(B)\frac{\mathrm{d}(\Delta a)}{\mathrm{d}t} = 0 \tag{3.49}$$

which includes N unknown phase angles $\Delta\varphi_k$ and N unknown amplitudes ΔA_k.

Eliminating $\mathrm{d}(\Delta\varphi)/\mathrm{d}t$ from Equations (3.48) and (3.49) yields the resulting first-order differential matrix equation to solve for Δa in the form

$$\frac{\mathrm{d}(\Delta a)}{\mathrm{d}t} + F\Delta a = 0 \tag{3.50}$$

where

$$F = \frac{\dfrac{\mathrm{Re}(D)}{\mathrm{Re}(B)} - \dfrac{\mathrm{Im}(D)}{\mathrm{Im}(B)}}{\dfrac{\mathrm{Im}(B)}{\mathrm{Re}(B)} + \dfrac{\mathrm{Re}(B)}{\mathrm{Im}(B)}}. \tag{3.51}$$

The oscillation system is stable if and only if all eigenvalues of the matrix F defined from the characteristic equation

$$\det(F + \lambda E) = 0 \tag{3.52}$$

where λ is an eigenvector of matrix F and E is a unit matrix, has positive real parts [6].

For a particular case of the single-frequency oscillator, by substituting the elements of the matrices B and D from Equation (3.45), the matrix F defined by Equation (3.51) can be rewritten as

$$F = \left(\frac{A\dfrac{\partial \mathrm{Re}W}{\partial A}}{\dfrac{\partial \mathrm{Re}W}{\partial \omega}} - \frac{A\dfrac{\partial \mathrm{Im}W}{\partial A}}{\dfrac{\partial \mathrm{Im}W}{\partial \omega}} \right) \Bigg/ \left(\frac{\dfrac{\partial \mathrm{Im}W}{\partial \omega}}{\dfrac{\partial \mathrm{Re}W}{\partial \omega}} + \frac{\dfrac{\partial \mathrm{Re}W}{\partial \omega}}{\dfrac{\partial \mathrm{Im}W}{\partial \omega}} \right) \tag{3.53}$$

or

$$F = \frac{A\left(\dfrac{\partial \mathrm{Re}W}{\partial A}\dfrac{\partial \mathrm{Im}W}{\partial \omega} - \dfrac{\partial \mathrm{Im}W}{\partial A}\dfrac{\partial \mathrm{Re}W}{\partial \omega} \right)}{\left(\dfrac{\partial \mathrm{Re}W}{\partial \omega}\right)^2 + \left(\dfrac{\partial \mathrm{Im}W}{\partial \omega}\right)^2} \tag{3.54}$$

which is similar to Equation (3.27) when the stable oscillations are established in the oscillator for $F > 0$, and also similar to the results of the stability analysis in the frequency domain using the harmonic balance technique [4, 7].

3.5.2 Two-frequency oscillation mode and its stability

In practice, the existence of the two-frequency oscillation mode in a single-frequency oscillator is absolutely undesirable. However, it may happen if the oscillation system with two coupled resonant circuits is used, or device and circuit parasitic elements compose an additional resonant circuit. Therefore, it is important to define the stability conditions for the two-frequency oscillation mode. In the case of a negative resistance oscillator with two coupled resonance circuits, the expression for the matrix F given in Equation (3.51) can be rewritten as

$$F = \frac{\begin{bmatrix} \frac{\partial R_1}{\partial I_1} & \frac{\partial R_1}{\partial I_2} \\ \frac{\partial R_2}{\partial I_1} & \frac{\partial R_2}{\partial I_2} \end{bmatrix}\begin{bmatrix} \frac{\partial X_1}{\partial \omega_1} & \frac{\partial X_1}{\partial \omega_2} \\ \frac{\partial X_2}{\partial \omega_1} & \frac{\partial X_2}{\partial \omega_2} \end{bmatrix} - \begin{bmatrix} \frac{\partial X_1}{\partial I_1} & \frac{\partial X_1}{\partial I_2} \\ \frac{\partial X_2}{\partial I_1} & \frac{\partial X_2}{\partial I_2} \end{bmatrix}\begin{bmatrix} \frac{\partial R_1}{\partial \omega_1} & \frac{\partial R_1}{\partial \omega_2} \\ \frac{\partial R_2}{\partial \omega_1} & \frac{\partial R_2}{\partial \omega_2} \end{bmatrix}}{\begin{bmatrix} \frac{\partial R_1}{\partial \omega_1} & \frac{\partial R_1}{\partial \omega_2} \\ \frac{\partial R_2}{\partial \omega_1} & \frac{\partial R_2}{\partial \omega_2} \end{bmatrix}^2 + \begin{bmatrix} \frac{\partial X_1}{\partial \omega_1} & \frac{\partial X_1}{\partial \omega_2} \\ \frac{\partial X_2}{\partial \omega_1} & \frac{\partial X_2}{\partial \omega_2} \end{bmatrix}^2} \begin{bmatrix} I_{10} & 0 \\ 0 & I_{20} \end{bmatrix} \tag{3.55}$$

where I_{10} and I_{20} are the steady-state current amplitudes corresponding to two independent frequencies.

The characteristic Equation (3.52) to define the matrix eigenvalues reduces to the quadratic form

$$\lambda^2 + (F_{11} + F_{12})\,\lambda + \Delta F = 0 \tag{3.56}$$

where $\Delta F = F_{11}F_{22} - F_{12}F_{21}$, the solution of which can be obtained by

$$\lambda_{1,2} = -\frac{F_{11} + F_{12}}{2} \pm \sqrt{\frac{(F_{11} + F_{12})^2}{4} - \Delta F} \tag{3.57}$$

As a result, the situation of two-frequency oscillation mode will be unstable and can be avoided if the conditions for the sum of the principal diagonal elements and determinant greater than zero are not fulfilled simultaneously [6]

$$F_{11} + F_{12} > 0 \qquad \Delta F > 0. \tag{3.58}$$

3.5.3 Single-frequency stability of oscillator with two coupled resonant circuits

Let us consider the single-frequency negative-resistance oscillator with a certain value of current $I_{10} \neq 0$ and zero value of current $I_{20} = 0$. As before, assume that $I_1 = I_{10} + \Delta I_1$, $I_2 = I_{20} + \Delta I_2$ and $\omega_1 = \omega_{10} + \Delta\omega_1, \omega_2 = \omega_{20} + \Delta\omega_2$. Then, the differential complex matrix Equation (3.47) for two independent frequencies can be rewritten as

$$Z + D\Delta i + B\frac{\mathrm{d}(\Delta\varphi)}{\mathrm{d}t} - jB\frac{\mathrm{d}(\Delta i)}{\mathrm{d}t} = 0 \tag{3.59}$$

where

$$\frac{\mathrm{d}(\Delta i)}{\mathrm{d}t} = \left[\frac{1}{I_{10}}\frac{\mathrm{d}(\Delta I_1)}{\mathrm{d}t}, \frac{1}{\Delta I_2}\frac{\mathrm{d}(\Delta I_2)}{\mathrm{d}t}\right]^{\mathrm{T}}$$

and, in view of the steady-state single-frequency operation defined by Equation (3.43),

$$Z = \begin{bmatrix} 0 & 0 \\ 0 & Z_2 \end{bmatrix}$$

where $Z_1 = R_1 + jX_1 = 0$ and $Z_2 = R_2 + jX_2$.

By equating the real and imaginary parts to zero and eliminating $\Delta\varphi$, Equation (3.59) can be rewritten in the form

$$\begin{aligned} &\mathrm{Re}^{-1}(B)\,\mathrm{Re}Z - \mathrm{Im}^{-1}(B)\,\mathrm{Im}Z + \left[\mathrm{Re}^{-1}(B)\,\mathrm{Re}D - \mathrm{Im}^{-1}(B)\,\mathrm{Im}D\right]\Delta i \\ &\qquad + \left[\mathrm{Re}^{-1}(B)\,\mathrm{Im}B + \mathrm{Im}^{-1}(B)\,\mathrm{Re}B\right]\frac{\mathrm{d}(\Delta i)}{\mathrm{d}t} = 0 \end{aligned} \tag{3.60}$$

If we limited ourselves to the first-order of values ΔI_1 and ΔI_2, the equation for ΔI_2 can be obtained as

$$\begin{aligned} &\left[\frac{R_2(I_{10},0)}{\det(\mathrm{Re}B)}\frac{\partial R_1}{\partial\omega_1} - \frac{X_2(I_{10},0)}{\det(\mathrm{Im}B)}\frac{\partial X_1}{\partial\omega_1}\right]\Delta I_2 \\ &\quad + \left[\frac{\dfrac{\partial X_1}{\partial\omega_1}\dfrac{\partial R_2}{\partial\omega_2} - \dfrac{\partial X_1}{\partial\omega_2}\dfrac{\partial R_1}{\partial\omega_2}}{\det(\mathrm{Im}B)} + \frac{\dfrac{\partial R_1}{\partial\omega_1}\dfrac{\partial X_2}{\partial\omega_2} - \dfrac{\partial R_1}{\partial\omega_2}\dfrac{\partial X_1}{\partial\omega_2}}{\det(\mathrm{Re}B)}\right]\frac{\mathrm{d}(\Delta I_2)}{\mathrm{d}t} = 0 \end{aligned} \tag{3.61}$$

where

$$\det(\mathrm{Re}B) = \frac{\partial R_1}{\partial\omega_1}\frac{\partial R_2}{\partial\omega_2} - \frac{\partial R_1}{\partial\omega_2}\frac{\partial R_2}{\partial\omega_1}$$

$$\det(\mathrm{Im}B) = \frac{\partial X_1}{\partial\omega_1}\frac{\partial X_2}{\partial\omega_2} - \frac{\partial X_1}{\partial\omega_2}\frac{\partial X_2}{\partial\omega_1}$$

Equation (3.61) can be integrated directly and its solution has the exponential form $\Delta I_2(t) = \Delta I_2(0)\exp(-\lambda_2 t)$, where $\Delta I_2(0)$ is the initial value for ΔI_2 determined at $t = 0$. Then, this solution is substituted into another equation for ΔI_1. As a result, the single-frequency oscillation stability require ΔI_1 and ΔI_2 to decay and this will take place if and only if the eigenvalues λ_1 and λ_2 are positive. Hence, the stability conditions for single-frequency oscillation can be written as

$$\frac{\dfrac{R_2(I_{10},0)}{\det(\mathrm{Re}B)}\dfrac{\partial R_1}{\partial\omega_1} - \dfrac{X_2(I_{10},0)}{\det(\mathrm{Im}B)}\dfrac{\partial X_1}{\partial\omega_1}}{\dfrac{1}{\det(\mathrm{Im}B)}\left(\dfrac{\partial X_1}{\partial\omega_1}\dfrac{\partial R_2}{\partial\omega_2} - \dfrac{\partial R_1}{\partial\omega_2}\dfrac{\partial X_1}{\partial\omega_2}\right) + \dfrac{1}{\det(\mathrm{Re}B)}\left(\dfrac{\partial R_1}{\partial\omega_1}\dfrac{\partial X_2}{\partial\omega_2} - \dfrac{\partial R_1}{\partial\omega_2}\dfrac{\partial X_1}{\partial\omega_2}\right)} > 0 \tag{3.62}$$

$$\frac{\dfrac{R_2(I_{10},0)}{\det(\mathrm{Re}B)}\left(\dfrac{\partial R_2}{\partial\omega_2}\dfrac{\partial R_1}{\partial I_1} - \dfrac{\partial R_2}{\partial\omega_1}\dfrac{\partial R_2}{\partial I_1}\right) - \dfrac{X_2(I_{10},0)}{\det(\mathrm{Im}B)}\left(\dfrac{\partial X_2}{\partial\omega_2}\dfrac{\partial X_1}{\partial I_1} - \dfrac{\partial R_2}{\partial\omega_1}\dfrac{\partial R_2}{\partial I_1}\right)}{\dfrac{1}{\det(\mathrm{Im}B)}\left(\dfrac{\partial R_1}{\partial\omega_1}\dfrac{\partial X_2}{\partial\omega_2} - \dfrac{\partial R_2}{\partial\omega_1}\dfrac{\partial X_2}{\partial\omega_1}\right) + \dfrac{1}{\det(\mathrm{Re}B)}\left(\dfrac{\partial R_2}{\partial\omega_2}\dfrac{\partial X_1}{\partial\omega_1} - \dfrac{\partial R_2}{\partial\omega_1}\dfrac{\partial X_2}{\partial\omega_1}\right)} > 0. \tag{3.63}$$

In the particular case of independent impedances, for instance, for oscillators with linear transfer characteristic of the active device, the stability conditions obtained by Equations (3.62) and (3.63) can be expressed in the simplified form

$$\frac{\partial R_1}{\partial I_1}\frac{\partial X_1}{\partial \omega_1} - \frac{\partial X_1}{\partial I_1}\frac{\partial R_1}{\partial \omega_1} > 0 \tag{3.64}$$

$$R_2(I_{10}, 0)\frac{\partial X_2}{\partial \omega_2} - X_2(I_{10}, 0)\frac{\partial R_2}{\partial \omega_2} > 0 \tag{3.65}$$

Equation (3.64) represents simply a stability condition for the single-resonant circuit oscillator. However, the case of two coupled resonant circuits, requires an additional stability condition given by Equation (3.65), which is necessary to guarantee the stable single-frequency oscillation mode.

3.5.4 Transistor oscillators with two coupled resonant circuits

In practical design, the oscillator loaded with an additional resonant circuit is widely used, for example, to improve linearity of the varactor frequency tuning bandwidth or to reduce the phase noise [8, 9]. If such an oscillator has a weak coupling between resonant circuits, only one frequency can be excited with a value close to the value of the resonant frequency of the main resonant circuit. For a strong coupling between resonant circuits, within the region where the resonant frequencies of the main and additional resonant circuits are very close to each other, the start-up conditions are satisfied for oscillation of two frequencies simultaneously.

Figure 3.7 shows the circuit examples of (a) the common source MOSFET oscillator with two transformer-coupled resonant circuits and (b) the common collector bipolar oscillator with two inductively coupled resonant circuits. Here, R_L is the load resistance, r_1 is the losses in

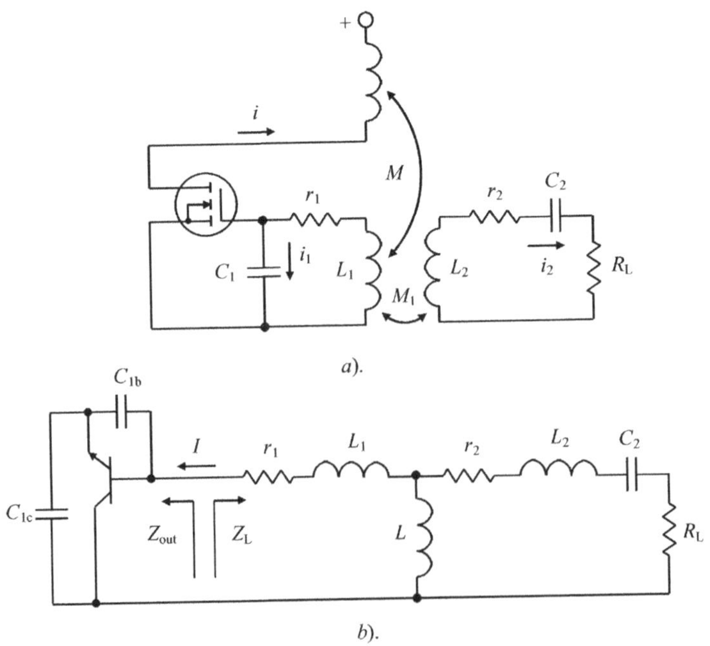

Figure 3.7 Schematics of transistor oscillators with two coupled resonant circuits

the main resonant circuit, r_2 is the losses in the additional resonant circuit, and inductances L_1, L_2 and L are assumed to be lossless.

Generally, a system with two degrees of freedom can be represented as two separate circuits with one degree of freedom linked to each other. The coupling between circuits results in the oscillations in one of them affecting the oscillations in the other and vice versa. These circuits with one degree of freedom into which a complex oscillating system can be divided are called *partial*. The frequencies of free-running oscillations of separate partial circuits are called *partial frequencies* of the entire oscillation system.

Let us consider the operation of the bipolar oscillator the equivalent circuit (without bias networks) of which is shown in Figure 3.7b. Here, the feedback capacitances C_{1b} and C_{1e} are used to realize an equivalent negative resistance for soft start-up mode. The steady-state oscillation condition is expressed as

$$Z_{\text{out}}(I,\omega) + Z_{\text{L}}(\omega) = 0 \tag{3.66}$$

where $Z_{\text{out}} = R_{\text{out}} + jX_{\text{out}}$ is the equivalent output impedance, $X_{\text{out}} = -1/\omega C_{\text{out}}$, and

$$Z_{\text{L}} = r_1 + j\omega L_1 + j\omega L \frac{1 + j\omega C_2(r_2 + R_{\text{L}}) - \omega^2 C_2 L_2}{1 + j\omega C_2(r_2 + R_{\text{L}}) - \omega^2 C_2(L + L_2)}$$

In order to analyse the frequency stability conditions in the steady-state operation mode, it is necessary to rewrite Equation (3.66) for real and imaginary parts in the form

$$\left(\omega_{01}^2 - \omega^2\right)\left(\omega_{02}^2 - \omega^2\right) - 4\delta_1\delta_2\omega^2 - k^2\omega^4 = 0 \tag{3.67}$$

$$\delta_1\left(\omega_{02}^2 - \omega^2\right) + \delta_2\left(\omega_{01}^2 - \omega^2\right) = 0 \tag{3.68}$$

where

$$\omega_{01}^2 = \frac{1}{C_{\text{out}}(L_1 + L)} \tag{3.69}$$

$$\omega_{02}^2 = \frac{1}{C_2(L_2 + L)} \tag{3.70}$$

are the partial frequencies of the main and additional resonant circuits,

$$\delta_1 = \frac{r_1 + R_{\text{out}}}{2(L_1 + L)} \tag{3.71}$$

$$\delta_2 = \frac{r_2 + R_{\text{L}}}{2(L_2 + L)} \tag{3.72}$$

are the attenuation factors of the main and additional resonant circuits,

$$k^2 = \frac{L^2}{(L_1 + L)(L_2 + L)} \tag{3.73}$$

is the coupling factor between the coupled resonant circuits, and

$$C_{\text{out}} = -\frac{1}{\omega X_{\text{out}}} \tag{3.74}$$

is the active device equivalent output capacitance.

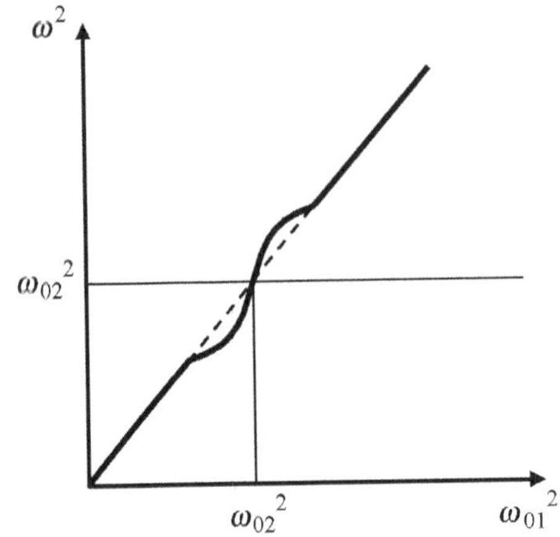

Figure 3.8 Frequency performance of two coupled resonant circuits with weak coupling

By substituting Equation (3.68) into Equation (3.67), the system oscillation frequency can be obtained from

$$\left(\omega_{01}^2 - \omega^2\right)\left(\omega_{02}^2 - \omega^2\right) + 4\delta_2^2\omega^2\frac{\omega_{01}^2 - \omega^2}{\omega_{02}^2 - \omega^2} - k^2\omega^4 = 0 \tag{3.75}$$

From Equation (3.75) it follows that connection of an additional resonant circuit contributes to changing the partial frequency provided by the main resonant circuit. In this case, if $\omega_{02} > \omega$, the equivalent partial frequency drops, i.e., an additional resonant circuit acts as a shunt capacitance connected to the main resonant circuit. If $\omega_{02} < \omega$, an additional resonant circuit acts as a shunt inductance connected to the main resonant circuit, and the equivalent partial frequency increases.

To simplify the analysis procedure, let us assume that the partial frequencies of two coupled resonant circuits are equal, i.e., $\omega_{01} = \omega_{02} = \omega_0$. Then,

$$(\omega_0^2 - \omega^2)\left[\left(\omega_0^2 - \omega^2\right)^2 + 4\delta_2^2\omega^2 - \omega^4k^2\right] = 0 \tag{3.76}$$

where one of the roots is $\omega = \omega_0$ and two other roots can be found from

$$\left(\omega_0^2 - \omega^2\right)^2 + \left(4\delta_2^2 - \omega^2k^2\right)\omega^2 = 0. \tag{3.77}$$

Equation (3.77) has two real solutions only for

$$k^2 > \frac{4\delta_2^2}{\omega^2} \tag{3.78}$$

that determines the existence of a critical coupling between resonant circuits.

In the case of weak coupling between resonant circuits when, for $\omega k < 2\delta_2$, Equation (3.77) has the only solution, the dependence of the oscillation frequency ω^2 versus partial frequency of the main resonant circuit ω_{01}^2 is shown in Figure 3.8. From Figure 3.8 it follows that the oscillation frequency is unambiguously related to the partial frequencies of the two coupled resonant circuits.

To define the amplitude of self-oscillations, let us consider the operation of the MOSFET transformer-coupled oscillator shown in Figure 3.7a [10]. Assume that the active device represents an ideal current source with infinite gate–source impedance. The equations for currents in the main and additional resonant circuits are written in the form

$$L_1\frac{di_1}{dt}+r_1i_1+\frac{1}{C_1}\int i_1dt+M_1\frac{di_2}{dt}=M\frac{di}{dt} \quad (3.79)$$

$$L_2\frac{di_2}{dt}+(r_2+R_L)\,i_2+\frac{1}{C_2}\int i_2dt+M_1\frac{di_1}{dt}=0 \quad (3.80)$$

where currents i_1 and i_2 are expressed through the voltages v_1 and v_2 across the capacitances C_1 and C_2, respectively, according to

$$v_1=\frac{1}{C_1}\int i_1dt \qquad v_2=\frac{1}{C_2}\int i_2dt$$

As a result, Equations (3.79) and (3.80) can be rewritten in the form of the second-order differential equations

$$\frac{d^2v_1}{dt^2}+2\delta_1\frac{dv_1}{dt}+\omega_{01}^2v_1+\alpha_1\frac{d^2v_2}{dt^2}=M\omega_{01}^2\frac{di}{dt} \quad (3.81)$$

$$\frac{d^2v_2}{dt^2}+2\delta_2\frac{dv_2}{dt}+\omega_{02}^2v_2+\alpha_2\frac{d^2v_1}{dt^2}=0 \quad (3.82)$$

where

$$\omega_{01}^2=\frac{1}{C_1L_1} \qquad \omega_{02}^2=\frac{1}{C_2L_2}$$

$$\delta_1=\frac{r_1}{2L_1} \qquad \delta_2=\frac{r_2+R_L}{2L_2}$$

$$\alpha_1=\frac{M_1C_2}{L_1C_1} \qquad \alpha_2=\frac{M_1C_1}{L_2C_2}$$

Suppose that the oscillator start-up conditions correspond to a soft mode and the transfer characteristic of the active device can be approximated by a third-order polynomial function

$$i=g_mv_1-\frac{a_3}{3}v_1^3 \quad (3.83)$$

where g_m is the device small-signal transconductance.

By applying the quasilinear method to the system of Equations (3.81) and (3.82) when the average transconductance g_{m1} is introduced, the solution is seeking in the form

$$v_1=V_1\cos(\omega t) \quad (3.84)$$

$$v_2=V_2\cos(\omega t+\varphi) \quad (3.85)$$

where V_1 and V_2 are the amplitudes of the oscillations in the main and additional resonant circuits, respectively, and φ is the phase shift between these oscillations.

Substituting voltages v_1 and v_2 from Equations (3.84) and (3.85) into the system of Equations (3.81) and (3.82) results in the following relationships between amplitudes V_1 and V_2 and phase φ

$$\frac{V_2}{V_1}=\frac{\alpha_2\omega^2}{\sqrt{\left(\omega_{02}^2-\omega^2\right)^2+4\delta_2^2\omega^2}} \quad (3.86)$$

$$\tan\varphi=-\frac{2\delta_2\omega}{\omega_{02}^2-\omega^2} \quad (3.87)$$

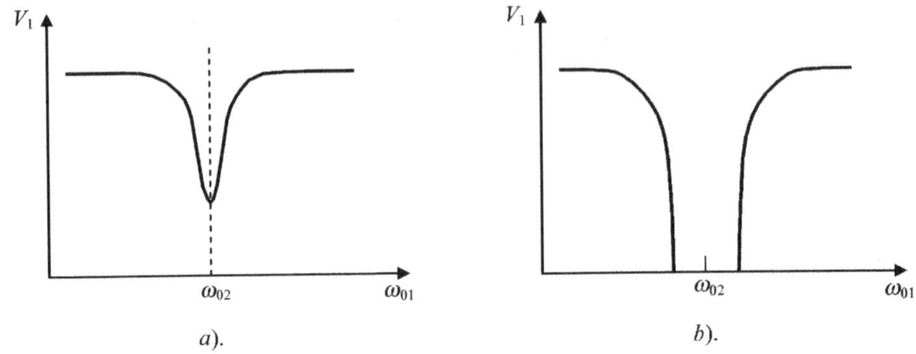

Figure 3.9 Oscillation amplitude versus partial frequency of the main circuit for different coupling

Finally, the relationship between the oscillation frequency ω and the amplitude of the oscillation V_1 can be obtained by

$$2\delta_1 + 2\delta_2 \frac{\alpha_1\alpha_2\omega^4}{\left(\omega_{02}^2 - \omega^2\right)^2 + 4\delta_2^2\omega^2} = M g_{m1}\omega_{01}^2 \tag{3.88}$$

$$\omega_{01}^2 - \omega^2 - \frac{\alpha_1\alpha_2\omega^4\left(\omega_{02}^2 - \omega^2\right)}{\left(\omega_{02}^2 - \omega^2\right)^2 + 4\delta_2^2\omega^2} = 0 \tag{3.89}$$

where

$$g_{m1} = g_m - \frac{a_3}{4}V_1^2 \tag{3.90}$$

is the large-signal transconductance of the active device averaged over the natural oscillation period at the frequency ω.

The amplitude of oscillations V_1 can be directly determined from Equation (3.88) as

$$V_1^2 = \frac{4}{a_3}\left[g_m - \frac{2\delta_1}{M\omega_{01}^2} - \frac{2\delta_2}{M\omega_{01}^2}\frac{\alpha_1\alpha_2\omega^4}{\left(\omega_{02}^2 - \omega^2\right)^2 + 4\delta_2^2\omega^2}\right] \tag{3.91}$$

From Equation (3.91) it follows that connection of the second resonant circuit results in additional dissipation whose value depends on the coupling coefficient and its partial frequency. Equation (3.89) is similar to Equation (3.76) obtained for the bipolar oscillator with two inductively coupled resonant circuits shown in Figure 3.7b, if one introduces the coupling coefficient between two transformer-coupled resonant circuits as

$$k^2 = \alpha_1\alpha_2 = \frac{M_1^2}{L_1L_2}. \tag{3.92}$$

Figure 3.9 shows the dependencies of the amplitude of self-oscillations on the partial frequency of the main resonant circuit. The greatest decrease of the amplitude (the energy is extracted from the main resonant circuit and delivered to the additional resonant circuit) is observed when the partial frequencies are equal, i.e., $\omega_{01} = \omega_{02} = \omega_0$. Then, when $\omega = \omega_0$, Equation (3.91) for the oscillation amplitude can be rewritten as

$$V_1^2 = \frac{8}{Ma_3\omega_0}\left(\frac{Mg_m\omega_0}{2} - \frac{\delta_1}{\omega_0} - \frac{\alpha_1\alpha_2\omega_0}{4\delta_2}\right) \tag{3.93}$$

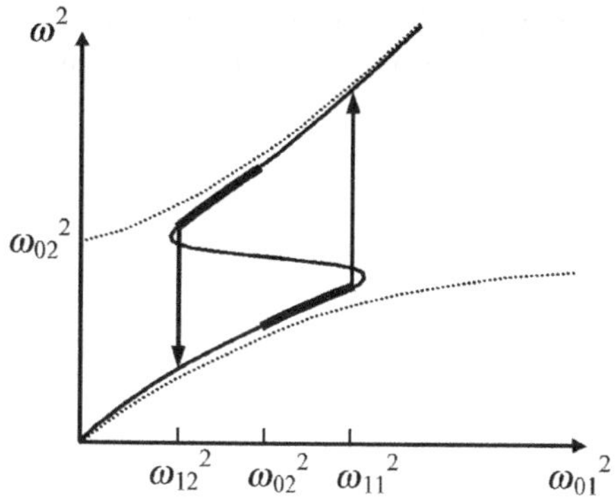

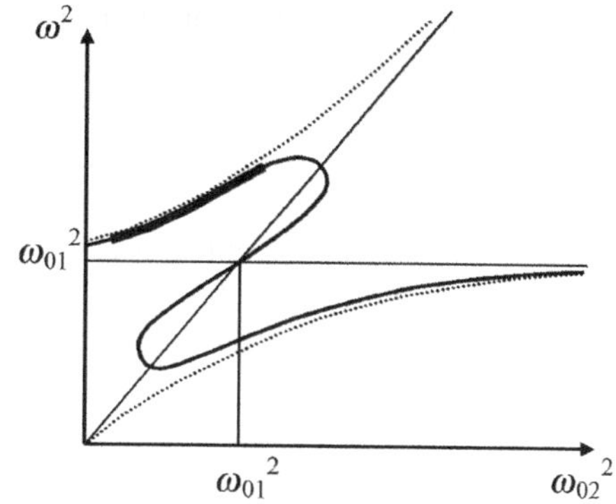

Figure 3.10 Frequency performance of two coupled resonant circuits with strong coupling

From Equation (3.93) it follows that the stronger the coupling M and the smaller losses of the additional resonant circuit δ_2 the greater decrease of the oscillation amplitude V_1, as shown in Figure 3.9a. For a sufficiently high quality factor of the additional resonant circuit, the self-oscillations near frequency synchronism of the resonant circuits can be entirely damped. The condition of the self-oscillations damping is provided when, in some frequency range, the increment of the main resonant circuit

$$\vartheta_0 = \frac{M g_m \omega_0}{2} - \vartheta_1 \tag{3.94}$$

where $\vartheta_1 = \delta_1/\omega_0$ is the decrement of the main circuit, becomes smaller than $\vartheta_2 = \alpha_1\alpha_2\omega_0/4\delta_2$ which is the loss (decrement) introduced into the main resonant circuit by the additional resonant circuit.

Figure 3.9b shows the oscillation amplitude V_1 versus partial frequency ω_{01} including the damping region. The borders of the damping region can be determined if we assume zero oscillation amplitude in Equation (3.91). When $\vartheta_0 = \vartheta_2$, the damping region represents just a point.

For the case of strong coupling between resonant circuits when $\omega k > 2\delta_2$, the dependence of the oscillation frequency ω^2 on partial frequency of the main resonant circuit ω_{01}^2 is shown in Figure 3.10a. If we maintain the resonant frequency ω_{01}^2 of the main circuit constant and vary only the resonant frequency of an additional circuit, then the oscillation frequency ω^2 will change, as shown in Figure 3.10b. It follows that, for $\omega_{01} \gg \omega_{02}$ and $\omega_{01} \ll \omega_{02}$, the oscillation system with two coupled resonant circuits generates a frequency close to the resonant frequency of the main circuit ω_{01}, i.e., the additional resonant circuit has no effect on oscillation frequency because of its weak coupling. Within the synchronism region when $\omega_{01} \approx \omega_{02}$, the dependence of ω^2 on ω_{01}^2 becomes three-valued. An analysis of the oscillation stability shows that the middle branch of the frequency characteristic is always unstable. As for the outer branches, their frequency behaviour is close to that of the normal frequencies of the linear conservative (lossless) system with two coupled resonant circuit shown by dashed lines (Wien graph). However, the presence of the loss factor $\delta_2 > 0$ in Equation (3.91) is equivalent to weakening of coupling between the resonant circuits so that the generated frequencies are closer to partial frequencies than in the conservative system.

Let us consider the situation within a synchronism or hysteresis region where two possible oscillation frequencies, ω_1 and ω_2, can exist simultaneously. In this case, the gate–source voltage v_1 in Figure 3.7a can be represented by

$$v_1 = V_1 \cos\omega_1 t + V_2 \cos\omega_2 t \tag{3.95}$$

where the phases can be omitted because these frequencies are independent.

Substituting Equation (3.95) into Equation (3.83) allows us to define the fundamental component of the drain current with a frequency ω_1 as

$$i_1 = \left(g_m V_1 - \frac{V_1^3 + 2V_1 V_2^2}{4} a_3 \right) \cos\omega_1 t \tag{3.96}$$

From Equation (3.96) it follows that, in the presence of both frequencies, the average transconductance of the device at a frequency ω_1 is equal to

$$g_{m1}(V_1, V_2) = g_m - \frac{V_1^2 + 2V_2^2}{4} a_3 \tag{3.97}$$

Similarly, the average transconductance of the device at a frequency ω_2 is equal to

$$g_{m2}(V_1, V_2) = g_m - \frac{2V_1^2 + V_2^2}{4} a_3 \tag{3.98}$$

From Equations (3.97) and (3.98) we can conclude that simultaneous coexistence of two frequencies established under soft start-up conditions reduces mutually the values of the averaged transconductance. To determine the steady-state amplitudes of the oscillations, V_1 and V_2, we need to substitute Equations (3.97) and (3.98) into Equation (3.88). As a result,

$$M\omega_{01}^2 \left(g_m - \frac{V_1^2 + 2V_2^2}{4} a_3 \right) = 2\delta_1 + 2\delta_2 \frac{\omega_{01}^2 - \omega_1^2}{\omega_{02}^2 - \omega_1^2} \tag{3.99}$$

$$M\omega_{01}^2 \left(g_m - \frac{2V_1^2 + V_2^2}{4} a_3 \right) = 2\delta_1 + 2\delta_2 \frac{\omega_{01}^2 - \omega_2^2}{\omega_{02}^2 - \omega_2^2} \tag{3.100}$$

where the left-hand sides of the equations describe the energy input into the main and additional resonant circuits while their right-hand sides define the energy losses in these resonant circuits. If for $V_1 = V_2 = 0$, the energy input is larger than energy losses in both resonant circuits, then the start-up conditions for the self-oscillations with frequencies ω_1 and ω_2 are satisfied and their amplitudes begin to grow. The voltage amplitudes V_1 and V_2 increase as long as the losses for any of the oscillations become equal to the incoming energy. Let it first be the case of the oscillations with frequency ω_2 when the voltage amplitude V_2 will no longer increase, while the voltage amplitude V_1 will continue growing. When the energy input becomes less than the losses at the frequency ω_2, the amplitude V_2 begins to decrease, resulting in increase of the amplitude V_1. As a result, V_2 decreases down to zero while V_1 can increase its value until the following conditions are satisfied:

$$M\omega_{01}^2 \left(g_m - \frac{V_1^2}{4} a_3 \right) = 2\delta_1 + 2\delta_2 \frac{\omega_{01}^2 - \omega_1^2}{\omega_{02}^2 - \omega_1^2} \tag{3.101}$$

$$M\omega_{01}^2 \left(g_m - \frac{V_1^2}{2} a_3 \right) < 2\delta_1 + 2\delta_2 \frac{\omega_{01}^2 - \omega_2^2}{\omega_{02}^2 - \omega_2^2} \tag{3.102}$$

where Equation (3.101) is the steady-state condition for the oscillations with frequency ω_1 and amplitude V_1. Equation (3.102) shows that the energy losses exceed the energy input at the frequency ω_2.

Similarly, the steady-state conditions for oscillations with frequency ω_2 and amplitude V_2 can be obtained in the form

$$M\omega_{01}^2\left(g_{\rm m} - \frac{V_2^2}{4}a_3\right) = 2\delta_1 + 2\delta_2\frac{\omega_{01}^2 - \omega_2^2}{\omega_{02}^2 - \omega_2^2} \tag{3.103}$$

$$M\omega_{01}^2\left(g_{\rm m} - \frac{V_2^2}{2}a_3\right) < 2\delta_1 + 2\delta_2\frac{\omega_{01}^2 - \omega_1^2}{\omega_{02}^2 - \omega_1^2} \tag{3.104}$$

It should be noted that oscillations with frequency ω_1 described by Equations (3.101) and (3.102) are stable. Let, for instance, the voltage amplitude V_1 increase slightly. In this case, the left-hand side of Equation (3.101) becomes less than its right-hand side, resulting in a decrease of the amplitude V_1. If the amplitude V_2 decreases slightly, the energy input becomes larger than losses and amplitude V_1 decreases again. Similarly, the oscillations with frequency ω_2 described by Equations (3.103) and (3.104) are stable as well.

Now consider the biharmonic mode described by Equations (3.99) and (3.100). For a small increase of amplitude V_1, the energy input at the frequency ω_2 decreases faster than the energy input at frequency ω_1, contributing to a fast decrease of amplitude V_2. As a result, the energy input at frequency ω_1 increases until only oscillations with frequency ω_1 are established. On the contrary, the small decrease of the amplitude V_1 results in the only oscillation frequency ω_2. Hence, a biharmonic mode with simultaneous existence of both frequencies is unstable.

The further analysis will be based on the behaviour of the oscillator with two coupled resonant circuits when partial frequencies of the main or additional resonant circuits are tuned. From Figure 3.10a it follows that, for the case of $\omega_{01} \ll \omega_{02}$, only oscillations with frequency ω_1 close to partial frequency ω_{01} can be established in the oscillator. With increasing ω_{01}, the oscillation system with two coupled resonant circuits creates a region where there is a possibility of oscillations at both frequencies ω_1 and ω_2. This region is called the *frequency pulling range*. In the pulling range, the generation mode for any of two possible frequencies depends on frequency, which exists first in the oscillation system. If the pulling range is entered from the side of small frequencies ω_{01}, then oscillations with frequency ω_1 and amplitude V_1 will exist. With further increasing ω_{01}, the oscillation system will suddenly change the generation mode into oscillations with frequency ω_2 and amplitude V_2 at $\omega_{01} = \omega_{11}$. If the pulling range is entered from the side of large frequencies ω_{01}, then oscillations with frequency ω_2 and amplitude V_2 will occur. The sudden change into a generation mode with frequency ω_2 and amplitude V_2 will take place at frequency ω_{12} which is much smaller than ω_{11}. The frequencies ω_{11} and ω_{12} determine the borders of the pulling range when the stability conditions are violated. The frequency and amplitude stability conditions can be analysed separately. So, the frequency stability conditions are violated at frequencies at which the curve $\omega^2 = f(\omega_1^2)$ shown in Figure 3.10a has a vertical tangent. The amplitude instability occurs when conditions given by Equations (3.102) and (3.104) are not satisfied. For instance, let the conditions given by Equations (3.101) and (3.103) be met at a certain frequency ω_{01}. With increasing ω_{01}, frequency ω_1 also increases and approaches ω_{02}. In this case, the right-hand side of Equation (3.101) increases and voltage amplitude V_1 decreases. At the same time, the right-hand side of Equation (3.102) decreases, while its left-hand side increases. Finally, at some value of ω_{01}, the inequality in Equation (3.102) changes sign. The energy input at frequency ω_2 becomes larger than the

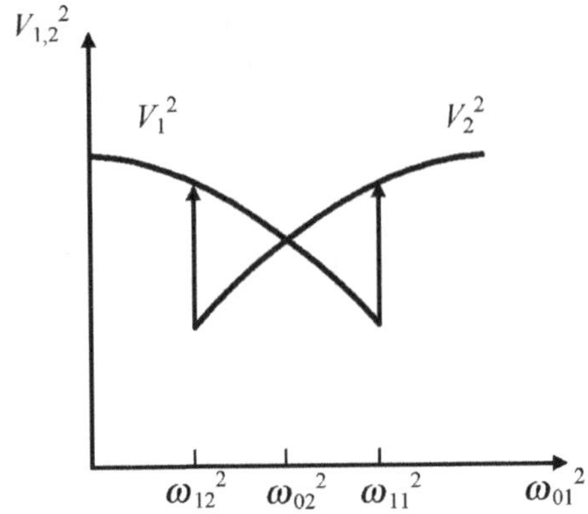

Figure 3.11 Amplitudes of oscillations versus partial frequency of the main resonant circuit

losses, and the oscillation system will change its mode into oscillations with frequency ω_2 and voltage amplitude V_2. The reverse transition to an oscillation mode with frequency ω_1 can take place only with decrease of ω_{01}, when, at $\omega_{01} = \omega_{12}$, the condition given by Equation (3.104) will be violated.

The behaviour of the oscillation amplitudes V_1 and V_2 versus partial frequency ω_{01} is shown in Figure 3.11. The ratio of the amplitudes at the border of the pulling range when $\omega_{01} = \omega_{11}$ can be obtained from Equations (3.100) and (3.101). The amplitude V_2, which is established in the oscillator after the jump, is $\sqrt{2}$ times greater the amplitude V_1 that existed in the oscillator before the jump. Accordingly, the amplitude V_1, which is established in the oscillator after the jump, is $\sqrt{2}$ times greater the amplitude V_2 that existed in the oscillator before the jump at another border when $\omega_{01} = \omega_{12}$.

Generally, for the case of strong coupling between resonant circuits when $\omega k > 2\delta_2$ (the left-hand side insignificantly exceeds the right-hand side), the pulling range is bounded by the frequency instability. However, when $\omega k \gg 2\delta_2$, the pulling range is bounded by the conditions of amplitude instability. When a self-excitation of the oscillations occurs within the pulling range, the steady-state operation mode depends on the initial conditions of the oscillation system.

Thus, an analysis of the oscillator with two coupled resonant circuit shows that, under the effect of the resonant load in a certain frequency range, it can change its oscillation frequency and amplitude, cease oscillation at all (damping mode) or enter the mode with sudden changes of its oscillation frequency and amplitude. Therefore, when designing a single-frequency oscillator using a resonant load, it is necessary to minimize the effect of such a load on the self-oscillating circuit. However, in some case, the pulling effect can be used to stabilize the oscillation frequency if the quality factor of an additional resonant circuit is very high. For example, the crystal resonator at RF frequencies and dielectric or cavity resonators at microwave frequencies can be used. On the other hand, if the quality factor of an additional circuit is small enough, it is possible to provide a linear varactor tuning of the oscillation frequency.

Assume that the partial frequency of an additional resonant circuit varies with the varactor tuning. In this case, the capacitances C_2 shown in Figure 3.7 are the equivalent varactor capacitances. Then, the oscillation frequency, the higher of two possible values, moves along the curve as shown in Figure 3.10b, where the bold section represents a linear tuning frequency bandwidth. The tuning linearity is improved by increasing the frequency sensitivity to the varactor capacitance variation at higher frequencies where the varactor capacitance has a smaller value and is less sensitive to the bias voltage. In order to avoid the frequency jumping, it is necessary to define the upper curve border when the dependence $\omega^2 = f(\omega_{02}^2)$ has a horizontal tangent and the derivative $\mathrm{d}(\omega^2)/\mathrm{d}(\omega_{02}^2)$ becomes equal to zero. Analysing Equation (3.89) in extremum, maximum varactor tuning frequency $\omega_{2\max}$ for stable operation conditions can be determined as

$$\omega_{2\max} = \omega_{01} \bigg/ \sqrt{1 - \frac{Q_{2\max}}{2} k^2} \tag{3.105}$$

where $Q_{2\max}$ is a quality factor of the additional resonant circuit at $\omega_{2\max}$. It follows from Equation (3.105) that the value of $\omega_{2\max}$ depends substantially on the coupling coefficient k, and, for a monotonic varactor tuning without frequency jumping, the quality factor $Q_{2\max}$ should be sufficiently small.

3.6 PHASE PLANE METHOD

The phase plane method is a qualitative method of an analysis of the dynamics of the oscillation systems described by second-order differential equations [11–13]. It is based on the representation of the totality of motions of such oscillation systems on the x, y plane, where x and y are orthogonal Cartesian coordinates. The x, y plane is called the *phase plane*, each point of which fully determines the state of the oscillation system at some moment in time t. It is assumed that the electrical behaviour of the oscillation systems with a resonant LC circuit is expressed through the system of the two first-order differential equations:

$$\frac{\mathrm{d}x}{\mathrm{d}t} = P(x, y) \tag{3.106}$$

$$\frac{\mathrm{d}y}{\mathrm{d}t} = Q(x, y) \tag{3.107}$$

By representing the second-order differential equations given by Equations (3.5) and (3.6) in a general form

$$\frac{\mathrm{d}^2x}{\mathrm{d}t^2} + 2\delta(x)\frac{\mathrm{d}x}{\mathrm{d}t} + \omega_0^2 x = 0 \tag{3.108}$$

Equation (3.108) can be rewritten as a system of the two first-order differential equations

$$\frac{\mathrm{d}x}{\mathrm{d}t} = y \tag{3.109}$$

$$\frac{\mathrm{d}y}{\mathrm{d}t} = -2\delta(x)y - \omega_0^2 x \tag{3.110}$$

Equations (3.109) and (3.110) are the particular case of Equations (3.106) and (3.107) when

$$P(x, y) = y \tag{3.111}$$

$$Q(x,y) = -2\delta(x)y - \omega_0^2 x \tag{3.112}$$

Plotting the values of the variables x and y along the proper coordinate axes yields the phase plane, whose points represent all possible states of the oscillation system. The instant state variation of the system is associated with the motion of a certain point on the phase plane, which is called the representative point. A path followed by the representative point is called the phase trajectory. The velocity of such representative point is called the phase velocity.

The direction of the phase velocity is a tangent to the phase trajectory at any point. Through its phase plane components

$$v_x = \frac{dx}{dt} = P(x, y) \tag{3.113}$$

$$v_y = \frac{dy}{dt} = Q(x, y) \tag{3.114}$$

the magnitude of the phase velocity can be written as

$$v_r = \sqrt{v_x^2 + v_y^2} = \sqrt{P^2(x, y) + Q^2(x, y)} \tag{3.115}$$

The phase plane mapped out by the phase trajectories gives an easily interpreted phase portrait, which determines the behaviour of the oscillation system at any initial conditions. To plot the phase trajectories in the form of the dependencies $y(x)$, it is necessary to eliminate time t as a variable from Equation (3.108). Dividing Equation (3.114) by Equation (3.113) results in an equation for the phase trajectories in the differential form

$$\frac{dy}{dx} = \frac{Q(x, y)}{P(x, y)} \tag{3.116}$$

Integrating Equation (3.116) for separate variables allows us to represent the equation for the phase trajectories in an analytical form of $y(x)$. Such an analytical dependence in an explicit form can be obtained in some simple cases, for example, for a lossless resonant LC circuit. However, in most cases, it is necessary to use numerical methods to calculate the phase trajectories.

Equation (3.116) determines directly a single tangent to the phase trajectory at each point of the phase plane, with the exception of the point x_0, y_0, where simultaneously

$$Q(x_0, y_0) = P(x_0, y_0) = 0 \tag{3.117}$$

i.e., the direction of the tangent becomes indeterminate. Such points are called singular points of the given differential equation. The phase velocity in singular points is equal to zero as $dx/dt = 0$ and $dy/dt = 0$. The singular points correspond to the state of equilibrium since the coordinates of the representative point do not change in time.

3.6.1 Free-running oscillations in lossless resonant LC circuits

Free-running oscillations in lossless resonant LC circuits can be described in the general form

$$\frac{d^2x}{dt^2} + \omega_0^2 x = 0 \tag{3.118}$$

Let us represent the second-order differential Equation (3.118) in the form of the two first-order differential equations given by Equations (3.109) and (3.110) as

$$\frac{dx}{dt} = y \tag{3.119}$$

$$\frac{dy}{dt} = -\omega_0^2 x \tag{3.120}$$

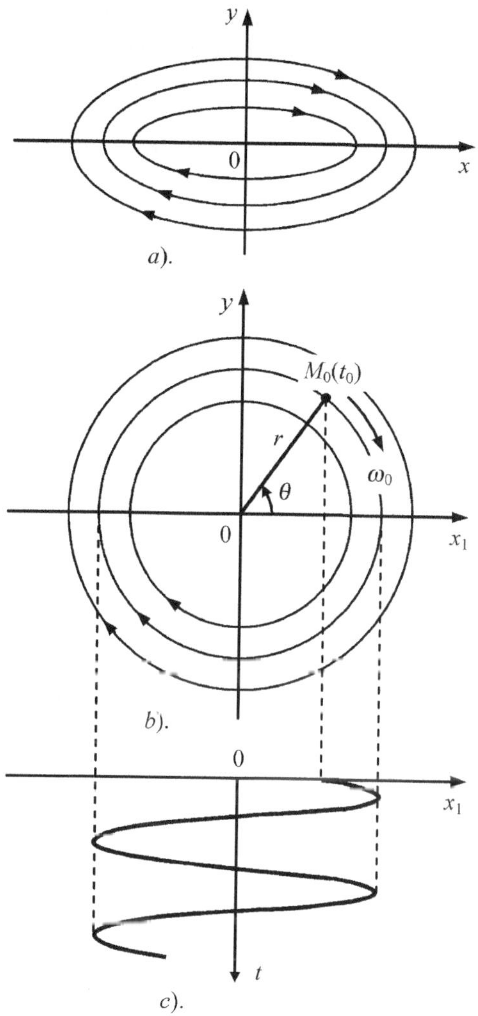

Figure 3.12 Free-running oscillations in a lossless resonant *LC* circuit

Dividing Equation (3.120) by Equation (3.119) allows us to obtain an equation for the phase trajectories in the differential form

$$\frac{dy}{dx} = -\omega_0^2 \frac{x}{y} \tag{3.121}$$

Separating variables and integrating Equation (3.121) resulting

$$\frac{x^2}{2} + \frac{y^2}{2\omega_0^2} = C \tag{3.122}$$

where C is the constant of integration defined by the initial conditions.

The phase trajectories corresponding to different values of C represent a family of nested ellipses, as shown in Figure 3.12a. In this case, there is only one singular point at $x = 0$, $y = 0$, which is called a centre. The phase velocity calculated according to Equations (3.115), (3.119) and (3.120) has nonzero value at any point on the elliptic trajectories. This means that each ellipse corresponds to the periodic oscillations with different amplitude, depending on an initial condition lying on the ellipse.

Changing the scale of one of the coordinate axis allows transforming the phase trajectories from elliptic to circular. For example, replacing x by $x_1 = \omega_0 x$ and its substituting to Equation (3.122) yields

$$x_1^2 + y^2 = 2\omega_0^2 C \tag{3.123}$$

resulting in circular trajectories enclosing each other, as shown in Figure 3.12b. Each circular trajectory has a radius of

$$r = \sqrt{2\omega_0^2 C} \tag{3.124}$$

with constant phase velocity corresponding to any point equal to

$$v_r = \sqrt{\left(\frac{dx_1}{dt}\right)^2 + \left(\frac{dy}{dt}\right)^2} = \omega_0 r \tag{3.125}$$

and the same angular velocity along any phase trajectory of $\omega = \omega_0$. Hence, at any deviation from the equilibrium state, sinusoidal or harmonic oscillations $x_1(t)$ arise in such an oscillation system with amplitude and phase determined by the initial conditions at t_0, as shown in Figure 3.12c.

3.6.2 Oscillations in lossy resonant *LC* circuits

The oscillations in lossy resonant *LC* circuits are described by a second-order differential equation in the general form

$$\frac{d^2x}{dt^2} + a\frac{dx}{dt} + bx = 0 \tag{3.126}$$

where $a = 2\delta$, $b = \omega_0^2$. The transient response in such a resonant *LC* circuit has an oscillating character if the roots of the characteristic equation

$$p_{1,2} = -\frac{a}{2} \pm \sqrt{\left(\frac{a}{2}\right)^2 - b} \tag{3.127}$$

have complex values when $b > a_2/4$. To simplify further analysis, let us assume that

$$\delta^2 \ll \omega_0^2 \tag{3.128}$$

Then, a solution of Equation (3.126) can be represented as

$$x = A\exp(-\delta t)\sin(\omega_1 t + \phi) \tag{3.129}$$

where the amplitude A and phase ϕ are determined by the initial conditions, and the oscillation frequency is equal to

$$\omega_1 = \sqrt{\omega_0^2 - \delta^2} \tag{3.130}$$

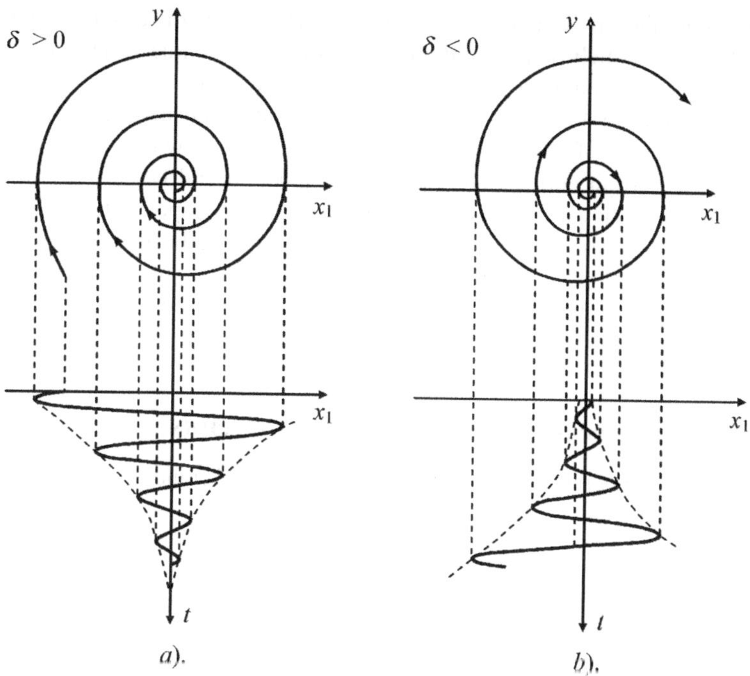

Figure 3.13 Oscillations in a lossy resonant *LC* circuit

To define the behaviour of the phase trajectories, let us differentiate Equation (3.129). Then,

$$y = \frac{\mathrm{d}x}{\mathrm{d}t} = A\omega_1 \exp(-\delta t)\left[\cos(\omega_1 t + \phi) - \frac{\delta}{\omega_1}\sin(\omega_1 t + \phi)\right] \tag{3.131}$$

By using an assumption given by Equation (3.128), Equation (3.131) is simplified to

$$y = A\omega_1 \exp(-\delta t)\cos(\omega_1 t + \phi) \tag{3.132}$$

Replacing x by $x_1 = \omega_1 x$ allows us to obtain Equation (3.129) in the form

$$x_1 = A\omega_1 \exp(-\delta t)\sin(\omega_1 t + \phi) \tag{3.133}$$

To plot the phase trajectories on the plane (x_1, y), let us rewrite Equations (3.132) and (3.133) in the polar coordinates $y = r\cos\psi$ and $x_1 = r\sin\psi$, where

$$\psi = \omega_1 t + \phi \tag{3.134}$$

$$r = \omega_1 A \exp(-\delta t) \tag{3.135}$$

By excluding the time t from Equations (3.134) and (3.135), the phase trajectories for the oscillations in the lossy resonant *LC* circuit can be calculated from

$$r = \omega_1 A \exp\left[-\frac{\delta}{\omega_1}(\psi - \phi)\right] \tag{3.136}$$

From Equation (3.136) it follows that the phase portrait on the plane x_1, y will be a family of shrinking logarithmic spirals with an asymptotic point at the origin of the coordinates, as shown in Figure 3.13a. Each spiral corresponds to a damped oscillation process tending to the position of equilibrium at the origin of the coordinates, which is a stable singular point. Such a singular point, which is an asymptotic point for all phase trajectories having the form of shrinking spirals enclosed in each other, is called a stable focus.

If the dissipation factor δ is negative, which is a result of the positive feedback or negative resistance, the magnitude of radius r in Equation (3.136) will grow exponentially in time to infinity. The phase portrait corresponding to such a situation will be a family of untwisted logarithmic spirals, as shown in Figure 3.13b. Each spiral corresponds to an exponentially growing oscillating process originating from the position of equilibrium at the origin of the coordinates, which becomes an unstable singular point. Such a singular point, which is a starting point for all phase trajectories having the form of untwisted spirals enclosed in each other, is called an unstable focus.

3.6.3 Aperiodic process in lossy resonant LC circuits

In a common case, a general solution for the second-order differential Equation (3.126) and its derivative is obtained as

$$x = A_1 \exp(p_1 t) + A_2 \exp(p_2 t) \tag{3.137}$$

$$y = \frac{dx}{dt} = A_1 p_1 \exp(p_1 t) + A_2 p_2 \exp(p_2 t) \tag{3.138}$$

where A_1 and A_2 are determined by the initial conditions.

For real roots of the characteristic equations, the process described by Equation (3.126) is aperiodic. In this situation, there are three possible cases depending on the values of parameters a and b:

- for any $a, b < 0$ and $p_1 > 0,\ p_2 < 0$ or $p_1 < 0,\ p_2 > 0$

  If $A_1 \neq 0$ and $A_2 \neq 0$, then one component in Equations (3.137) and (3.138) tends to zero while another tends to infinity. Consequently, the system moves away from the state of equilibrium in due course. The phase portrait represents a family of equilateral hyperbolas referred to their principal axes, shown in Figure 3.14a. The two asymptotes of this family pass through the origin of the coordinates, which is the only singular point. Such an unstable singular point is called a saddle.

- $a > 0, 0 < b < (0.5a)^2$ and $p_1 < 0,\ p_2 < 0$

  Both components in Equations (3.137) and (3.138) tend to zero. This means that any representative point aperiodically approaches a singular point in due course. The velocity of its motion reduces to zero only at the origin. The phase portrait represents a family of parabolas passing through the origin of the coordinates, which is a stable singular point, as shown in Figure 3.14b. Such a singular point is called a stable node.

- $a < 0, 0 < b < (0.5a)^2$ and $p_1 > 0,\ p_2 > 0$

  Both components in Equations (3.137) and (3.138) tend to infinity. This means that any representative point aperiodically moves away from a singular point in due course. The phase portrait represents a family of parabolas passing through the origin of the coordinates, which is an unstable singular point, as shown in Figure 3.14c. Such a singular point is called an unstable node.

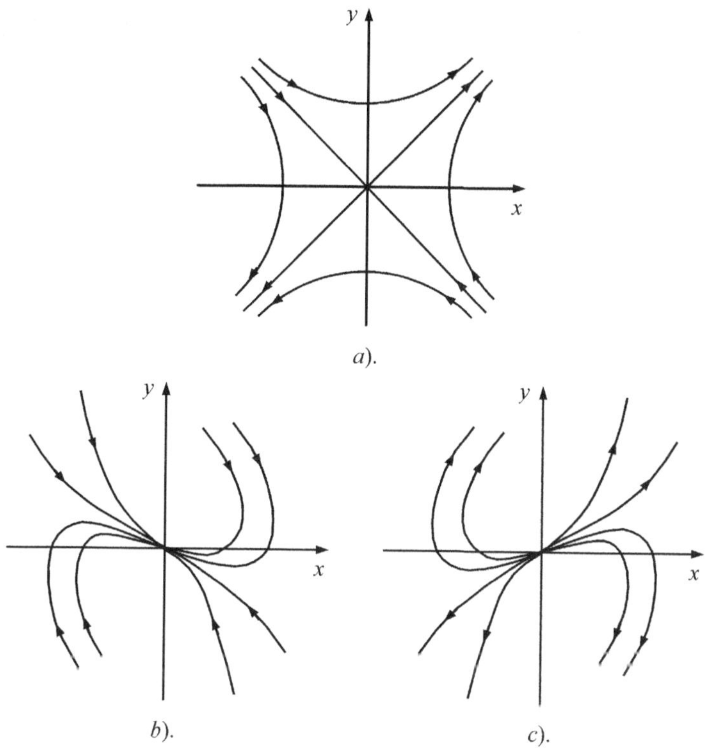

Figure 3.14 Aperiodic process in a lossy resonant LC circuit

Figure 3.15 shows the complete stability diagram determining the location of all possible singular points on the plane of a, b, which are the coefficients of the second-order differential equation given by Equation (3.126). The behaviour of the phase trajectories in the vicinity of the singular point completely determines its stability. In particular, if the representative point

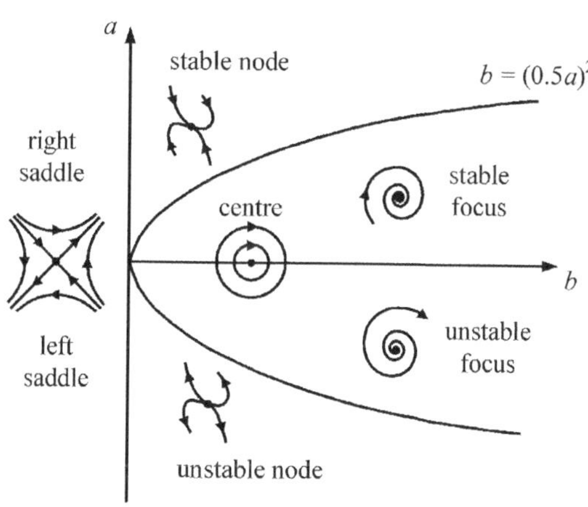

Figure 3.15 Complete stability diagram

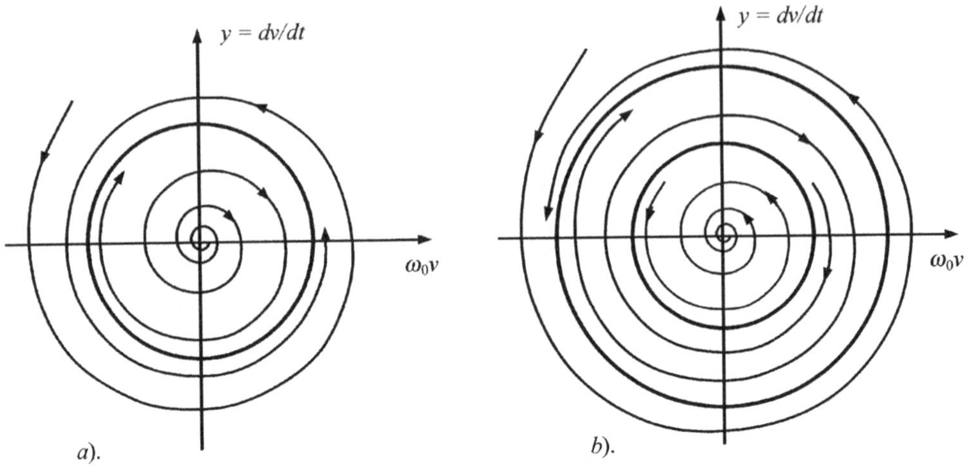

Figure 3.16 Phase portraits of self-oscillations

moves away from the state of equilibrium, the latter is considered unstable. However, if the representative point approaches the state of equilibrium, it is considered asymptotically stable.

If the representative point is moving along the closed phase trajectory and its phase velocity never becomes zero, i.e., such a phase trajectory never passes a singular point, then, on expiry of a certain period of time, the process of motion on the phase plane will recur. Such a phase trajectory characterizes the periodic free-running oscillations shown in Figure 3.12 and is called a limit cycle. If the phase trajectories are approaching the limit cycle in time as shown in Figure 3.16a, then such a limit cycle is called stable, corresponding to a stable oscillation process. On the contrary, the limit cycle, with phase trajectories moving away from it, is called unstable, corresponding to an unstable oscillation process.

3.6.4 Transformer-coupled MOSFET oscillator

Consider the transformer-coupled MOSFET oscillator, the simplified equivalent circuit of which is shown in Figure 2.1b (see Chapter 2). Assume that the transfer function of the MOSFET device can be approximated by Equation (3.83). Then, the second-order differential equation given by Equation (1.53) (see Chapter 1), which describes the behaviour of such a transformer-coupled oscillator, can be rewritten in the form

$$\frac{d^2 v}{dt^2} + (2\delta + \gamma v^2)\frac{dv}{dt} + \omega_0^2 v = 0 \tag{3.139}$$

where

$$\gamma = \frac{a_3 M}{4LC} \tag{3.140}$$

Figure 3.16a shows the phase portrait of the oscillations for a soft start-up condition corresponding to a coupling coefficient of $M = M_3$ (see Figure 2.6a in Chapter 2), when only two steady-state operation modes are possible: a state of equilibrium (point 0) and a state of periodic oscillations (point A_3). For such a case, let us represent Equation (3.139) by

$$\frac{dv}{dt} = y \quad (3.141)$$

$$\frac{dy}{dt} = -\left[(2\delta + \gamma v^2)y + \omega_0^2 x\right] \quad (3.142)$$

where $\delta < 0$ and $\gamma > 0$.

The phase velocity can be obtained as

$$v_r = \sqrt{y^2 + \left[(2\delta + \gamma v^2)y + \omega_0^2 x\right]^2} \quad (3.143)$$

which allows us to define the coordinates of the singular points when $v_r = 0$. Since both components under the square root sign on the right-hand side of Equation (3.143) are positive, the only singular point is zero. For small voltages v, Equation (3.139) is similar to Equation (3.126) with negative coefficient a when a singular point behaves as an unstable focus with the corresponding phase trajectories shown in Figure 3.13b. For large voltages v, the solution of Equation (3.126) will contain the exponentially decreasing amplitude resulting in the damped oscillation process. For a particular value of the oscillation amplitude corresponding to the situation when the positive feedback in the oscillator (or induced negative resistance) is compensated by the losses in its resonant circuit, the periodic oscillation process is established in the oscillator, as shown in Figure 3.12. The phase portrait of such periodic oscillations is shown in Figure 3.16a where the stable periodic oscillations are characterized by the stable limit cycle.

The phase portrait of the oscillator characterized by a hard start-up condition corresponding to a coupling coefficient $M = M_3$ (see Figure 2.7a in Chapter 2) is shown in Figure 3.16b with three possible steady-state operation modes: a stable state of equilibrium (point 0), an unstable state of the periodical oscillations (point B_3) characterized by an unstable limit cycle, and a stable state of periodic oscillations (point A_3) with the stable limit cycle. The limit cycles shown in Figure 3.16 separate the regions of different character of the phase trajectories corresponding to growing or damping oscillation processes.

3.7 NYQUIST STABILITY CRITERION

The Nyquist criterion is intended to test for stability of a closed-loop system based on the analysis of the transfer characteristic of its open-loop equivalent [12, 13]. In the case of complicated systems described by high-order differential equations when it is very difficult to obtain the solutions in an analytical form, a Nyquist graphic technique can help to determine the circuit stability in a much easier and clear way. Strictly saying, such an approach is applied to the stability analysis of the only linear time-invariant circuits. However, it can also be useful to determine the steady-state oscillation modes and their stability for the oscillation systems characterized by slowly varying amplitudes [14].

Figure 3.17a shows the two-port network with external feedback or closed-loop two-port network, where Z_{in} is its input impedance. Now let us transform this circuit into the open-loop two-port network, as shown in Figure 3.17b, keeping its operating capability the same by connecting the input impedance Z_{in} as load impedance. The transfer characteristic of such a

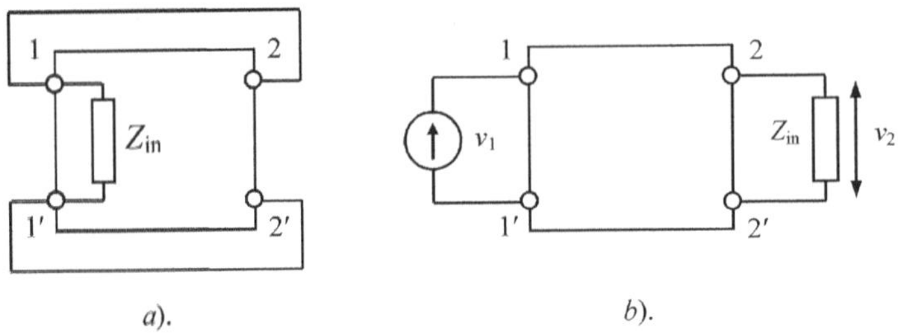

Figure 3.17 Closed-loop and open-loop two-port networks

open-loop two-port network is defined by

$$\boldsymbol{T}(j\omega) = T\exp(j\varphi) = \frac{\boldsymbol{V}_2}{\boldsymbol{V}_1} \tag{3.144}$$

where $\boldsymbol{V}_2$ and $\boldsymbol{V}_1$ are the magnitudes of the output and input voltages, respectively. The trajectory of the phasor $\boldsymbol{T}(j\omega)$ on the complex plane by varying frequency ω from 0 to ∞ is called a *magnitude–phase characteristic* or *Nyquist plot*.

For small oscillation amplitudes when all circuit elements can be considered as linear, the electrical behaviour of the closed-loop two-port network shown in Figure 3.17a can generally be described by a linear nth-order differential equation

$$a_0\frac{d^n v_2}{dt^n} + a_1\frac{d^{n-1} v_2}{dt^{n-1}} + \ldots + a_{n-1}\frac{dv_2}{dt} + a_n v_2 = 0 \tag{3.145}$$

A solution of Equation (3.145) represents a sum of the components in the form

$$v_2 = V_2\exp(pt) \tag{3.146}$$

Substituting Equation (3.146) into Equation (3.145) allows us to obtain the characteristic equation

$$D(p) = a_0 p^n + a_1 p^{n-1} + \ldots + a_{n-1}p + a_n = 0 \tag{3.147}$$

which can be written as

$$v_2 = V_2^{(1)}\exp(p_1 t) + V_2^{(2)}\exp(p_2 t) + \ldots + V_2^{(n)}\exp(p_n t) \tag{3.148}$$

where $p_1, p_2, \ldots, p_n$ are the roots of the characteristic equation given by Equation (3.147).

As a result, Equation (3.145) can be rewritten in the form

$$DV_2\exp(pt) = 0 \tag{3.149}$$

The left-hand side of Equation (3.147) can be represented by

$$D(p) = a_0(p - p_1)(p - p_2)\ldots(p - p_n) \tag{3.150}$$

By introducing a substitution of $p = j\omega$ into the right-hand side of Equation (3.150), we can write the characteristic polynomial $D(p)$ in the complex form

$$\boldsymbol{D}(j\omega) = a_0(j\omega - p_1)(j\omega - p_2)\ldots(j\omega - p_n) \tag{3.151}$$

Each factor in Equation (3.151) can be represented in the complex plane by a phasor

$$\boldsymbol{r}_k = j\omega - p = r_k \exp(j\varphi_k) \tag{3.152}$$

whereas, the characteristic polynomial can be represented by a phasor

$$\boldsymbol{D}(j\omega) = a_0 \exp(j\varphi) \prod_{k=1}^{n} r_k \tag{3.153}$$

where

$$r_1 r_2 \ldots r_n = \prod_{k=1}^{n} r_k \qquad \sum_{k=1}^{n} \varphi_k = \varphi$$

When frequency ω varies, each of the phasors $\boldsymbol{r}_k$ changes its position and total phase angle φ determines the direction of the resultant phasor $\boldsymbol{D}(j\omega)$. Let us consider the possible variations of the phase angles φ_k when frequency ω varies from zero to infinity for the following different values of the roots $p_{k,k+1} = \alpha_k \pm j\beta_k$:

- $p_k = \alpha_k < 0$ is real and negative, and the phasor $\boldsymbol{r}_k$ rotates counterclockwise by

$$\varphi_k = \pi/2$$

- $p_{k,k+1}$ are conjugate with negative real part $\alpha_k < 0$, and the phasors $\boldsymbol{r}_k$ and r_{k+1} rotate counterclockwise by $\varphi_k = \pi/2 + \gamma$ and $\varphi_{k+1} = \pi/2 - \gamma$, where $\gamma = \tan^{-1}(\beta_k/\alpha_k)$, with total phase rotation of

$$\varphi_k + \varphi_{k+1} = \pi$$

- $p_k = \alpha_k > 0$ is real and positive, and the phasor $\boldsymbol{r}_k$ rotates clockwise by

$$\varphi_k = -\pi/2$$

- $p_{k,k+1}$ are conjugate with positive real part $\alpha_k > 0$, and the phasors $\boldsymbol{r}_k$ and r_{k+1} rotate clockwise by $\varphi_k = -(\pi/2 + \gamma)$ and $\varphi_{k+1} = -(\pi/2 - \gamma)$, respectively, with total phase rotation of

$$\varphi_k + \varphi_{k+1} = -\pi$$

In this situation, the first two cases correspond to the stable systems when the average variation of the phase angle is equal to $\pi/2$ for the frequency variation from 0 to ∞.

Let us assume that the input impedance of the closed-loop two-port network shown in Figure 3.17a is real, i.e., $Z_{\rm in} = {\rm Re}Z_{\rm in}$. In this case, the electrical behaviour of the open-loop two-port network shown in Figure 3.17b can generally be described by the linear differential equation of the same order

$$b_0 \frac{d^n v_2}{dt^n} + b_1 \frac{d^{n-1} v_2}{dt^{n-1}} + \ldots + b_{n-1} \frac{dv_2}{dt} + b_n v_2$$
$$= c_0 \frac{d^m v_1}{dt^m} + c_1 \frac{d^{m-1} v_1}{dt^{m-1}} + \ldots + c_{m-1} \frac{dv_1}{dt} + c_m v_1 \tag{3.154}$$

Let $v_1 = V_1 \exp(pt)$ and $v_2 = V_2 \exp(pt)$. Then, Equation (3.154) can be rewritten in the form

$$D_0 V_2 \exp(pt) = D_1 V_1 \exp(pt) \tag{3.155}$$

where D_0 and D_1 are the characteristic polynomials

$$D_0(p) = b_0 p^n + b_1 p^{n-1} + \ldots + b_{n-1} p + b_n$$

$$D_1(p) = c_0 p^m + c_1 p^{m-1} + \ldots + c_{m-1} p + c_m$$

For a closed-loop circuit, we have $v_1 = v_2$. Consequently, Equation (3.155) can be rewritten as

$$(D_0 - D_1) V_2 \exp(pt) = 0 \tag{3.156}$$

Comparison of Equations (3.149) and (3.156) shows that, for any p,

$$D = D_0 - D_1 \tag{3.157}$$

For the case of the harmonic voltage $v_1 = V_1 \exp(j\omega t)$ at the input of the open-loop two-port network, the complex transfer function $\boldsymbol{T}(j\omega)$ defined by Equation (3.144) can be rewritten as

$$\boldsymbol{T}(j\omega) = \frac{\boldsymbol{D}_1(j\omega)}{\boldsymbol{D}_0(j\omega)} = 1 - \boldsymbol{F}(j\omega) \tag{3.158}$$

where

$$\boldsymbol{F}(j\omega) = \frac{\boldsymbol{D}(j\omega)}{\boldsymbol{D}_0(j\omega)} \tag{3.159}$$

From the definition of polynomial $D(p)$ given by Equation (3.154), it follows that, from a total number of n roots of the characteristic equation $D(p) = 0$, n_1 roots have negative real parts and n_2 roots have positive real parts. Then, as frequency ω varies from zero to infinity, the phasor $\boldsymbol{D}(j\omega)$ rotates in the angle of $(n_1 - n_2)\pi/2$. By assuming the stable operation of the open-loop two-port network when all n roots of the characteristic equation $D_0(p) = 0$ have negative real parts, the total rotation angle φ_F of the phasor $\boldsymbol{F}(j\omega)$ can be calculated from

$$\varphi_F = (n_1 - n_2 - n)\frac{\pi}{2} \tag{3.160}$$

The system is stable if all n roots of its characteristic equation have negative real parts, i.e., $n_1 = n$ and $n_2 = 0$. Hence, it follows from Equation (3.160) that the state of equilibrium of the closed-loop system will be stable, if, for variation of frequency ω from 0 to ∞, the total clockwise rotating angle φ_F of the phasor $\boldsymbol{F}(j\omega)$ is equal to zero, i.e., $\varphi_F = 0$. If $n_2 \neq 0$, for the stable open-loop system, the closed-loop system is always unstable as $n_1 - n_2 - n \neq 0$.

Figure 3.18 shows the magnitude–phase characteristic of the phasor $\boldsymbol{F}(j\omega)$ on the complex plane. Here, Figure 3.18a illustrates an unstable state of equilibrium as the total phasor angle φ_F rotating in the clockwise direction is equal to $\varphi_F = -2\pi$. Figure 3.18b,c demonstrate the stable states of equilibrium as $\varphi_F = 0$ for them.

Now transform the Nyquist plot of the phasor $\boldsymbol{F}(j\omega)$ on the $\boldsymbol{F}$-plane to the Nyquist plot of the phasor $\boldsymbol{T}(j\omega)$ on the $\boldsymbol{T}$-plane according to Equation (3.158). First, it is necessary to map a plot of the phasor $-\boldsymbol{F}(j\omega)$, and then move it to the right by 1, resulting in appropriate plots shown in Figure 3.19a,b,c. The origin (0, 0) of the $\boldsymbol{F}$-plane becomes the point (1, 0) on the $\boldsymbol{T}$-plane. The phase angle on the $\boldsymbol{T}$-plane is defined as a rotation angle of the phasor $\boldsymbol{T}(j\omega)$ drawn from the point (1, 0) to the point (T, φ) when the frequency ω varies from 0 to ∞.

Thus, a Nyquist stability criterion can be formulated as follows: the state of equilibrium of the closed-loop system is stable if the locus of the phasor, which determines the magnitude–phase characteristic of the open-loop system, does not go through nor encircle the point $(1, 0)$.

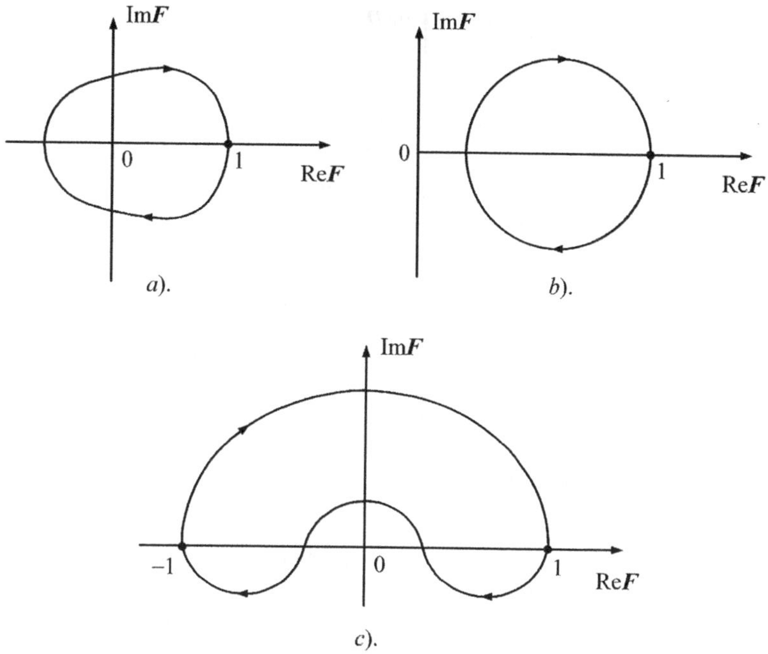

Figure 3.18 Nyquist plots of phasor $\boldsymbol{F}(j\omega)$

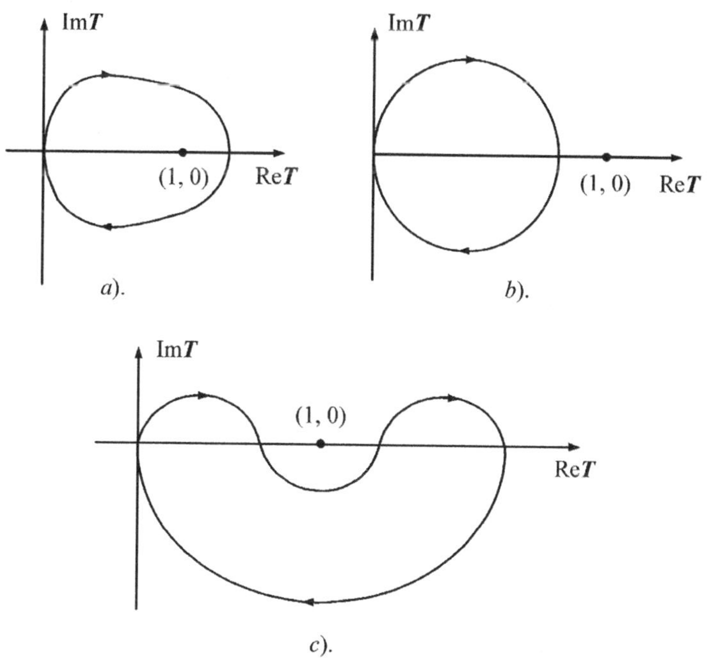

Figure 3.19 Nyquist plots of phasor $\boldsymbol{T}(j\omega)$

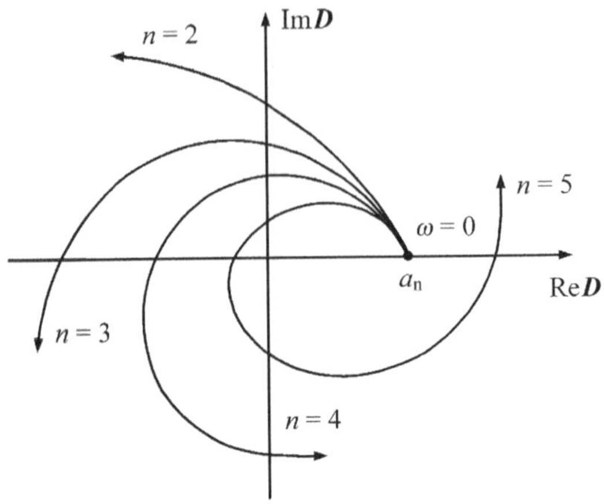

Figure 3.20 Mikhailov plots of phasor $\boldsymbol{D}(j\omega)$

Consequently, Figure 3.19a corresponds to an unstable state of equilibrium, whereas Figure 3.19b,c correspond to the stable states of equilibrium.

In some cases, when using a Nyquist stability criterion, it is difficult to predict exactly whether an oscillator circuit can be stable at the state of equilibrium or not; for example, for an unstable open-loop circuit or when the order of the characteristic polynomial for the closed-loop system is higher than that of for the open-loop system. In this case, it is possible to characterize the stability conditions directly from the characteristic polynomial of the closed-loop system in the complex form given by Equation (3.151) [12, 15].

The derivation of the system stability conditions are based on an analysis of the motion of the phasor $\boldsymbol{D}(j\omega)$, the magnitude–phase characteristic of which is defined in Equation (3.153). Consideration of the frequency behaviour of the phasors $\boldsymbol{r}_k$ results in the following criterion which is called the *Mikhailov stability criterion*: the state of equilibrium of the system described by the nth-order differential equation is stable if its phasor $\boldsymbol{D}(j\omega)$ rotates by the total phase angle $\varphi = n\pi/2$ in the counterclockwise direction as frequency ω varies from 0 to ∞. If the phase angle φ differs from the values of $n\pi/2$, the state of equilibrium is unstable.

Figure 3.20 shows the plots of the phasor $\boldsymbol{D}(j\omega)$ corresponding to the stable conditions of the systems described by differential equations of different orders. It follows from Equation (3.147) that the initial point when $\omega = 0$ is always located on the right-hand side of the complex plane having a real number, because $D(0) = a_n$ when $p = 0$ and all coefficients of the characteristic equation $D(p)$ should be positive for the stable system. Each n curve turns counterclockwise around the origin, moving to infinity in an appropriate part of the complex plane. The system becomes unstable if any trajectory intersects the right-hand side of the real axis instead of the top part of the imaginary axis, which means that such a trajectory does not encircle the origin and one of the roots has a positive real part.

3.8 START-UP AND STABILITY

To illustrate a Nyquist stability criterion, consider the start-up condition of the transformer-coupled MOSFET oscillator shown in Figure 3.21a. Disconnecting the circuit at the points

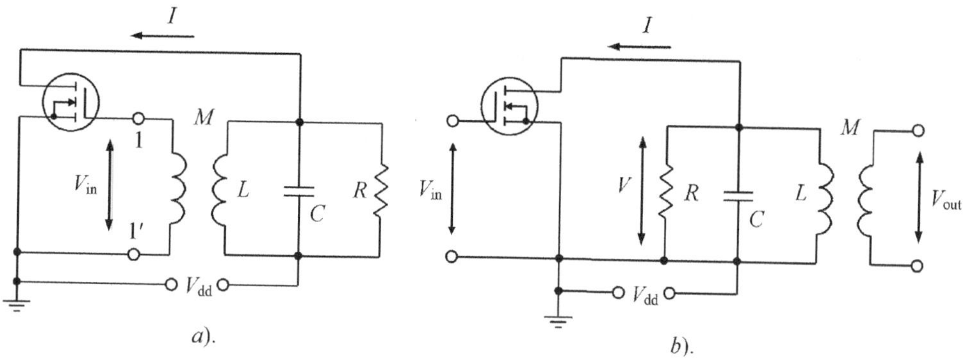

Figure 3.21 Schematics of transformer-coupled MOSFET oscillators

1–1′ results in the modified equivalent oscillator circuit shown in Figure 3.21b. Let a MOSFET device be represented by an ideal voltage-controlled current source. Then, under an assumption of small-signal conditions, the drain current fundamental complex amplitude $\boldsymbol{I}$ can be obtained as

$$\boldsymbol{I} = g_{\mathrm{m}}\boldsymbol{V}_{\mathrm{in}} \tag{3.161}$$

where g_{m} is the small-signal device transconductance.

The driving-point impedance $Z(j\omega)$ of the resonant circuit is written as

$$Z = \frac{\boldsymbol{V}}{\boldsymbol{I}} = \left(\frac{1}{R} + \frac{1}{j\omega L} + j\omega C\right)^{-1} = \frac{R}{1 + jQ\left(\dfrac{\omega}{\omega_0} - \dfrac{\omega_0}{\omega}\right)} \tag{3.162}$$

where

$$Q = \omega_0 RC = \frac{R}{\omega_0 L} \tag{3.163}$$

is the quality factor of the resonant circuit at the resonant frequency ω_0 [16].

The output voltage magnitude $\boldsymbol{V}_{\mathrm{out}}$ is defined through the voltage magnitude $\boldsymbol{V}$ as

$$\boldsymbol{V}_{\mathrm{out}} = \frac{M}{L}\boldsymbol{V} \tag{3.164}$$

It should be noted that the drain–source fundamental voltage is out-of-phase with the fundamental voltage across the parallel resonant circuit as an increase in the drain current results in the decrease of the drain–source voltage and an increase of the voltage across the parallel resonant circuit.

As a result, the complex transfer function $\boldsymbol{T}(j\omega)$ of the circuit in Figure 3.21b can be obtained using Equations (3.161–3.164) as

$$\boldsymbol{T}(j\omega) = \frac{\boldsymbol{V}_{\mathrm{out}}}{\boldsymbol{V}_{\mathrm{in}}} = \frac{Mg_{\mathrm{m}}R}{L}\frac{1}{1 + jQ\left(\dfrac{\omega}{\omega_0} - \dfrac{\omega_0}{\omega}\right)} \tag{3.165}$$

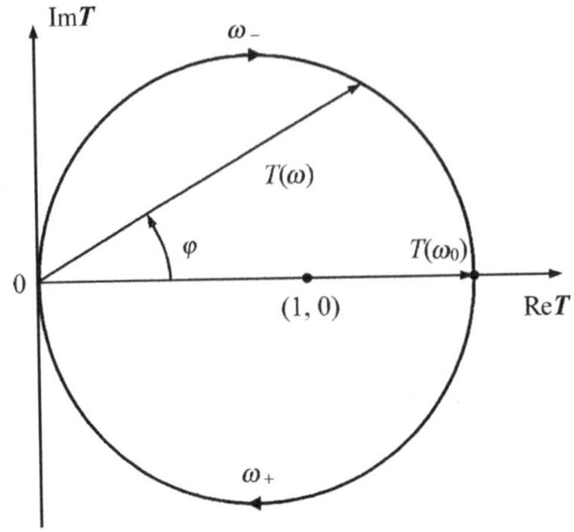

Figure 3.22 Nyquist locus of a MOSFET transformer-coupled oscillator

Hence, the magnitude and phase of the complex transfer function $\boldsymbol{T}(j\omega)$ can be written, respectively, in the form

$$T(\omega) = \frac{Mg_mR}{L\sqrt{1+\tan^2\varphi}} = T_0\cos\varphi \tag{3.166}$$

$$\varphi(\omega) = -\tan^{-1}\left[Q\left(\frac{\omega}{\omega_0} - \frac{\omega_0}{\omega}\right)\right] \tag{3.167}$$

where

$$T_0 = T(\omega_0) = \frac{Mg_mR}{L} \tag{3.168}$$

is the magnitude of the transfer function at the resonant frequency ω_0 when $\omega = \omega_0$ and $\varphi = 0$.

The locus of the phasor $\boldsymbol{T}(j\omega)$ plotted in the complex plane as ω varies from 0 to ∞ is shown in Figure 3.22. It is a circle with centre $(T_0/2, 0)$ and radius $T_0/2$ according to

$$\left[\mathrm{Re}T(j\omega) - \frac{T_0}{2}\right]^2 + \mathrm{Im}^2T(j\omega) = \left(\frac{T_0}{2}\right)^2 \tag{3.169}$$

which can be obtained by a rotation of the phasor $\boldsymbol{T}(j\omega)$ with different values of its magnitude T and phase φ calculated from Equations (3.166) and (3.167).

According to a Nyquist stability criterion, the oscillation start-up condition is provided when $T_0 > 1$ or, using Equation (3.166),

$$M > \frac{L}{g_mR} \tag{3.170}$$

which is identical to that of derived from the second-order differential equation given by Equation (1.53) (see Chapter 1). The oscillation frequency ω is determined by setting Equation (3.167) to zero, corresponding to a condition of complete phase compensation in the oscillator, from

$$\omega = \omega_0 = \frac{1}{\sqrt{LC}} \tag{3.171}$$

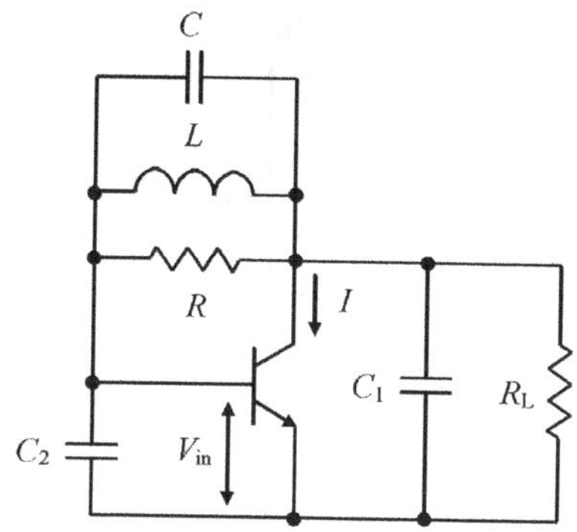

Figure 3.23 Circuit schematic of the parallel feedback bipolar oscillator

In most cases, the locus of the phasor $\boldsymbol{T}(j\omega)$ represents a closed curve going through the origin because $T(\omega) = 0$ for $\omega = 0$ and $\omega = \infty$. For $\omega = 0$, this occurs when one part of the electrical circuit is connected to another part through a blocking capacitor with a certain capacitance value. For $\omega = \infty$, it is a result of a shunt output capacitance of the device or resonant circuit capacitance connected in parallel to the load.

Now consider the examples of the parallel feedback and negative-resistance bipolar oscillators to illustrate the start-up predictions based on a stability analysis [17]. For a parallel feedback oscillator with a complex transfer function (or loop gain) $\boldsymbol{T}(j\omega)$, the start-up conditions can be written as

$$\text{Mag}\,[\boldsymbol{T}\,(j\omega)] > 1 \tag{3.172}$$

$$\text{Phase}\,[\boldsymbol{T}\,(j\omega)] = 0 \tag{3.173}$$

which are similar to Equations (2.8) and (2.9), respectively (see Chapter 2). Figure 3.23 shows the circuit schematic of the parallel feedback bipolar oscillator. For the idealized transistor, which is considered simply as a voltage-controlled current source with the small-signal transconductance $g_{\rm m} = I/V_{\rm in}$, the loop gain of the oscillator circuit in symbolic form can be written as

$$T(p) = -T_0 \frac{p^2 LC + p\dfrac{L}{R} + 1}{a_3 p^3 + a_2 p^2 + a_1 p + a_0} \tag{3.174}$$

where

$$T_0 = g_{\rm m} R_{\rm L}$$

$$a_3 = LC_2 R_{\rm L} C_1 \left(1 + \frac{C}{C_1} + \frac{C}{C_2}\right)$$

$$a_2 = L(C + C_2) + \frac{L}{R} R_{\rm L}(C_1 + C_2)$$

$$a_1 = \frac{L}{R} + R_{\rm L}(C_1 + C_2)$$

$$a_0 = 1$$

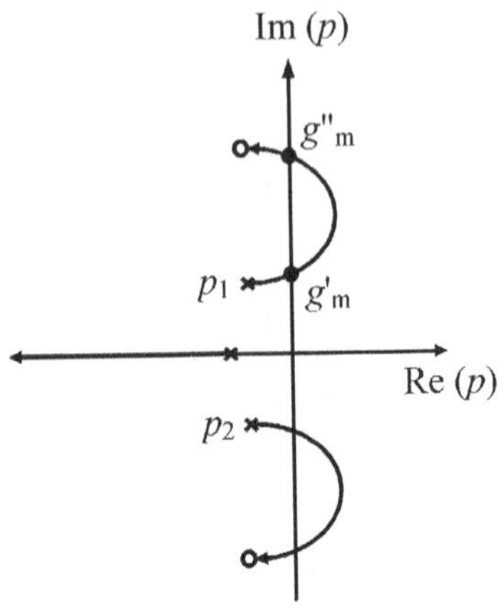

Figure 3.24 Root locus for the parallel feedback bipolar oscillator

where R_L is the load resistance, R, L and C are the parallel resonant circuit elements, C_1 and C_2 are the feedback capacitances.

Analysing the roots of the characteristic equation $1 - T(p) = 0$ enables determination of the stability properties of the oscillator. Figure 3.24 shows the root locus on the p-plane as a function of g_m, where values g'_m and g''_m correspond to the boundary conditions when the roots p_1 and p_2 are purely reactive. Hence, the roots p_1 and p_2 have positive real parts within the range $g'_m < g_m < g''_m$, corresponding to the circuit instability with a growing oscillation signal. At the same time, the circuit is stable when $g_m < g'_m$ and $g_m > g''_m$, because the loop gain is either too small or too large, producing a decaying oscillation signal in response to an excitation.

Another way to determine the stability conditions is to generate a Nyquist plot for the oscillator circuit. Figure 3.25 shows the Nyquist plots for a parallel feedback bipolar oscillator for bias conditions corresponding to (*a*) $g'_m < g_m < g''_m$ and (*b*) $g_m > g''_m$, respectively. Figure 3.25a shows the two clockwise encirclements of the point (1, 0) that indicates that two roots of the circuit characteristic equation have positive real parts, showing the circuit instability.

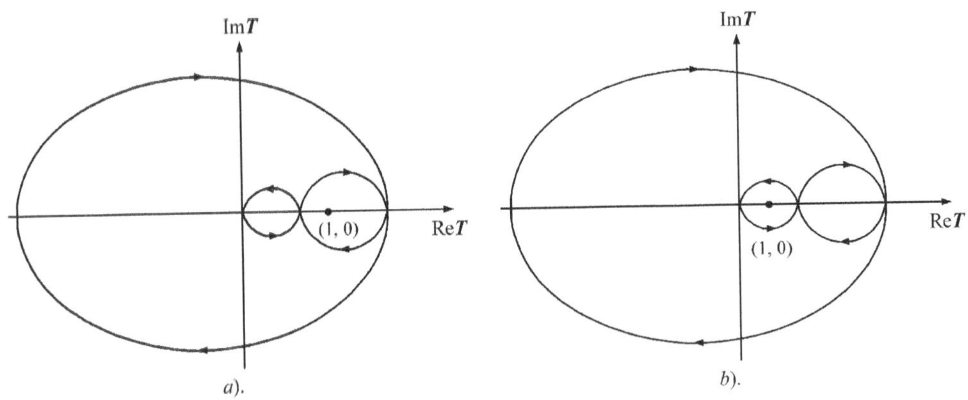

Figure 3.25 Nyquist plots of phasor $T(j\omega)$ for the parallel feedback bipolar oscillator

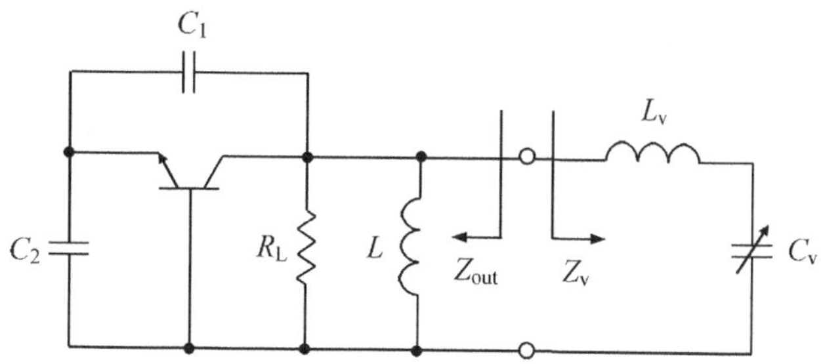

Figure 3.26 Circuit schematic of the common base bipolar oscillator

In Figure 3.25b, the net clockwise encirclement of the point (1, 0) is zero, confirming that, for $g_m > g''_m$, the circuit is stable. All methods of the stability analysis are valid if the start-up conditions (or circuit instability) hold at only one frequency and the phase characteristic (or circuit reactance plot) is monotonic over frequency. However, Equations (3.172) and (3.173) in some cases can be satisfied at a certain frequency, but the circuit will nevertheless be stable if we examine its root locus or Nyquist plot. Let us consider such a situation for the example of the negative-resistance bipolar oscillator.

Figure 3.26 shows the circuit schematic of the negative-resistance common base bipolar oscillator where R_L is the load resistance, L is the feedback inductance, C_1 and C_2 are the feedback capacitances, C_v is the varactor junction capacitance and L_v is the varactor parasitic series inductance. For such a negative resistance oscillator with a varactor as a frequency-tuning element, which is modelled by the equivalent output impedance $Z_{out} = R_{out} + jX_{out}$ and varactor impedance $Z_v = R_v + jX_v$, the start-up conditions can be written as

$$R_{out}(\omega) + R_v(\omega) < 0 \tag{3.175}$$

$$X_{out}(\omega) + X_v(\omega) = 0 \tag{3.176}$$

which are similar to Equations (2.84) and (2.85) (see Chapter 2), respectively.

Equations (3.175) and (3.176) are the analytical start-up conditions, the fulfilment of which, however, cannot always guarantee the oscillation build-up and corresponding stable steady-state oscillation process. For example, Figure 3.27a shows the oscillator impedance characteristics versus frequency for load resistance $R_L = 200\ \Omega$ and varactor capacitance $C_v = 4$ pF, where the start-up conditions are satisfied at a frequency of 1.8 GHz [17]. However, the simulated transient response shows a decaying oscillation process as well as no clockwise encirclement of the point (1, 0) by the Nyquist plot. The root locus shown in Figure 3.27b indicates by points 1–4 that the roots of the circuit characteristic equation have the negative real parts. For a negative resistance oscillator with constant load, the frequency stability conditions of the steady-state self-oscillation process are given by Equation (3.31), requiring the growth of the real and imaginary parts of the circuit impedance or positive slope of their derivatives over frequency. However, it follows from Figure 3.27a that the derivative $\partial(X_{out} + X_v)/\partial\omega$ is negative at the zero point corresponding to the potential oscillation frequency of 1.8 GHz. This means that the steady-state self-oscillation process at this frequency is unstable, producing a decaying transient response.

Figure 3.28 shows the case when, for the increased value of the load resistance $R_L = 500\ \Omega$ and reduced value of the varactor capacitance $C_v = 2$ pF, the start-up conditions are satisfied

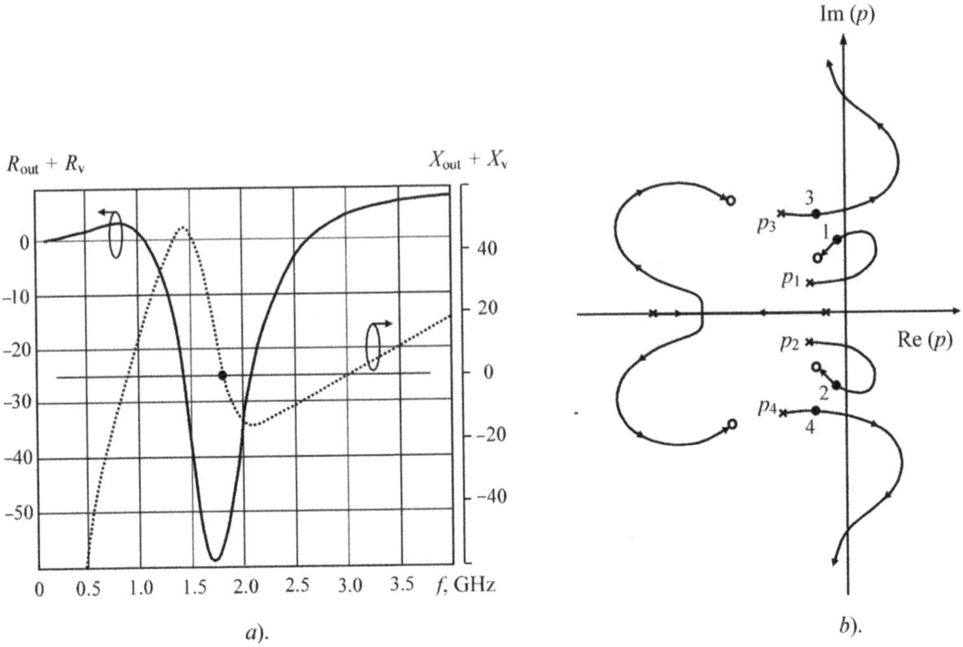

Figure 3.27 Impedance plot and root locus for the common base bipolar oscillator ($R_L = 200\ \Omega$, $C_v = 4$ pF) © 1992 IEEE

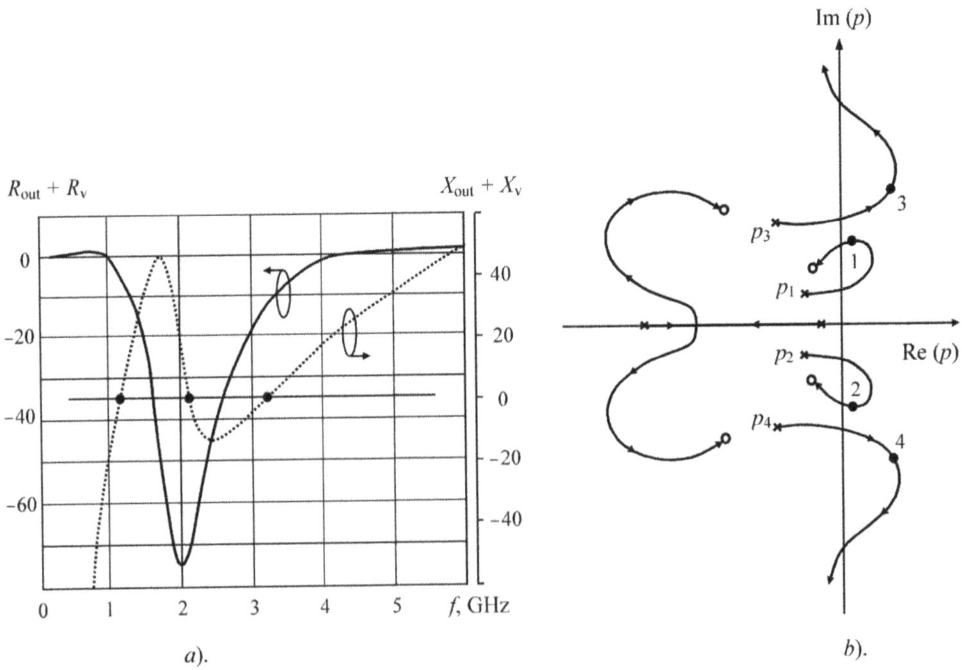

Figure 3.28 Impedance plot and root locus for the common base bipolar oscillator ($R_L = 500\ \Omega$, $C_v = 2$ pF) © 1992 IEEE

at three frequencies, as shown in Figure 3.28a, and root locus demonstrates the positive real parts of the roots of the circuit characteristic equation by points 1–4, as shown in Figure 3.28b. We can see that the behaviour of the reactance curve $X_{out} + X_v$ over frequency is very similar to the behaviour of two coupled resonant circuit oscillators with strong coupling having a hysteresis region or pulling range. In this case, two possible oscillation frequencies at 1.2 GHz and 3.2 GHz can exist simultaneously, having positive slopes of the reactance curve. The oscillation frequency corresponding to the middle branch of the reactance curve with negative slope at 2.1 GHz is always unstable. Moreover, a biharmonic mode with simultaneous existence of both frequencies is normally also unstable.

However, for varactor frequency tuning within a pulling range, the oscillation process can be accompanied by sudden changes of its oscillation frequency and amplitude. The generation mode for any of two possible frequencies depends on the original frequency of the oscillation system. To eliminate the potential multi-oscillation phenomenon, it is necessary to minimize the effect of the parasitic reactive elements in the oscillator circuit. In particular, for the negative resistance common base bipolar oscillator shown in Figure 3.26, it is necessary to minimize the effect of the series bondwire parasitic varactor inductance.

REFERENCES

1. A. V. Grebennikov, 'Stability of Negative Resistance Oscillator Circuits', *Int. J. Electrical Engineering Education*, **36**, 242–254 (1999).
2. A. V. Grebennikov, 'Stability Analysis of Wideband Linear Bipolar Varactor-Controlled Oscillator with Two-Coupled Resonant Circuits', *Int. J. Numerical Modelling: Electronic Networks, Devices and Fields*, **12**, 465–482 (1999).
3. K. Kurokawa, 'Some Basic Characteristics of Broadband Negative Resistance Oscillator Circuits', *Bell Sys. Tech. J.*, **48**, 1937–1955 (1969).
4. V. Rizzoli and A. Lipparini, 'General Stability Analysis of Periodic Steady-State Regimes in Nonlinear Microwave Oscillators', *IEEE Trans. Microwave Theory Tech.*, **MTT-33**, 1059–1063 (1985).
5. B. D. Bates and P. J. Khan, 'Stability of Multifrequency Negative-Resistance Oscillators', *IEEE Trans. Microwave Theory Tech.*, **MTT-32**, 1310–1318 (1984).
6. S. Barnett, *Introduction of Mathematical Control Theory*, London: Oxford University Press (1975).
7. M. Fillebock and P. Russer, 'Efficient Stability Analysis of Free Runing Microwave Oscillators in the Frequency Domain', *Int. J. Numerical Modelling: Electronic Networks, Devices and Fields*, **9**, 171–179 (1996).
8. W. El-Kamali, J.-P. Grimm, R. Meierer and C. Tsironis, 'New Design Approach for Wide-Band FET Voltage-Controlled Oscillators', *IEEE Trans. Microwave Theory Tech.*, **MTT-34**, 1059–1063 (1986).
9. V. Gunderich, F. Zinkler, W. Anzill and P. Russer, 'Noise Calculation and Experimental Results of Varactor Tunable Oscillators with Significantly Reduced Phase Noise', *IEEE Trans. Microwave Theory Tech.*, **MTT-43**, 278–285 (1995).
10. V. V. Migulin, V. I. Medvedev, E. R. Mustel and V. N. Parygin, *Basic Theory of Oscillations*, Moscow: Mir Publishers (1983).
11. A. A. Andronov, A. A. Vitt and S. E. Khaikin, *Theory of Oscillations*, New York: Dover Publications (1987).
12. V. S. Andreev, *Theory of Nonlinear Electrical Circuits* (in Russian), Moskva: Radio i Svyaz (1982).
13. L. O. Chua, C. A. Desoer and E. S. Kuh, *Linear and Nonlinear Circuits*, Singapore: McGraw-Hill (1987).

14. K. F. Teodorchik, *Self-Oscillatory Systems* (in Russian), Moskva: Gostekhizdat (1952).
15. M. V. Kapranov, V. N. Kuleshov and G. M. Utkin, *Theory of Oscillations in Radio Engineering* (in Russian), Moskva: Nauka (1984).
16. D. M. Pozar, *Microwave Engineering*, New York: John Wiley & Sons (1998).
17. N. M. Nguyen and R. G. Meyer, 'Start-up and Frequency Stability in High-Frequency Oscillators', *IEEE Trans. Solid-State Circuits*, **SC-27**, 810–819 (1992).

4

Optimum design and circuit technique

Generally, RF and microwave transistor oscillator design is a complex problem. Depending on the technical requirements, it is necessary to define the configuration of the oscillator circuit, choose a proper transistor type, evaluate and measure the parameters of the transistor nonlinear model under small- and large-signal conditions. Finally, an appropriate nonlinear simulator must be used to simulate the oscillator performance in time and frequency domains. An oscillator analysis can be based on using the two-port network approach to describe the active device and feedback circuit. In this case, the basic parameters of the transistor equivalent circuit can be directly measured, or approximated on the basis of experimental data, with sufficient accuracy across a wide frequency range. However, the values of the external feedback circuit elements are initially unknown. The process of determining the optimum values of the feedback and load parameters can be time-consuming and, in a typical case, calls for much simulation. Consequently, it is convenient to use an analytic method of optimizing oscillator design. This method should incorporate the explicit expressions for feedback elements and load impedance in terms of the transistor equivalent circuit elements and its static volt–ampere (I–V) and voltage–capacitance (C–V) characteristics.

Use of analytic method involves a two-step procedure. First, the optimum combination of feedback elements required to achieve a small-signal maximum negative resistance to permit oscillations at the largest amplitude is defined. Second, taking into account the nonlinearities of the transistor equivalent circuit elements, the maximum negative resistance condition under large-signal operation is characterized to determine the optimum load impedance when maximum output power is generated for a given oscillator circuit configuration. This procedure calls for the use of different time-domain and frequency-domain computer-aided design methods, including the most popular quasilinear and harmonic balance methods.

This chapter presents both the empirical and analytic optimum design approaches applied to series and parallel feedback oscillators, including circuit design and simulation aspects, and high-efficiency design techniques as well. Typical practical examples of RF and microwave oscillators using MOSFET, MESFET, HEMT, and bipolar devices, including the descriptions of their circuit configurations, are given.

RF and Microwave Transistor Oscillator Design A. Grebennikov

4.1 EMPIRICAL OPTIMUM DESIGN APPROACH

Optimum oscillator design with maximum output power generally requires extensive small- and large-signal measurements [1]. The small-signal S-parameter measurements can be made at several frequencies, along with estimated device equivalent circuit parameters, including package parasitics. Then, a computer optimization program is used to match the measured S-parameters to the S-parameters computed for the active device from its equivalent circuit. The next step is to vary those elements of the equivalent circuit, which can vary under large signals. By varying the active device nonlinear equivalent circuit parameters, the set of large-signal S-parameters can be obtained corresponding to the saturation condition where maximum oscillator output power can be achieved.

The simplified design approach assumes that all the S-parameters except the magnitude of S_{21} are constant under large signals. For example, for a MESFET device up to saturation with small limitation in accuracy, it is possible to derive the large-signal behaviour of the main active device nonlinear elements as functions of the device transconductance g_m [1]. Consequently, the S-parameters become functions of g_m only and, at each incremental reduction of g_m, are recomputed and optimized along with power gain. If one were interested in the 1-dB compression point, the S-parameters used would be those at that point.

For a common source (emitter) power oscillator, it is helpful to characterize the maximum power through the saturated output power P_{sat} and small-signal transducer power gain G_T of the corresponding power amplifier. An empirical expression for output power of the common source MESFET power amplifier can be written as

$$P_{out} = P_{sat}\left[1 - \exp\left(-\frac{G_T P_{in}}{P_{sat}}\right)\right] \tag{4.1}$$

where P_{in} is the input power [1].

The objective is to maximize the oscillator output power $P_{osc} = P_{out} - P_{in}$, that is, to provide a condition

$$\frac{dP_{osc}}{dP_{in}} = \frac{d(P_{out} - P_{in})}{dP_{in}} = 0 \tag{4.2}$$

Substituting Equation (4.1) into Equation (4.2) yields

$$\frac{dP_{out}}{dP_{in}} = G_T \exp\left(-\frac{G_T P_{in}}{P_{sat}}\right) = 1 \tag{4.3}$$

From Equation (4.3) it follows that

$$\exp\left(\frac{G_T P_{in}}{P_{sat}}\right) = G_T \tag{4.4}$$

or

$$\frac{P_{in}}{P_{sat}} = \frac{\ln G_T}{G_T} \tag{4.5}$$

As a result,

$$P_{out} = P_{sat}\left(1 - \frac{1}{G_T}\right) \tag{4.6}$$

and the maximum output power of the oscillator can be approximated by

$$P_{\text{osc}} = P_{\text{sat}}\left(1 - \frac{1}{G_{\text{T}}} - \frac{\ln G_{\text{T}}}{G_{\text{T}}}\right) \tag{4.7}$$

Consequently, the oscillator maximum power can be predicted from the saturated power and small-signal transducer power gain of a common source power amplifier. From Equation (4.7) it follows that the oscillator output power P_{osc} will approach P_{sat} at low frequencies where the small-signal transducer power gain G_{T} is large. As the transducer power gain approaches unity, the oscillator output power approaches zero.

In this case, the maximum efficiency power gain G_{ME} can be calculated from

$$G_{\text{ME}} = \frac{G_{\text{T}} - 1}{\ln G_{\text{T}}} \tag{4.8}$$

For example, a MESFET device, which is characterized by the small-signal transducer power gain $G_{\text{T}} = 7.5$ dB and saturated amplifier output power of 1 W, would be capable of a maximum oscillator power of 515 mW. The maximum efficient gain at this point calculated from Equation (4.8) is equal to $G_{\text{ME}} = 4.3$ dB. The expression for maximum efficient gain given by Equation (4.8) can be used to determine at what gain level the large-signal S-parameters are to be used for optimum oscillator design [1].

In a common case, there are a multitude of possible circuit configurations with feedback inductive or capacitive elements depending on the load connection points. Having computed the large-signal S parameters following the procedure outlined before, the embedding elements for six oscillator topologies with three parallel feedback configurations shown in Figure 4.1a,b,c and three series feedback configurations shown in Figure 4.1d,e,f may be computed [2]. To represent the design equations for the feedback elements and load in terms of admittance Y-parameters or impedance Z-parameters, it is necessary to transform the S-parameters by the usual S to Y and Z conversion formulas.

Based on a MSX801G device, the six MESFET oscillator circuits embedding element values are shown in Figure 4.2 [1]. It should be noted that not all of these circuit configurations can be readily realized in practice. In particular, all the configurations with parallel feedback are difficult to realize with a coax packaged device, because it is not so easy to physically locate a capacitor between the gate and drain since the source ring gets in the way. The circuit configurations with series feedback are more convenient for practical implementation.

To overcome some limitations of the large-signal S-parameter measurement approach, the method based on the large-signal active device model for de-embedding the dominant nonlinearities using Equations (4.5) and (4.6) can be used [3]. Due to the similarity between nonlinear device operation in power amplifiers and oscillators, the results obtained for power amplifier design can be applied to the oscillator design as well. The device model used in this approach recognizes only nonlinearities associated with the transconductance g_{m} and output differential drain–source resistor R_{ds}. These two elements produce the nonlinear fundamental effects of practical significance [4]. The values of the linear model elements and the small-signal values of the nonlinear elements are derived from measured small-signal S-parameters using curve-fitting techniques. The main issue is the assessment of the large-signal behaviours of these nonlinear elements and finding their optimum operating conditions. The instantaneous values of g_{m} and R_{ds} can be described as time-invariant functions of the voltage across the intrinsic gate–source capacitor and the voltage across the intrinsic drain–source terminals. Consequently, the oscillation mode is uniquely determined by three basic parameters, the fundamental frequency

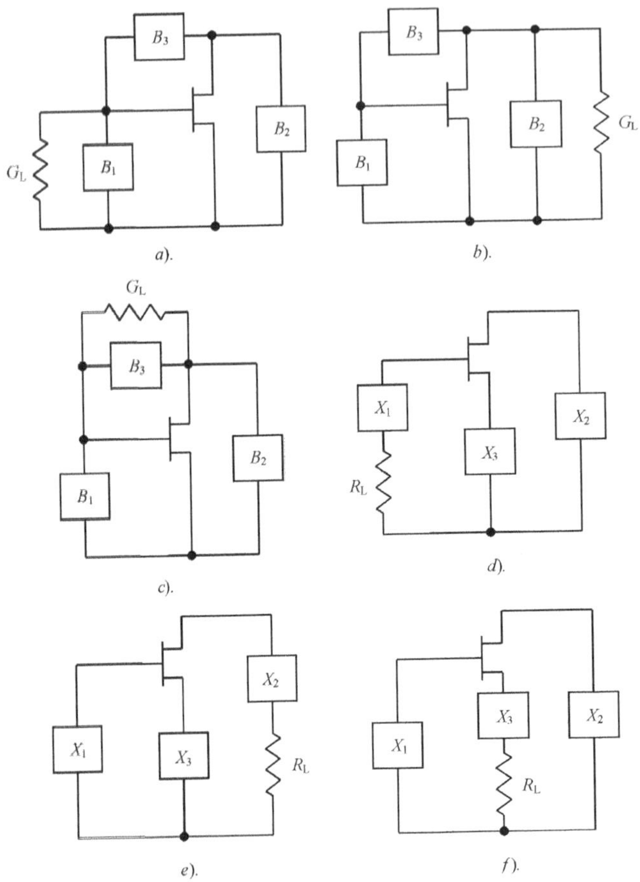

Figure 4.1 Oscillator topologies with parallel and series feedback

Fourier components of these two voltages and the intrinsic load impedance. Although all these three parameters are generally interrelated, due to the resistive nature of both nonlinear model elements, an independent approximation for intrinsic load impedance can be employed without introducing errors of practical importance. The optimum large-signal solution for intrinsic load resistance (the reactive part of load impedance is compensated in a steady-state mode) is determined from the large-signal load line plotted on the static output voltage–ampere characteristic of the active device. As a result, such an approach is based on the empirical equations for the maximum oscillator output power, the measured static output voltage–ampere device characteristic, the similarity between the static and dynamic characteristics and the knowledge of the small-signal values of resistive model elements at the nominal bias point.

A similar quasilinear design approach based on the measured small-signal S-parameters and static output voltage–ampere device characteristic to predict the oscillator output power has been proposed [5]. However, this method assumes that the significant nonlinear effects in GaAs MESFET are the rapid increase in the gate–source conductance due to the forward-biasing effect of the model gate–source Schottky diode and output drain–source conductance [6]. To maximize the output power of a series feedback MESFET oscillator, an analytic procedure was derived using the input and output fundamental voltages as independent variables. The sys-

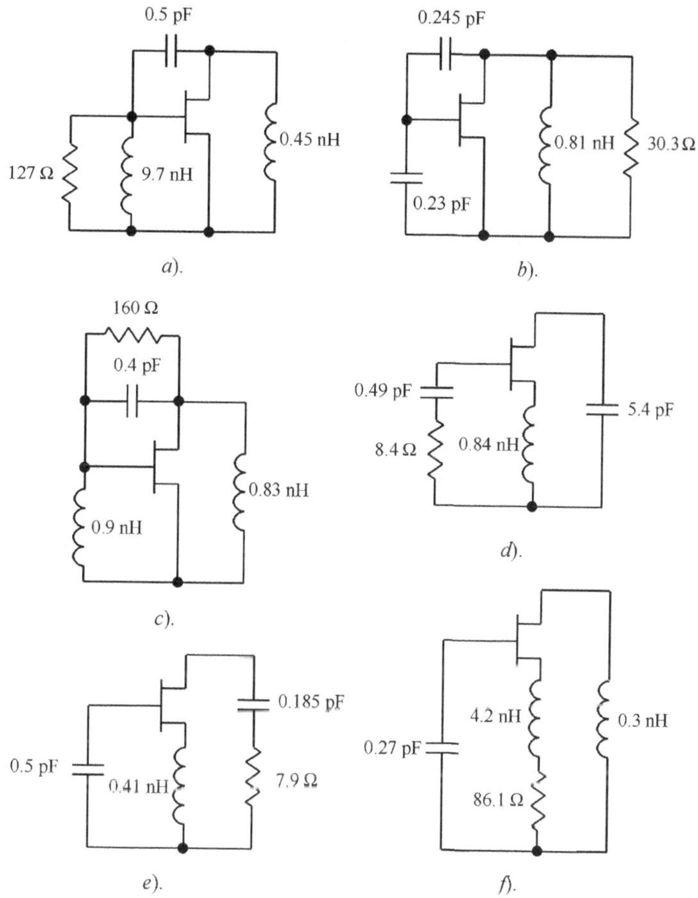

Figure 4.2 MESFET oscillator circuit schematics with parallel and series feedback (© 1979 IEEE)

tematic approach to oscillator design using large-signal S-parameters, which may be measured under high drive conditions or obtained through the use of an active device nonlinear model, is described elsewhere [7]. The device can be represented in a quasilinear approximation by its large-signal S-parameters, each assumed to be a function of a single variable, i.e., S_{11} and S_{21} are functions of the input fundamental voltage amplitude while S_{12} and S_{22} are functions of the output fundamental voltage amplitude. As a result, the set of four equations describes the oscillation condition that requires standard computer-based nonlinear root finding methods to determine the optimum feedback parameters providing a delivery of maximum output power into the load.

A technique for the design of microwave transistor oscillators, in which measurements made on an experimentally optimized power amplifier, has been presented [8]. Generally, the transistor power amplifier is more easily analysed and optimized than the corresponding oscillator because, in quasilinear approximation at a given frequency and bias point, only two parameters need be varied: the input RF drive level and the output load impedance. Once these parameters are experimentally optimized, measuring the input impedance and constructing an input matching circuit complete the design. In the case of the transistor oscillator, however, there are a multitude of possible oscillator configurations and a large number of interacting

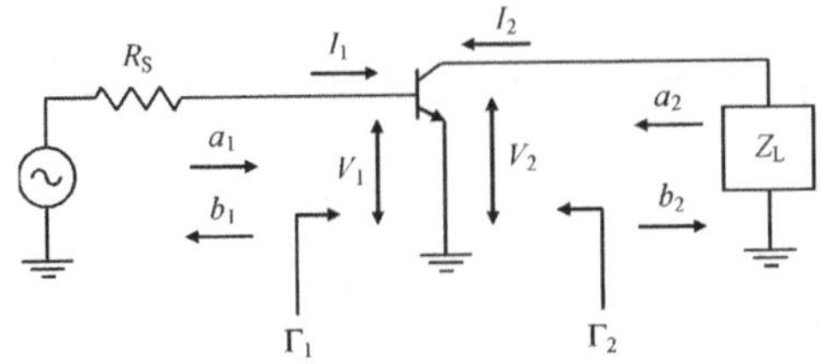

Figure 4.3 Power amplifier schematic to be experimentally optimized

circuit elements, which must be varied to optimize an oscillator for maximum power. But since we know that a transistor operates under the same set of RF voltages and currents when delivering its maximum output power, it is possible to take information obtained from an easily optimized power amplifier and use this information to calculate optimum oscillator configurations.

The first step in this procedure is to experimentally optimize the large-signal behaviour of the transistor by varying the load impedance and RF drive level for the maximum output power delivered into the load. The incident and reflected waves a_1, b_1, a_2, and b_2 shown in Figure 4.3 can be measured with calculation of the output power delivered into the load according to

$$P_L = |b_2|^2 \left(1 - \frac{1}{|S'_{22}|^2} - \frac{1 - |S'_{11}|^2}{|S'_{21}|^2}\right) \tag{4.9}$$

where

$$S'_{11} = \frac{b_1}{a_1} \qquad S'_{22} = \frac{b_2}{a_2} \qquad S'_{21} = \frac{b_2}{a_1}$$

S'_{11} is the input reflection coefficient Γ_1 and S'_{22} is the output reflection coefficient Γ_2 [9].

Using the three measured parameters S'_{11}, S'_{21} and S'_{22}, it is then possible to calculate the required ratios of the transistor terminal voltages and currents

$$\frac{V_2}{V_1} = \frac{S'_{21}\left(\dfrac{1}{S'_{22}} + 1\right)}{1 + S'_{11}} \tag{4.10}$$

$$\frac{I_2}{I_1} = \frac{S'_{21}\left(\dfrac{1}{S'_{22}} - 1\right)}{1 - S'_{11}} \tag{4.11}$$

$$\frac{V_1}{I_1} = R_S \frac{1 + S'_{11}}{1 - S'_{11}} \tag{4.12}$$

$$\frac{V_2}{I_1} = R_S \frac{S'_{21}\left(\dfrac{1}{S'_{22}} + 1\right)}{1 - S'_{11}} \tag{4.13}$$

$$\frac{I_2}{V_1} = \frac{1}{R_S} \frac{S'_{21}\left(\dfrac{1}{S'_{22}} - 1\right)}{1 + S'_{11}} \tag{4.14}$$

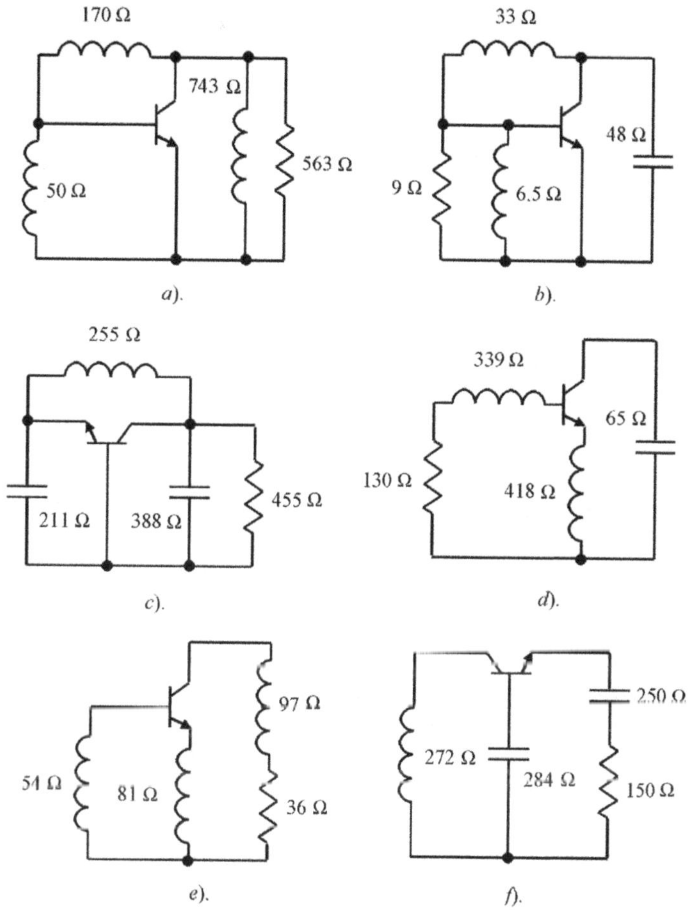

Figure 4.4 Bipolar oscillator schematic with parallel and series feedback (© 1984 IEEE)

where V_1 and I_1 are the transistor input port voltage and current, V_2 and I_2 are the transistor output port voltage and current, and R_S is the source resistance considered as the characteristic impedance of the measuring system, as shown in Figure 4.3.

Based on these ratios, the values of all elements of six basic bipolar oscillator configurations at 1 GHz based on silicon bipolar transistor NE73435, three with parallel feedback shown in Figure 4.4a,b,c and three with series feedback shown in Figure 4.4d,e,f, are obtained [8]. For maximum output power, the output resistive load can vary from very high to very low, depending on the circuit configuration. As a result, for a parallel feedback oscillator, a load with maximum value of 563 Ω is required for a fully inductive oscillator when it is connected between the collector and the emitter, as shown in Figure 4.4a. To use a load with minimum value of 9 Ω, it should be connected between the base and the emitter. In this case, the required external collector–emitter reactance must be capacitive, as shown in Figure 4.4b. By contrast for a series feedback oscillator, a high load condition can be provided by its connection between the base and the emitter, as shown in Figures 4.4d,f, whereas a load with minimum value of 36 Ω is connected between the collector and the emitter, as shown in Figure 4.4e.

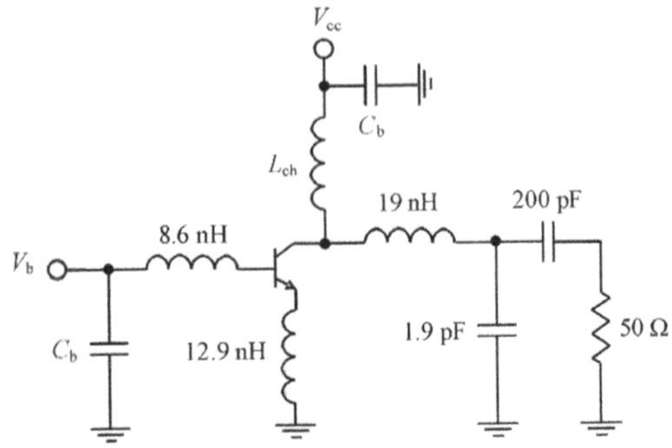

Figure 4.5 Circuit schematic of the 1-GHz test oscillator (© 1984 IEEE)

The series feedback oscillator configuration shown in Figure 4.4e, which was selected for practical implementation, yielded a stable oscillation at 1.03 GHz with a measured output power of 23 mW and efficiency of 36% at a bias point of $V_{cc} = 8$ V with dc collector current of 8 mA. Figure 4.5 shows the practical oscillator schematic where L_{ch} is the choke inductance and C_b are the bypass capacitors. The use of a simple L-section output matching circuit with a series inductor and a shunt capacitor provides the impedance transformation of the resistance 36 Ω into a standard load resistance of 50 Ω.

For a general nonlinear active device, the problem of finding the conditions under which the power delivered to a load is maximized can be also solved in a quasilinear approximation by using the large-signal impedance or admittance two-port parameters which are functions of a single port voltage [10]. Provided that operation in the saturation region is excluded, the linearized large-signal parameters can be functions of the dc bias voltage and port 1 voltage amplitude V_1 shown in Figure 4.6a. For fixed voltage amplitude V_1, there is a freedom to choose the complex voltage $\boldsymbol{V}_2$ to maximize the output power delivered by the active device

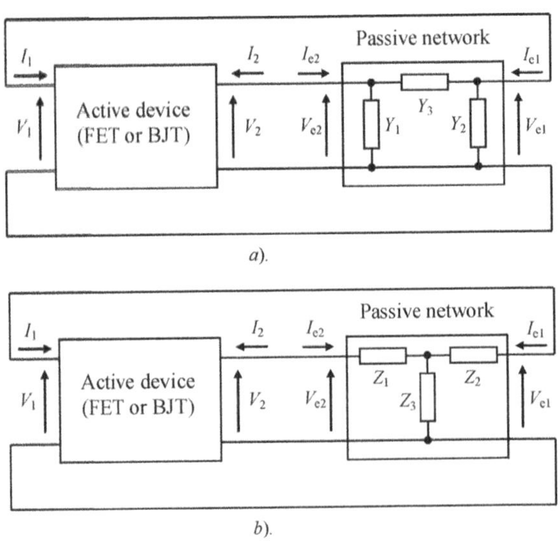

Figure 4.6 Equivalent circuits of two-port oscillators

into the load. Therefore, it is convenient to define the voltage complex ratio of port 2 voltage $\boldsymbol{V}_2$ and the port 1 voltage $\boldsymbol{V}_1$ as

$$\boldsymbol{A} = \frac{\boldsymbol{V}_2}{\boldsymbol{V}_1} = A_r + jA_i \tag{4.15}$$

which determines the device output power. In this case, various embeddings for constant $\boldsymbol{A}$ will produce oscillators with the same power delivered to the embeddings.

Optimization of $\boldsymbol{A} = \boldsymbol{V}_2/\boldsymbol{V}_1$ keeping the voltage amplitude V_1 constant requires that complex voltage $\boldsymbol{V}_2$ be varied in both magnitude and phase until the real part of the power dissipated in the active device is a minimum. If this minimum is negative then the active device is generating real power flowing into a load. The total power P dissipated by the two-port network can be found by

$$P = \frac{1}{2}\text{Re}\left(\boldsymbol{V}_1^*\boldsymbol{I}_1 + \boldsymbol{V}_2^*\boldsymbol{I}_2\right) \tag{4.16}$$

The active device can be described by the admittance two-port parameters in matrix form

$$\begin{bmatrix} \boldsymbol{I}_1 \\ \boldsymbol{I}_2 \end{bmatrix} = \begin{bmatrix} Y_{11} & Y_{12} \\ Y_{21} & Y_{22} \end{bmatrix} \begin{bmatrix} \boldsymbol{V}_1 \\ \boldsymbol{V}_2 \end{bmatrix} \tag{4.17}$$

In a steady-state oscillation mode, the following relationships between the voltages and currents of the active device and the voltages and currents of the embedding network must be satisfied:

$$\boldsymbol{V}_1 = \boldsymbol{V}_{e1} \qquad \boldsymbol{V}_2 = \boldsymbol{V}_{e2} \tag{4.18}$$

$$\boldsymbol{I}_1 = -\boldsymbol{I}_{e1} \qquad \boldsymbol{I}_2 = -\boldsymbol{I}_{e2} \tag{4.19}$$

Rewriting the total power in terms of Y-parameters and voltage complex ratio $\boldsymbol{A}$ results in

$$P = \left[G_{11} + G_{22}\left(A_r^2 + A_i^2\right) + (G_{11} + G_{22})A - (B_{12} - B_{21})A_i\right]|\boldsymbol{V}_1|^2 \tag{4.20}$$

where G_{ij} and B_{ij} are the real and imaginary parts of Y_{ij}.

The condition for the minimum value of $P = P_{\min}$ gives the optimum value of $\boldsymbol{A} = \boldsymbol{A}_{\text{opt}}$, which can be found from the partial derivatives of Equation (4.20) with respect to A_r and A_i, in the form of

$$\boldsymbol{A}_{\text{opt}} = -\frac{Y_{21} + Y_{12}^*}{2G_{22}} \tag{4.21}$$

Substituting Equation (4.21) into Equation (4.20) yields

$$P_{\min} = \frac{4G_{11}G_{22} - |Y_{21} + Y_{12}^*|^2}{4G_{22}}|\boldsymbol{V}_1|^2 \tag{4.22}$$

Once $\boldsymbol{A}_{\text{opt}}$ is determined, one can generate an infinite number of embeddings (characterized by the admittance parameters $Y_k = G_k + jB_k$, $k = 1, 2, 3$) with positive conductances G_1, G_2, and G_3. In a common case, each conductance can represent a load. For the extreme case when the load is confined to G_1 with $G_2 = G_3 = 0$, it can be determined from

$$G_1 = \frac{\left|Y_{21} + Y_{12}^*\right|^2 - 4G_{11}G_{22}}{4G_{22}} \tag{4.23}$$

which is the maximum load conductance for port 1 at a given amplitude of port 1 voltage.

When load is connected to port 2 and $G_1 = G_3 = 0$, the load conductance G_2 can be calculated from

$$G_2 = G_{22}\left[1 - \frac{4G_{11}G_{22}}{\left|Y_{21} + Y_{12}^*\right|^2}\right] \tag{4.24}$$

An analysis of the oscillation conditions through Y-parameters with determination of the maximum output power corresponding to optimum voltage ratio $\boldsymbol{A}_{\mathrm{opt}}$ given by Equation (4.21) has been presented [11]. Here, the embedding circuit impedances of an oscillator can be directly calculated from the measured two-port small-signal parameters, and the terminal voltage ratio should be chosen inside the circle in the complex $\boldsymbol{A}$-plane

$$\left|\boldsymbol{A}_0 - \boldsymbol{A}_{\mathrm{opt}}\right| = r$$

where $\boldsymbol{A}_0$ is the voltage gain value when the circuit is at the verge of oscillations with $P = 0$, $\boldsymbol{A}_{\mathrm{opt}}$ is the centre and r is the radius given by

$$r = \frac{\sqrt{\left|Y_{21} + Y_{12}^*\right|^2 - 4G_{11}G_{22}}}{2G_{22}} \tag{4.25}$$

When $P \geq 0$, there is no power flowing from the active device to the external passive circuit, which means that the circuit does not constitute an oscillator. The condition when $\boldsymbol{A} = \boldsymbol{A}_0$ represents the values of $\boldsymbol{A}$ which lie on the circle described by Equation (4.25). The values of $\boldsymbol{A}$ within the circle correspond to a circuit that will oscillate. To analyse the oscillator with passive two-port T-type network shown in Figure 4.6b, it is possible to use the well-known equivalence between admittances for a π-circuit and impedances for a T-circuit.

However, it is hardly to be expected that port 1 voltage amplitude V_1 will remain constant for all values of $\boldsymbol{A}$. Therefore, the better way to maximize the power delivered into the load is to determine the optimum value of $\boldsymbol{A}$, which gives the minimum positive (power is going in) input power delivered to the port 1 for any arbitrary negative (power is going out) constant output power delivered to the port 2 [12]. As a result,

$$\boldsymbol{A}_{\mathrm{opt}} = \frac{-2G_{11}Y_{21}}{\sqrt{N} + 2G_{11}G_{22} + j(G_{12}B_{21} + G_{21}B_{12})} \tag{4.26}$$

where

$$N = 4G_{11}G_{22}(G_{11}G_{22} + B_{12}B_{21} - G_{12}G_{21}) - (G_{12}B_{21} + G_{21}B_{12})^2$$

Numerical results using 20-GHz gain-compressed Y-parameters of typical GaAs MESFETs show that the values of $\boldsymbol{A}_{\mathrm{opt}}$ obtained by Equation (4.26) are different from those obtained by Equation (4.21). However, such a difference is not extremely large, which accounts for the experimental success of oscillators constructed under the assumption of constant port 1 voltage amplitude V_1.

4.2 ANALYTIC OPTIMUM DESIGN APPROACH

Depending on the type of the transistor used, frequency band, required output power level and spectral characteristics, most of the oscillator schematics can be reduced to two basic arrangements with external positive (a) parallel or (b) series feedback shown in Figure 4.7. One model may be preferred over another, depending on the oscillator configuration and characteristics. The two-port network representing an active device can be characterized in a

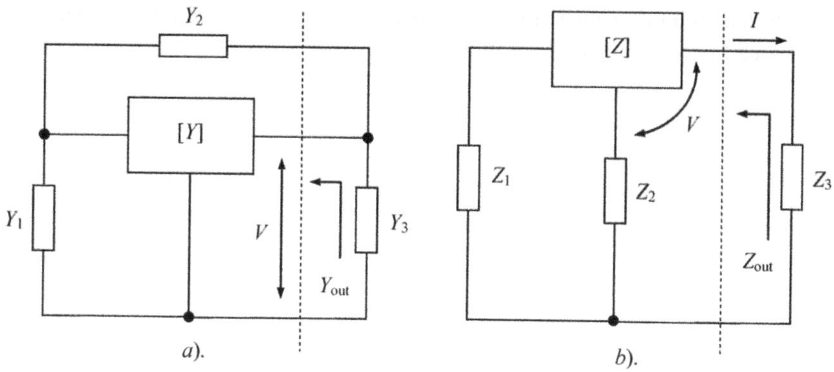

Figure 4.7 Two-port oscillator circuits with (a) parallel and (b) series feedback

common case by a system of immittance W-parameters. This describes parameter systems of the two-port network, for example, a system of Y-parameters for parallel feedback or a system of Z-parameters for series feedback oscillator circuits, respectively. The oscillator circuit and the output load are characterized by immitances W_1, W_2 and $W_3 = W_L$, respectively.

Then, the steady-state oscillation condition for a single frequency of oscillation can be expressed as

$$W_{out} + W_L = 0 \tag{4.27}$$

where the active two-port network, together with the feedback elements W_1 and W_2, will be considered as a one-port negative resistance oscillator circuit.

To optimize the oscillator circuit in terms of the maximum value of the negative real part of the equivalent one-port network output immitance, the expression for output immitance W_{out} should be written as

$$W_{out} = W_{22} + W_2 - \frac{(W_{12} \mp W_2)(W_{21} \mp W_2)}{W_{11} + W_1 + W_2} \tag{4.28}$$

where the minus signs in the factors correspond to output admittance of the circuit shown in Figure 4.7a, and the plus signs correspond to the output impedance of the circuit shown in Figure 4.7b. Such an optimization approach was first applied to a bipolar transformer-coupled oscillator to define maximum output power [13]. Later it was used to determine optimum feedback elements with reference to a series feedback MESFET oscillator using a simplified device equivalent circuit without the intrinsic feedback gate–drain capacitance [14].

It is advisable to represent this optimum approach in a generalized form regardless of the type of the active device, for commonly used parallel and series feedback oscillators [15]. The first step in the design is to determine the optimum combination of the values of the feedback reactive elements $\text{Im}W_1$ and $\text{Im}W_2$, which maximizes the absolute value of the real part of the output immittance $\text{Re}W_{out}$ at the desired frequency of oscillation. Such a condition will permit us to obtain self-sustained oscillations with the largest amplitude that implies maximum output power delivered to the load [16]. Analysing Equation (4.28) in extremum, we can find the optimum values $\text{Im}W_1^o$ and $\text{Im}W_2^o$, at which the negative value $\text{Re}W_{out}$ is maximal, by solving

$$\frac{\partial \text{Re}W_{out}}{\partial \text{Im}W_1} = 0 \qquad \frac{\partial \text{Re}W_{out}}{\partial \text{Im}W_2} = 0 \tag{4.29}$$

As a result, the optimum values $\text{Im}W_1^{\text{o}}$ and $\text{Im}W_2^{\text{o}}$ depend on the immittance parameters of the active device two-port network as follows:

$$\text{Im}W_1^{\text{o}} = \mp\frac{\text{Re}(W_{21} - W_{12})}{\text{Im}(W_{21} - W_{12})}\left[\frac{\text{Re}(W_{12} + W_{21})}{2} \pm \text{Re}(W_{11} + W_1)\right] - \text{Im}W_{11} \mp \frac{\text{Im}(W_{12} + W_{21})}{2} \quad (4.30)$$

$$\text{Im}W_2^{\text{o}} = \pm\frac{\text{Re}(W_{21} + W_{12} \mp 2W_2)\text{Re}(W_{21} - W_{12})}{2\text{Im}(W_{21} - W_{12})} \pm \frac{\text{Im}(W_{12} + W_{21})}{2} \quad (4.31)$$

The next step is to determine the optimum load immittance W_{L} in a steady-state operation mode defined by Equation (4.27). By substituting expressions for $\text{Im}W_1^{\text{o}}$ and $\text{Im}W_2^{\text{o}}$ into Equation (4.28), the optimum real and imaginary parts of the output immittance $W_{\text{out}}^{\text{o}}$ will be, respectively, defined by

$$\text{Re}W_{\text{out}}^{\text{o}} = \text{Re}W_{22} + \text{Re}W_2 - \frac{\text{Re}^2(W_{12} + W_{21} \mp 2W_1) + \text{Im}^2(W_{21} - W_{12})}{4\text{Re}(W_{11} + W_1 + W_2)} \quad (4.32)$$

$$\text{Im}W_{\text{out}}^{\text{o}} = \text{Im}W_{22} + \text{Im}W_2^{\text{o}} - \frac{\text{Re}(W_{21} - W_{12})}{\text{Im}(W_{21} - W_{12})}\text{Re}\left(W_{\text{out}}^{\text{o}} - W_{22} - W_2\right) \quad (4.33)$$

In a large-signal operation mode, all immittance parameters generally become functions of the voltage amplitudes across the elements of the active device equivalent circuit. For the negative resistance (conductance) one-port oscillators shown in Figure 4.7, the output power can be written as $P_{\text{out}} = V^2G_{\text{out}}(V)/2$ in terms of Y-parameters or $P_{\text{out}} = I^2R_{\text{out}}(I)/2$ in terms of Z-parameters, where V is voltage amplitude across the load resistance and I is the amplitude of the output current flowing into the load. Since, for the same output voltage or current amplitude, the combination of the optimum feedback parameters obtained by Equations (4.30) and (4.31) provides a maximum negative real part of the output immittance according Equation (4.32), then the maximum power will be delivered to the load. Thus, such an analytic approach presumes that, once the nonlinear model of the active device is developed, the elements of the optimum feedback parameters and the load can be easily and explicitly calculated under both small-signal and large-signal conditions corresponding to the start-up and steady-state operation modes, respectively.

4.3 PARALLEL FEEDBACK OSCILLATORS

4.3.1 Optimum oscillation condition

For the parallel feedback two-port oscillator shown in Figure 4.7a, a steady-state single-frequency oscillation condition is defined by

$$Y_{\text{out}} + Y_3 = 0. \quad (4.34)$$

In this case, the active two-port network together with the feedback elements Y_1 and Y_2 is considered as a one-port negative conductance oscillator circuit, the output admittance of which is written as

$$Y_{\text{out}} = Y_{22} + Y_3 - \frac{(Y_{12} - Y_3)(Y_{21} - Y_3)}{Y_{11} + Y_3 + Y_2} \quad (4.35)$$

By analysing Equation (4.35) on extremum, it is possible to obtain the optimum values B_1 and B_2, at which the negative value $\mathrm{Re}Y_{out}$ is maximal, by solving the following two equations [17]:

$$\frac{\partial \mathrm{Re}Y_{out}}{\partial B_1} = 0 \qquad \frac{\partial \mathrm{Re}Y_{out}}{\partial B_2} = 0 \tag{4.36}$$

As a result, for $Y_1 = jB_1$ and $Y_2 = jB_2$, the optimum values of feedback susceptances B_1^o and B_2^o can be expressed through the two-port active device Y-parameters as

$$B_1^o = -\frac{G_{21} - G_{12}}{B_{21} - B_{12}}\left(\frac{G_{12} + G_{21}}{2} + G_{11}\right) - B_{11} - \frac{B_{12} + B_{21}}{2} \tag{4.37}$$

$$B_2^o = \frac{(G_{21} + G_{12})(G_{21} - G_{12})}{2(B_{21} - B_{12})} + \frac{B_{12} + B_{21}}{2} \tag{4.38}$$

The optimum values of the real and imaginary parts of output admittance $Y_{out}^o = G_{out}^o + jB_{out}^o$ are written as

$$G_{out}^o = G_{22} - \frac{(G_{21} + G_{12})^2 + (B_{21} - B_{12})^2}{4G_{11}} \tag{4.39}$$

$$B_{out}^o = B_{22} + \frac{G_{21} - G_{12}}{B_{21} - B_{12}}\left(\frac{G_{21} + G_{12}}{2} + G_{22} - G_{out}^o\right) + \frac{B_{21} + B_{12}}{2} \tag{4.40}$$

Thus, in the steady-state stationary oscillation mode with admittance $Y_3 = G_L + jB_3$, where $G_L = 1/R_L$ and R_L is the load resistance, Equation (4.34) can be written in the form of the following amplitude and phase balance conditions:

$$G_{out}^o + G_L = 0 \tag{4.41}$$

$$B_{out}^o + B_3^o = 0 \tag{4.42}$$

The output power can be calculated using Equation (4.41) according to

$$P_{out} = \frac{1}{2}V_L^2 G_L \tag{4.43}$$

where V_L is the voltage amplitude across the load, which should be considered as the voltage amplitude across the output and common terminals of the active device.

4.3.2 Optimum MOSFET oscillator

The electrical behaviour of the MOSFET device in the high frequency ranges can be described with sufficient accuracy by the equivalent circuit shown in Figure 4.8 [18]. According to this equivalent circuit, the transistor Y-parameters in common source small-signal operation are written as

$$Y_{11} = \frac{j\omega C_{gs}}{1 + j\omega C_{gs}R_{gs}} + j\omega C_{gd} \qquad Y_{12} = -j\omega C_{gd}$$
$$Y_{21} = \frac{g_m}{1 + j\omega C_{gs}R_{gs}} - j\omega C_{gd} \qquad Y_{22} = \frac{1}{R_{ds}} + j\omega(C_{ds} + C_{gd}) \tag{4.44}$$

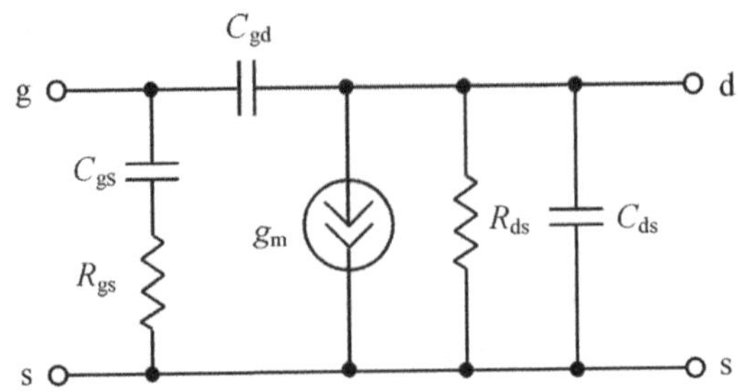

Figure 4.8 Small-signal intrinsic MOSFET model

By substituting the expressions for real and imaginary parts of Y-parameters from Equation (4.44) into Equations (4.37) and (4.38), the optimum values of the feedback susceptances B_1^o and B_2^o expressed through the parameters of the MOSFET equivalent circuit can be calculated as

$$B_1^o = \frac{g_m}{2\omega C_{gs} R_{gs}} \tag{4.45}$$

$$B_2^o = -\omega C_{gd} - \frac{g_m}{2\omega C_{gs} R_{gs}} \tag{4.46}$$

According to Equations (4.45) and (4.46), the optimum values of the susceptances B_1^o and B_2^o should be capacitive and inductive, respectively. Taking into account that the value of the feedback capacitance C_{gd} is sufficiently small and variations of the transition frequency $\omega_T = g_m/C_{gs}$ and gate–source resistance R_{gs} in large-signal operation are not significant, Equations (4.45) and (4.46) give a close approximation for the feedback elements in a steady-state oscillation mode based on the small-signal parameters of the MOSFET device at the operating bias point. The same conclusion is valid for the optimum oscillator using a JFET device, the large-signal behaviour of the elements of the equivalent circuit of which is similar to that of the MOSFET device [19].

The optimum values of the real and imaginary parts of the output admittance Y_{out}^o can be obtained as

$$G_{out}^o = \frac{1}{R_{ds}} - \frac{1}{R_{gs}} \left(\frac{g_m}{2\omega C_{gs}} \right)^2 \tag{4.47}$$

$$B_{out}^o = \omega C_{gd} - \frac{g_m}{2\omega C_{gs} R_{gs}} \left(1 - \frac{1}{\omega C_{gs} R_{gs}} \frac{g_m}{2\omega C_{gs}} \right) \tag{4.48}$$

From Equation (4.48) it follows that the absolute value of the negative conductance G_{out}^o reduces as the frequency increases. At the maximum oscillation frequency f_{max}, the negative conductance becomes zero. Then, the expression for f_{max} is given by

$$f_{max} = \frac{g_m}{4\pi C_{gs}} \sqrt{\frac{R_{ds}}{R_{gs}}} \tag{4.49}$$

that coincides with well-known expression for maximum frequency, at which the unilateral power gain of the FET power amplifier becomes equal to unity. The value of the optimum output susceptance B_{out}^o, positive or negative, depends in an optimal oscillator mostly on the values of the device transconductance g_m and gate time constant $\tau_{gs} = R_{gs} C_{gs}$.

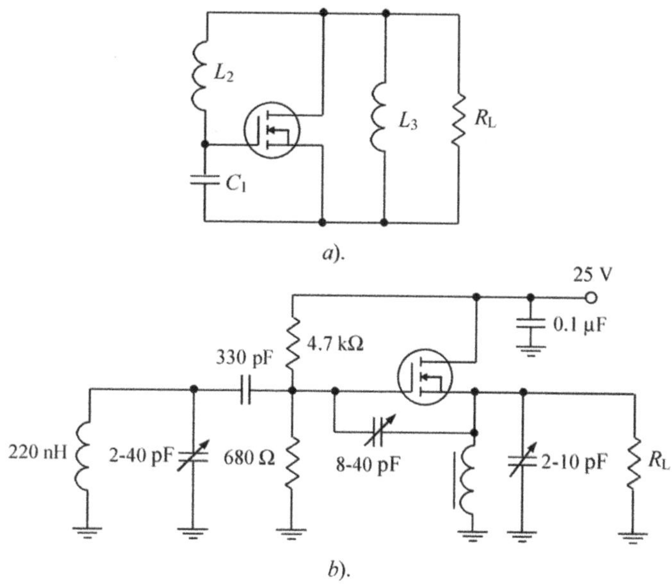

Figure 4.9 Schematics of MOSFET oscillators

As an example, consider a 200-MHz optimum MOSFET power oscillator with the following electrical characteristics and small-signal parameters of the device equivalent circuit: transconductance $g_m = 30$ mA/V, $R_{gs} = 70\ \Omega$, $C_{gs} = 8$ pF, $C_{gd} = 0.4$ pF, $C_{ds} = 2$ pF, maximum drain current $I_{dmax} = 0.3$ A, transition frequency $f_T = 1$ GHz. According to Equations (4.45) and (4.46), the feedback elements of the optimum oscillator shown in Figure 4.9a must have the approximate values of $C_1^o = 17$ pF and $L_2^o = 36$ nH. In this case, the value of B_{out}^o is positive and can be represented by the output capacitance $C_{out}^o = 26$ pF. Consequently, to satisfy the phase balance condition, an inductive character of the load must be provided. In the case of the parallel inductor, its inductance is determined by

$$L_3^o = -\frac{1}{\omega B_3^o} = \frac{1}{\omega B_{out}^o} = \frac{1}{\omega^2 C_{out}^o}$$

For a common drain MOSFET oscillator with varactor wideband tuning, it is easy to achieve the wideband tuning of the oscillation frequency using a variable capacitor connected in parallel to a feedback inductor. The equivalent circuit of such an oscillator is shown in Figure 4.9b. By tuning the parallel resonant circuit capacitor in the capacitance range from 2 to 40 pF, it is possible to easily tune the oscillator in the frequency range from 50 to 200 MHz with output power of 100–200 mW delivered to the load $R_L = 200\ \Omega$. Maximum flatness of the amplitude characteristic can be achieved by choosing a proper value of the feedback capacitances. This can be accomplished by tuning the variable capacitors in the capacitance ranges of 2–10 pF and 8–40 pF, respectively.

The oscillator performance is invariant to the point of grounding, which is considered as a common point, provided there are no changes in connection of the feedback elements and the load. This occurs, for example, when the load for the oscillators with a common source configuration shown in Figure 4.10a or with a common drain configuration shown

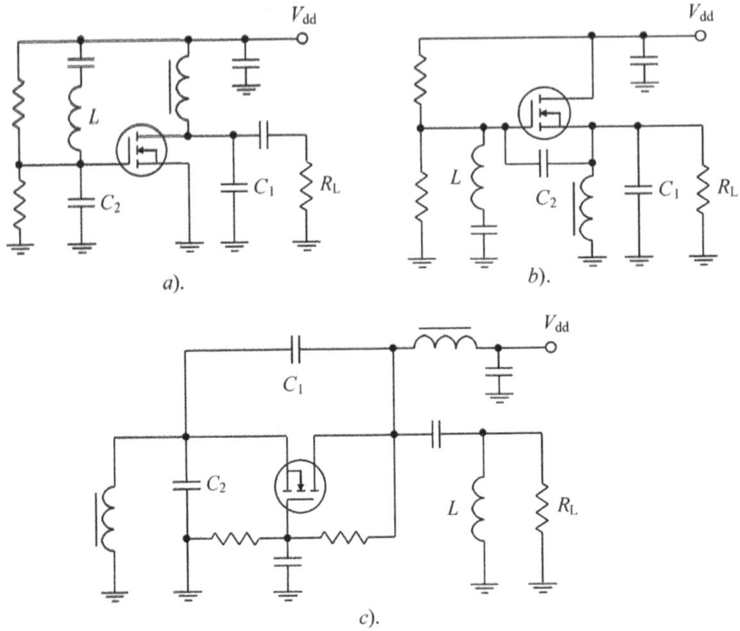

Figure 4.10 MOSFET oscillators with different grounded terminals

in Figure 4.10b is connected between the drain and the source, having in each case a direct ground connection. However, when the gate is grounded, the load has no connection with the ground. Therefore, for a common gate oscillator shown in Figure 4.10c, when the load is connected between the drain and the gate, the electrical characteristics — such as output power, frequency spectrum or circuit stability — can change because of the different effect of the intrinsic MOSFET device parameters on the amplitude and phase oscillation conditions. The attractiveness of using the common drain or common gate oscillator is attributable to the schematic simplicity, realizing wideband frequency tuning. The wide frequency tuning in a common source oscillator is limited by a negative influence of the source lead inductance and the inconvenience of the parallel feedback circuit implementation, especially at higher frequencies.

4.4 SERIES FEEDBACK BIPOLAR OSCILLATORS

4.4.1 Optimum oscillation condition

The steady-state single-frequency oscillation condition for a series feedback oscillator shown in Figure 4.7b can be written as

$$Z_{out}(I, \omega) + Z_L(\omega) = 0 \tag{4.50}$$

where I is the load current amplitude, $Z_{out}(I, \omega) = R_{out}(I, \omega) + jX_{out}(I, \omega)$ is the equivalent one-port network output impedance, $Z_L(\omega) = R_L(\omega) + jX_L(\omega)$ is the frequency-dependent load impedance. The output impedance Z_{out} can be expressed through the transistor equivalent circuit and feedback parameters by

$$Z_{out} = Z_{22} + Z_2 - \frac{(Z_{12} + Z_2)(Z_{21} + Z_2)}{Z_{11} + Z_2 + Z_1} \tag{4.51}$$

According to the optimum criterion given by Equation (4.29) and written in impedance form

$$\frac{\partial R_{out}}{\partial X_1} = 0 \qquad \frac{\partial R_{out}}{\partial X_2} = 0 \tag{4.52}$$

the optimum values X_1^o, X_2^o depend on the impedance parameters of the active two-port network as

$$X_1^o = \frac{R_{21} - R_{12}}{X_{21} - X_{12}} \left(\frac{R_{12} + R_{21}}{2} - R_{11} - R_1 \right) - X_{11} + \frac{X_{12} + X_{21}}{2} \tag{4.53}$$

$$X_2^o = \frac{(2R_2 + R_{12} + R_{21})(R_{21} - R_{12})}{2(X_{21} - X_{12})} - \frac{X_{12} + X_{21}}{2} \tag{4.54}$$

By substituting X_1^o and X_2^o into Equation (4.51), the optimum real and imaginary parts of the output impedance $Z_{out}^o = R_{out}^o + jX_{out}^o$ can be defined by

$$R_{out}^o = R_2 + R_{22} - \frac{(2R_2 + R_{12} + R_{21})^2 + (X_{21} - X_{12})^2}{4(R_{11} + R_2 + R_1)} \tag{4.55}$$

$$X_{out}^o = X_2^o + X_{22} - \frac{R_{21} - R_{12}}{X_{21} - X_{12}} \left(R_{out}^o - R_2 - R_{22} \right) \tag{4.56}$$

Thus, in a steady-state operation mode, Equation (4.50) can be written in the form of the following amplitude and phase balance conditions:

$$R_{out}^o + R_L = 0 \tag{4.57}$$

$$X_{out}^o + X_L^o = 0 \tag{4.58}$$

The ratio of the voltage $\boldsymbol{V}$ between output and common terminal of the transistor and the load current $\boldsymbol{I}$ is expressed through the transistor Z-parameters and oscillator feedback elements by

$$\boldsymbol{I} = \frac{Z_{11} + Z_1 + Z_2}{Z_{22}(Z_{11} + Z_1 + Z_2) - Z_{21}(Z_{12} + Z_2)} \boldsymbol{V} \tag{4.59}$$

As a result, by using Equation (4.59), one can easily calculate the maximum output power of the oscillator $P_{out} = 0.5I^2 \mathrm{Re} Z_L$ delivered to the load in a steady-state operation mode.

4.4.2 Optimum common base oscillator

Figure 4.11 shows the generalized two-port circuit of the series feedback bipolar oscillator commonly used with RF and microwave oscillator design. The equivalent Gummel–Poon SPICE model of the bipolar transistor, providing sufficiently accurate simulation the device dc and high-frequency behaviour up to its transition frequency $f_T = g_m/2\pi C_\pi$, is also shown [20, 21].

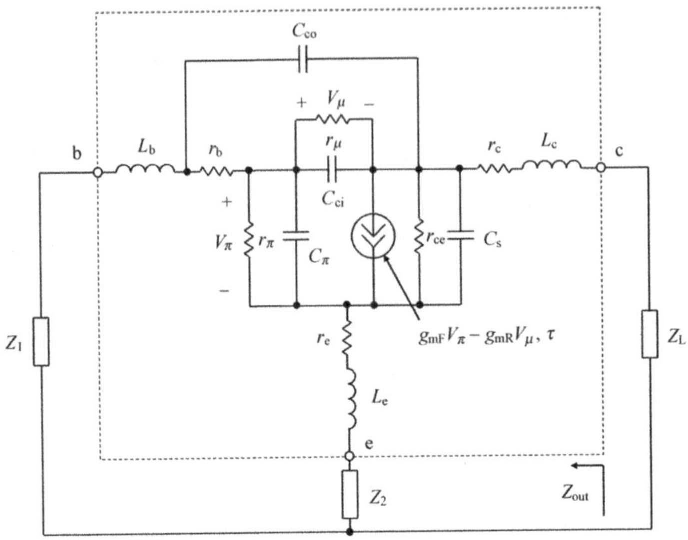

Figure 4.11 Generalized bipolar oscillator series feedback circuit

To evaluate analytically the start-up and steady-state oscillation conditions, it is best to simplify the transistor equivalent circuit. Figure 4.12 shows another, simplified version of the bipolar transistor equivalent circuit, where $C_c = C_{co} + C_{ci}$, $r_{b1} = r_b C_{ci}/C_c$, $r_{b2} = r_b C_{co}/C_c$. This equivalent circuit becomes possible because the condition $r_b \ll (C_{ci} + C_{co})/\omega C_{ci} C_{co}$ exists over the entire frequency range up to a maximum oscillation frequency f_{max}. In this case, it is advisable to ignore the effect of a base-width modulation (Early effect), so that the value of the resistance r_{ce} can be assumed as infinite. The parasitic lead inductances and resistances as well as substrate capacitance can be considered amongst the external feedback circuit. The condition $r_\pi \gg 1/\omega C_\pi$ simplifies the analytical and numerical calculations substantially without a significant decrease in accuracy at high frequencies.

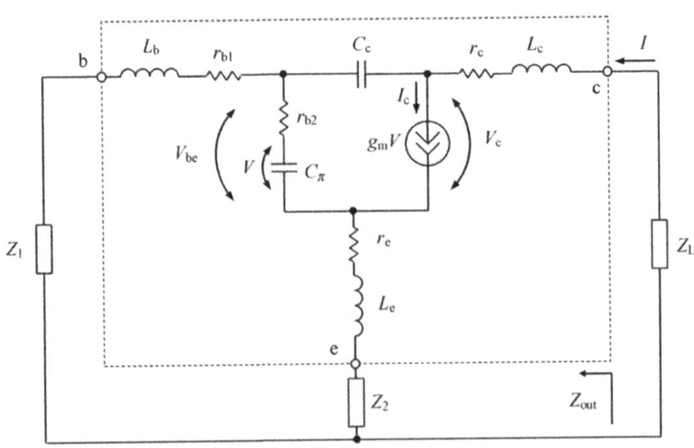

Figure 4.12 Simplified bipolar oscillator series feedback circuit

As a result, the following real and imaginary parts of the impedance Z-parameters will characterize the intrinsic bipolar transistor in a common emitter small-signal operation:

$$R_{11} = R_{12} = a\left[\frac{1}{g_m} + r_{b2}\left(\frac{\omega}{\omega_T}\right)^2\right]$$
$$R_{21} = R_{22} = R_{11} + \frac{a}{\omega_T C_c}$$
$$X_{11} = X_{12} = -a\frac{\omega}{\omega_T}\left(\frac{1}{g_m} - r_{b2}\right) \tag{4.60}$$
$$X_{21} = X_{11} + \frac{a}{\omega C_c}$$
$$X_{22} = X_{11} - \frac{a}{\omega_T C_c}\left(\frac{\omega}{\omega_T}\right)^2$$

where

$$a = 1\Bigg/\left[1 + \left(\frac{\omega}{\omega_T}\right)^2\right]$$

Let us consider the transistor parasitic series resistances and lead inductances among the external feedback elements Z_1, Z_2 and Z_L. In addition, loss in the feedback elements should be taken into account in elements r_{b1} and r_e. As a result, the optimum values of imaginary parts of the feedback elements X_1^o and X_2^o, expressed through the parameters of the bipolar transistor equivalent circuit, can be calculated as

$$X_1^o = \frac{1}{2\omega C_c} - r_{b1}\frac{\omega}{\omega_T} \tag{4.61}$$

$$X_2^o = -\frac{1}{2\omega C_c} - (r_{b2} + r_c)\frac{\omega}{\omega_T} \tag{4.62}$$

The real and imaginary parts of the optimum output impedance $Z_{out}^o = R_{out}^o + jX_{out}^o$ with respect to the collector series resistor r_c and lead inductance L_c are

$$R_{out}^o = r_c + \frac{1}{r_{b1} + r_e + R_{11}}\left[r_{b1}\left(r_e + R_{11} + \frac{1}{a\omega_T C_c}\right) - \frac{1}{a}\left(\frac{1}{2\omega C_c}\right)^2\right] \tag{4.63}$$

$$X_{out}^o = \omega L_c - \frac{1}{2\omega C_c} + \left(R_{out}^o - r_c\right)\frac{\omega}{\omega_T} \tag{4.64}$$

From Equation (4.64) it follows that, as frequency increases, the absolute value of the negative resistance R_{out}^o reduces and becomes zero at maximum oscillation frequency f_{max}. Without accounting for the parasitic parameters of the bipolar transistor equivalent circuit, the expression for f_{max} is

$$f_{max} = \sqrt{\frac{f_T}{8\pi r_{b1} C_c}} = \sqrt{\frac{f_T}{8\pi r_b C_{ci}}} \tag{4.65}$$

This coincides with the well-known expression for a maximum oscillation frequency f_{max} of the bipolar transistor, for which a maximum available power gain becomes equal to unity and a steady-state oscillation condition can be established only for lossless feedback elements. Such a condition corresponds to the three-terminal mode of transistor operation. However, the circuit may be capable of oscillating at frequencies greater than the f_{max} obtained by Equation (4.65), when it is operated in two-terminal mode, by making Z_1 an open circuit. In this configuration, no RF current is flowing into the base terminal to produce negative resistance (though the base terminal can still be used for biasing purposes). The transistor is now behaving as a two-terminal negative resistance, and the extrinsic base resistance does not directly affect the maximum oscillation frequency, which will be determined by the parasitic resistances in the emitter and collector circuits [22].

4.4.3 Quasilinear approach [23]

The oscillator design procedure generally involves at least three steps: a choice of an appropriate circuit topology; the determination of the transistor small- and large-signal characteristics; and the optimization of the oscillator circuit to achieve the desired performance. As mentioned in Chapter 1, the derivation of equivalent linear elements in terms of signal voltages is based on static voltage–ampere and voltage–capacitance bipolar transistor characteristic. It is enough to be limited to the main nonlinear elements g_m, ω_T and C_c, because the base resistance r_b poorly depends on the bias conditions. In addition, to calculate the dc and fundamental collector currents, it is possible to use a linear approximation of transition frequency ω_T and the small-signal value C_c at an operating point. This bipolar transistor model does not represent the effect of the forward-rectified current across the collector–base junction and the collector voltage-breakdown phenomenon under the assumption of sufficiently low collector–emitter bias voltage value. In that case, it is assumed that the soft build-up of the self-sustained oscillations is mainly caused by the nonlinear characteristic of the fundamentally averaged large-signal transconductance g_{m1}.

The following exponential model can describe the bipolar transistor transfer I–V characteristic when the base–emitter current is sufficiently small:

$$i_c = I_{sat}\left[\exp\left(\frac{v_b}{V_T}\right) - 1\right] \tag{4.66}$$

where I_{sat} is the reverse saturation current and V_T is the temperature voltage. If our interest is restricted to the fundamental frequency, then $v_b = V_b - V_e - I_{c0}R_e + V\cos\omega t$, where I_{c0} is the dc collector current, V_b is the base bias voltage, V_e is the emitter bias voltage, R_e is the self-bias resistor. The voltage drops across the collector and emitter series resistors r_c and r_e, are negligible.

The values of the large-signal transconductance g_{m1} and the dc collector current I_{c0} as the functions of the fundamental voltage amplitude V across the capacitance C_π are defined by

$$g_{m1} = \frac{2I_{sat}}{V}\exp\left(\frac{V_b - V_c - I_{c0}R_c}{V_T}\right) I_1\left(\frac{V}{V_T}\right) \tag{4.67}$$

$$I_{c0} = I_{sat}\left[I_0\left(\frac{V}{V_T}\right)\exp\left(\frac{V_b - V_e - I_{c0}R_e}{V_T}\right) - 1\right] \tag{4.68}$$

where $I_0(V/V_T)$, $I_1(V/V_T)$ are modified first-order Bessel functions of the first kind.

Table 4.1 Microwave bipolar transistor equivalent circuit parameters

C_{co} (pF)	$C_{ci}(V_{\mu} = 0V)$ (pF)	$C_{\pi}(V_{\pi} = 0V)$ (pF)	L_e (nH)	L_b (nH)	L_c (nH)	r_e (Ω)	r_b (Ω)	r_c (Ω)	f_T (GHz)
0.35	0.34	2.0	0.3	0.3	0.5	0.3	4.0	1.75	6.0

An analytic ratio between the voltage amplitudes V_{be} and V is obtained by

$$V_{be} = V\sqrt{1 + \left(\frac{\omega}{\omega_T} r_{b2} g_{m1}\right)^2} \tag{4.69}$$

Equation (4.67) implies that the large-signal transconductance g_{m1} is maximal in the absence of the RF signal and, near $V = 0$, takes the small-signal value of

$$g_m = \frac{I_{sat}}{V_T} \exp\left(\frac{V_b - V_e - I_{c0}R_e}{V_T}\right) \tag{4.70}$$

For a steady-state oscillation mode, the analytic relationships between the amplitude I of the fundamental output current flowing into the load and collector voltage fundamental amplitude V_c as functions of the voltage amplitude V_{be} expressed by the feedback and transistor Z-parameters can be written in the form

$$I = \frac{Z_{11} + Z_2 + Z_1}{Z_{11}Z_2 - Z_{12}(Z_2 + Z_1)} V_{be} \tag{4.71}$$

$$V_c = \frac{Z_{22}(Z_{11} + Z_2 + Z_1) - Z_{21}(Z_2 + Z_{12})}{Z_{12}(Z_1 + Z_2) - Z_{11}Z_2} V_{be} \tag{4.72}$$

As a result, in a steady-state operation mode, the output power of the oscillator as a function of the input voltage amplitude V and parameters of the bipolar transistor equivalent circuit can be written as

$$P_{out} = -\frac{a g_{m1}^2 (r_{b1} + r_e + R_{11}) R_{out}^o}{r_{b1} + r_c - R_{out}^o} \frac{V^2}{2} \tag{4.73}$$

This analytical approach was applied to microwave bipolar oscillator design with the transistor equivalent circuit parameters listed in Table 4.1. The numerical calculations were performed for oscillation frequency 4 GHz with bias voltages $V_b = 0$ V and $V_e = -2$ V, collector supply voltage $V_{cc} = 9$ V, self-bias resistor $R_e = 100$ Ω, maximum dc collector current $I_{c0max} = 50$ mA and reverse saturation current $I_{sat} = 10$ μA.

Figure 4.13 shows the amplitude dependencies of the large-signal transconductance g_{m1}, dc collector current I_{c0} and fundamental collector voltage amplitude V_c. From Figure 4.13 it follows that, with the increase of the base–emitter amplitude V, the large-signal transconductance g_{m1} decreases significantly, whereas the values of V_c and I_{c0} increase practically linearly, beginning with sufficiently small values.

Figure 4.14 shows the amplitude dependencies of the optimum feedback parameters X_1^o and X_2^o, real and imaginary parts of the output resistance R_{out}^o and X_{out}^o, and output power P_{out}. As the negative resistance R_{out}^o reduces with the increase of the amplitude V, the stable oscillations are established in the oscillator. In that case, the value of the large-signal transconductance g_{m1} is significantly reduced. From Figure 4.14 it follows that the output power P_{out} increases monotonically with increase of amplitude V. Consequently, the maximum output power in the

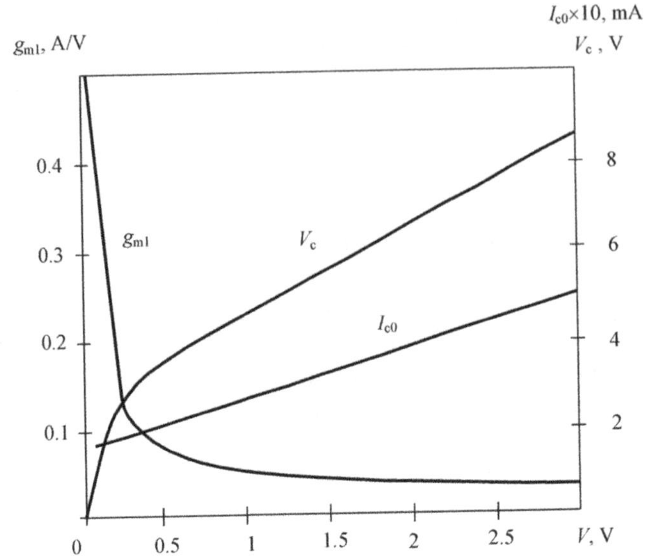

Figure 4.13 Amplitude dependencies of g_{m1}, I_{c0} and V_c

bipolar self-bias oscillator is realized under some compression condition due to the saturation effect when the collector–base junction is forward-biased. Therefore, to realize maximum output power in a linear operation region without saturation effect, it is necessary to choose the proper load resistance or use a diode. This microwave bipolar transistor model does not include a collector–base diode because such an operation region is inappropriate for the oscillator

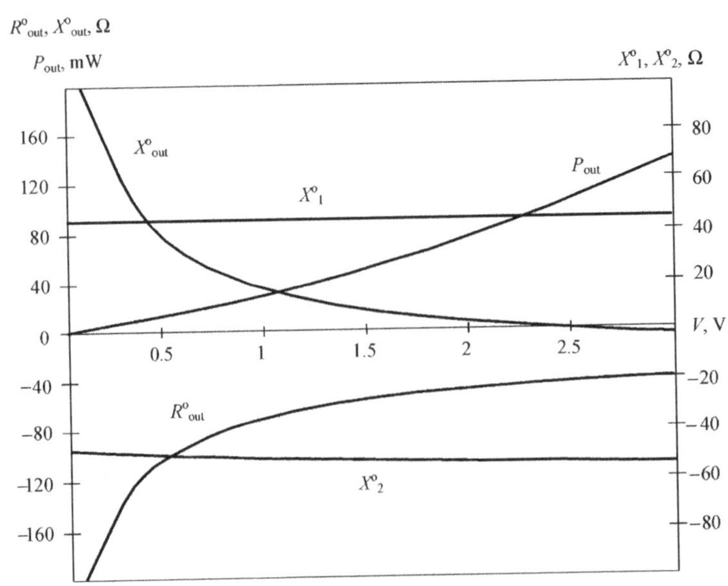

Figure 4.14 Amplitude dependencies of output power and feedback elements

because of the significant deterioration of its noise properties. The amplitude restriction at the numerical calculations is taken into consideration by means of the condition $V_c \leq V_{cc}$.

Numerical calculations show that the values of V_c and I_{c0} increase linearly with increase of the base–emitter junction amplitude V, beginning with sufficiently small values. Let us consider the influence of the emitter self-bias resistor R_e on the character of amplitude dependence of the fundamental collector current amplitude I_c, which can be written as

$$I_c = 2I_{sat}\left[\exp\left(\frac{V_b - V_e - I_{c0}R_e}{V_T}\right) I_1\left(\frac{V}{V_T}\right)\right] \tag{4.74}$$

In view of

$$\frac{dI_0\left(\frac{V}{V_T}\right)}{d\left(\frac{V}{V_T}\right)} = I_1\left(\frac{V}{V_T}\right) \tag{4.75}$$

Equation (4.68) can be rewritten, being differentiated as

$$\frac{dI_{c0}}{dV} = \frac{I_{c0} + I_{sat}}{V_T + (I_{c0} + I_{sat})\,R_e}\,\frac{I_1\left(\frac{V}{V_T}\right)}{I_0\left(\frac{V}{V_T}\right)} \tag{4.76}$$

From the definition of the first-order modified Bessel functions it follows that, for $(V/V_T) \geq 5$, the following inequality is satisfied:

$$0.9 \leq \frac{I_1\left(\frac{V}{V_T}\right)}{I_0\left(\frac{V}{V_T}\right)} \leq 1 \tag{4.77}$$

Hence, from Equation (4.76) it follows that the value of the dc collector current I_{c0} varies practically linearly with increase in V under the condition $V \geq 5V_T$. The slope of the dependence $I_{c0}(V)$ is determined by the value of R_e. Substituting Equation (4.68) into Equation (4.74) yields

$$I_c(V) = 2\left[I_{c0}\,(V) + I_{sat}\right]\frac{I_1\left(\frac{V}{V_T}\right)}{I_0\left(\frac{V}{V_T}\right)} \tag{4.78}$$

From Equations (4.76) and (4.78) it follows that, in the bipolar oscillator with emitter self-bias resistor, the collector constant bias current I_{c0} and the fundamental output current amplitude I_c depend linearly on the base–emitter voltage amplitude V, beginning from $V/V_T \geq 5$. As a result, the linearizing influence of the emitter self-bias resistor is similar to the effect of the negative feedback resistor on linearization of the power amplifier transfer characteristics, contributing to phase noise improvement.

In addition, sufficiently high output power levels can be achieved without use of special load matching circuit, when a standard load $R_L = 50\ \Omega$ is directly connected. Furthermore, on the basis of optimum bipolar oscillator, it is easy to realize wideband varactor-tuned microwave oscillator when the varactors are placed both in the base and emitter circuits.

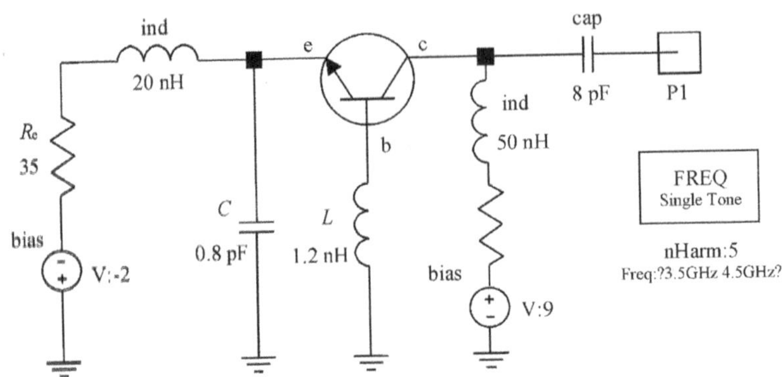

Figure 4.15 Simulated series feedback bipolar oscillator equivalent circuit

4.4.4 Computer-aided design [24]

One of the most popular approaches to nonlinear free-running oscillator analysis is to use the harmonic balance equations and to consider the oscillation frequency as an additional optimization variable. Such an algorithm is used in a Microwave Harmonica that is a part of the computer circuit simulator Serenade 7.5 [25]. Let us verify the accuracy of an analytic approach for the series feedback microwave bipolar oscillator shown in Figure 4.15. For a preliminary chosen oscillation frequency $f = 4$ GHz, the optimum oscillator feedback parameters according to the theoretical predictions given by Equations (4.61) and (4.62) must be equal to $L = 1.2$ nH and $C = 0.8$ pF, respectively, for the transistor equivalent circuit parameters listed in Table 4.1.

The starting oscillation conditions were determined by sweeping the frequency f of the external test source injecting signal into the oscillator circuit from 3.5 to 4.5 GHz. The curves satisfy the starting oscillation conditions under linear small-signal operation where $\mathrm{Re}(I_1) < 0$ and $\mathrm{Im}(I_1) = 0$ at 4.24 GHz, as shown in Figure 4.16.

Figures 4.17 and 4.18 show the oscillation frequency and output power in a steady-state mode for various values of the emitter bias resistor R_e, load resistance R_L and

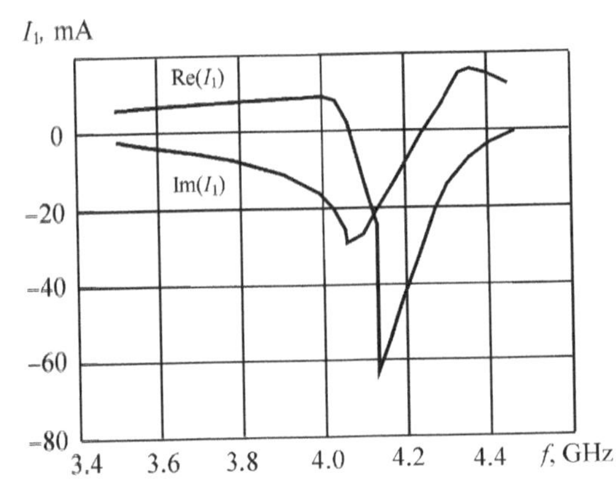

Figure 4.16 Simulated start-up oscillation conditions

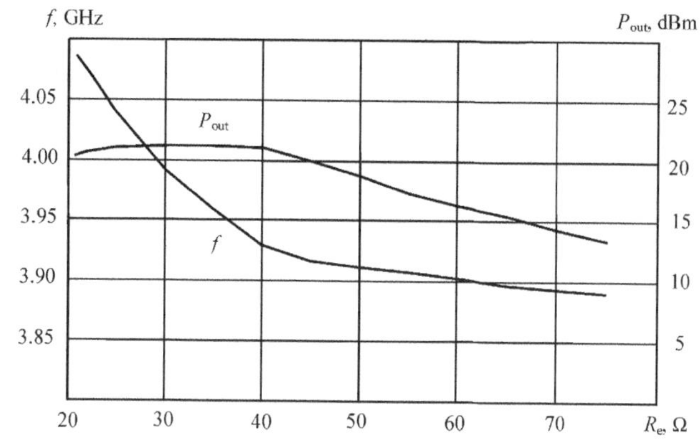

Figure 4.17 Output power and oscillation frequency versus emitter bias resistance

feedback elements L and C. Despite preliminary simplification, the experimental results indicate minimal discrepancy between the chosen oscillation frequency and simulated oscillation frequency using an analytic calculation of the oscillator feedback parameters (taking into account the standard load resistance $R_L = 50\ \Omega$). The exact value of the oscillation frequency $f = 4$ GHz in a steady-state mode is realized at $R_e = 28\ \Omega$ when the output power P_{out} is close to the maximum value, as shown in Figure 4.17. Moreover, the output power $P_{out} = 21.5$ dBm is very close to a maximum value of 21.9 dBm at optimal load $R_L^o = 45\ \Omega$ (Figure 4.18).

To verify the validity of the analytically calculated feedback parameters $L = 1.2$ nH and $C = 0.8$ pF, circuit simulations with other sets of feedback parameters that satisfy the oscillation conditions at the resonant frequency $f = 4$ GHz were performed. From the simulation results it follows that a deviation of the feedback parameters from an optimal set leads first to a deterioration of the phase conditions and then to oscillation failure (the broken lines in Figure 4.19 define the borders of the oscillation region). The dc collector current did not exceed the value of 32 mA under the simulation procedure. Hence, the theoretical small-signal predictions for the oscillator feedback elements based on an analytic optimum approach provide

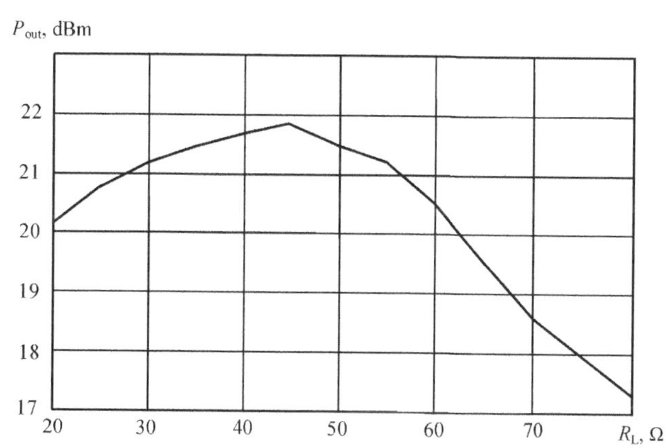

Figure 4.18 Output power versus load resistance

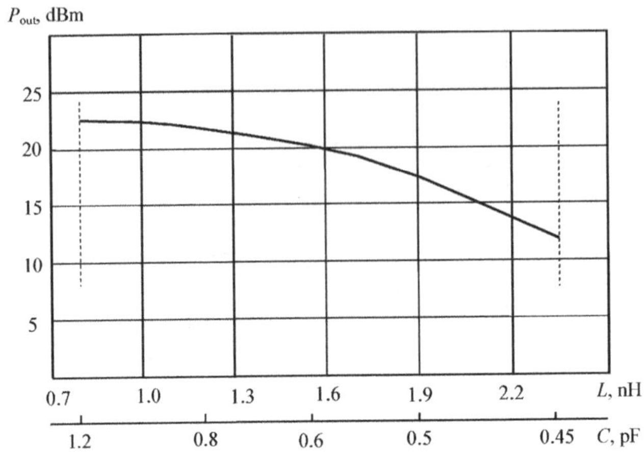

Figure 4.19 Output power versus oscillator feedback parameters

values of the oscillation frequency and maximum output power very close to their simulated values under large-signal operation, thus simplifying and significantly shortening the entire oscillator design procedure.

4.5 SERIES FEEDBACK MESFET OSCILLATORS

4.5.1 Optimum common gate oscillator

A common gate MESFET oscillator configuration with series feedback between the gate and the ground is shown in Figure 4.20. Such a circuit configuration was selected because of its inherent broadband negative resistance. If the correct feedback reactance is added, oscillations can occur from very low frequencies to approximately maximum oscillation frequency f_{max}.

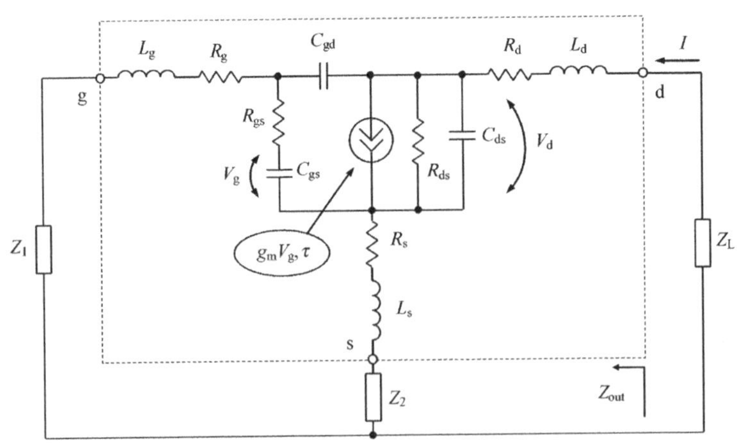

Figure 4.20 Series feedback MESFET oscillator equivalent circuit

Figure 4.20 also shows the small-signal MESFET equivalent circuit, which characterizes with good accuracy the device performance up to 50 GHz [26].

The internal GaAs MESFET device (without taking into account the lead inductances and parasitic series resistances) in a common source small-signal operation is characterized by the following admittance Y-parameters:

$$Y_{11} = \frac{j\omega C_{gs}}{1 + j\omega C_{gs} R_{gs}} + j\omega C_{gd} \qquad Y_{12} = -j\omega C_{gd}$$
$$Y_{21} = \frac{g_m \exp(-j\omega\tau)}{1 + j\omega C_{gs} R_{gs}} - j\omega C_{gd} \qquad Y_{22} = \frac{1}{R_{ds}} + j\omega(C_{ds} + C_{gd}) \tag{4.79}$$

where τ is the transit time in the MESFET channel.

For analytical calculation of the series feedback FET oscillator parameters, it is more convenient to use the impedance Z-parameters in a common source configuration by applying a standard Y- to Z-parameters transformations given by Table 2.1 (Chapter 2). As a result, the optimum values of the imaginary parts of the feedback elements and output impedance X_1^o, X_2^o and X_{out}^o, expressed through the parameters of the MESFET equivalent circuit with $\tau = 0$, can be obtained by [15]

$$X_1^o = -(R_1 + R_g)\frac{b}{a} + \frac{1}{a}\left[\frac{1}{\omega C_{gs}} - R_{ds}\left(\omega R_{gs} C_{ds} - \frac{g_m}{2\omega C_{gs}}\right)\right] \tag{4.80}$$

$$X_2^o = -(R_2 + R_s)\frac{b}{a} + \frac{R_{ds}}{a}\left(\omega R_{gs} C_{gd} - \frac{g_m}{2\omega C_{gs}}\right) \tag{4.81}$$

$$X_{out}^o = -R_{out}^0 \frac{b}{a} + \frac{R_{ds}}{a}\frac{g_m}{2\omega C_{gs}} \tag{4.82}$$

where

$$a = 1 + \frac{C_{gd}}{C_{gs}}(1 - \omega^2 R_{gs} R_{ds} C_{gs} C_{ds}) + R_{ds} C_{gd} \frac{g_m}{C_{gs}}$$

$$b = \omega R_{ds}(C_{ds} + C_{gd}) + \omega \frac{C_{gd}}{C_{gs}}(R_{ds} C_{ds} + R_{gs} C_{gs}).$$

The obtained analytic expression to calculate the optimum value of the output resistance R_{out}^o is rather bulky. However, in some cases and in particular when neglecting the capacitance C_{gd}, it becomes simpler and can be written as

$$R_{out}^o = R_s + R_2 + \frac{R_{ds}}{1 + (\omega R_{ds} C_{ds})^2}\left[1 - \frac{R_{ds}}{R_g + R_s + R_1 + R_2 + R_{gs}}\left(\frac{g_m}{2\omega X_{gs}}\right)^2\right] \tag{4.83}$$

For a GaAs MESFET oscillator with the gate length of less than 1 μm, such a simplification allows us to calculate the oscillation frequency and output power in a frequency range up to 18 GHz with good accuracy compared with the experimental results [14, 15].

4.5.2 Quasilinear approach [15]

The theoretical and experimental results show that, to develop a sufficiently accurate nonlinear model of the MESFET device, it is enough to be limited to the nonlinear behaviour of four major elements. They are the gate capacitance C_{gs}, the gate charging resistance R_{gs}, the drain–source differential resistance R_{ds} and the transconductance g_m. In a simplified case, such a MESFET model does not represent the effect of the forward-rectified current across the gate–source junction and the voltage–breakdown phenomenon in the gate–drain region under the assumption of low gate–source bias value and moderate gate–drain voltage when the gate forward current is negligibly small. From Equations (4.80–4.83) it follows that the part of the MESFET equivalent circuit elements represents some combinations that significantly simplifies the design procedure. For example, the charging time constant $\tau_g = R_{gs}C_{gs}$ can be considered to be a constant parameter. For modern GaAs MESFETs with gate length of the order of 1 μm or less, it is possible to assume the drift carrier velocity to be equal to the saturated velocity, resulting in a practically constant ratio g_m/C_{gs}.

The gate–source capacitance C_{gs} can be found from Schottky barrier theory as

$$C_{gs} = C_{gso} \Big/ \sqrt{1 - \frac{v_g}{\varphi}} \tag{4.84}$$

where C_{gso} is a zero-bias gate–source capacitance, φ is the built-in potential of the Schottky barrier. If our interest is restricted to the fundamental frequency only when $v_g = V_g + V \sin(\omega t + \phi)$, then the effective averaged large-signal gate–source capacitance C_{gs1} can be written as

$$C_{gs1} = \frac{C_{gso}}{\pi\sqrt{1 - \dfrac{V_g}{\varphi}}} \int_0^{2\pi} \frac{\cos^2\psi}{\sqrt{1 - \dfrac{V}{\varphi - V_g}\sin\psi}} d\psi \tag{4.85}$$

where $\psi = \omega t + \phi$.

To approximate the family of experimental GaAs MESFET output I–V curves, it is advisable to use the popular Materka model:

$$i_d = I_{dss}\left(1 - \frac{v_g}{V_p}\right)^2 \tanh\left(\frac{\alpha v_d}{v_g - V_p}\right) \tag{4.86}$$

where I_{dss} is the drain saturation current, $V_p = V_{po} + \gamma v_d$ is the gate pinch-off voltage, α, γ, V_{po} are the model parameters [27]. Further analytical calculations are significantly simplified for the case of $\gamma = 0$.

Assuming $v_d = V_{dd} + V_d \sin\omega t$ and applying a Taylor series expansion of Equation (4.86) in the vicinity of V_{dd} equating the fundamental frequency component terms, the large-signal differential drain–source resistance R_{ds1} as a function of the fundamental drain voltage amplitude V_d can be obtained by

$$R_{ds1} = \frac{R_{ds}}{1 + AV_d^2} \tag{4.87}$$

where

$$R_{ds} = \frac{\cosh^2\left(\dfrac{\alpha V_{dd}}{V_g - V_p}\right)}{I_{dss}\left(1 - \dfrac{V_g}{V_p}\right)^2} \frac{V_g - V_p}{\alpha}$$

$$A = \frac{3\tanh^2\left(\dfrac{\alpha V_{dd}}{V_g - V_p}\right) - 1}{4\left(\dfrac{V_g - V_p}{\alpha}\right)^2}$$

Equation (4.87) implies that the large-signal value of R_{ds1} is maximal in the absence of the oscillation signal near $V_d = 0$, becoming equal to its small-signal value of R_{ds} and reducing as the oscillation signal amplitude V_d increases. Consequently, according to such a simplified analytic approach, the values of the feedback elements, output impedance and output power are functions of the varying drain–source differential resistance under large-signal operation.

The relationship between the complex drain voltage $\boldsymbol{V}_d$ and output current $\boldsymbol{I}$ expressed through the MESFET device Z-parameters and oscillator feedback parameters can be written as

$$\boldsymbol{I} = \frac{Z_{11} + Z_1 + Z_2}{Z_{22}(Z_{11} + Z_1 + Z_2) - Z_{21}(Z_{12} + Z_2)} \boldsymbol{V}_d \tag{4.88}$$

Consequently, the output power of the oscillator $P_{out} = I^2 \mathrm{Re} Z_L/2$ in a steady-state operation mode, taking into account the drain series resistance R_d, can be calculated from

$$P_{out} = -\frac{1 + \left(\dfrac{R_{21} - R_{12}}{X_{21} - X_{12}}\right)^2}{(R_{22} + R)^2 + (X_{22} + X)^2} (R_{out} + R_d) \frac{V_d^2}{2} \tag{4.89}$$

where

$$R = \frac{X_{21}(X_{12} + X_2^o) - R_{21}(R_{12} + R_2 + R_s) - X_{22}(X_{11} + X_1^o + X_2^o)}{R_{11} + R_1 + R_2 + R_g + R_s}$$

$$X = \frac{R_{21}(X_{11} + X_1^o + X_2^o) - R_{21}(X_{12} + X_2^o) - X_{21}(R_{12} + R_2 + R_s)}{R_{11} + R_1 + R_2 + R_g + R_s}$$

The given analytic approach in a quasilinear approximation was applied to the microwave 12 GHz series feedback oscillator design using a GaAs MESFET device with the gate geometry of 0.8×300 μm.

The amplitude dependencies of the optimum feedback parameters X_1^o and X_2^o, large-signal drain–source differential resistance R_{ds1}, real and imaginary parts of the output resistance R_{out}^o and X_{out}^o and output power P_{out} are shown in Figures 4.21 and 4.22, respectively. Numerical calculations show that, under maximum output power, complete phase compensation with no need to use an additional external reactive element is realized. For optimum feedback elements

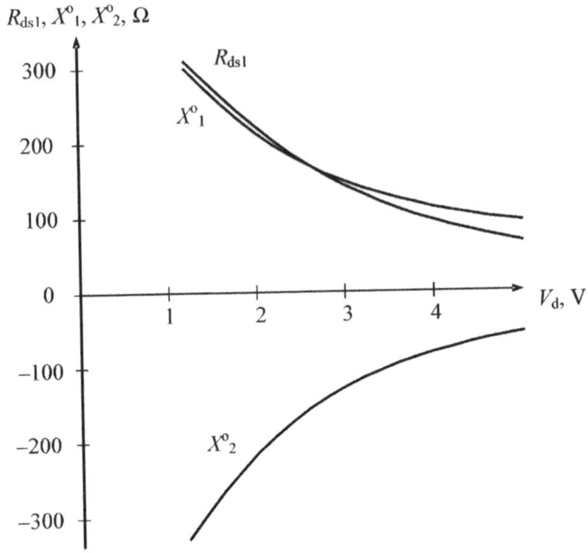

Figure 4.21 Amplitude dependencies of drain–source resistance and feedback parameters

$L = X_1^o/\omega = 1.3$ nH and $C = -1/X_2^o\omega = 0.2$ pF, the maximum output power level of 14 mW is achieved. It should be noted that, in contrast to the series feedback bipolar oscillator, the maximum output power of the series feedback MESFET oscillator is obtained in the active operation region without influence of the saturation effect, due to the different active device nonlinear model.

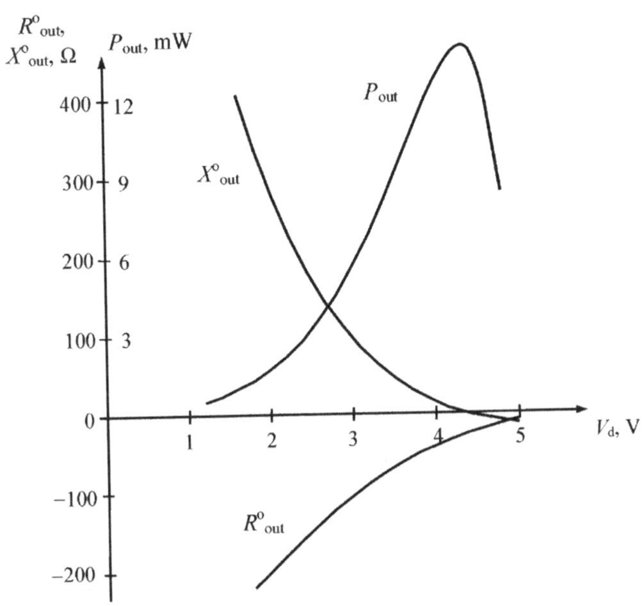

Figure 4.22 Amplitude dependencies of output power and output impedance components

4.5.3 Computer-aided design [28]

The results of microwave bipolar oscillator analytic design based on a quasilinear approach show the attractiveness and high effectiveness of preliminary analytical calculations of the oscillator feedback parameters according to the simple optimum analytical expressions. Therefore, one can assume that, to speed up the design process for microwave MESFET oscillators, a simplification of the analytical expressions for feedback elements is required. For example, influence of the parasitic series resistances, gate–drain capacitance C_{gd} and transit time τ can be ignored. Then, the optimum values of imaginary parts of the lossless feedback elements X_1^o, X_2^o and output reactance X_{out}^o can be obtained from Equations (4.80–4.82) as

$$X_1^0 = \frac{1}{\omega C_{gs}} + R_{ds}\left[-\omega C_{ds}(R_{gs} + R_g) + \frac{g_m}{2\omega C_{gs}}\right] \tag{4.90}$$

$$X_2^0 = -R_{ds}\left(\omega C_{ds} R_s + \frac{g_m}{2\omega C_{gs}}\right) \tag{4.91}$$

$$X_{out}^0 = -R_{ds}\left[\omega C_{ds} R_{out}^o + \frac{g_m}{2\omega C_{gs}}\right] \tag{4.92}$$

As a result, for an optimum series feedback MESFET oscillator, according to Equations (4.90) and (4.91), the optimum values of the reactances X_1^o and X_2^o should be inductive and capacitive, respectively. An analytical equation to calculate the optimum output resistance R_{out}^o in a small-signal operation mode can be written as

$$R_{out}^0 \cong R_s + \frac{R_{ds}}{1 + (\omega C_{ds} R_{ds})^2}\left[1 - \frac{R_{ds}}{R_g + R_s + R_{gs}}\left(\frac{g_m}{2\omega C_{gs}}\right)^2\right] \tag{4.93}$$

To determine the differential drain–source resistance R_{ds} as a function of the optimum output resistance R_{out}^o, it is enough to solve the quadratic equation for R_{ds} obtained from Equation (4.93). As a result,

$$R_{ds} = \frac{1 + \sqrt{1 - 4\left(R_{out}^o - R_s\right) G_{dso}}}{2G_{dso}} \tag{4.94}$$

where

$$G_{dso} = \frac{1}{R_g + R_s + R_{gs}}\left(\frac{g_m}{2\omega C_{gs}}\right)^2 + \left(R_{out}^o - R_s\right)(\omega C_{ds})^2$$

Let us verify the accuracy of the analytic approach using the power microwave MESFET ($l = 1$ μm, $w = 4 \times 200$ μm). To determine a large-signal value of the output resistance R_{out}^o for a certain value of the load resistance R_L, it is necessary to use the amplitude balance equation $R_{out}^o + R_d + R_L = 0$. In such a situation, R_{ds} is considered as a fundamentally averaged drain–source resistance R_{ds1} under large-signal operation. For preliminary defined oscillation frequency $f = 4$ GHz, the optimum oscillator feedback parameters — according to theoretical calculations — must be $L = 5.0$ nH and $C = 0.4$ pF, respectively (for load resistance $R_L = 50\ \Omega$). In this case, the value of 50 Ω is not strictly an exact optimum load resistance corresponding to a maximum available output power; to define an optimum value, it can be considered as a parameter under computer optimization. To satisfy the phase balance condition $X_{out}^o + \omega L_d + X_L = 0$, where L_d is the drain lead inductance, the value of X_L should be

Table 4.2 Small-signal parameters of GaAs MESFET equivalent circuit

L_g (pH)	L_s (pH)	L_d (pH)	R_g (Ω)	R_{gs} (Ω)	R_s (Ω)	R_d (Ω)	C_{gs} (pF)	C_{gd} (pF)	C_{ds} (pF)	g_m (mS)	τ (ps)
50.4	0.1	60.1	2.0	2.0	0.93	1.1	1.2	0.087	0.199	97.4	4.8

inductive with $L_L = 4.0$ nH. The small-signal parameters of the GaAs MESFET equivalent circuit are listed in Table 4.2 [26].

Using the circuit simulator Serenade 7.5, the nonlinear circuit simulation was performed for a microwave series feedback MESFET oscillator, whose equivalent circuit is shown in Figure 4.23.

The starting oscillation conditions were found by sweeping the frequency f of the external test source from 2 to 6 GHz. The curves satisfy starting oscillation conditions under linear small-signal operation where $\text{Re}(I_1) < 0$ and $\text{Im}(I_1) = 0$ at 4.4 GHz, as shown in Figure 4.24. In a steady-state operation mode, the value of the oscillation frequency becomes equal to 3.85 GHz. This is in a good agreement with the predicted theoretical value of 4 GHz. Neglecting the gate–drain capacitance C_{gd} and transit time τ leads to the inductive value of load reactance. In optimum oscillator, the maximum output power is realized under the conditions of complete phase compensation with zero load reactance.

For a 12 GHz MESFET oscillator which circuit schematic is shown in Figure 4.25, the optimum oscillator feedback parameters — according to the theoretical predictions — must be equal to $L = 0.35$ nH and $C = 0.5$ pF (Figure 4.25). In this case, to satisfy the phase balance condition, the value of the load reactance X_L should be capacitive of $C_L = 1$ pF. A simulated value of the oscillation frequency is 10.72 GHz, which differs from the theoretical value by only 11%.

Figure 4.26 shows the dependencies of the oscillation frequency f and output power P_{out} on the load resistance R_L. From this it follows that maximum output power $P_{out} = 22.9$ dBm can be realized for load values in limits of 20–30 Ω. The dc drain current did not exceed the value of 96 mA under the simulation procedure.

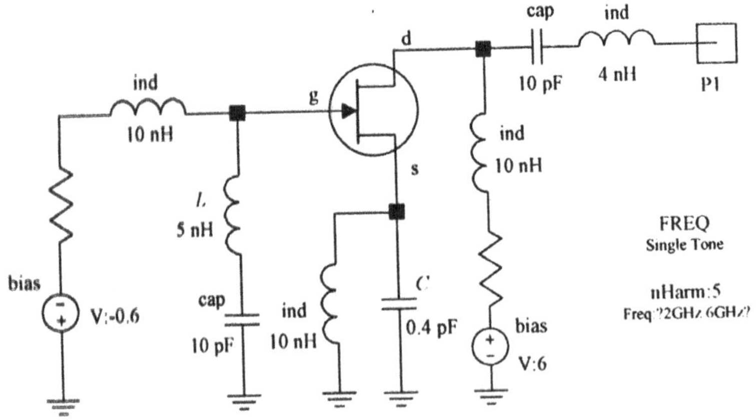

Figure 4.23 Simulated series feedback 4 GHz MESFET oscillator

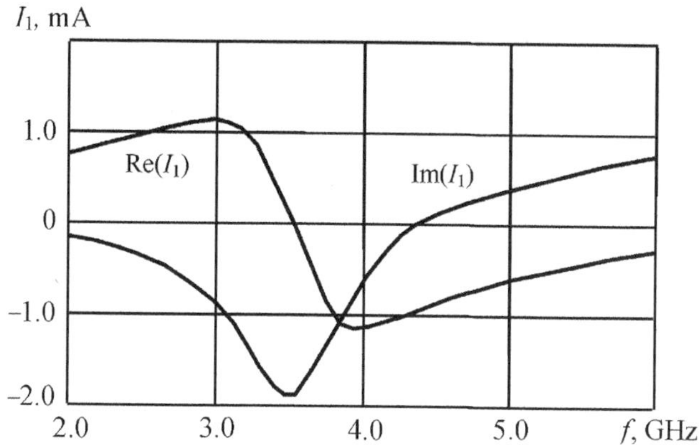

Figure 4.24 Simulated start-up oscillation conditions

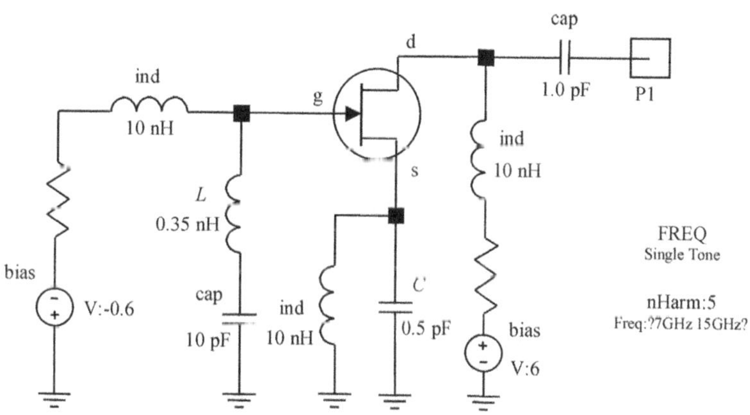

Figure 4.25 Simulated series feedback 12 GHz MESFET oscillator

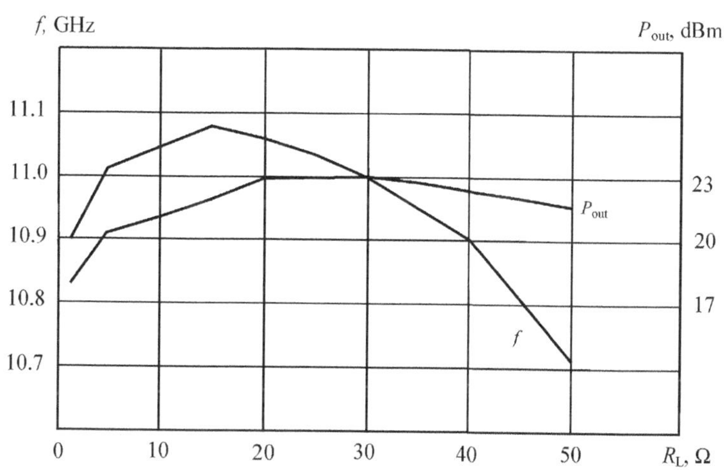

Figure 4.26 Output power and oscillation frequency versus load resistance

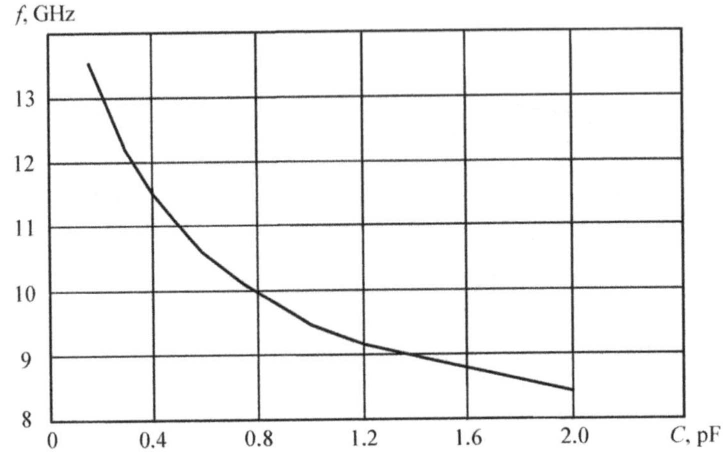

Figure 4.27 Oscillation frequency versus feedback capacitance

Further computer simulation shows that the large-signal oscillation conditions for different values of the feedback elements L and C are fulfilled in a very wide frequency range. Figure 4.27 shows the dependence of the oscillation frequency f on the source feedback capacitance C when $R_L = 25\ \Omega$ and $L = 0.35$ nH. Capacitance tuning from 0.15 to 2.0 pF leads to the appropriate frequency tuning of 8.3–13.6 GHz with approximately constant level of the output power of 23–24 dBm.

Improved frequency tuning characteristics can be achieved by inductance tuning. Figure 4.28 shows the dependence of the oscillation frequency f on the gate feedback inductance L when $R_L = 25\ \Omega$ and $C = 0.5$ pF. In this case, the inductance tuning from 7.0 to 0.2 nH leads to the appropriate frequency tuning from 3.9 to 13.3 GHz with output power level of 19–23 dBm. Moreover, the oscillation conditions are satisfied under the further increase of the gate feedback

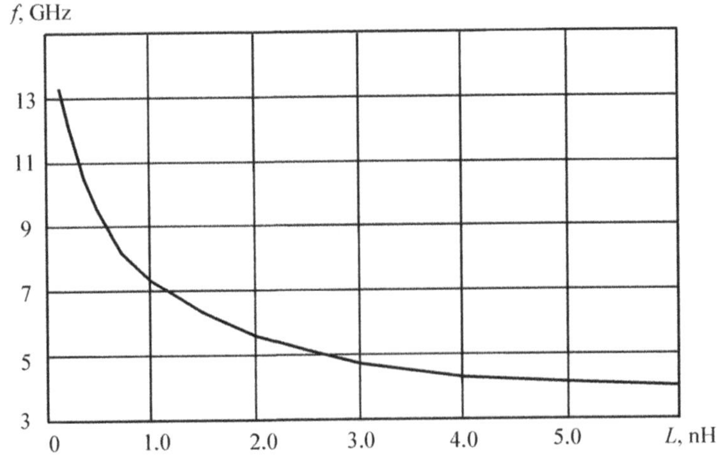

Figure 4.28 Oscillation frequency versus feedback inductance

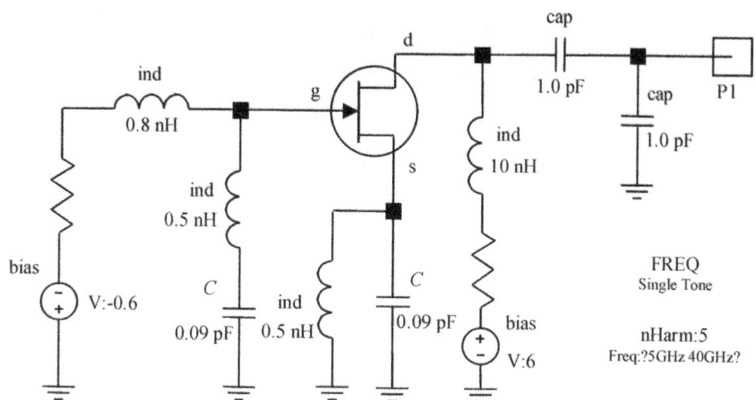

Figure 4.29 Simulated series feedback-tuned MESFET oscillator

inductance up to the value of 100 nH with the appropriate decrease of the oscillation frequency to 2.8 GHz.

A series feedback MESFET oscillator can be successfully used as an octave-band voltage-controlled oscillator in the frequency range 6–18 GHz [29]. In the case of the given power GaAs MESFET, both the upper bandwidth frequency and output power of the oscillator can be increased significantly. Figure 4.29 shows an example of the equivalent oscillator circuit for wideband frequency tuning.

The sufficiently small values of the source and gate bias inductances were initially chosen to maximize the frequency tuning bandwidth. An additional capacitor of 1 pF in parallel to the load resistor was used to provide stable oscillation conditions across the entire tuning range when $R_L = 50\ \Omega$. As a result, a simultaneous gate and source capacitance tuning from 2.0 to 0.09 pF with overlap factor of 22.2 leads to a wide frequency tuning bandwidth of 10.1–26.0 GHz with high value of the output power, as shown in Figure 4.30.

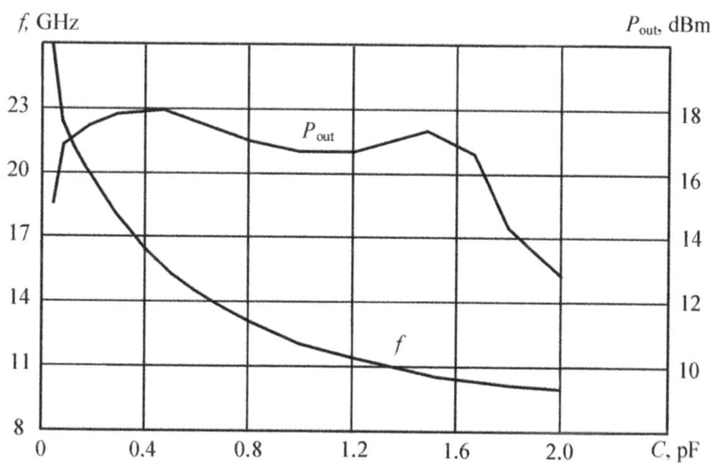

Figure 4.30 Output power and oscillation frequency versus tuned capacitances

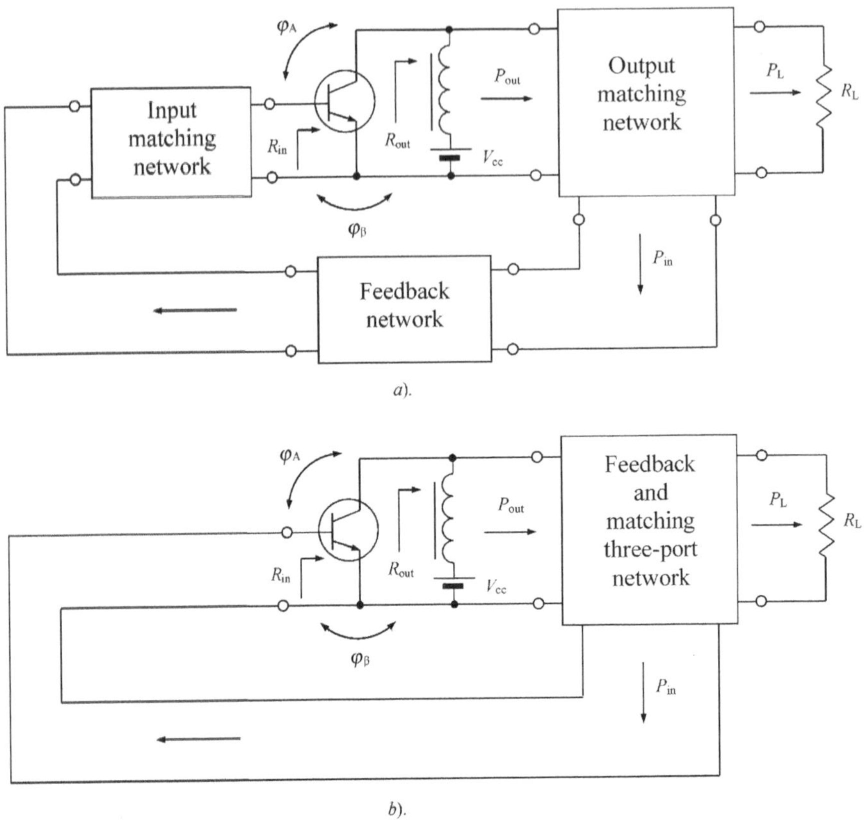

Figure 4.31 Basic circuits of Class E power amplifier with shunt capacitance

4.6 HIGH-EFFICIENCY DESIGN TECHNIQUE

4.6.1 Class C operation mode

Achieving high efficiency in oscillators is very important in order to increase battery lifetime, output power with minimum requirements for heat dissipation and reliability of measurement equipment and radio transmitters including active antennas and space power combiners. The popular design approach is based on the method of equivalent two-port networks representing the active device and parallel feedback circuit. The oscillator can be designed in a Class C operation mode so that the phase oscillation condition is fulfilled at the resonant frequency of the output matching circuit [30]. Therefore, the output power and collector efficiency of the oscillator is close to their high values in the corresponding tuned power amplifiers when high operating power gain is provided.

Figure 4.31a shows a block diagram of the tuned power oscillator where the feedback network is connected in between output and input matching circuits providing the amplitude and phase conditions for steady-state oscillations. However, the feedback network elements can be simultaneously a part of the output and input matching networks. In this case, the oscillator block diagram can be simplified to the form shown in Figure 4.31b. The resulting feedback

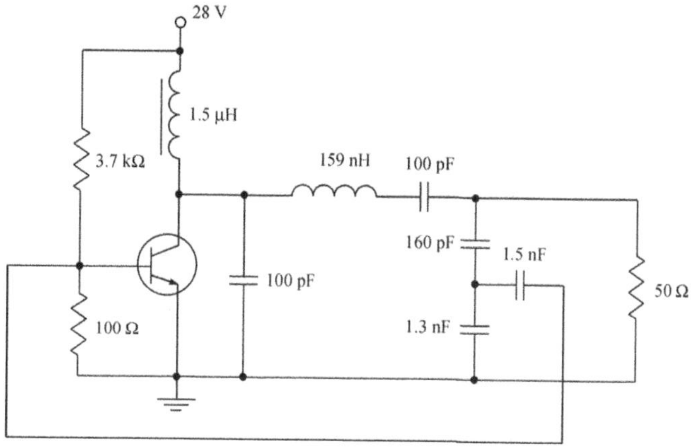

Figure 4.32 Schematic of high-efficiency bipolar oscillator with parallel feedback (© 1972 IEEE)

and matching three-port network should fulfil the amplitude and phase oscillation conditions for optimum oscillator operation when loop gain is equal to unity and $\varphi_A + \varphi_\beta = 2\pi n$, where $n = 0, 1, 2, \ldots$. The phase oscillation condition should result in an oscillation frequency equal to the resonant frequency of the output matching circuit.

According to Figure 4.31, the output power of the oscillator is written as

$$P_{out} = P_L + P_{in} \tag{4.95}$$

where P_L is the power delivered to the load and P_{in} is the power at the input of the active device. The operating power gain is defined by

$$G_p = \frac{P_{out}}{P_{in}} \tag{4.96}$$

Hence, the output power and output resistance can be obtained as

$$P_{out} = \frac{P_L}{1 + \dfrac{1}{G_p}} \tag{4.97}$$

$$R_{out} = \frac{(V_{cc} - V_{sat})^2}{2P_{out}} \tag{4.98}$$

where V_{cc} is the supply voltage and V_{sat} is the saturation voltage determined from the output volt–ampere characteristic of the active device. It is assumed that the collector voltage is sinusoidal. To calculate the values of the circuit elements, the method of equivalent two-terminal networks connected in parallel and in series is used.

Figure 4.32 shows the circuit schematic of the Class C tuned power oscillator operating at resonant frequency of 60 MHz and providing load power of 5 W [30]. The input of the active device is connected to the output matching circuit of the tuned power amplifier through the series capacitance of 1.5 nF and capacitive divider formed by the capacitances of 160 pF and 1.3 nF. At higher frequencies, it is convenient to use a directional coupler with proper coupling to provide the amplitude oscillation condition and a transmission line with electrical length required to satisfy the phase balance condition.

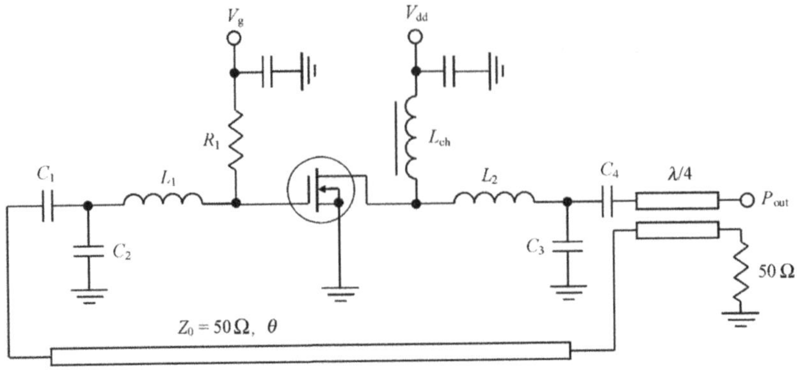

Figure 4.33 Schematic of high-efficiency MOSFET oscillator with parallel feedback

The circuit schematic of such a high-efficiency MOSFET power oscillator is shown in Figure 4.33, where the input and output matching circuits represent T-type lumped transformers, both matched to the source and load impedances of 50 Ω of the corresponding power amplifier. The directional coupler and feedback transmission line can easily be implemented using microstrip lines with the characteristic impedance of 50 Ω. Such a design method has an important advantage because the same transistor parameters and operating frequency as well as the same parameters of the input and output matching circuits are used to design both high-efficiency tuned power amplifier and oscillator. The additional transistor parameter is its phase shift between the fundamental-frequency components of the collector–emitter and base–emitter voltages, which should be compensated by a proper choice of the electrical length of the feedback microstrip line.

It was experimentally found long ago that by applying a biharmonic driving signal containing the fundamental and second harmonic components produces the dc-to-RF signal transformation more efficiently because of the much steeper driving waveform resulting in a lower power loss in the active device [31]. In this case, the resultant driving waveform shown in Figure 4.34a consists of the fundamental and second harmonic components being in phase at their maximum amplitudes, and the amplitude of the second harmonic is preferably chosen

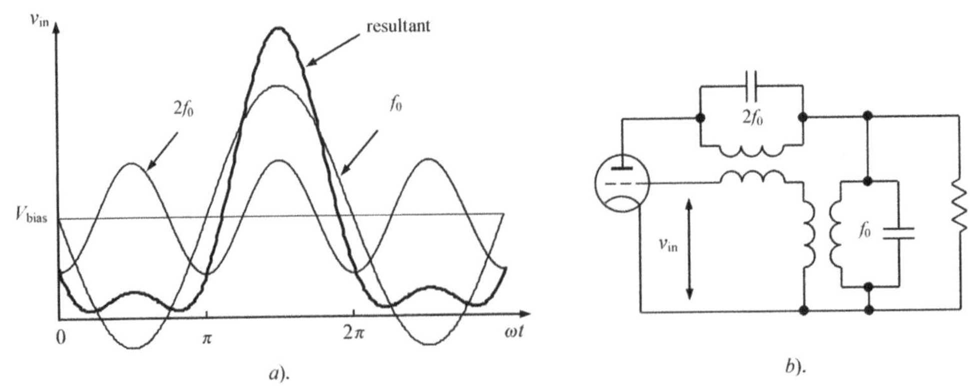

Figure 4.34 Oscillator with fundamental-frequency and second-harmonic resonators

to have approximately one-quarter the amplitude of the fundamental. Figure 4.34b shows a simplified circuit of a vacuum tube power oscillator containing the fundamental-frequency (f_0) and second-harmonic ($2f_0$) LC resonators in the anode circuit inductively coupled with the grid circuit. As a result, the driving feedback voltage v_{in} represents a sum of the fundamental-frequency and second-harmonic voltages, while the only fundamental-frequency output signal flows to the load resistance R_L.

4.6.2 Class E power oscillators

The switched-mode tuned Class E power amplifiers are widely used in different frequency ranges and output power levels, ranging from several kilowatts at low RF frequencies up to about one watt at microwaves. In the Class E power amplifier, the transistor operates as an on–off switch and the shapes of the current and voltage waveforms provide a condition where the high current and high voltage do not overlap simultaneously. This minimizes the power dissipation and maximizes the power amplifier efficiency. The possibility of increasing the efficiency of the single-ended power amplifier by mistuning the output matching circuit was found quite long ago [32]. However, the single-ended switched-mode power amplifier with shunt capacitance as a Class E power amplifier was first introduced in 1975 [33].

The characteristics of a Class E power amplifier can be determined by finding its steady-state collector voltage and current waveforms. The load network consists of a capacitance C shunting the transistor, a series inductance L, a series fundamentally tuned L_0C_0 circuit and a load resistance R. In a common case, a shunt capacitance C can represent the intrinsic device output capacitance and external circuit capacitance added by the load network. The collector of the transistor is connected to the supply voltage by RF choke with high reactance at the fundamental frequency. The active device is considered to be an ideal switch that is driven in such a way in order to provide the device switching between its on-state and off-state operation conditions. As a result, the collector voltage waveform is determined by the switch when it is turned on and by the transient response of the load network when the switch is turned off. To minimize the collector power losses, the following optimum conditions for voltage across the switch just prior to the start of switch on at the moment $\omega t = 2\pi$, when the transistor is saturated, should be provided:

$$v(\omega t)|_{\omega t=2\pi} = 0 \tag{4.99}$$

$$\left.\frac{dv(\omega t)}{d\omega t}\right|_{\omega t=2\pi} = 0 \tag{4.100}$$

where v is the voltage across the switch.

To achieve Class E operation for a power oscillator based on the corresponding Class E power amplifier, it is necessary to design a three-port matching and feedback reactive network satisfying the following conditions [34, 35]:

- the collector–emitter voltage should be the same as in the corresponding Class E power amplifier;
- the energy transmission from collector to load should be provided in the same way as in the corresponding Class E power amplifier, taking into account the driving power;
- the energy stored in the load network should be high enough to ensure the required frequency stability;

- the feedback voltage should have the same amplitude and phase (with respect to the phase of the collector voltage) as the input voltage of the corresponding Class E power amplifier.

One of the important requirements to the output spectrum of the oscillator is a sufficiently small level of harmonics. The loaded quality factor of the resonant circuit must be not less than 5.1 to obtain the second harmonic level below −20 dBc without additional filtering [36]. Therefore, the almost sinusoidal output voltage is the input voltage of the feedback network or driving voltage of the oscillator. This means that the active device cannot operate either in pinch-off or saturation regions, and there is a time period when it is operated in the active region, which is characterized by power losses. However, the effect of finite switching time is not so significant compared with losses provided by the finite value of the saturation resistance [18].

The design of Class E power oscillator is divided into two parts. The first step is to design the load network of the corresponding Class E power amplifier. Then, the second step is to provide the proper phase shift between input and output of the feedback network. A part of the power amplifier output power is transmitted to the input, which means that the load resistance R of the power amplifier will be equal to a parallel combination of the load resistance R_L and the input resistance of the feedback network R_{in}. Figure 4.35a shows an example of the MOSFET Class E power oscillator with series inductance 48 μH in the feedback network having efficiency of 86% and output power of 1 W at oscillation frequency 800 kHz and supply voltage 4.5 V [35]. Similar results with the description of a complete design procedure of a Class E MOSFET tuned power oscillator have been reported [37]. To simplify the analysis of such an oscillator, let us represent its equivalent circuit using a switch instead of the active device, as shown in Figure 4.35b. The input resistance R_{in} and capacitance C_{in} represent a parallel equivalent of the input circuit of the MOSFET device equivalent circuit consisting of the gate resistance R_{gs} and gate–source capacitance C_{gs} connected in series. The optimum parameters of the Class E load network with shunt capacitance C, series inductance L and load resistance R are given by [38]

$$L = \frac{1.1525R}{\omega} \tag{4.101}$$

$$C = \frac{0.1836}{\omega R} \tag{4.102}$$

$$R = 0.5768\frac{V_{cc}^2}{P_{out}} \tag{4.103}$$

Thus, the phase angle of the load network (between the fundamental voltage and current phasors) seen from the device collector and required for idealized optimum Class E with shunt capacitance can be determined through the load network parameters using Equations (4.101) and (4.102) as

$$\phi = \tan^{-1}\left(\frac{\omega L}{R}\right) - \tan^{-1}\left(\frac{\omega C R}{1 - \frac{\omega L}{R}\omega C R}\right) = 35.945^\circ \tag{4.104}$$

which means that the load network is slightly mistuned with respect to the fundamental frequency. The values of the load network parameters should be chosen to create an inductive reac-

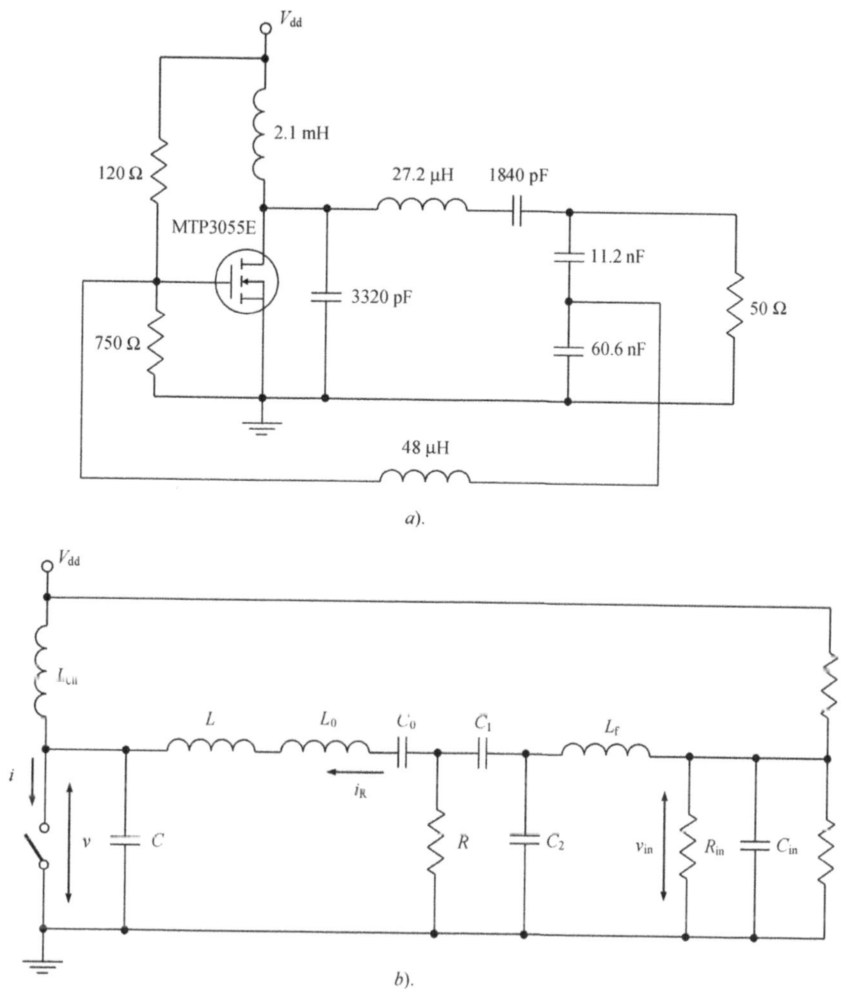

Figure 4.35 Basic circuits of Class E power oscillators (© 2002 IEEE)

tance at the fundamental frequency and the capacitive reactances at other harmonic components. The feedback inductance L_f is necessary to provide greater than 180° phase shift between the fundamental components of the gate and drain voltages.

Figure 4.36 shows the normalized (a) load current, (b) collector voltage and (c) current and (d) input sinusoidal voltage waveforms for idealized optimum Class E with shunt capacitance. From the collector voltage and current waveforms it follows that, when the transistor is turned on, there is no voltage across the switch and the current i, consisting of the load sinusoidal current and dc current, flows through the active device. However, when the transistor is turned off, this current now flows through the parallel capacitance C. When the feedback driving voltage v_{in} is larger than the device threshold voltage V_{th}, the active device is turned on. Otherwise, when $v_{in} < V_{th}$, the MOSFET device is turned off, corresponding to off-state operation mode.

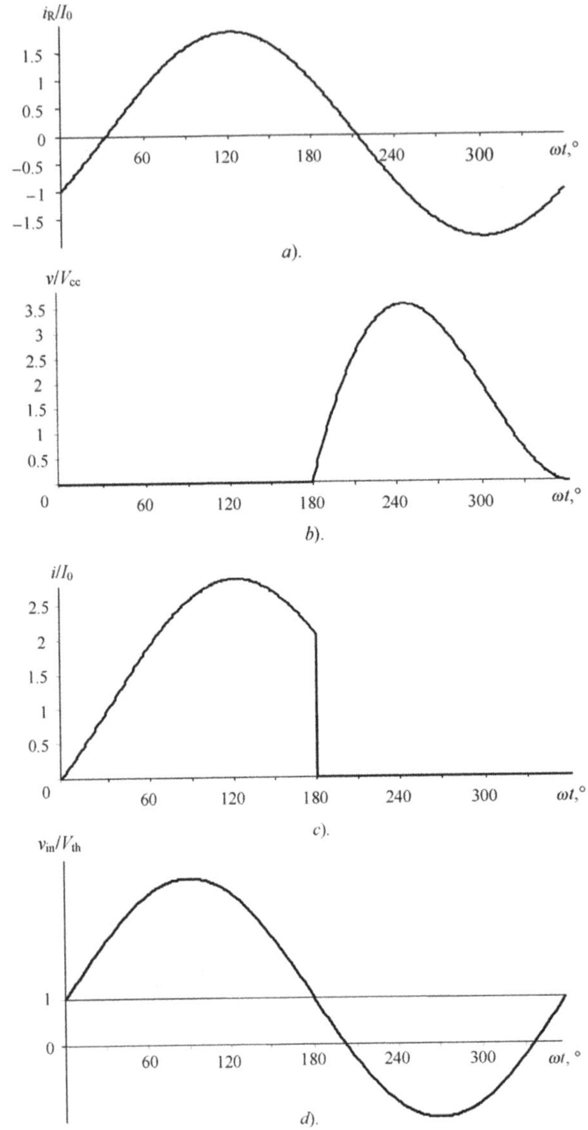

Figure 4.36 Normalized (a) load current, collector (b) voltage and (c) current and (d) input voltage waveforms for the idealized optimum Class E oscillator

In practical design, it is impossible to realize choke inductance L_{ch} with infinite impedance at the fundamental and any harmonics. Moreover, using a finite dc feed inductance has the advantage of minimizing size, cost and complexity of the overall circuit. When $H = L/L_{ch}$ is small corresponding to only dc current flowing through the choke inductance L_{ch}, the optimum load network parameters defined by Equations (4.101–4.103) are almost constant. However, in the range of large H, the optimum load network parameters vary since inductance L_{ch} no longer works as a finite dc feed inductance [39]. The experimental results for the MOSFET Class E

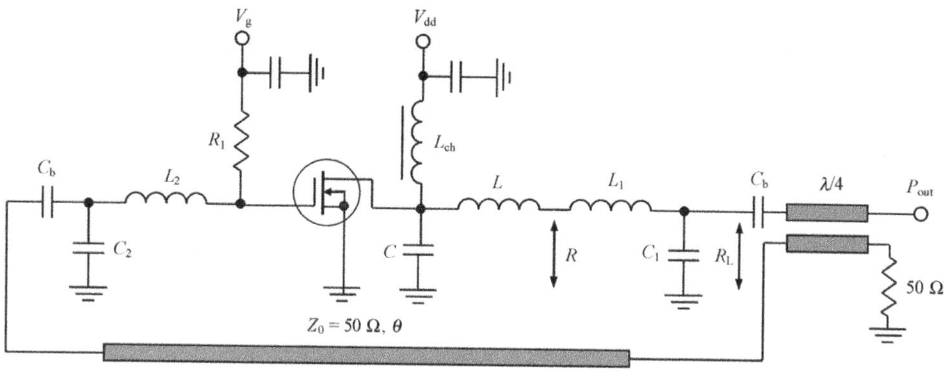

Figure 4.37 Schematic of a Class E power oscillator with impedance transformation

power oscillator show the efficiency of 90.7% and output power of 6.8 W at the oscillation frequency of 2.02 MHz for high quality factor $Q = \omega L/R$ and low H. For the case of low Q and high H, the efficiency was 89.7% with the output power 2.8 W at oscillation frequency 1.97 MHz. By using the same topology for the bipolar Class E power oscillator, the collector efficiency over 95% with the output power of 3 W at 2 MHz was measured [34].

If a theoretical value of the resistance R for the optimum Class E load network given by Equation (4.103) is too small or differs significantly from the required load impedance (typically 50 Ω), it is necessary to use a matching circuit instead of a series L_0C_0 resonant circuit to deliver maximum output power to the load, as shown in Figure 4.37. In a practical oscillator, the inductors L and L_1 are combined together and considered as a single inductor. To provide a sufficient level of harmonic suppression, the loaded quality factor of such a transformer must be high enough. In addition, the first series element of the matching circuit should be an inductor with sufficiently large inductance value to realize high impedance conditions for harmonics.

The simple L-type output matching circuit with the shunt capacitance C_1 and series inductance L_1, the input impedance of which should be sufficiently high at second- and higher-order harmonics, can be used. The output matching circuit can also be designed using any combination of the lumped capacitors and inductors with a series inductor as the first element. To determine the circuit parameters of the L-type matching circuit, we first calculate the loaded quality factor from

$$Q_L = \sqrt{\left(\frac{R_L}{R}\right) - 1} \tag{4.105}$$

Then, the matching circuit parameters can be obtained by

$$L_1 = \frac{Q_L R}{\omega} \tag{4.106}$$

$$C_1 = \frac{Q_L}{\omega R_L} \tag{4.107}$$

At microwave frequencies, to minimize the insertion losses, the transmission lines should replace any lumped inductors in output matching circuit. For example, any type of transmission line, including open-circuited or short-circuited stubs to provide the required matching and

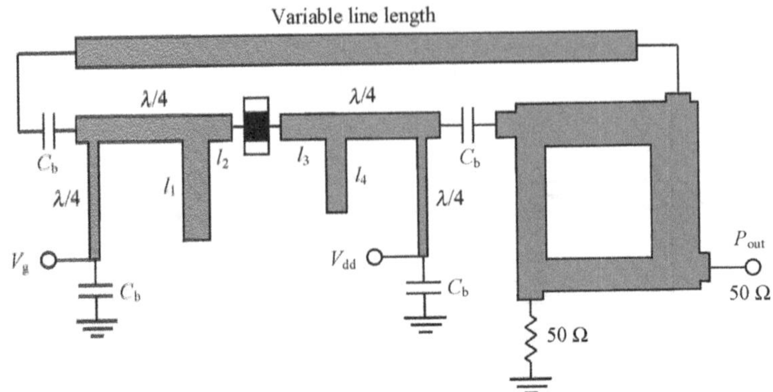

Figure 4.38 High-efficiency microwave MESFET oscillator topology

harmonic suppression conditions can compose the matching circuit. As a result, to approximate the idealized Class E with shunt capacitance, it is necessary to design the transmission-line load network satisfying the required idealized optimum impedance Z_{net1} at the fundamental frequency given by

$$Z_{net1} = R(1 + j\tan 49.052^\circ) \tag{4.108}$$

which can be obtained from Equations (4.101) and (4.102). At the same time, the open-circuited conditions should be realized for all higher-order harmonics. However, as shown from the Fourier analysis, a good approximation to Class E mode may be obtained with only two harmonics (fundamental and second) of the voltage waveform across the switch [40].

The parallel feedback can be used to design Class E microwave oscillators when the load network has impedance at the fundamental frequency of

$$Z_{net1} = \frac{0.28015}{\omega C_{out}} \exp(j49.052^\circ) \tag{4.109}$$

where C_{out} is the transistor switch output capacitance [41]. Equation (4.109) establishes a relationship between the load network impedance seen by the device collector and transistor switch capacitance, and is similar to Equation (4.108). Here, the output device capacitance fully represents the shunt capacitance required for optimum switched-mode Class E operation. Such a Class E oscillator with a Fujitsu FLK052WG MESFET device shown in Figure 4.38 achieved a maximum efficiency of 59% with output power of 300 mW at the operating frequency of 5.0 GHz and supply voltage of 6.5 V. The matching circuits and variable feedback microstrip lines of 50 Ω (1.6 mm wide) were fabricated on the 0.508-mm-thick substrate with dielectric permittivity $\varepsilon_r = 2.2$, when $l_1 = 9$ mm, $l_2 = 1.8$ mm, $l_3 = 5.3$ mm and $l_4 = 6.2$ mm. An asymmetric microstrip branch-line coupler is used to provide the feedback through a microstrip line of the appropriate length.

4.6.3 Class DE power oscillators

Figure 4.39 shows the circuit schematic of Class DE RF power oscillator representing a Class D inverter with Class E switching conditions [42]. A Class D inverter is composed of two switch devices conducting alternate half cycles and a series resonant circuit tuned to the fundamental

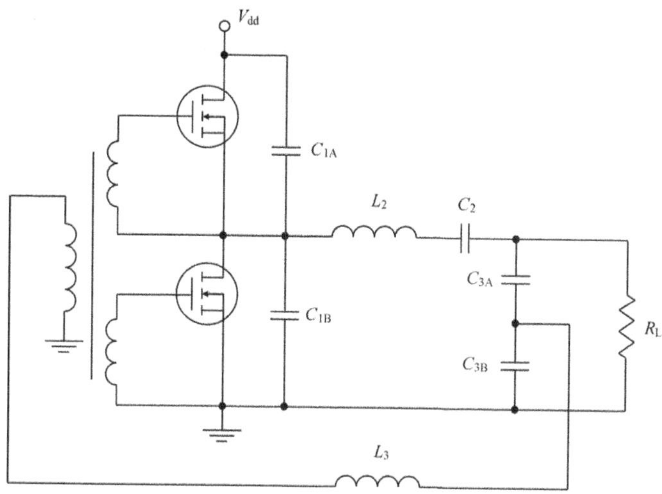

Figure 4.39 Circuit schematic of a Class DE power oscillator

frequency, converting the square collector (drain) voltage waveforms into a sinusoidal load current. At high frequencies, its efficiency reduces because of the power losses due to the device internal parasitic capacitance (collector capacitance in the case of the bipolar transistor or drain–source capacitance in the case of MOSFET device). Thus, applying Class E switching conditions can improve efficiency since the output device capacitance can be added to the required shunt capacitance of the Class E load network. The necessary modification of the Class DE power amplifier to the corresponding Class DE power oscillator can be accomplished by providing an additional phase-shifting feedback circuit, consisting of the capacitive divider and series inductance, and a centre-tapped transformer at the input.

The two switches are driven by the sinusoidal signal, which generates a dead time between the instants when one switch has turned off and another switch will turn on. The gate sinusoidal voltages are out of phase by 180°. Consequently, one device will be turning off if the gate voltage decreases below the threshold value and remaining off until the gate voltage exceeds the threshold level. At the same time, another device will also be turned off when the gate voltage is below the threshold level. Hence, during this period, the two devices are turned off, resulting in a dead time. During the dead time, the sinusoidal output current charges one shunt capacitor and discharges the other shunt capacitor, and the midpoint voltage between two switching conditions becomes V_{dd} or zero at the end of the dead time. The load network parameters with a dead time duty ratio of 0.25 can be calculated as

$$C_{1A} = C_{1B} = \frac{1}{2\pi \omega R_L} \tag{4.110}$$

$$L_2 = \frac{Q_L R_L}{\omega} \tag{4.111}$$

$$C_2 = \frac{1}{\omega (Q_L - 0.5\pi) R_L} \tag{4.112}$$

$$R_L = \frac{V_{dd}^2}{2\pi^2 P_{out}} \tag{4.113}$$

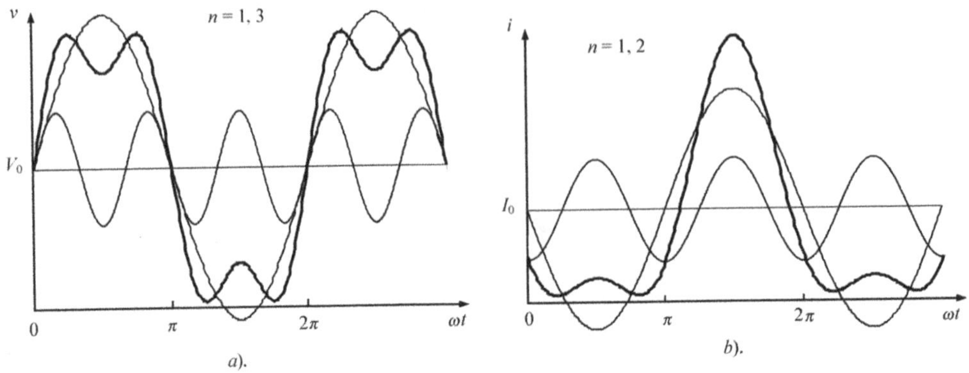

Figure 4.40 Fourier voltage and current waveforms with third and second harmonics

where Q_L is the loaded quality factor [43]. The feedback reactive elements C_{3A}, C_{3B} and L_3 together with the gate–source capacitance C_{gs} provide the required phase shift. As a result, the drain efficiency of 93.3% with output power of 2.3 W was measured at the oscillation frequency of 1.0 MHz.

4.6.4 Class F mode and harmonic tuning

Another way to design a high-efficiency oscillator, based on the active device common source (emitter) configuration, is to use a corresponding high-efficiency power amplifier with Class F harmonic conditions. To understand the basic approach, let us evaluate the contribution of each harmonic component to an ideal half-sinusoidal current waveform and a square voltage waveform. In this case, it is useful to calculate the partial Fourier series of current $i(\omega t)$ and voltage $v(\omega t)$ in normalized form according to

$$\frac{i(\omega t)}{I_0} = 1 - \frac{\pi}{2}\sin\omega t - 2\sum_{n=2,4,6,...}^{N}\frac{\cos n\omega t}{n^2 - 1} \quad (4.114)$$

$$\frac{v(\omega t)}{V_0} = 1 + \frac{4}{\pi}\sin\omega t + \frac{4}{\pi}\sum_{n=3,5,7,...}^{N}\frac{\sin n\omega t}{n} \quad (4.115)$$

where I_0 and V_0 are the dc current and voltage components, respectively.

Figure 4.40 shows that the shapes of the voltage and current waveforms can be significantly changed with increasing fundamental voltage amplitude by adding even one additional harmonic component if properly phased. For example, the combination of the fundamental and third harmonic components being out of phase at centre point results in a flattened voltage waveform with depression in its centre. It is clearly seen from Figure 4.40a that the proper ratio between the amplitudes of the fundamental-frequency and third-harmonic components can provide the flattened voltage waveform with minimum depression and maximum difference between its peak amplitude and amplitude of the fundamental harmonic. Similarly, the combination of the fundamental-frequency and second-harmonic components being in phase at the centre point flattens the current waveform corresponding to the maximum values of the voltage waveform and sharpens the current waveform corresponding to the minimum values of

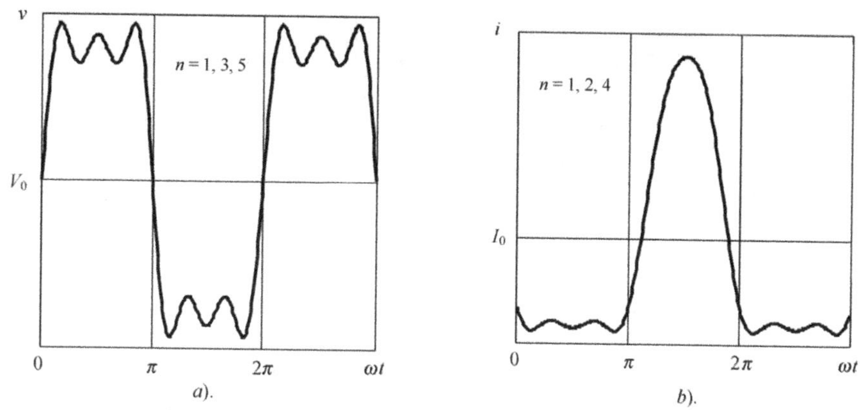

Figure 4.41 Fourier voltage and current voltage waveforms with three harmonics

the voltage waveform, as shown in Figure 4.40b. The optimum ratio between the amplitudes of the fundamental-frequency and second-harmonic current components can maximize the peak value of the current waveform with its minimized value determined by the device saturation resistance in a practical circuit. Thus, power loss due to the active device can be minimized since the results of the integration over the period when minimum voltage corresponds to maximum current will give a small value compared with the power delivered to the load.

Ideally, the half-sinusoidal current waveform does not contain the third harmonic component because its third-harmonic Fourier current coefficient is equal to zero, i.e., $\gamma_3(\theta) = 0$. However, a load line analysis of a Class B power amplifier with sinusoidal output voltage waveform under overdriven condition when device operates in pinch-off, active and saturation modes during one oscillation period shows that operation in the saturation mode is characterized by a depression in the output current waveform. From Fourier analysis it follows that such a current waveform includes the third harmonic component, which is out of phase with the fundamental component at the point of symmetry of $\omega t = \pi/2$. Therefore, when an additional resonant circuit tuned to the third harmonic is included into the anode circuit operating in a saturation mode, the voltage drop with opposite phase will appear across this resonant circuit, resulting in a similar depressed voltage waveform shown in Figure 4.40a by a solid line. Hence, for the increased fundamental voltage amplitude, the output power at the fundamental frequency and anode efficiency can be increased for the same input drive. Physically, an efficiency improvement can be explained by the fact that fundamental voltage or fundamental current have negative values during some part of the period corresponding to the negative power as an integration of a product of the instantaneous fundamental voltage and current. This means that the power loss for the active device is partly compensated by the reactive power provided by the harmonic resonator. Adding one or more high-order harmonic components can further improve the voltage or current waveform. Figure 4.41a shows the voltage waveform with third and fifth harmonic peaking, which is close to an ideal rectangular waveform. Figure 4.42b shows the current waveform with second and fourth harmonic peaking resulting in a close-to-ideal half-sinusoidal waveform.

The possibility of improving efficiency by approximation of the anode voltage waveform to a square wave in order to minimize the values of the saturation voltage compared with the supply voltage over the entire interval $0 \le \omega t \le 2\pi$ has been known for a long time [44]. In

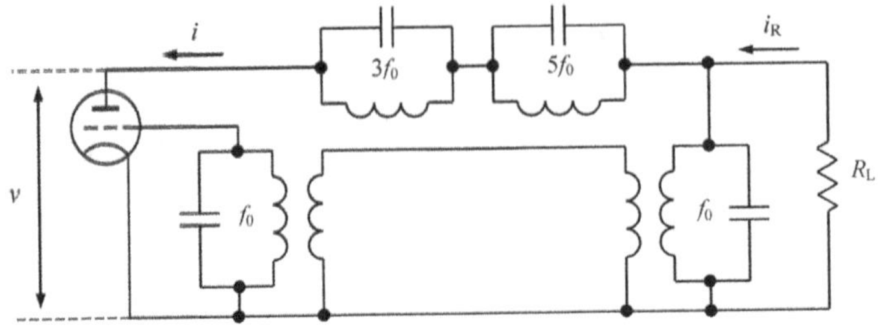

Figure 4.42 Polyharmonic power oscillator

this case, it was suggested to use the additional resonators tuned to the third and fifth harmonic components in order to maximize efficiency of the vacuum tube power oscillator. Figure 4.42 shows a simplified circuit of a vacuum tube power oscillator containing the fundamental-frequency (f_0), third-harmonic ($3f_0$) and fifth-harmonic ($5f_0$) LC resonators in the anode circuit. The anode fundamental-frequency resonator is inductively coupled with the tank grid circuit. As a result, the anode voltage $v(\omega t)$ represents a sum of the fundamental-frequency, third-harmonic and fifth-harmonic voltages, while the only fundamental-frequency output signal flows to the load resistance R_L.

In an ideal Class F power amplifiers, the fundamental and harmonic load impedances are optimized by short-circuit termination and open-circuit peaking in order to control the voltage and current waveforms at the collector (drain) of the active device to obtain maximum efficiency. To provide 100% idealized collector efficiency, the required impedance conditions at the device collector must be

$$Z_1 = R_L = \frac{8}{\pi^2}\frac{V_{cc}}{I_0}$$

$$Z_n = 0 \quad \text{for even } n$$

$$Z_n = \infty \quad \text{for odd } n$$

where I_0 is the dc current, V_{cc} is the dc supply voltage and n is the harmonic number [18]. Figure 4.43 shows the ideal voltage and current waveforms corresponding to a Class F operation mode. Here, a sum of odd harmonics gives a square voltage waveform, a sum of fundamental-frequency and even harmonics approximates a half-sinusoidal current shape and sinusoidal current flowing to the load resistance R_L.

In practical implementation at microwave frequencies, the matching circuit with a series lumped inductance is replaced by the matching circuit with a series transmission line, as shown in Figure 4.44a. The directional coupler at the output of the oscillator with coupling coefficient less than the operating power gain of the equivalent power amplifier and the transmission line with proper electrical length in the feedback circuit provide the necessary phase shift. By using an additional third harmonic impedance peaking circuit, the total oscillator efficiency can be increased. The ideal class F power amplifier with all even harmonic short-circuit termination due to the quarter-wave transmission line and third harmonic peaking achieves a maximum drain efficiency of 88.4% [18].

It is easy to realize such impedance conditions using the series transmission line with electrical length of θ_1 and open-ended stub with electrical length of θ_2, as shown in Figure 4.44b. An appropriate control of short-circuit termination at the second harmonic and open-circuit

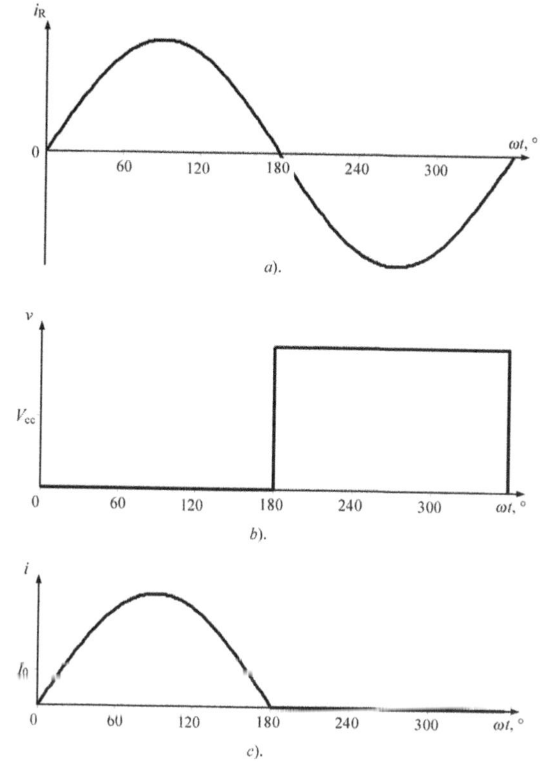

Figure 4.43 Ideal Class F collector current and voltage waveforms

termination at the third harmonic results in a drain (collector) voltage waveform close to a square wave, thus approximating an ideal Class F operation. In this case, it is necessary to provide the following electrical lengths of the transmission lines at the fundamental frequency:

$$\theta_1 = \frac{1}{3}\tan^{-1}\left(\frac{1}{3Z_0\omega C_{out}}\right)$$

$$\theta_2 = \frac{\pi}{6}$$

where C_{out} is the device output capacitance, Z_0 is the characteristic impedance of the microstrip line [18]. The Class F load network then must be followed by the output matching circuit to match the realized fundamental-frequency output impedance to a standard 50-Ω load.

At very high frequencies, it is necessary to take into account the parasitic elements of the device equivalent circuit since their effect on the impedance conditions becomes significant at higher-order harmonics. In this case, it is not enough to consider the drain–source capacitance C_{ds} as the only element contributing to the device output reactance, unlike at lower frequencies where it can be considered as the output device capacitance C_{out}. Figure 4.45 shows the equivalent circuit of the HEMT device with series parasitic gate, drain and source inductances (L_g, L_d and L_s) and pad gate and drain capacitances (C_{gp} and C_{dp}). The device output admittance (assuming negligible value of the feedback gate–drain capacitance C_{gd}) to compensate for the

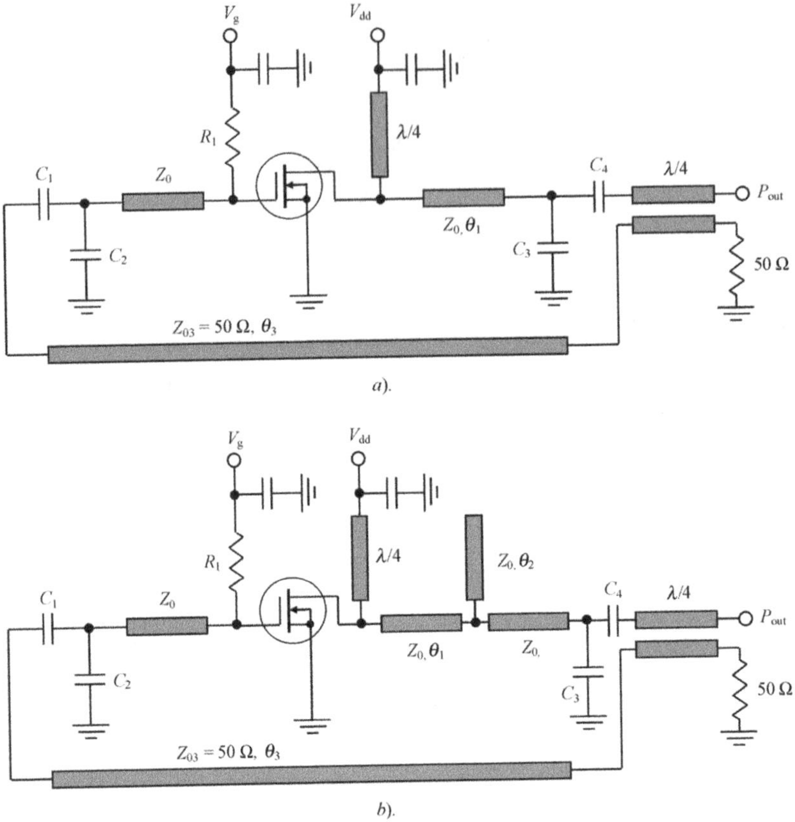

Figure 4.44 High-efficiency Class F MOSFET oscillator circuits

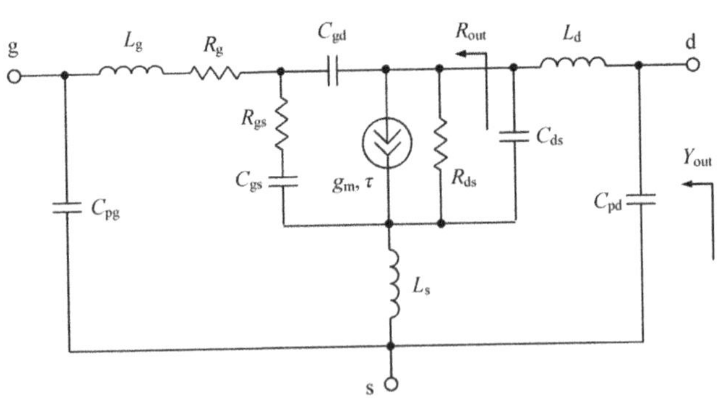

Figure 4.45 HEMT equivalent circuit with parasitic elements

parasitic elements can be written as

$$Y_{out}(n\omega) = \frac{1}{\dfrac{1}{\dfrac{1}{G_{out} - jn\omega C_{ds}} - jn\omega C_{ds}} - jn\omega(L_d + L_s)} - jn\omega C_{pd} \tag{4.116}$$

where $G_{out} = 1/R_{out}$ and n is a harmonic number.

By providing a short-circuit impedance condition at the second harmonic, a drain efficiency of 61% with output power 14.46 dBm at operating frequency 1.86 GHz for low supply voltage of 2 V was obtained for a HEMT power oscillator [45]. Such an approach was based on the measurement and optimization of the input incident-wave power and load reflection coefficients at the fundamental frequency and second harmonic using an active load measurement system with input and output reflectometers [46]. To design a high-efficiency Class F MESFET power oscillator with second and third harmonic control, a parallel feedback topology, similar to that shown in Figure 4.44, was chosen [47]. The attenuation of the entire feedback loop was adjusted to 10.5 dB. The measured efficiency was found to be 67% for supply voltage of 5.5 V and output power of 24 dBm at the operating frequency of 1.6 GHz.

4.7 PRACTICAL OSCILLATOR SCHEMATICS

Figure 4.46a shows a common source GaAs MESFET oscillator circuit configuration with a parallel feedback circuit between the drain and the gate using lumped elements. This feedback

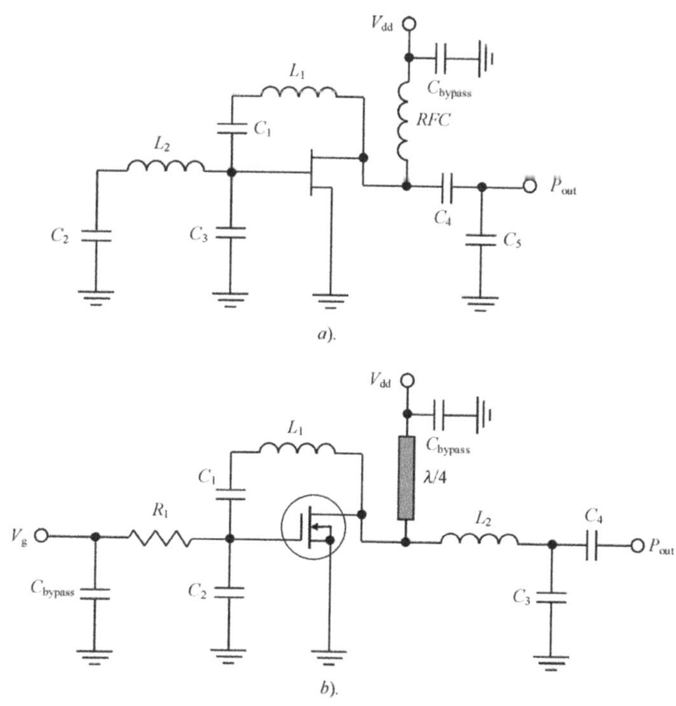

Figure 4.46 Parallel feedback common source oscillator circuit configurations

circuit represents a series connection of an inductor L_1 and a capacitor C_1 [48]. The capacitance C_3, which is required to satisfy the amplitude and phase oscillation conditions for the oscillator large-signal operation, is connected between the gate and the source. The drain terminal is connected to the load through the blocking capacitor C_4 and load capacitor C_5. This connection compensates for the inductive reactance at the device output. To provide the start-up oscillation conditions at 5.96 GHz, the external circuit feedback parameters were chosen as $L_1 = 1.1$ nH, $C_1 = 1.05$ pF, $C_3 = 0.63$ pF, and $C_5 = 1.28$ pF. The series circuit consisting of an inductor L_2 and a capacitor C_2 is connected to the gate terminal to maintain a self-bias voltage. The circuit resonant frequency $f_1 = 1/2\pi\sqrt{L_1C_1}$ is low enough for it to be considered as an open circuit at the oscillation frequency.

The operational principle of such a self-bias circuit is as follows. At the first moment when the drain supply voltage V_{dd} is switched on, the gate–source voltage and the voltage across the capacitance C_3 are zero. Due to the Shottky barrier rectifying property, the RF gate–source voltage generates a dc gate current component that discharges capacitance C_3 to a negative voltage. This leads to a decrease of the device effective output resistance and transconductance which, in turn, contributes to a decrease of the negative conductance and the fulfilment of the large-signal steady-state oscillation conditions. For a given oscillator, the oscillation build-up is established at around 25 ns.

For a MOSFET oscillator, stable oscillations are provided by using an external gate biasing. This procedure is required because the gate leakage current is too small to provide the self-bias mode oscillation conditions. For such an operation mode, it is possible to use a self-bias series resistor between the source and the ground. In addition, for high-efficiency operation, the quarter-wave transmission line can be connected to the drain (instead of the RF choke) together with the series lumped inductor L_2, as shown in Figure 4.46b. This creates high impedance at the third and higher odd harmonics. The series inductor L_2 can be considered as an element of a T-type output matching circuit.

A GaAs MESFET self-bias mode oscillator, which was implemented in a microwave integrated circuit using alumina substrate, is shown in Figure 4.47 [48]. In this case, the parallel lumped feedback circuit between the drain and gate terminals is realized using the two short microstrip lines, whose total length is less than a quarter-wavelength, to create an inductive feedback impedance. An open-ended microstrip line replaces the feedback capacitance C_3 connected to the gate terminal. The matching circuit, which is necessary to convert the optimum device impedance to the 50-Ω load, is realized with a quarter-wave microstrip transformer

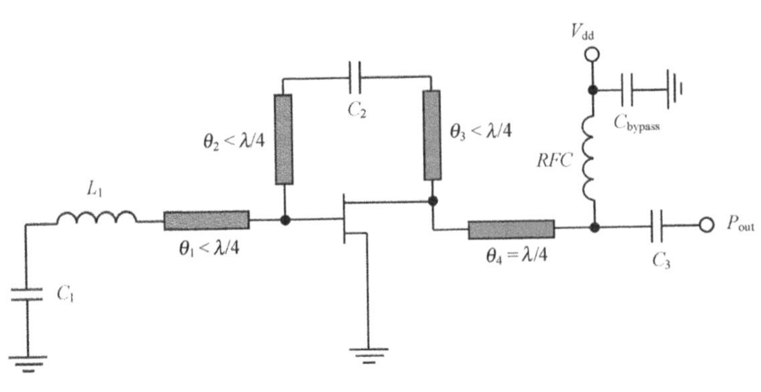

Figure 4.47 Microwave self-bias mode GaAs MESFET oscillator

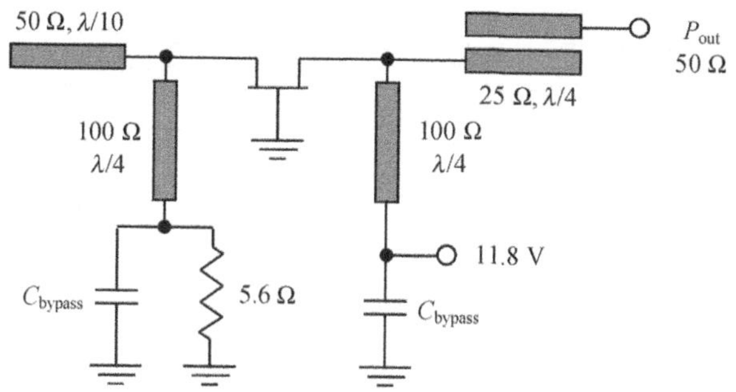

Figure 4.48 Common gate MESFET oscillator circuit configuration (© 1978 IEEE)

as a replacement for the capacitance C_5. As a result, using an active device with gate width 2500 μm and gate length 1.5 μm, under supply voltage of 4.5 V, the oscillation frequency was 5.51 GHz with output power 110 mW and efficiency 17%.

Figure 4.48 shows a common gate oscillator circuit configuration. By using a DXL-4640-P100 MESFET device, such an oscillator produced 1.2 W output power at 8 GHz operating frequency [49]. The transition from self-oscillating small-signal conditions to large-signal steady-state operation mode is provided by a self-bias resistor of 5.6 Ω connected to the source circuit in parallel with a bypass capacitor. Both RF short-circuited quarter-wave microstrip lines connected to the source and drain, respectively, eliminate an influence of device bias circuits on the electrical properties of the resonant circuit. An open-ended microstrip line provides the capacitive reactance seen by the source terminal. Since the output resonator is a quarter-wave microstrip stub, a simple adjacent line coupling section was chosen and adjusted for required coupling with maximum output power. A quarter-wave microstrip line connected to the drain improves the oscillator efficiency and spectral characteristics attributable to the short-circuit conditions for the even voltage harmonics.

Generally, it is easy to achieve the negative resistance start-up conditions by using an inductive feedback element located between the gate and the ground, like the inductance L_g in Figure 4.49a [50]. In this case, the imaginary part of the impedance Z_s provided by the feedback circuit, which is connected between the source terminal and ground, should be capacitive, with minimum real part. A matching circuit connected to the drain terminal is used to maximize the output power delivered to a 50-Ω load. Experimental results show that microwave MESFET oscillators having such a configuration can produce 50–60 mW output powers at efficiencies of more than 20%. These oscillators have soft build-up of the oscillations without showing any frequency jumping or spurious outputs. Low sensitivity of output power and oscillation frequency to drain bias variation is observed.

The series feedback common gate MMIC oscillator configuration operating at 75 GHz is shown in Figure 4.49b. Transmission lines of different length were used to provide inductive impedance in the gate circuit and capacitive impedance in the source circuit [51]. All transmission lines are realized in the form of the coplanar lines with the characteristic impedance of 50 Ω. The gate length of the active device is 0.16 μm and the gate width is 2 × 25 μm. Because of the geometry of an active device, the source is grounded through two symmetric coplanar lines. The drain bias is supplied by a coplanar line, which also serves as the oscillator output. Output power 8 mW with drain bias 3 V was achieved. The similar common gate oscillator topology using a power HEMT device with gate width 1.5 mm based on coplanar transmission

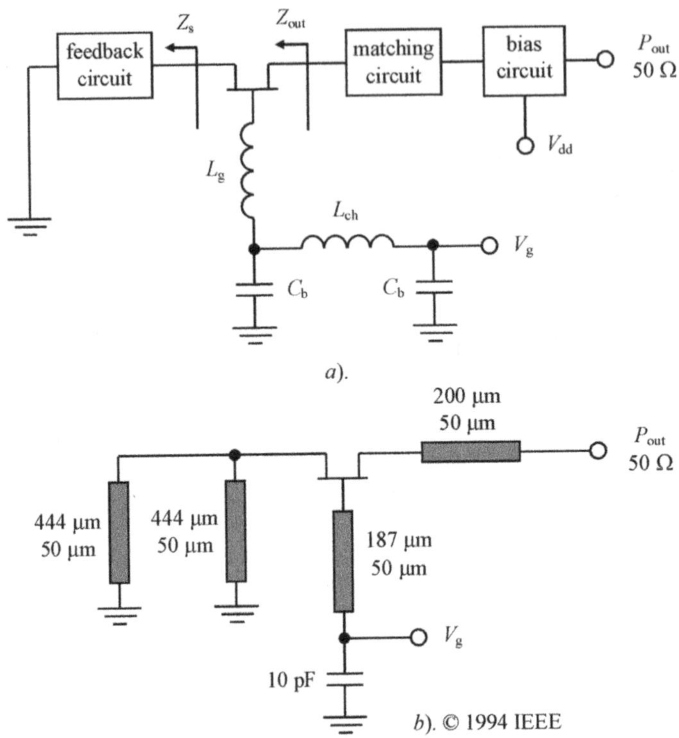

Figure 4.49 Series feedback common gate oscillator circuits

lines could provide a delivery of 1.7 W at 9.556 GHz into 50-Ω load with drain bias 30 V [52]. The output matching circuit was realized in the form of an L-transformer with series transmission line and open-circuited stub.

Figure 4.50 shows the circuit layout of the microstrip pHEMT *Ku*-band oscillator having a series feedback topology [53]. The oscillator circuit uses a packaged GaAs pHEMT device (model KH1032-C02) where one source lead is connected to an open-ended microstrip line and the second source lead is removed. S-parameters measurements with one source lead removed and another source lead connected to the ground revealed the device to be highly unstable around 16.2 GHz. Two open-circuited microstrip lines are placed on the gate and

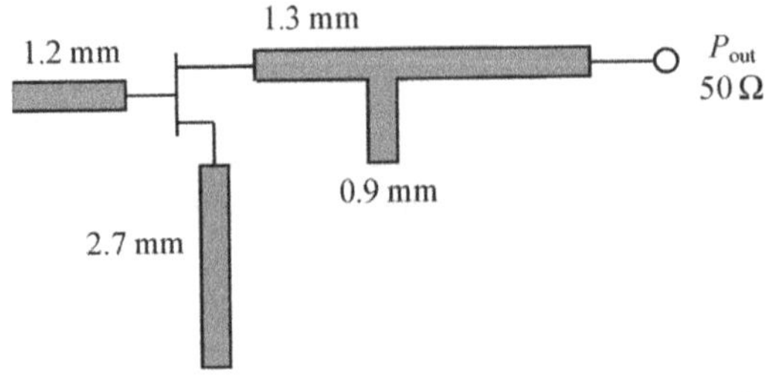

Figure 4.50 Series feedback pHEMT oscillator topology (© 1998 IEEE)

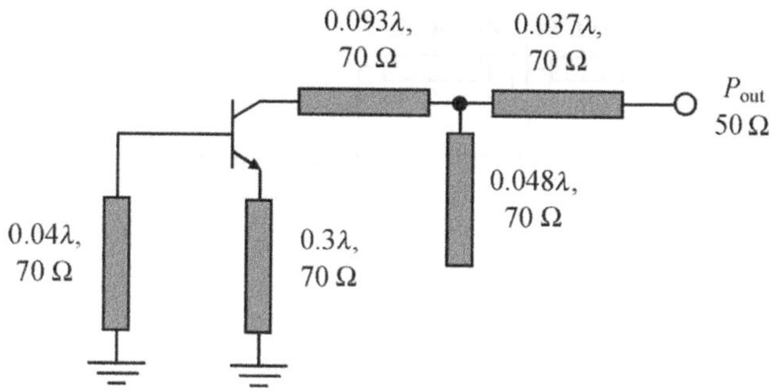

Figure 4.51 Series inductive feedback microwave HBT oscillator schematic (© 1998 IEEE)

source, and the drain is connected to the microstrip output impedance matching circuit. The source open-ended microstrip provides capacitive series feedback, and the gate open-ended microstrip is used to control the oscillation frequency. The substrate used for the oscillator circuit implementation was Rogers RO4350 (thickness = 0.254 mm, $\varepsilon_r = 3.48$, loss tangent = 0.004). Using the small-signal S-parameters of pHEMT with one source lead removed, measured near device pinch-off, the initial design predicted the oscillation frequency around 16.7 GHz. The highest efficiency of 41% was achieved at 17.1 GHz with output power 13.4 mW at supply voltage 2 V.

An example of a common base series inductive feedback microwave HBT oscillator using microstrip lines is shown in Figure 4.51 [54]. Optimizing the oscillator for maximum output power at an oscillation frequency of 20 GHz yields the base inductive reactance with an inductance of 1.4 nH and the emitter capacitive reactance with a capacitance of 40 fF. The AlGaAs/GaAs HBT device with geometry of 2.0 × 10 μm exhibits dc current gain $\beta = 30$, transconductance $g_m = 80$ mS and transition frequency $f_T = 54$ GHz. Changing the value of the emitter capacitance in the range 20–40 fF provides the oscillation conditions within a 20–28 GHz frequency range. The inductive base and capacitive emitter reactances are easy to realize using short-circuited microstrip lines with an electrical length of less than a quarter-wavelength and with an electrical length of more than a quarter-wavelength, respectively. The output matching circuit represents a T-type network consisting of the series inductive and parallel capacitive elements. All feedback elements, as well as the matching network, were realized using microstrip lines with characteristic impedance of 70 Ω. Oscillations started at a collector voltage around 2.6 V. For a 5.2 V collector voltage, an output power as high as 9 dBm with 19% efficiency at oscillation frequency 20.9 GHz was achieved. To provide base–emitter biasing, it is necessary to apply the positive voltage to the base terminal or the negative voltage to the emitter terminal with the appropriate short-circuited condition for the microstrip line using a bypass capacitor.

In order to enhance the negative resistance at higher frequencies, the oscillator circuit configuration can include both the parallel feedback circuit and the series feedback element. For example, each of two oscillators with a common source configuration shown in Figure 4.52 uses a dual feedback scheme consisting of a series feedback element from the source terminal to the ground and a parallel feedback circuit from the drain terminal to the gate terminal. The MESFET oscillator configuration shown in Figure 4.52a is composed of an input open-ended microstrip line connected to the gate terminal, a parallel feedback circuit with two series microstrip lines

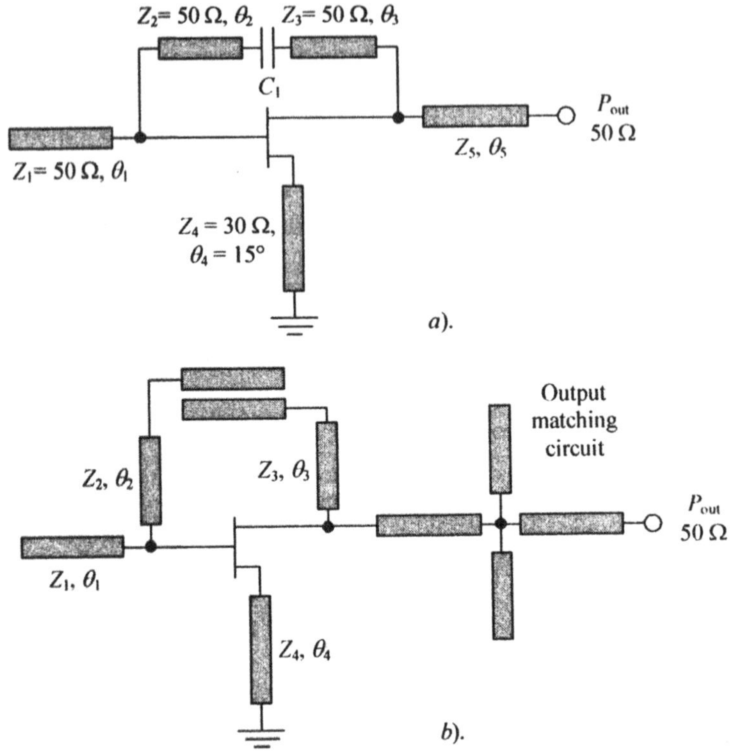

Figure 4.52 Microstrip parallel feedback MESFET oscillator circuit schematics

and blocking capacitor, and a short-circuited microstrip line connected to the source terminal [55]. The output matching circuit consists of the series microstrip line connected between the drain terminal and the load. With the device gate geometry of 1 × 1500 μm, an output power of 350 mW with 26% efficiency can be achieved at 6.5 GHz. The parallel feedback circuit can be designed as a combination of the phase-delay microstrip lines and edge-coupled microstrip lines shown in Figure 4.52b [56]. The edge-coupled lines are used for dc blocking instead of a metal–insulator–metal capacitor having high losses at high frequencies, above 100 GHz. The output matching circuit, consisting of two series microstrip lines and two microstrip stubs, provides the optimum impedance conditions at the drain terminal. The open-ended microstrip line connected to the gate terminal determines the oscillation frequency as a main frequency-tuning element. This monolithic fundamental HEMT oscillator with the device geometry of 0.1 × 90 μm can operate at 130.7 GHz with an output power of −7.9 dBm at a drain voltage of 1.3 V.

REFERENCES

1. K. M. Johnson, 'Large Signal GaAs MESFET Oscillator Design', *IEEE Trans. Microwave Theory Tech.*, **MTT-27**, 217–227 (1979).
2. K. L. Kotzebue and W. J. Parrish, 'The Use of Large Signal *S*-Parameters in Microwave Oscillator Design', *Proc. 1975 IEEE Int. Symp. Circuits and Systems*, pp 487–490 (1975).

3. C. Rauscher, 'Large-Signal Technique for Designing Single-Frequency and Voltage-Controlled Oscillators', *IEEE Trans. Microwave Theory Tech.*, **MTT-29**, 293–304 (1981).
4. C. Rauscher and H. A. Willing, 'Design of Broadband GaAs FET Power Amplifiers', *IEEE Trans. Microwave Theory Tech.*, **MTT-28**, 1054–1059 (1980).
5. H. Abe, 'A GaAs MESFET Oscillator Using Quasi-Linear Design Method', *IEEE Trans. Microwave Theory Tech.*, **MTT-34**, 19–25 (1986).
6. Y. Tajima, B. Wrona and K. Mishima, 'GaAs FET Large-Signal Model and its Application to Circuit Design', *IEEE Trans. Electron Devices*, **ED-28**, 171–175 (1981).
7. R. J. Gilmore and F. J. Rosenbaum, 'An Analytic Approach to Optimum Oscillator Design Using *S*-Parameters', *IEEE Trans. Microwave Theory Tech.*, **MTT-31**, 633–639 (1983).
8. K. L. Kotzebue, 'A Technique for the Design of Microwave Transistor Oscillators', *IEEE Trans. Microwave Theory Tech.*, **MTT-32**, 719–721 (1984).
9. G. D. Vendelin, A. M. Pavio and U. L. Rohde, *Microwave Circuit Design Using Linear and Nonlinear Techniques*, New York: John Wiley & Sons (1990).
10. M. Vehovec, L. Houselander and R. Spense, 'On Oscillator Design for Maximum Power', *IEEE Trans. Circuit Theory*, **CT-15**, 281–283 (1968).
11. V. M. T. Lam, P. C. L. Yip and C. R. Poole, 'Microwave Oscillator Design with Power Prediction', *Electronics Lett.*, **27**, 1574–1575 (1991).
12. B. K. Kormanyos and G. M. Rebeiz, 'Oscillator Design for Maximum Added Power', *IEEE Microwave and Guided Wave Lett.*, **4**, 205–207 (1994).
13. D. F. Page, 'A Design Basis for Junction Transistor Oscillator Circuits', *Proc. IRE*, **46**, 1271–1280 (1958).
14. M. Madihian and T. Noguchi, 'An Analytical Approach to Microwave GaAs FET Oscillator Design', *NEC Res. & Develop.*, **82**, 49–56 (1986).
15. A. V. Grebennikov and V. V. Nikiforov, 'An Analytic Method of Microwave Transistor Oscillator Design', *Int. J. Electronics*, **83**, 849–858 (1997).
16. M. Maeda, K. Kimura and H. Kodera, 'Design and Performance of *X*-Band Oscillators with GaAs Schottky-Gate Field-Effect Transistors', *IEEE Trans. Microwave Theory Tech.*, **MTT-23**, 661–667 (1975).
17. A. V. Grebennikov, 'A Dual Mode Wideband Bipolar Microwave Transistor Voltage-Controlled Oscillator Design', *Int. J. RF & Microwave Computer-Aided Eng.*, **9**, 424–434 (1999).
18. A. Grebennikov, *RF and Microwave Power Amplifier Design*, New York: McGraw-Hill (2004).
19. B. J. Koster, A. Beyer, P. Waldow and I. Wolf, 'An Improved CAD Procedure for Oscillator Circuit Design Using Fast-Approximation Formulas Describing the Voltage-Dependent Nonlinear Characteristics of the Intrinsic Equivalent Circuit Elements of the Active Devices', *Int. J. RF & Microwave Computer-Aided Eng.*, **14**, 550–563 (2004).
20. P. Antognetti and G. Massobrio, *Semiconductor Device Modeling with SPICE*, New York: McGraw-Hill (1993).
21. N. M. Rohringer and P. Kreuzgruber, 'Parameter Extraction for Large-Signal Modeling of Bipolar Junction Transistors', *Int. J. Microwave & Millimeter-Wave Computer-Aided Eng.*, **5**, 161–172 (1995).
22. J. F. Gibbons, 'An Analysis of the Modes of Operation of a Simple Transistor Oscillator', *Proc. IRE*, **49**, 1383–1390 (1961).
23. A. V. Grebennikov, 'An Analytic Method of Microwave Bipolar Single-Frequency and Voltage-Controlled Oscillator Design', *Int. J. RF & Microwave Computer-Aided Eng.*, **9**, 403–414 (1999).
24. A. V. Grebennikov, 'Microwave Transistor Oscillators: Analytic Approach to Simplify Computer-Aided Design', *Microwave J.*, **42**, 292–300 (1999).
25. Serenade 7.5, *Reference Volume*. Ansoft Corporation, New Jersey (1998).
26. J.-M. Dortu, J.-E. Muller, M. Pirola and G. Ghione, 'Accurate Large-Signal GaAs MESFET and HEMT Modeling for Power MMIC Amplifier Design', *Int. J. Microwave & Millimeter-Wave Computer-Aided Eng.*, **5**, 195–209 (1995).

27. A. Materka and T. Kacprzak, 'Computer Calculation of Large-Signal GaAs FET Amplifier Characteristics', *IEEE Trans. Microwave Theory Tech.*, **MTT-33**, 129–135 (1985).
28. A. V. Grebennikov, 'Microwave FET Oscillators: Analytic Approach to Simplify Computer-Aided Design', *Microwave J.*, **43**, 100–110 (2000).
29. J. Kitchen, 'Octave Bandwidth Varactor-Tuned Oscillators', *Microwave J.*, **30**, 347–353 (1987).
30. M. Kazimierczuk, 'A New Approach to the Design of Tuned Power Oscillators', *IEEE Trans. Circuits and Systems*, **CAS-29**, 261–267 (1972).
31. J. C. Shelleng, Amplifying System, U.S. Patent 1,484,967 (February 1924).
32. A. V. Grebennikov, 'Switched-Mode Tuned High-Efficiency Power Amplifiers: Historical Aspect and Future Prospect', *2002 IEEE RFIC Symp. Dig.* 49–52 (2002).
33. N. O. Sokal and A. D. Sokal, 'Class E — A New Class of High-Efficiency Tuned Single-Ended Switching Power Amplifiers', *IEEE J. Solid-State Circuits*, **SC-10**, 168–176 (1975).
34. J. Ebert and M. Kazimierczuk, 'Class E High-Efficiency Tuned Power Oscillator', *IEEE J. Solid-State Circuits*, **SC-16**, 62–66 (1981).
35. D. V. Chernov, M. Kazimierczuk and V. G. Krizhanovski, 'Class-E MOSFET Low-Voltage Power Oscillator', *Proc. 2002 IEEE Int. Symp. Circuits and Systems*, **5**, 509–512 (2002).
36. N. O. Sokal and F. H. Raab, 'Harmonic Output of Class-E RF Power Amplifiers and Load Coupling Network Design', *IEEE J. Solid-State Circuits*, **SC-12**, 86–88 (1977).
37. M. K. Kazimierczuk, V. G. Krizhanovski, J. V. Rassohina and D. V. Chernov, 'Class-E MOSFET Tuned Power Oscillator Design Procedure', *IEEE Trans. Circuits and Systems — I: Regular Papers*, **CAS-I-52**, 1138–1147 (2005).
38. F. H. Raab, 'Idealized Operation of the Class E Tuned Power Amplifier', *IEEE Trans. Circuits and Systems*, **CAS-24**, 725–735 (1977).
39. H. Hase, H. Sekiya, J. Lu and T. Yahagi, 'Novel Design Procedure for MOSFET Class E Oscillator', *IEICE Trans. Fundamentals*, **E87-A**, 2241–2247 (2004).
40. T. B. Mader, E. W. Bryerton, M. Marcovic, M. Forman and Z. Popovic, 'Switched-Mode High-Efficiency Microwave Power Amplifiers in a Free-Space Power-Combiner Array', *IEEE Trans. Microwave Theory Tech.*, **MTT-46**, 1391–1398 (1998).
41. E. W. Bryerton, W. A. Shiroma and Z. B. Popovic, 'A 5-GHz High-Efficiency Class-E Oscillator', *IEEE Microwave and Guided Wave Lett.*, **6**, 441–443 (1996).
42. M. Matsuo, H. Sekiya, T. Suetsugu, K. Shinoda and S. Mori, 'Design of a High-Efficiency Class DE Tuned Power Oscillator', *IEEE Trans. Circuits and Systems — I: Fundamental Theory Appl.*, **CAS-I-47**, 1645–1649 (2000).
43. H. Koizumi, T. Suetsugu, M. Fujii, K. Shinoda, Mori and K. Ikeda, 'Class DE High-Efficiency Tuned Power Amplifier', *IEEE Trans. Circuits and Systems — I: Fundamental Theory Appl.*, **CAS-I-43**, 51–60 (1996).
44. H. J. Round, Wireless Telegraph and Telephone Transmission, U.S. Patent 1,564,627 (December 1925).
45. M.-Q. Lee, S. Nam, Y. Kwon and K.-W. Yeom, 'Analytic Design of High Efficiency Harmonic Two Signal Method', *1997 IEEE MTT-S Int. Microwave Symp. Dig.*, **3**, 1495–1498 (1997).
46. P. Berini, M. Desgagne, F. M. Ghannouchi and R. G. Bosisio, 'An Experimental Study of the Effects of Harmonic Loading on Microwave MESFET Oscillators and Amplifiers', *IEEE Trans. Microwave Theory Tech.*, **MTT-42**, 943–950 (1994).
47. M. Prigent, M. Camiade, G. Pataut, D. Reffet, J. M. Nebus and J. Obregon, 'High Efficiency Free Running Class F Oscillators', *1995 IEEE MTT-S Int. Microwave Symp. Dig.*, **3**, 1317–1320 (1995).
48. H. Abe, 'A GaAs MESFET Self-Bias Mode Oscillator', *IEEE Trans. Microwave Theory Tech.*, **MTT-34**, 167–172 (1986).
49. R. M. Rector and G. Vendelin, 'A 1.0 Watt GaAs MESFET Oscillator at *X*-Band', *1978 IEEE MTT-S Int. Microwave Symp. Dig.*, 145–146 (1978).

50. H. J. Finlay, J. S. Joshi and S. C. Cripps, 'An X Band F.E.T. Oscillator with Low F.M. Noise', *Electronics Lett.*, **14**, 198–199 (1978).
51. A. Bangert, M. Schlechtweg, W. Reinhert, W. H. Haydl, A. Hulsmann and K. Kohler, 'Monolithic Integrated 75 GHz Oscillator with High Output Power Using a Pseudomorphic HFET', *1994 IEEE MTT-S Int. Microwave Symp. Dig.*, 135–138 (1994).
52. V. S. Kaper, V. Tilak, H. Kim, A. V. Vertiatchikh, R. M. Thompson, T. R. Prunty, L. F. Eastman and J. R. Shealy, 'High-Power Monolithic AlGaN/GaN HEMT Oscillator', *IEEE J. Solid-State Circuits*, **SC-38**, 1457–1461 (2003).
53. J. O. McSpadden, L. Fan and K. Chang, 'High-Efficiency *Ku*-Band Oscillators', *IEEE Trans. Microwave Theory Tech.*, **MTT-46**, 1566–1570 (1998).
54. M. Madihian and H. Takahashi, 'A Low-Noise *K*-*Ka* Band Oscillator Using AlGaAs/GaAs Heterojunction Bipolar Transistors', *IEEE Trans. Microwave Theory Tech.*, **MTT-39**, 133–136 (1991).
55. Y. Mitsui, M. Nakatani and S. Mitsui, 'Design of GaAs MESFET Oscillator Using Large-Signal *S*-Parameters', *IEEE Trans. Microwave Theory Tech.*, **MTT-25**, 981–984 (1977).
56. Y. Kwon, D. Pavlidis, T. L. Brock and D. C. Streit, 'A *D*-Band Monolithic Fundamental Oscillator Using InP-Based HEMTs', *IEEE Trans. Microwave Theory Tech.*, **MTT-41**, 2336–2344 (1993).

5

Noise in oscillators

It is very important for the oscillator noise model to express a clear relationship between the resonant circuit and active device noise model parameters. The simple Leeson linear model for a feedback oscillator, which was derived empirically, is based on the expectation that the contribution to real oscillator output spectrum is provided by two basic processes. The first process is a result of the phase fluctuations due to the additive white noise at frequency offsets close to the carrier. The second process is a result of the low-frequency fluctuations or flicker noise up-converted to the carrier region because of the active device nonlinear effects. The nonlinear Kurokawa analysis based on the sinusoidal representation of the current in the negative-resistance oscillator extends the oscillator noise model by introducing relationships between the noise power, stability conditions and amplitude-to-phase conversion. However, such a noise generation mechanism does not consider the mixing effect from the inherent nonlinear behaviour of the active device when the current at the output of the active device must be represented by a Fourier series expansion. Thus, the phase noise generated around the fundamental frequency of the oscillation is generally an equal contribution of two simultaneous and correlated phenomena: additive phase noise due to phase modulation process and converted phase noise due to conversion from one sideband to another.

5.1 NOISE FIGURE

There are several primary noise sources in the oscillator circuit. Thermal or white noise is created by random motion of charge carriers due to thermal excitation, being always found in any conducting medium whose temperature is above absolute zero whatever the nature of the conduction process or the nature of the mobile charge carriers [1]. This random motion of carriers creates a fluctuating voltage on the terminals of each resistive element which increases with temperature. However, if the average value of such a voltage is zero, then the noise power on its terminal is not zero being proportional to the resistance of the conductor and to its absolute temperature. The resistor as a thermal noise source can be represented by either of the noise sources shown in Figure 5.1. The noise voltage source and noise current source can be respectively described by Nyquist equations through their mean-square noise voltage and

RF and Microwave Transistor Oscillator Design A. Grebennikov

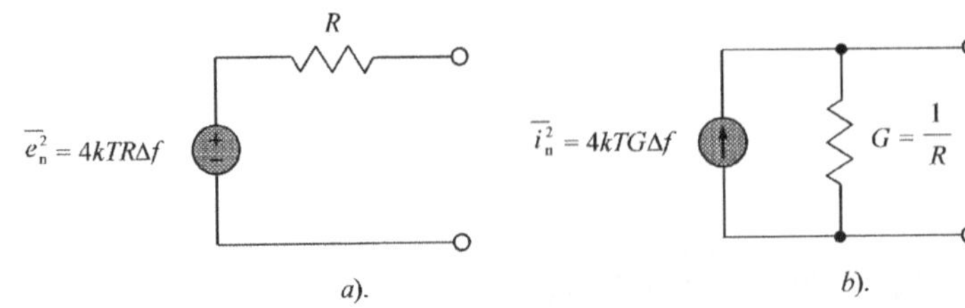

Figure 5.1 Equivalent circuits to represent thermal noise sources

noise current values as

$$\overline{e_n^2} = 4kTR\Delta f \tag{5.1}$$

$$\overline{i_n^2} = \frac{4kT\Delta f}{R} \tag{5.2}$$

where $k = 1.38 \times 10^{-23}$ J/K is the Boltzmann constant, T is the absolute temperature, and $kT = 4 \times 10^{-21}$ W/Hz $= -174$ dBm/Hz at ambient temperature $T = 290$ K. The thermal noise is proportional to the frequency bandwidth Δf. It can be represented by the voltage source in series with resistor R, or by the current source in parallel to the resistor R. The maximum noise power can be delivered to the load when $R = R_L$, where R_L is the load resistance, being equal to $kT\Delta f$. Hence, the noise power density when the noise power is normalized by Δf is independent of frequency and is considered as white noise. The root-mean-square noise voltage and current are proportional to the square root of the frequency bandwidth Δf.

Shot noise is associated with the carrier injection through the device *p–n* junction, being generated by the movement of individual electrons within the current flow. In each forward-biased junction, there is a potential barrier that can be overcome by the carriers with higher thermal energy. Such a process is random and mean-square noise current can be given by

$$\overline{i_n^2} = 2qI\Delta f \tag{5.3}$$

where q is the electron charge and I is the direct current flowing through the *p–n* junction. The shot noise depends on the thermal energy of the carriers near the potential barrier and its power density is independent of frequency. It has essentially a flat spectral distribution and can be treated as the thermal or white type of noise with current source $\overline{i_n^2}$ connected in parallel to the small-signal junction resistance. In a voltage noise representation, when the noise voltage source is connected in series with such a resistor, it can be written as

$$\overline{e_n^2} = 2kTr\Delta f \tag{5.4}$$

where $r = kT/qI$ is the junction resistance.

Circuits containing more than one resistor can be analysed by reducing their number to the only one (Thevenin) equivalent resistance to obtain the mean-square noise voltage in the form of Equation (5.1) [2]. As an example, the noise equivalent of a circuit shown in Figure 5.2a, where a signal source V_S is driving a hypothetical noise-free load resistor R_{in} (which can be considered an input of the power amplifier) through three noise resistors R_1, R_2, and R_3, is a noise voltage source $\overline{e_n^2} = 2kTR_T\Delta f$ connected in series with an ideal (noise-free) resistor equal to the Thevenin resistance R_T, as shown in Figure 5.2b.

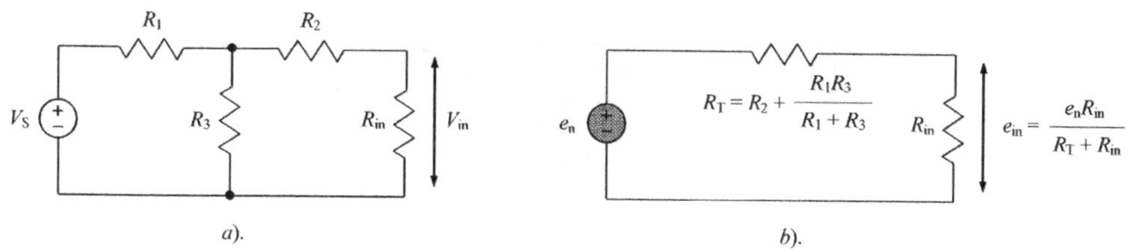

Figure 5.2 Circuit with three resistors and its equivalent with a noise voltage source

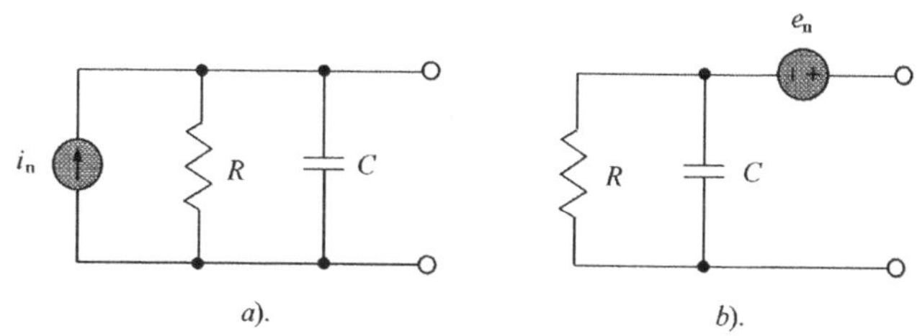

Figure 5.3 Noise characterization of a two-port *RC* network

Consider now a simple parallel *RC* circuit shown in Figure 5.3a, where the thermal noise due to the parallel resistor is represented by a parallel noise current source i_n. Nyquist has determined the thermal noise output of a two-port network containing both resistive and reactive elements, as shown in Figure 5.3b. In this case, the mean-square thermal noise voltage is given by

$$\overline{e_n^2} = 4kT \int_{\Delta f} R(f)\,df \tag{5.5}$$

where integration is performed over the frequency bandwidth of interest Δf and

$$R(f) = \frac{R}{1 + (2\pi fCR)^2} \tag{5.6}$$

is the real part of the output circuit impedance at frequency f.

Hence, the parallel current noise source can be equivalently transformed to the series noise voltage source by integration over infinite frequency bandwidth with the total mean square noise voltage given by

$$\overline{e_n^2} = \frac{4kT}{2\pi} \int_0^\infty \frac{R\,d\omega}{1 + (\omega CR)^2} = \frac{2kTR}{\pi} \int_0^\infty \frac{d\omega}{1 + (\omega CR)^2} = \frac{kT}{C} \tag{5.7}$$

where the resistance R has no effect on the noise voltage which depends on the value of the capacitance C and temperature T only [2, 3].

It is well known that any linear noisy two-port network can be represented as a noise-free two-port part with noise sources at the input and the output connected in different way [4, 5]. For example, the noisy linear two-port network with internal noise sources shown in Figure 5.4a can be redrawn, either in the impedance form with external series voltage noise sources shown

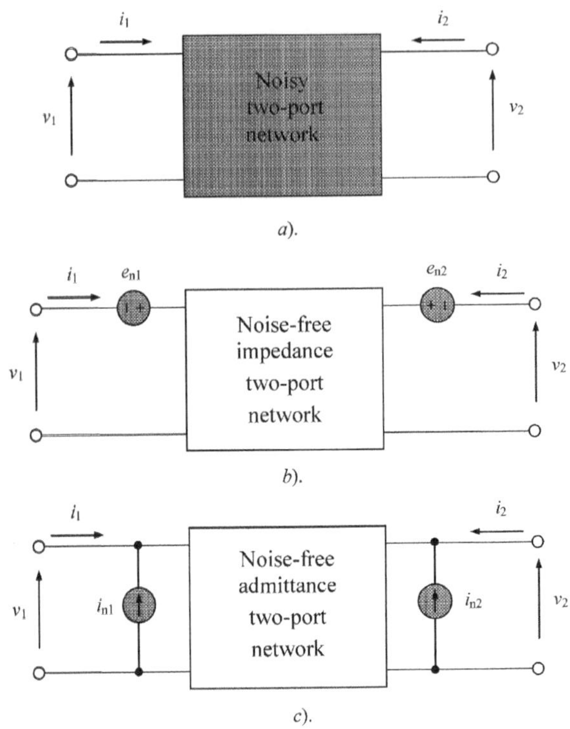

Figure 5.4 Linear two-port network with noise sources

in Figure 5.4b, or in the admittance form with external parallel current noise sources shown in Figure 5.4c.

However, to fully describe the noise properties of the two-port network at fixed frequency, sometimes it is convenient to represent it through the noise-free two-port part and the noise sources equivalently located at the input. Such a circuit is equivalent to the configurations with noise sources located at the input and the output [6]. In this case, it is enough to use four parameters: the noise spectral densities of both noise sources and the real and imaginary parts of its correlation spectral density. These four parameters can be defined by measurements at the two-port network terminals. The two-port network current and voltage amplitudes are related to each other through a system of two linear algebraic equations. By taking into account the noise sources at the input and the output, these equations in the impedance and admittance forms can be respectively written as

$$V_1 = Z_{11}I_1 + Z_{12}I_2 - V_{n1} \tag{5.8}$$

$$V_2 = Z_{21}I_1 + Z_{22}I_2 - V_{n2} \tag{5.9}$$

and

$$I_1 = Y_{11}V_1 + Y_{12}V_2 - I_{n1} \tag{5.10}$$

$$I_2 = Y_{21}V_1 + Y_{22}V_2 - I_{n2} \tag{5.11}$$

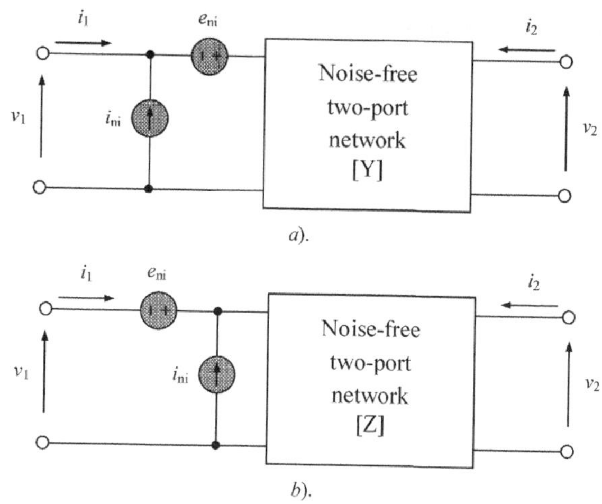

Figure 5.5 Linear two-port network with noise sources at input

where the voltage and current noise amplitudes represent the Fourier transforms of noise fluctuations.

The equivalent two-port network with voltage and current noise sources located at its input is shown in Figure 5.5a, where $[Y]$ is the two-port network admittance matrix and ratios between current and voltage amplitudes can be written as

$$I_1 = Y_{11}(V_1 + V_{ni}) + Y_{12}V_2 - I_{ni} \tag{5.12}$$

$$I_2 = Y_{21}(V_1 + V_{ni}) + Y_{22}V_2 \tag{5.13}$$

From comparison of Equations (5.10) and (5.11) with Equations (5.12) and (5.13) it follows that

$$V_{ni} = -\frac{I_{n2}}{Y_{21}} \tag{5.14}$$

$$I_{ni} = I_{n1} - \frac{Y_{11}}{Y_{21}}I_{n2} \tag{5.15}$$

representing the relationships between the current noise sources at the input and the output corresponding to the circuit shown in Figure 5.4c and voltage and current noise sources at the input only corresponding to the circuit shown in Figure 5.5a. In this case, Equations (5.14) and (5.15) are valid only if $Y_{21} \neq 0$ that always takes place in practice. Similar equations can be written for the circuit with the series noise voltage source followed by a parallel noise current source shown in Figure 5.5b in terms of impedance Z-parameters to represent the relationships between the voltage noise sources at the input and the output corresponding to the circuit shown in Figure 5.4b. The use of voltage and current noise sources at the input enables the combination of all internal two-port network noise sources.

To evaluate a quality of the two-port network, it is important to know the amount of noise added to a signal passing through it. Usually this can be done by introducing such an important parameter as the noise figure or noise factor. The noise figure of the two-port network is intended as an indication of its noisiness. The lower the noise figure, the less noise is contributed by the two-port network. The noise figure is defined as

$$F = \frac{(S/N)_{\text{in}}}{(S/N)_{\text{out}}} \tag{5.16}$$

where $(S/N)_{\text{in}}$ is the signal-to-noise ratio available at the input and $(S/N)_{\text{out}}$ is the signal-to-noise ratio available at the output.

For a two-port network characterized by the power gain G, the noise figure can be rewritten as

$$F = \frac{S_{\text{in}}/N_{\text{in}}}{GS_{\text{in}}/G(N_{\text{in}} + N_{\text{int}})} = 1 + \frac{N_{\text{int}}}{N_{\text{in}}} \tag{5.17}$$

where N_{int} is the additional noise power added by the two-port network referred to the input. From Equation (5.17) it follows that the noise figure depends on the source impedance Z_S shown in Figure 5.6a, but not on the circuit connected to the output of the two-port network.

Hence, if the two-port network is driven from the source with impedance $Z_S = R_S + jX_S$, the noise figure F of this two-port network in terms of the model shown in Figure 5.6b with

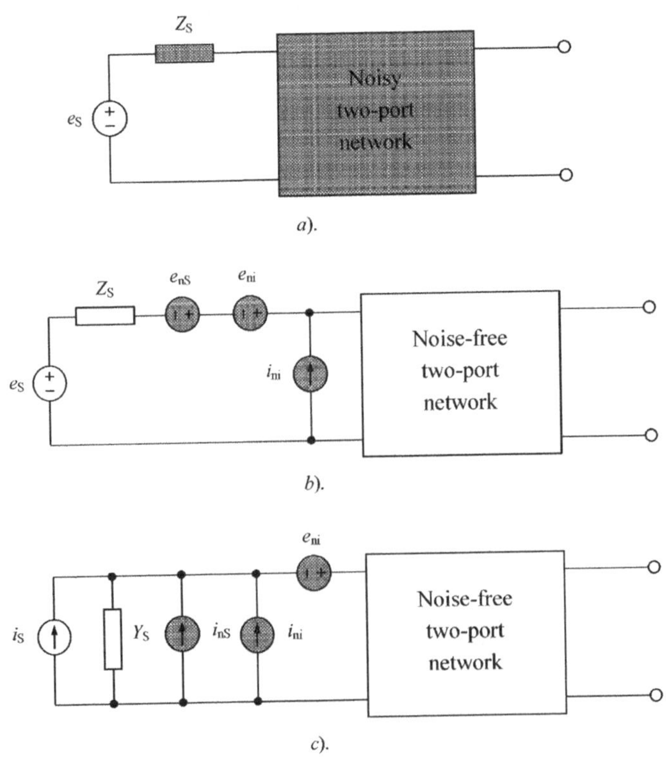

Figure 5.6 Linear two-port networks to calculate the noise figure

input voltage and current noise sources and noise-free two-port network can be obtained by

$$F = 1 + \frac{\overline{|e_n + Z_S i_n|^2}}{4kTR_S\Delta f} = 1 + \frac{R_n + |Z_S|^2 G_n + 2\sqrt{R_n G_n}\mathrm{Re}(CZ_S)}{R_S} \tag{5.18}$$

where

$$R_n = \frac{\overline{e_{ni}^2}}{4kT\Delta f} \tag{5.19}$$

is the equivalent input-referred noise resistance corresponding to the noise voltage source,

$$G_n = \frac{\overline{i_{ni}^2}}{4kT\Delta f} \tag{5.20}$$

is the equivalent input-referred noise conductance corresponding to the noise current source, and

$$C = \frac{\overline{i_{ni}e_{ni}^*}}{\sqrt{\overline{i_{ni}^2}\,\overline{e_{ni}^2}}} \tag{5.21}$$

is the correlation coefficient representing a complex number less than or equal to 1 in magnitude [5].

Here, G_n and R_n generally do not represent the particular circuit immitances, but depend on the bias level, resulting in a dependence of the noise figure on the operating bias point of the active device.

As the source impedance Z_S is varied over all values with positive R_S, the noise figure F has a minimum value of

$$F_{min} = 1 + 2\sqrt{R_n G_n}\left[\sqrt{1 - (\mathrm{Im}C)^2} + \mathrm{Re}C\right] \tag{5.22}$$

which occurs for the optimum source impedance $Z_{Sopt} = R_{Sopt} + jX_{Sopt}$ given by

$$|Z_{Sopt}|^2 = \frac{R_n}{G_n} \tag{5.23}$$

$$X_{Sopt} = \sqrt{\frac{R_n}{G_n}}\mathrm{Im}C \tag{5.24}$$

As a result, the noise figure F for nonoptimum input impedance Z_S can be expressed in terms of F_{min} as

$$F = F_{min} + |Z_S - Z_{Sopt}|^2 \frac{G_n}{R_S} = F_{min} + \left[(R_S - R_{Sopt})^2 + (X_S - X_{Sopt})^2\right]\frac{G_n}{R_S} \tag{5.25}$$

Similarly, the noise figure F can be equivalently expressed using a model shown in Figure 5.6c with source admittance $Y_S = G_S + jB_S$ as

$$F = F_{min} + \left|Y_S - Y_{Sopt}\right|^2 \frac{R_n}{G_S}$$
$$= F_{min} + \left[(G_S - G_{Sopt})^2 + (B_S - B_{Sopt})^2\right]\frac{R_n}{G_S} \quad (5.26)$$

where F_{min} is the minimum noise figure of the two-port network which can be realized with respect to the source admittance Y_S, $Y_{Sopt} = G_{Sopt} + jB_{Sopt}$ is the optimal source admittance, and R_n is the equivalent noise resistance which measures how rapidly the noise figure degrades when the source admittance Y_S deviates from its optimum value Y_{Sopt} [7]. Since the admittance Y_S is generally complex, then its active and imaginary parts can be controlled independently. To obtain the minimum value of the noise figure, the two matching conditions of $G_S = G_{Sopt}$ and $B_S = B_{Sopt}$ must be satisfied.

The physical interpretation of the noise sources which are assumed to be stationary random processes is given by their self- and cross-power spectral densities which are defined as the Fourier transform of their auto- and cross-correlation functions. These spectral densities in two-port matrix form leads to the so-called correlation matrices with their admittance, impedance, or chain representations [8]. The correlation matrix C belonging to the noise sources s_{n1} and s_{n2} can be written as

$$C = \frac{1}{\Delta f}\begin{bmatrix} \overline{s_{n1}s_{n1}^*} & \overline{s_{n1}s_{n2}^*} \\ \overline{s_{n2}s_{n1}^*} & \overline{s_{n2}s_{n2}^*} \end{bmatrix} \quad (5.26)$$

where the asterisk denotes the complex conjugate. For example, the admittance correlation matrix for the circuit shown in Figure 5.4c with two parallel current noise sources is obtained as

$$C_Y = \frac{1}{\Delta f}\begin{bmatrix} \overline{i_{n1}i_{n1}^*} & \overline{i_{n1}i_{n2}^*} \\ \overline{i_{n2}i_{n1}^*} & \overline{i_{n2}i_{n2}^*} \end{bmatrix} \quad (5.27)$$

Determination of the noise correlation matrix is based on the following procedure:

- each element in the diagonal matrix is equal to the sum of the noise current of each element connected to the corresponding node: the first diagonal element is the sum of noise currents connected to the node 1, while the second diagonal element is the sum of noise currents connected to node 2;
- the off-diagonal elements are the negative noise current of the element connected to the pair of the corresponding node: therefore, a noise current source between nodes 1 and 2 goes into the matrix at locations (1, 2) and (2, 1);
- if a noise current source is grounded, it will only contribute to one entry in the noise correlation matrix — at the appropriate location on the diagonal: if it is not grounded, it will contribute to four entries in the matrix — two diagonal entries corresponding to the two nodes and two off-diagonal entries.

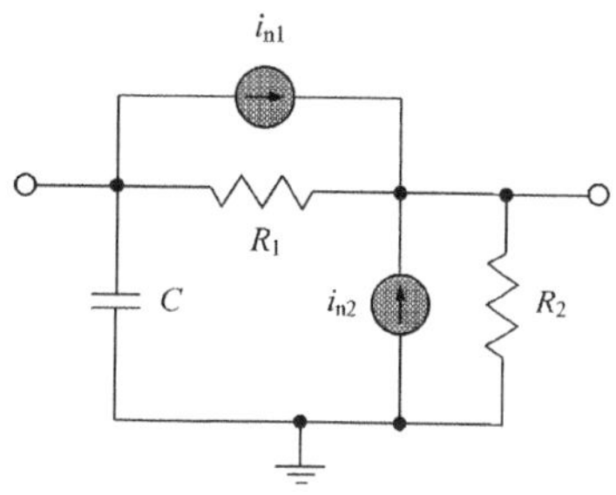

Figure 5.7 Circuit with two noise current sources

By applying these rules for the circuit with two current noise sources shown in Figure 5.7, the admittance noise correlation matrix C_Y can be defined as

$$C_Y = \frac{1}{\Delta f}\begin{bmatrix} \overline{i_{n1}^2} & -\overline{i_{n1}^2} \\ -\overline{i_{n1}^2} & \overline{i_{n1}^2} + \overline{i_{n2}^2} \end{bmatrix} \tag{5.28}$$

To form the impedance noise correlation matrix with voltage noise sources, we can write

$$C_Z = \frac{1}{\Delta f}\begin{bmatrix} \overline{e_{n1}e_{n1}^*} & \overline{e_{n1}e_{n2}^*} \\ \overline{e_{n2}e_{n1}^*} & \overline{e_{n2}e_{n2}^*} \end{bmatrix} = [Z]\,[C_Y]\,[Z]^{T} \tag{5.29}$$

where $[Z]$ is the impedance Z-matrix of the two-port network and T denotes the Hermitian or transposed conjugation.

In the case where the correlation matrix cannot be theoretically derived, the measurements can be used for its determination. Such measurements are usually done by defining the equivalent noise resistance R_n, the optimal source admittance Y_{Sopt} and the minimum noise figure F_{min}. As a result, the chain representation of the noise correlation matrix is obtained as

$$C_A = 4kT\begin{bmatrix} R_n & \dfrac{F_{min}-1}{2} - R_n Y_{Sopt} \\ \dfrac{F_{min}-1}{2} - R_n Y_{Sopt}^* & R_n\left|Y_{Sopt}\right|^2 \end{bmatrix} \tag{5.30}$$

where T is the absolute temperature [8]. However, if the correlation matrix has been determined, the noise parameters can be calculated analytically from

$$R_n = \frac{C_{11}^A}{4kT} \tag{5.31}$$

$$Y_{Sopt} = \sqrt{\frac{C_{22}^A}{C_{11}^A} - \mathrm{Im}^2\left(\frac{C_{12}^A}{C_{11}^A}\right)} - j\mathrm{Im}\left(\frac{C_{12}^A}{C_{11}^A}\right) \tag{5.32}$$

$$F_{min} = 1 + \frac{C_{12}^A + C_{11}^A Y_{Sopt}}{2kT} \tag{5.33}$$

where C_{11}^A, C_{12}^A, C_{21}^A and C_{22}^A are the elements of the chain correlation matrix C_A.

5.2 FLICKER NOISE

The flicker or $1/f$ noise is a low-frequency noise associated with a fluctuation in the conductance with a power spectral density proportional to $f^{-\gamma}$, where $\gamma = 1.0 \pm 0.1$ in a wide frequency range, usually measured from 1 Hz to 10 kHz [9]. Its spectrum cannot be exactly f^{-1} at offset frequencies from $f = 0$ to $f \to \infty$, since neither the integral over the power density nor the Fourier transform would be able to have finite values. Unlike the thermal or shot noise sources, the origin of the $1/f$ noise is not exactly clear and open to debate despite its predictable behaviour. Generally, it is a result of both surface and bulk effects in the semiconductor material and is not generated by the current. In series experiments it was shown that there is a type of $1/f$ noise that is a fluctuation in the carrier mobility due to lattice scattering.

A significant contribution to the low-frequency noise is made by the generation–recombination and burst noises [10]. The generation–recombination noise associated with the fluctuations in the number of the carriers rather than their mobility is due to trap centres within the bandgap of a semiconductor. It may have any frequency behaviour between f^0 and f^{-2}. If not masked by thermal noise, the low-frequency noise generated from these trap centres becomes f^{-2} at very high frequencies. However, if the lifetime of the carriers in the semiconductor is finite, the noise spectral density reaches a plateau at very low frequencies. Burst noise (random telegraph noise) is a special kind of generation–recombination noise due to a single trap in the active device region. It is often observed in submicrometre devices or in devices with very poor crystalline quality. In such devices, a trap level with a certain energy and at a specific location in the active device region (a single localized trap) traps and detraps the carriers causing an on–off time-dependent signal similar to a telegraph signal [11].

Flicker noise in bipolar transistors is associated mainly with generation–recombination centres which contribute to random trapping and release of free carriers [12, 13]. This relatively slow process is always associated with flowing current the mean-square value of which $\overline{i_n^2}$ as a function of the offset frequency f can be approximated by

$$\overline{i_n^2} = K_F I^k \frac{\Delta f}{f} \tag{5.34}$$

where K_F is the flicker noise coefficient and k is the flicker noise exponent. Both these coefficients are device dependent. In conventional high-quality silicon bipolar devices, the low-frequency noise is determined by $1/f$ noise in the base current due to carriers injected from the base into the emitter, since the emitter series resistance can be neglected, the base series resistance is low and the collector current has an ideality factor of 1. In downscaled polysilicon bipolar transistors with lower emitter area, at low bias currents the $1/f$ noise in the base current is dominant, while at higher bias currents the influence of the series resistances on the noise becomes noticeable. In GaAs/AlGaAs HBT devices, the $1/f$ noise in the base current can be described by Equation (5.34) with $K_F \approx 10^{-10}\mathrm{A}^{2-k}$ and $k \approx 1.6$, whereas, for the $1/f$ noise in the collector current, $K_F \approx 10^{-12}\mathrm{A}^{2-k}$ and $k \approx 1.3$ [14]. The contribution of the extrinsic base resistance noise becomes more important with scaling, especially for devices with very high transition frequency f_T [15].

There are two major theories to explain the physical origin of $1/f$ noise in MOSFET devices, one is based on the carrier number fluctuation theory when the flicker noise is attributed to the random trapping and detrapping processes of charges near the Si–SiO$_2$ interface, the other is based on mobility fluctuation theory considering the flicker noise as a result of the fluctuations in bulk mobility [3, 16, 17]. Assuming that the channel can exchange charges with the oxide traps through tunnelling, the charge fluctuation results in fluctuation of the surface potential, which in turn modulates the channel carrier density. At the same time, it is considered that the fluctuation of bulk mobility is induced by fluctuations in phonon population through phonon scattering. Generally, the measured noise power in MOSFET devices has a more complicated dependence on the gate bias and oxide thickness than predictions based on the number or bulk mobility fluctuation theory only. Also, the surface mobility fluctuation mechanism should be taken into account, attributed to the scattering effect of fluctuating oxide charge [18]. The implementation of oxide and interface trapping noise into a partial differential equation-based semiconductor device simulator shows the correct prediction of $1/f$ noise spectral densities for submicrometre MOSFET devices operating at subthreshold and strong inversion in saturation [19].

The dependence of flicker noise power on gate bias and oxide thickness for an n-channel MOSFET in terms of the equivalent gate noise power $\overline{e^2_{\rm ng}}$ can be modelled by the following empirical expression:

$$\overline{e^2_{\rm ng}} = \frac{\Delta f}{LWf}\left[K_1\left(\frac{q}{C_{\rm ox}}\right)^{\rm m}(V_{\rm gs} - V_{\rm th}) + K_2\left(\frac{q}{C_{\rm ox}}\right)^2\right] \tag{5.35}$$

where $V_{\rm gs}$ is the gate–source voltage, $V_{\rm th}$ is the threshold voltage, L is the channel length, W is the channel width, $C_{\rm ox}$ is the gate–oxide capacitance per unit area, q is the electron charge, K_1 and K_2 are empirical constants, and $m = 0.7$–1.2 [17]. In this case, it was proposed that the term with K_1 represents the contribution from the mobility fluctuation whereas the term with K_2 represents the contribution from the number fluctuation.

Based on the experimental results for different n-channel MOSFET devices, Equation (5.34) for relative fluctuations of the drain current can be rewritten as

$$\frac{\overline{i^2_{\rm nd}}}{I^2_{\rm d}\Delta f} = \begin{cases} \dfrac{K_{\rm F}}{I_{\rm d}^{3/2} f} & V_{\rm gs} < V^*_{\rm gs} \\ \dfrac{K_{\rm F} I^*_{\rm d}}{I_{\rm d}^{5/2} f} & V_{\rm gs} \geq V^*_{\rm gs} \end{cases} \tag{5.36}$$

where $V^*_{\rm gs}$ and $I^*_{\rm d}$ are the gate–source voltage and drain current, respectively, at which the device transconductance $g_{\rm m}$ begins to drop, i.e., $\partial g_{\rm m}/\partial I_{\rm d} \leq 0$, and $K_{\rm F} = 3.1$–$6.6 \times 10^{-14}{\rm A}^{3/2}$ [20]. It should be noted that, for constant $V_{\rm gs}$, the relative fluctuations of the drain current remain almost constant. Using the negative feedback resistor $R_{\rm s}$ connected between the device source terminal and the ground results in flicker noise reduction according to $1/(1 + g_{\rm m}R_{\rm s})^2$. To determine the parameters of the equivalent noise voltage source at the gate of the device, we can write

$$\overline{e^2_{\rm ng}} = \frac{\overline{i^2_{\rm nd}}}{I^2_{\rm d}}\frac{I^2_{\rm d}}{g^2_{\rm m}} \tag{5.37}$$

whose dependence on the drain current becomes significant only at high values when the device transconductance reduces.

Generally, the noise behaviour of the n-channel and p-channel MOSFET devices is different, since the pMOS transistor, being less noisy, usually has a channel at a larger distance from the interface [21]. Unlike MOSFET device, the flicker noise of its JFET counterpart is negligible. At low temperatures the noise spectrum of the JFET indicates the presence of several types of generation–recombination process, but the $1/f$ noise component is absent [22]. However, the component related to $1/f$ noise can appear at high temperatures, above 200 K [23]. Such a situation can be explained by the fact that the noise behaviour of a JFET device having a p–n junction cannot be characterized by semiconductor–oxide surface effects since its channel is separated by the depletion region localized along the device channel. On the contrary, the GaAs MESFET devices are characterized by a significant value of flicker noise. This is a combined result of the gate leakage current, fluctuations of the Schottky barrier space charge region, carrier number fluctuations in the channel and at the interface between the channel and substrate due to trapping phenomena [24]. The trapping mechanism is especially pronounced in GaAs material where the trapping centres can arise from a variety of causes such as trace impurities and crystal defects [25]. It is shown that the low-frequency noise in GaAs MESFET devices on InP substrate is directly related to the structural quality of GaAs active layers when increasing the buffer layer thickness for GaAs lattice mismatched on InP substrate improves the noise performance [26].

5.3 ACTIVE DEVICE NOISE MODELLING

5.3.1 MOSFET devices

The noise behaviour of the MOSFET device can be described based on its equivalent circuit representation which includes the main elements responsible for the device electrical behaviour and noise sources. The noise generated by a circuit element can be modelled as a result of a small-signal electrical excitation. Each noise source is considered as statistically uncorrelated to the other noise sources in the circuit and the contribution of each noise source to the total noise is determined on an individual basis. The total noise represents the root-mean-square sum of these individual noise contributions. Since a device channel material is resistive, it exhibits thermal noise as a major source of noise, which can be represented by a noise current source $\overline{i_{\rm nd}^2}$ connected between the drain and the source in the MOSFET small-signal equivalent circuit shown in Figure 5.8a, where the flicker noise is also included too. The induced gate current noise is modelled by the gate noise current source $\overline{i_{\rm ng}^2}$ connected across the gate–source capacitance $C_{\rm gs}$.

The noise voltage and current sources can be given through their mean-square values as

$$\overline{e_{\rm ng}^2} = 4kT R_{\rm g}\Delta f \tag{5.38}$$

$$\overline{i_{\rm ng}^2} = \frac{4kT\eta(\omega C_{\rm gs})^2\Delta f}{g_{\rm m}} \tag{5.39}$$

$$\overline{i_{\rm ns}^2} = \frac{4kT\Delta f}{R_{\rm s}} \tag{5.40}$$

$$\overline{e_{\rm nd}^2} = 4kT R_{\rm d}\Delta f \tag{5.41}$$

$$\overline{i_{\rm nd}^2} = \frac{4kT\gamma}{R_{\rm ds0}}\Delta f + K_{\rm F}\frac{I_{\rm d}^{A_{\rm F}}}{C_{\rm ox}L_{\rm eff}^2 f}\Delta f \tag{5.42}$$

where η is the induced gate noise coefficient, $g_{\rm m}$ is the device transconductance, $I_{\rm d}$ is the drain

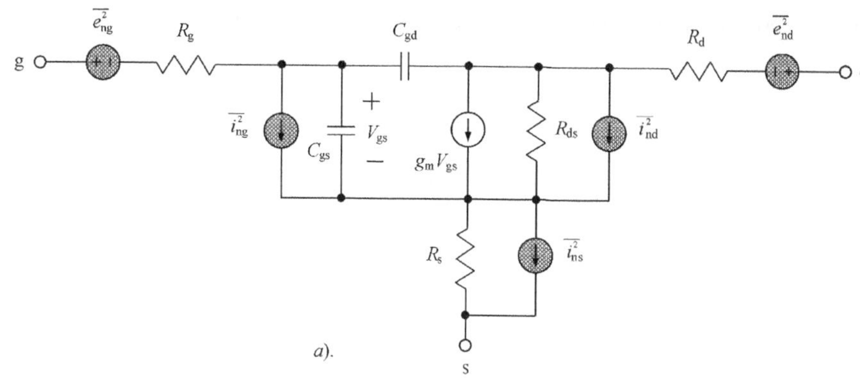

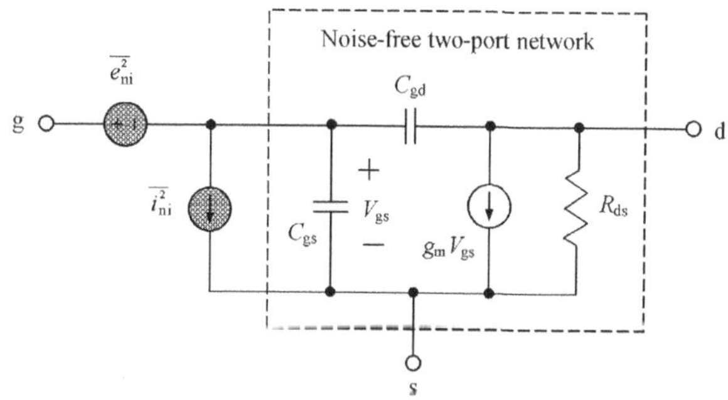

Figure 5.8 MOSFET equivalent circuits with noise sources

bias current, A_F is the flicker noise exponent, K_F is the flicker noise coefficient, C_{ox} is the oxide capacitance per unit area, L_{eff} is the effective channel length, R_{ds0} is the differential drain–source resistance at $V_{ds} = 0$, γ is the channel noise coefficient [27, 28].

In the long-channel devices $R_{ds0} = 1/g_m$ and $\gamma = 2/3$, while both R_{ds0} and γ are complicated functions of the device parameters in the short-channel MOSFETs [29]. The equation for $\overline{i_{nd}^2}$ (with excluded flicker noise) valid for both short-channel and long-channel devices expressed through the device parameters can be written in a simple form as

$$\overline{i_{nd}^2} = 4kT\beta I_d \Delta f \tag{5.43}$$

where

$$\beta = \frac{1}{V_{dsat}} + \frac{\alpha^2 V_{dsat}}{3(V_{gs} - V_{th})^2}$$

V_{dsat} is the saturation drain-source voltage and α is the bulk-charge effect coefficient [30].

The required minimum noise figure F_{min}, noise resistance R_n and optimum source admittance Y_{Sopt} using the noise correlation C_A-matrix parameters as functions of the input referred noise voltage $\overline{v_{ni}^2}$ and noise current $\overline{i_{ni}^2}$ as well as Y-parameters of the simplified noise-free two-port network shown in Figure 5.8b can be approximately estimated as

$$R_n = R_g + \frac{\beta I_d}{g_m^2} \tag{5.44}$$

$$Y_{Sopt} = \frac{\sqrt{\beta I_d R_g}}{R_n}\left(\frac{f}{f_T}\right)\left(1 - j\frac{1}{g_m}\sqrt{\frac{\beta I_d}{R_g}}\right) \tag{5.45}$$

$$F_{min} = 1 + \left(\frac{f}{f_T}\right)\sqrt{\beta I_d R_g}\left[1 + \left(\frac{f}{f_T}\right)\sqrt{\beta I_d R_g}\right] \tag{5.46}$$

where f is the operation frequency and $f_T = g_m/2\pi(C_{gs} + C_{gd})$ [30].

5.3.2 MESFET devices

The noise properties of a MESFET device can be described based on both its physical and equivalent circuit models. The dominant intrinsic noise of a microwave GaAs MESFET device is the diffusion noise introduced by electrons experiencing velocity saturation. In a device two-zone model, a portion of the channel near the source end is assumed to be in the constant mobility operation mode (zone I), while the remaining portion near the drain end is postulated to be in velocity saturation (zone II). The position of the boundary between these zones is a strong function of the source–drain bias with weak dependence on the gate–source bias. It is assumed that the noise in zone I is thermally enhanced by hot electron effects [31, 32]. However, zone II cannot be treated as an ohmic conductor. Its contribution must be represented as a high-field diffusion noise, being dominant in microwave devices [33]. This diffusion noise is proportional to the high-field diffusion coefficient and is linearly dependent on drain current. On the other hand, the thermal noise of zone I decreases with increasing drain current. As a result, a strong correlation exists between the drain noise and induced gate noise which leads to a high degree of cancellation in the noise output of the GaAs MESFET [34].

The noise equivalent equivalent circuit of the MESFET device with both intrinsic and extrinsic noise sources is shown in Figure 5.9a [31, 34]. The noise source $\overline{i_{ng}^2}$ represents the noise induced on the gate electrode by the passing thermal fluctuations in the drain current. The intrinsic drain noise source $\overline{i_{nd}^2}$ has a flat spectrum. The resistance R_{gs} represents the resistive charging path for the gate–source capacitance C_{gs}, and noise associated with this resistor is imbedded in the gate noise source. The series gate, source and drain resistances are represented by the voltage thermal noise sources $\overline{e_{ng}^2}$, $\overline{e_{ns}^2}$ and $\overline{e_{nd}^2}$, respectively.

The noise voltage and current sources can be given through their mean-square values as

$$\overline{e_{ng}^2} = 4kTR_g\Delta f \tag{5.47}$$

$$\overline{i_{ng}^2} = \frac{4kT(\omega C_{gs})^2 R\Delta f}{g_m} \tag{5.48}$$

$$\overline{e_{ns}^2} = 4kTR_s\Delta f \tag{5.49}$$

$$\overline{e_{nd}^2} = 4kTR_d\Delta f \tag{5.50}$$

$$\overline{i_{nd}^2} = 4kTg_mP\Delta f \tag{5.51}$$

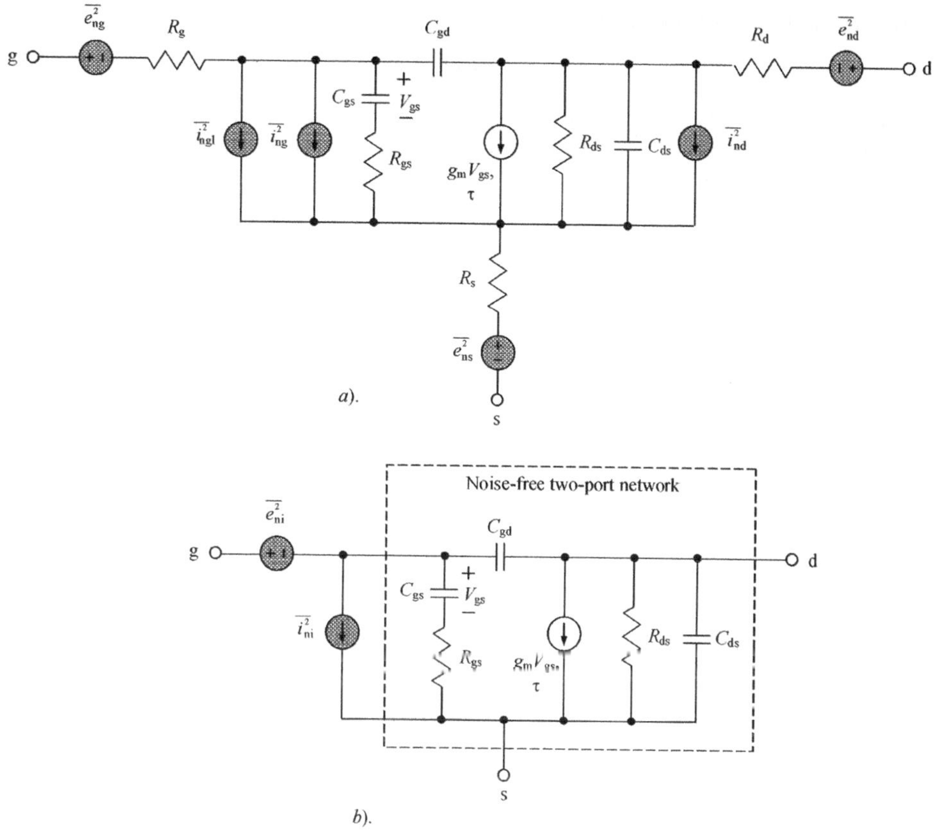

Figure 5.9 MESFET equivalent circuits with noise sources

where R and P are the gate and drain noise model parameters depending upon the implementation technology and biasing conditions [35, 36].

The cross-correlation between the gate drain noise current sources $\overline{i_{ng}^2}$ and $\overline{i_{nd}^2}$ can be written as

$$\overline{i_{nd}i_{ng}} = 4kT\omega C_{gs}C\sqrt{PR}\Delta f \tag{5.52}$$

where C is the correlation coefficient. The quantities P, R and C are bias-dependent empirical correction factors, which may be obtained by noise de-embedding techniques [37]. Their typical values based on measurement and calculation of the noise figure for different devices can be chosen as $P = 1$, $R = 0.5$ and $C = 0.9$ [38]. It should be noted that $\overline{i_{nd}^2}$ and C increase in the ohmic region and tend to saturate at high drain voltage, while $\overline{i_{ng}^2}$ increases with a near constant slope versus drain voltage.

The noise current source $\overline{i_{ngl}^2}$ is responsible for the effect of the gate leakage current which should be taken into account when using a submicrometre gate-length HEMT device. It can be written as a shot noise source in the form

$$\overline{i_{\mathrm{ngl}}^2} = 2\alpha q I_{\mathrm{gl}} \Delta f \tag{5.53}$$

where I_{gl} is the gate leakage current and α is the fitting parameter [39, 40].

The admittance Y-parameters of the MESFET intrinsic small-signal equivalent circuit can be written in matrix form

$$Y = \begin{bmatrix} \dfrac{j\omega C_{\mathrm{gs}}}{1 + j\omega R_{\mathrm{gs}} C_{\mathrm{gs}}} + j\omega C_{\mathrm{gd}} & -j\omega C_{\mathrm{gd}} \\ \dfrac{g_{\mathrm{m}} \exp(-j\omega\tau)}{1 + j\omega R_{\mathrm{gs}} C_{\mathrm{gs}}} - j\omega C_{\mathrm{gd}} & \dfrac{1}{R_{\mathrm{ds}}} + j\omega(C_{\mathrm{ds}} + C_{\mathrm{gd}}) \end{bmatrix} \tag{5.54}$$

The corresponding admittance noise correlation matrix to calculate the equivalent noise resistance R_{n}, the optimal source admittance Y_{Sopt} and the minimum noise figure F_{min} is given by

$$C_{\mathrm{Y}} = 4kT \begin{bmatrix} \dfrac{\alpha I_{\mathrm{gl}}}{2V_{\mathrm{T}}} + \dfrac{(\omega C_{\mathrm{gs}})^2 R}{g_{\mathrm{m}}} & j\omega C_{\mathrm{gs}} C\sqrt{PR} \\ -j\omega C_{\mathrm{gs}} C\sqrt{PR} & g_{\mathrm{m}} P \end{bmatrix} \tag{5.55}$$

where $V_{\mathrm{T}} = kT/q$ is the thermal voltage. However, if the correlation matrix has been determined, the noise parameters can be analytically calculated by substituting the elements of the chain correlation matrix

$$C_{11}^{\mathrm{A}} = \frac{C_{22}^{\mathrm{Y}}}{|Y_{21}|^2} \tag{5.56}$$

$$C_{12}^{\mathrm{A}} = \frac{Y_{11}^{*} C_{22}^{\mathrm{Y}}}{|Y_{21}|^2} - \frac{C_{12}^{\mathrm{Y}}}{Y_{12}} \tag{5.57}$$

$$C_{22}^{\mathrm{A}} = C_{11}^{\mathrm{Y}} + \frac{|Y_{11}|^2 C_{22}^{\mathrm{Y}}}{|Y_{21}|^2} - \frac{Y_{11} C_{21}^{\mathrm{Y}}}{Y_{21}} - \frac{Y_{11}^{*} C_{12}^{\mathrm{Y}}}{Y_{21}^{*}} \tag{5.58}$$

into Equations (5.31–5.33) [41].

In a first approximation, the gate noise source $\overline{i_{\mathrm{ng}}^2}$, feedback capacitance C_{gd} and series drain resistance R_{d} can be neglected. As a result, the simple approximate expressions based on measurements can be obtained in terms of the device equivalent circuit elements

$$F_{\mathrm{min}} = 1 + 0.016 f C_{\mathrm{gs}} \sqrt{\frac{R_{\mathrm{g}} + R_{\mathrm{s}}}{g_{\mathrm{m}}}} \tag{5.59}$$

$$R_{\mathrm{n}} = \frac{0.8}{g_{\mathrm{m}}} \tag{5.60}$$

$$R_{\mathrm{Sopt}} = 2.2 \left(\frac{1}{4 g_{\mathrm{m}}} + R_{\mathrm{g}} + R_{\mathrm{s}} \right) \tag{5.61}$$

$$X_{\mathrm{Sopt}} = \frac{160}{f C_{\mathrm{gs}}} \tag{5.62}$$

provided that R_{n}, R_{Sopt}, X_{Sopt}, R_{g} and R_{s} are in ohms, transconductance g_{m} in mhos, capacitance C_{gs} in picofarads, and operating frequency f is in gigahertz [42, 43].

The minimum noise figure F_{min} given by Equation (5.59) can also be expressed in terms of the device geometric parameters as

$$F_{min} = 1 + 0.27Lf\sqrt{g_m(R_g + R_s)} \tag{5.63}$$

where the effective gate length L is in micrometres [44].

A comparison of the noise performance of both HEMT and conventional MESFET devices demonstrates the HEMT superiority, mainly related to its higher transition frequency and correlation coefficient [38]. The transition frequency of HEMT device is greater for two main reasons: the higher carrier mobility results in a higher average velocity and, therefore, a higher transconductance; the small epilayer thickness yields higher transconductance and less effect of the parasitic capacitances. In addition, the correlation coefficient C is close to 0.7–0.8 for short-gate-length MESFETs, but becomes close to 0.8–0.95 for HEMTs.

5.3.3 Bipolar transistors

The noise in a bipolar transistor is assumed to arise from three basic sources: diffusion fluctuations, recombination fluctuations in the base region, and thermal noise in the base resistance [45]. The noise behaviour of the bipolar transistor can be described based on its equivalent circuit representation shown in Figure 5.10a, which includes the main elements responsible for the device electrical behaviour and noise sources. Since the process of the carrier drifting into

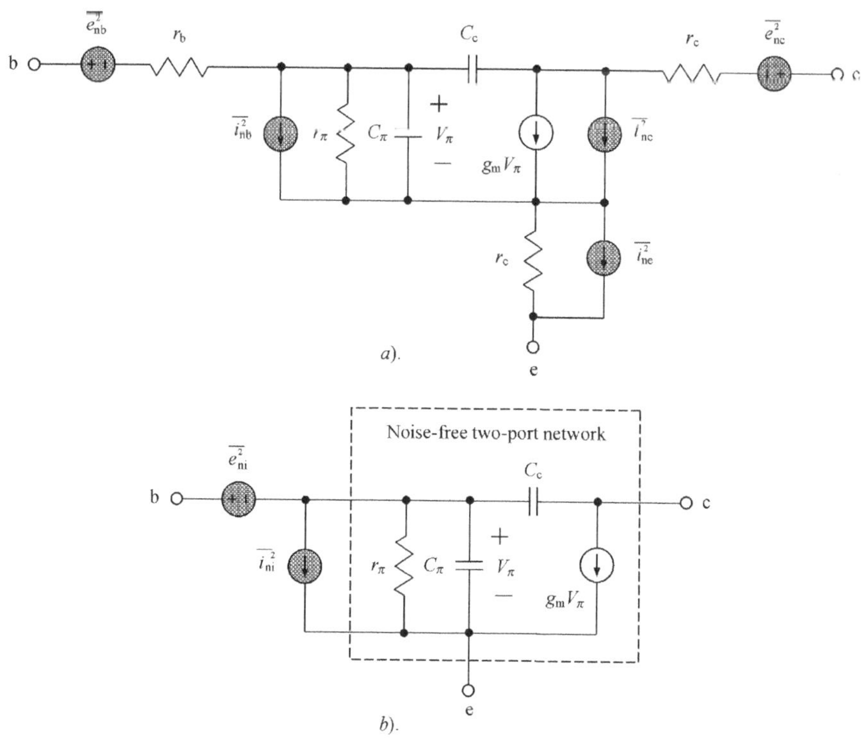

Figure 5.10 Bipolar equivalent circuits with noise sources

the collector–base depletion region is a random process, the collector current I_c demonstrates shot noise and is represented by a shot-noise collector current source $\overline{i_{nc}^2}$. The base current I_b is a result of the carrier injection from the base to the emitter and generation–recombination effect in the base and base–emitter depletion regions. Because all these components are independent, representing a random process, the base current also demonstrates a shot-noise behaviour and is represented by a shot-noise base current source $\overline{i_{nb}^2}$. Flicker noise is represented by a current source across the internal base–emitter junction combined with the base current source $\overline{i_{nb}^2}$. The series base, emitter and collector resistances are represented by the voltage and current thermal noise sources $\overline{e_{nb}^2}$, $\overline{i_{ne}^2}$ and $\overline{e_{nc}^2}$, respectively.

The noise voltage and current sources can be given through their mean-square values as

$$\overline{e_{nb}^2} = 4kTr_b\Delta f \tag{5.64}$$

$$\overline{i_{ne}^2} = \frac{4kT\Delta f}{r_e} \tag{5.65}$$

$$\overline{i_{nc}^2} = 2qI_c\Delta f \tag{5.66}$$

$$\overline{e_{nc}^2} = 4kTr_c\Delta f \tag{5.67}$$

$$\overline{i_{nb}^2} = 2qI_b\Delta f + K_F\frac{I_b^{A_F}}{f}\Delta f \tag{5.68}$$

where q is the electron charge, A_F is the flicker noise exponent, K_F is the flicker noise coefficient calculated as $2qf_c$, f_c is the flicker noise corner frequency [28, 46].

The required minimum noise figure F_{min}, noise resistance R_n and optimum source admittance Y_{Sopt} using the noise correlation C_A-matrix parametres as functions of the input referred noise voltage $\overline{v_{ni}^2}$ and noise current $\overline{i_{ni}^2}$ as well as Y-parametres of the simplified noise-free two-port network shown in Figure 5.10b can be calculated for a sufficiently high value of the low-frequency current gain $\beta = g_m r_\pi$ from

$$R_n = r_b\left(1 + \frac{r_b}{2r_\pi}\right) + \frac{r_\pi}{2\beta} + \frac{r_b^2}{2\beta r_\pi}\left(\frac{f}{f_T}\right)^2 \tag{5.69}$$

$$Y_{Sopt} = \sqrt{\frac{1}{2r_\pi R_n}\left[1 + \frac{1}{\beta}\left(\frac{f}{f_T}\right)^2\right] - \left(\frac{1}{2\beta R_n}\frac{f}{f_T}\right)^2} - j\frac{1}{2\beta R_n}\frac{f}{f_T} \tag{5.70}$$

$$F_{min} = 1 + \frac{r_b}{r_\pi}\left[1 + \frac{1}{\beta}\left(\frac{f}{f_T}\right)^2\right] + \sqrt{\frac{2r_b}{r_\pi}\left(1 + \frac{r_b}{2r_\pi}\right) + \frac{2r_b}{\beta r_\pi}\left(1 + \frac{r_b}{r_\pi}\right)\left(\frac{f}{f_T}\right)^2} \tag{5.71}$$

where f is the operation frequency and $f_T = g_m/2\pi C_\pi$ is the bipolar transition frequency (the effect of the feedback collector capacitance C_c is not taken into account) [47]. It should be noted that if the optimum noise conductance G_{Sopt} is insensitive to the collector capacitance C_c, then it can severely affect the other noise parametres at microwave frequencies, mostly due to the reduction of the device gain capability.

A noise model for a HBT device operated at very high frequencies should include the contribution of both space-charge layers (at the emitter–base junction and the base–collector junction) to the shot noise. These two noise sources related to the collector current I_c are the result of the same electrons, which are injected from the emitter into the base, cross this layer, and then reach the collector. Therefore, their correlation can be given by a time delay function $\exp(-j\omega\tau)$, where τ is the transit time through the base and the collector–base junction [48–50]. Thus, to extend the HBT noise model valid up to its transition frequency, the base and collector noise current sources are rewritten as

$$\overline{i_{nb}^2} = 2q(I_b + |1 - \exp(-j\omega\tau)|^2 I_c)\Delta f \tag{5.72}$$

$$\overline{i_{nc}^2} = 2qI_c\Delta f \tag{5.73}$$

$$\overline{i_{nb}i_{nc}^*} = 2q[\exp(-j\omega\tau) - 1]I_c\Delta f. \tag{5.74}$$

5.4 OSCILLATOR NOISE SPECTRUM: LINEAR MODEL

The instantaneous output signal of an oscillator can be represented by

$$v(t) = A_0 \left[1 + \frac{\Delta A(t)}{A_0}\right] \cos[2\pi f_0 t + \phi_0 + \Delta\phi(t)] \tag{5.75}$$

where A_0 is the voltage amplitude of the steady-state oscillations, f_0 is the oscillation frequency, ϕ_0 is the initial phase of the oscillations at $t = 0$, $\Delta A(t)$ and $\Delta\phi(t)$ are the amplitude and phase deviations of the corresponding amplitude and phase fluctuations, respectively. Generally, the nature of the fluctuations may be discrete or random where the discrete signals are called spurious, appearing as distinct spectral components, while the random fluctuations are called phase noise.

The instantaneous frequency as a function of time can be written as

$$f(t) = \frac{1}{2\pi}\frac{d}{dt}[2\pi f_0 t + \phi_0 + \Delta\phi(t)] = f_0 + \Delta f(t) \tag{5.76}$$

where

$$\Delta f(t) = \frac{1}{2\pi}\frac{d\Delta\phi(t)}{dt}$$

Since the process associated with frequency fluctuations $\Delta f(t)$ is stationary, then

$$\Delta\phi(t) = 2\pi \int_0^t \Delta f(\tau)\,d\tau \tag{5.77}$$

which is in general a nonstationary process. However, as the phase fluctuation process is sufficiently slow during the natural period of the oscillations, the stationary model to describe the phase noise performance of free-running oscillators can be used. It should be noted that an autonomous free-running oscillator does not have a reference plane and the initial phase ϕ_0 in an autonomous system can be chosen arbitrarily, for example, of zero value.

5.4.1 Parallel feedback oscillator

It is very important for the oscillator noise model to express a clear relationship between the oscillator spectral noise power density and resonant circuit and active device parametres. The

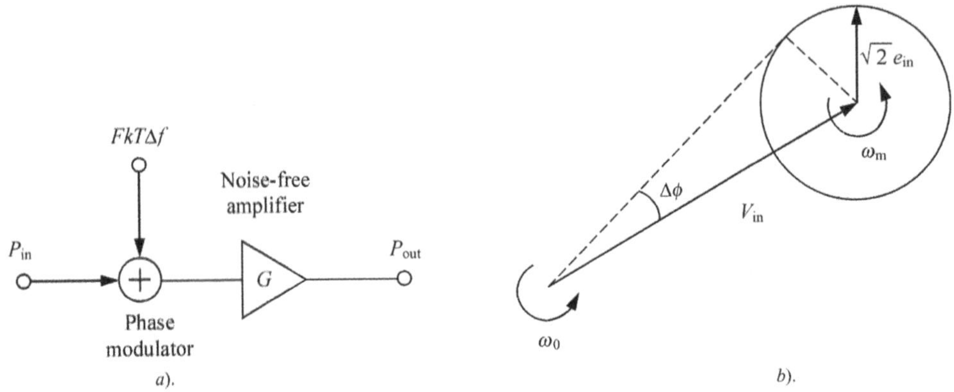

Figure 5.11 Simplified feedback oscillator noise model

simple Leeson linear model for a feedback oscillator, which was derived empirically, is based on the expectation that the real oscillator has two basic components [51]. The first component is caused by the phase fluctuations due to the additive white noise at frequency offsets close to the carrier, as well as due to the noise having a mixing nature resulting from the circuit nonlinearities. The second component is a result of the low-frequency fluctuations or flicker noise up-converted to the carrier region because of the active device nonlinear effects.

The phase noise at the input of the power amplifier is added to a signal as the sum of every bandwidth $\Delta f = 1$ Hz, each producing an available noise power at the input of the noise-free amplifier. Maximum power delivery can be achieved when the source internal impedance is conjugate-matched to the input impedance of the amplifier. As a result, only one-half of the root-mean-square noise voltage appears across the amplifier input and is equal to

$$e_{\text{in}} = \frac{e_{\text{n}}}{2} = \frac{\sqrt{4FkTR}}{2} = \sqrt{FkTR} \tag{5.78}$$

where R is the equivalent resistance, which can be represented as the input resistance for the input root-mean-square noise voltage [52]. The input phase noise produces a root-mean-square phase deviation $\Delta\phi_{\text{rms}} = \Delta\phi/\sqrt{2}$ at each offset frequency $\pm f_{\text{m}}$ from the carrier f_0, as shown in Figure 5.11, for which a total power-wise sum can be written for a small phase perturbation as

$$\Delta\phi = \Delta\phi_{\text{rms}}\sqrt{2} = \frac{e_{\text{in}}\sqrt{2}}{V_{\text{in}}} = \sqrt{\frac{FkT}{P_{\text{in}}}} \tag{5.79}$$

where $V_{\text{in}} = \sqrt{2P_{\text{in}}R}$ is the signal voltage amplitude at the power amplifier input.

As a result, the double sideband spectral power density of the thermal phase noise in a frequency bandwidth $\Delta f = 1$ Hz can be written as

$$S_{\Delta\phi} = \Delta\phi^2 = \frac{FkT}{P_{\text{in}}} \tag{5.80}$$

The Leeson model consists of an amplifier with a noise figure F and a resonator (or filter) in the feedback loop, as shown in Figure 5.12 [52]. The oscillator phase noise is modelled by assuming a noise-free power amplifier and adding a phase modulator to its input. Based on empirical predictions, the phase noise level of the oscillator at an offset frequency f_{m} from the

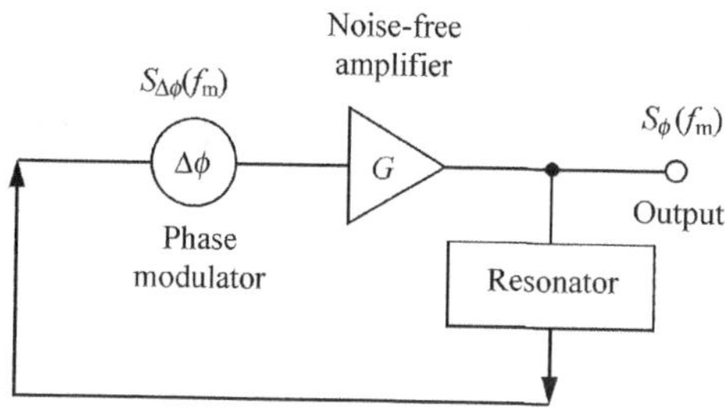

Figure 5.12 Equivalent model of parallel feedback oscillator

carrier f_0 can be described by

$$S_\phi(f_m) = S_{\Delta\phi}(f_m)\left(\frac{f_0}{2Q_L f_m}\right)^2 \qquad \text{for } f_m < \frac{f_0}{2Q_L} \tag{5.81}$$

$$S_\phi(f_m) = S_{\Delta\phi}(f_m) \qquad \text{for } f_m > \frac{f_0}{2Q_L} \tag{5.82}$$

where $S_{\Delta\phi}(f_m)$ is determined, using Equation (5.80), as

$$S_{\Delta\phi}(f_m) = \frac{FkT}{P_{in}}\left(1 + \frac{f_c}{f_m}\right) \tag{5.83}$$

taking into account the effect of the signal purity degradation due to the low-frequency flicker noise close to the carrier, described empirically by the corner frequency f_c. It should be noted that the empirical Leeson equation for $S_{\Delta\phi}(f_m)$ contains a multiplication factor of two in the numerator. Moreover, accurate agreement was achieved between the model and experimental results when the power density $S_{\Delta\phi}(f_m)$ was expressed in terms of the compressed (or large-signal) power gain G and output power P_{out} as $S_{\Delta\phi}(f_m) = 2GFkT/P_{out}$ [53].

The parameter F in Equation (5.83) is associated with the active device noise figure and can be called an effective noise factor because, generally, it should represent the effects of the active device noise sources and the cyclostationary noise resulting from periodically varying processes in practical oscillators. Due to the inherent nonlinear nature of the active device, the effects of intermodulation between the wideband white noise and various harmonics of fundamental frequency (for example, nonlinear transformation of the noise near the third harmonic down-converted to the near carrier region due to mixing effect with second harmonic) must be included [51]. Also, the effect of low-frequency noise modulation of the current, resulting in the reactance modulation of the input impedance of the circuit (for example, variation of the phase angle of the device forward transfer function versus emitter current), cannot be neglected [54]. Hence, it is impossible to calculate F accurately without taking into account the effect of the oscillator resonant circuit. Therefore, for such a linear model, the effective noise factor F as well as the corner frequency f_c can be considered more like fitting parametres, based on measured data.

The corresponding combined expression to calculate the normalized double-sideband phase noise power spectral density or the double-sideband noise-to-carrier ratio at the input of the feedback oscillator can be obtained from

$$S_\phi(f_m) = \frac{FkT}{P_{in}}\left(1 + \frac{f_c}{f_m}\right)\left[1 + \left(\frac{f_0}{2Q_L f_m}\right)^2\right] \tag{5.84}$$

which gives an asymptotic model showing generally the noise reduction of 9 dB/octave in the offset region with predominant low-frequency $1/f$ noise, 6 dB/octave in the offset region due to feedback loop and 0 dB/octave representing the thermal or white noise spectrum.

The single sideband noise-to-carrier ratio at the input of the feedback oscillator can be described by

$$L(f_m) = \frac{1}{2}\frac{FkT}{P_{in}}\left(1 + \frac{f_c}{f_m}\right)\left[1 + \left(\frac{f_0}{2Q_L f_m}\right)^2\right] \tag{5.85}$$

whose idealized sideband spectral behaviour for different values of the loaded quality factors is illustrated in Figure 5.13. For the low-Q_L case, there are regions with $1/f_m^3$ and $1/f_m^2$ dependencies for spectral power density close to carrier, as shown in Figure 5.13a. For the moderate-Q_L case, Figure 5.13b demonstrates only $1/f_m^3$ dependence as far as intersection with thermal noise floor. For the high-Q_L case, the regions with $1/f_m^3$ and $1/f_m$ dependencies are observed near the carrier, as shown in Figure 5.13c. Closest to the carrier, $1/f_m^3$ phase noise behaviour is a result of random frequency modulation of the oscillator by low-frequency $1/f$ noise. In the region of $1/f_m^2$ phase noise behaviour, the white noise causes random frequency modulation. The $1/f_m$ dependence is due to the mixing up of the $1/f$ noise with the oscillation frequency. Finally, the phase noise becomes constant, which is a result of the mixing up of white noise around the oscillation frequency.

To calculate the same phase noise power spectral density at the oscillator output, it is necessary to replace the input power P_{in} by the power available at the output P_{out} and to multiply the numerator of Equation (5.85) by the power gain G. As a result, neglecting the effect of flicker noise and considering the case of $f_m \ll f_0$, one can obtain

$$L(f_m) = \frac{GFkT}{8Q_L^2 P_{out}}\left(\frac{f_0}{f_m}\right)^2 \tag{5.86}$$

where

$$G = \frac{1}{\left(1 - \frac{Q_L}{Q_0}\right)^2} \tag{5.87}$$

is considered the transducer power gain and Q_0 is the unloaded quality factor [55]. From Equations (5.85) and (5.86) it follows that, to minimize the oscillator phase noise, it is necessary to reduce the noise figure F and to increase the input power P_{in} (or the output power P_{out} for a fixed power gain G of the amplifier) as much as possible. In addition, for frequency offsets inside the resonator bandwidth, it is desirable to maximize the oscillator loaded quality factor Q_L. However, the resonator insertion loss and loaded Q_L are interrelated, and one cannot arbitrarily increase Q_L without increasing the insertion loss, otherwise a larger power gain G is needed. The two competing effects result in an optimum loaded Q_L of approximately one-half the unloaded Q_0 and insertion loss of about 6 dB [56, 57]. Thus, the minimum noise

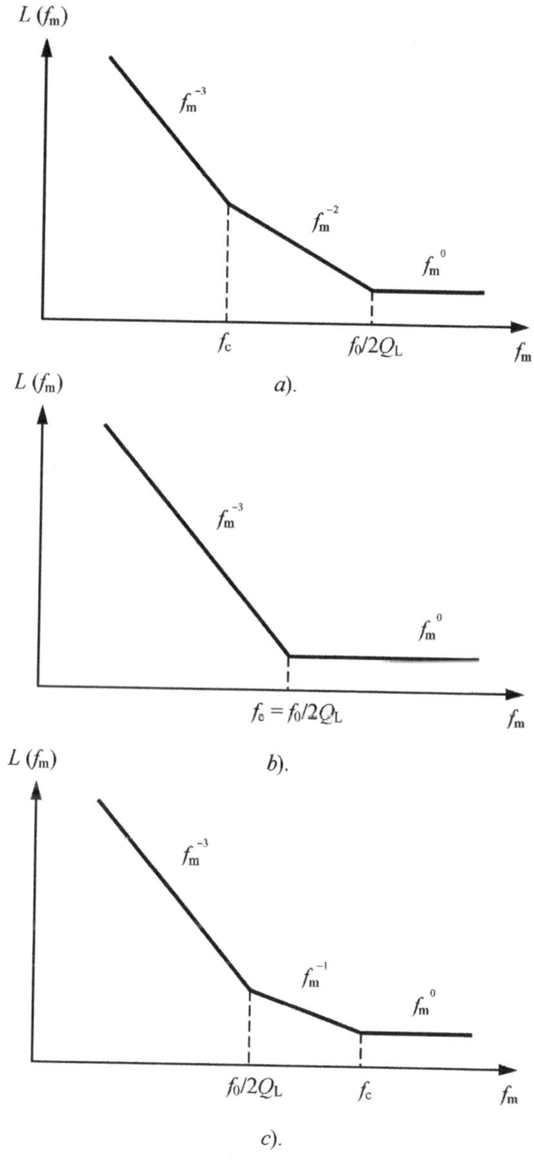

Figure 5.13 Single sideband oscillator phase noise behaviour

occurs when $Q_L/Q_0 = 0.5$ resulting in

$$L(f_m) = \frac{2FkT}{P_{out}} \left(\frac{f_0}{2Q_L f_m} \right)^2 \tag{5.88}$$

Note that the difference in the optimum noise performance predicted by different definitions of the output power (power dissipated in the resonant circuit or power available at the amplifier output) is small [55].

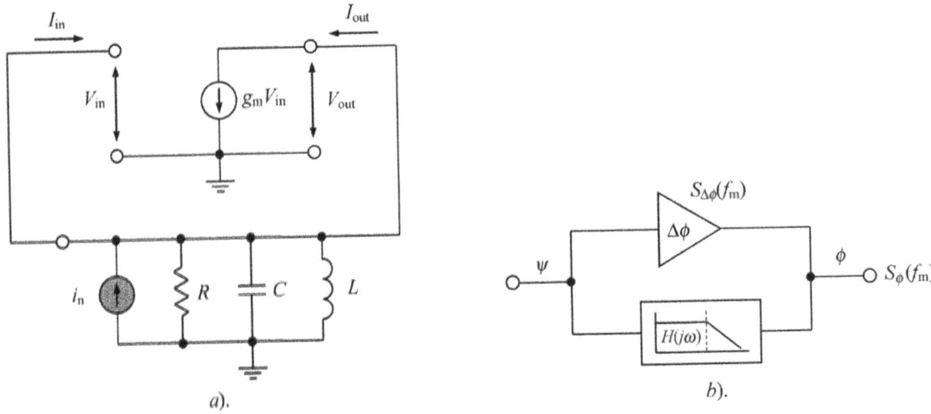

Figure 5.14 Equivalent models of feedback oscillator

Now let us represents a parallel feedback oscillator model using the circuit schematic, with an active device and a parallel resonant circuit, shown Figure 5.14a. Here, the active device is an ideal voltage-control current source with transconductance g_m. At operating frequencies $f \ll f_T$, where f_T is the transition frequency, such a simplified transistor model can describe the behaviour of a FET device with the input gate–source capacitance C_{gs} (it can be included in the resonant circuit capacitance C), the gate–drain capacitance C_{gd} (its value is typically sufficiently small) and the drain–source resistance R_{ds} (its value is assumed to be infinite). The impedance of the parallel resonant circuit for $\Delta\omega \ll \omega_0$, where $\Delta\omega$ is the offset from the carrier, can be calculated from

$$Z_L(j\omega) = \frac{R}{1 + j2\dfrac{\Delta\omega}{\omega_0}Q_L} \tag{5.89}$$

where $\omega_0 = 1/\sqrt{LC}$ is the resonant frequency and $Q_L = \omega CR$ is the loaded quality factor.

For a given open-loop voltage transfer function $H(j\omega) = g_m Z_L(j\omega)$, the closed-loop voltage transfer function $T(j\omega)$ can be written as

$$T(j\omega) = \frac{V_{out}(j\omega)}{V_{in}(j\omega)} = \frac{g_m Z_L}{1 - g_m Z_L} \tag{5.90}$$

By substituting Equation (5.89) into Equation (5.90) and using the steady-state oscillation condition corresponding to the *Bakhausen criterion* as $g_m R_L = 1$, the magnitude $T(f_m)$ of the complex transfer function $\boldsymbol{T}(j\omega)$ can be obtained

$$T(f_m) = \frac{V_{out}(f_m)}{V_{in}(f_m)} = \frac{f_0}{2Q_L f_m} \tag{5.91}$$

where $f_m = \Delta\omega$.

Since noise current i_n is produced by the resistor R only, it can be transformed to the equivalent voltage noise with amplitude $V_{in} = \sqrt{8kTR\Delta f}$ at the device input. In this case,

Equation (5.91) can be rewritten as

$$V_{\mathrm{out}}^2(f_{\mathrm{m}}) = 8kTR\Delta f\left(\frac{f_0}{2Q_{\mathrm{L}}f_{\mathrm{m}}}\right)^2 \tag{5.92}$$

As a result, the single sideband spectral noise power density in a frequency bandwidth $\Delta f = 1$ Hz, normalized to the total power dissipated in the oscillator $P = V_{\mathrm{in}}^2/2R$, can be calculated from

$$L(f_{\mathrm{m}}) = \frac{1}{2}\frac{V_{\mathrm{out}}^2(f_{\mathrm{m}})}{V_{\mathrm{in}}^2} = \frac{2kT}{P}\left(\frac{f_0}{2Q_{\mathrm{L}}f_{\mathrm{m}}}\right)^2 \tag{5.93}$$

where the factor $^1/_2$ arises from neglecting the contribution of amplitude noise since, for a totally uncorrelated noise, one-half of the total noise power contributes to AM sidebands and the other half of the total noise power is converted into PM sidebands. Equation (5.93) is similar to Equation (5.88) with only the difference in power definitions. Since P represents the total or dc power dissipated in the oscillator with an ideal lossless active device (for example, a device operating in switching Class E mode), the power delivered to the load is $P_{\mathrm{L}} = \eta P$, where η is the efficiency of the oscillator. The load resistance can represent a part of the resonant circuit resistance R. Thus, Equation (5.93) can be given in the form

$$L(f_{\mathrm{m}}) = \frac{2\eta kT}{P_{\mathrm{L}}}\left(\frac{f_0}{2Q_{\mathrm{L}}f_{\mathrm{m}}}\right)^2 \tag{5.94}$$

Figure 5.14b shows another representation of the Leeson model with a phase feedback loop. Suppose that the phase noise modulation occurs in the oscillator active element as $\Delta\phi(t)$, and it is necessary to define how the oscillator reacts to this internal noise. As is known from transmission theory, the transfer function of a modulated high-frequency signal, passing through a bandpass filter, equals the transfer function of the modulating signal passing through an equivalent low-pass filter prototype. Thus, the phase relationship resulting from the null-phase condition upon the loop and from the filtering of $\phi(t)$ by the selective filter can be written as

$$\psi(t) + \Delta\phi(t) = \phi(t) \tag{5.95}$$

where

$$\psi(t) = \int_{-\infty}^{+\infty} \phi(t)h(\tau - t)\,\mathrm{d}\tau = \phi(t) * h(t) \tag{5.96}$$

$h(t)$ is the impulse response of the equivalent low-pass filter and the asterisk denotes a convolution product. The integral in Equation (5.96) converges for nearly all samples of $\phi(t)$ provided that the filter is linear and time invariant, and that the stationary random process $\phi(t)$ possesses a finite second-order moment.

In this case, the Leeson formula for the double-sideband phase noise power spectral density of the feedback oscillator given by Equation (5.84) can be expressed in a more general form

$$S_{\phi}(\omega_{\mathrm{m}} = S_{\Delta\phi}(\omega_{\mathrm{m}})\left\{[H(j\omega_{\mathrm{m}}) - 1][H^*(j\omega_{\mathrm{m}}) - 1]\right\}^{-1} \tag{5.97}$$

where $H(j\omega_{\mathrm{m}})$ is the equivalent low-pass transfer function and the asterisk denotes the complex-conjugate value [58].

Thus, by representing the transfer function of the first order low-pass filter as

$$H(j\omega_{\mathrm{m}}) = \frac{\alpha}{\alpha + j\omega_{\mathrm{m}}} \tag{5.98}$$

where $\alpha = \omega_0/2Q_{\mathrm{L}}$ is the half-bandwidth of the resonator and $\omega_{\mathrm{m}} = 2\pi f_{\mathrm{m}}$, Equation (5.97) can be rewritten as

$$S_{\phi}(\omega_{\mathrm{m}} = S_{\Delta\phi}(\omega_{\mathrm{m}}) \left[1 + \left(\frac{\alpha}{\omega_{\mathrm{m}}} \right)^2 \right] \tag{5.99}$$

which is similar to Equation (5.84).

Now consider the second-order low-pass filter case based on two coupled resonators having the transfer function

$$H(j\omega_{\mathrm{m}}) = \frac{\alpha_1}{\alpha_1 + j\omega_{\mathrm{m}}} \frac{\alpha_2}{\alpha_2 + j\omega_{\mathrm{m}}} = \frac{1}{1 - \left(\frac{\omega_{\mathrm{m}}}{\alpha} \right)^2 + j2\omega_{\mathrm{m}}\delta} \tag{5.100}$$

where $\alpha = \sqrt{\alpha_1\alpha_2}$ and $\delta = (\alpha_1 + \alpha_2)/2\sqrt{\alpha_1\alpha_2}$ [59]. In this case, Equation (5.97) can be rewritten as

$$S_{\phi}(\omega_{\mathrm{m}}) = S_{\Delta\phi}(\omega_{\mathrm{m}}) \left[1 + \frac{1 - 2\left(\frac{\omega_{\mathrm{m}}}{\alpha} \right)^2}{\left(\frac{\omega_{\mathrm{m}}}{\alpha} \right)^2 + 4\delta^2} \left(\frac{\alpha}{\omega_{\mathrm{m}}} \right)^2 \right] \tag{5.101}$$

which shows substantially better phase noise performance in the near vicinity of the carrier frequency for the case of loosely coupled resonators when $\delta > 1$. However, since the available output power becomes low, it is necessary to provide post-low-noise output signal amplification.

Let us quantitatively compare both cases of the Leeson phase noise models using the first-order and second-order low-pass filters in the oscillator feedback loop for the same arbitrary chosen technical data:

- oscillation (carrier) frequency $f_0 = 2$ GHz;
- offset frequency $f_{\mathrm{m}} = 10$ kHz;
- oscillator resonant circuit loaded quality factor $Q_{\mathrm{L}} = 10$;
- noise figure of the active device $F = 6$ dB;
- input power delivered to the device $P_{\mathrm{in}} = 10$ mW;
- corner frequency for flicker noise $f_{\mathrm{c}} = 3$ kHz.

Substituting these parametres into Equation (5.85) for a single-sideband phase noise spectral density of the oscillator with the first-order low-pass filter results in

$$L(f_{\mathrm{m}}) = 10 \log(1.05128 \times 10^{-10}) = -99.78 \text{ dBc/Hz}$$

For the case of the oscillator with two coupled resonators, let us assume that the loaded Q_{L} of the second resonator is five times as much as the first one, that is $\alpha_1 = 6.283 \times 10^8$ and $\alpha_2 = 1.2566 \times 10^8$. Then,

$$L(f_{\mathrm{m}}) = 10 \log(2.918 \times 10^{-12}) = -115.35 \text{ dBc/Hz}$$

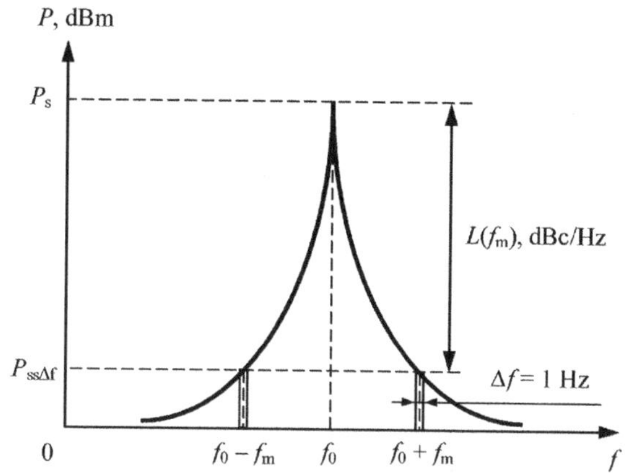

Figure 5.15 Oscillator output power spectrum

which clearly shows the significant phase noise improvement compared with the oscillator having a first-order low-pass filter in a feedback loop.

The typical oscillator output power spectrum is shown in Figure 5.15. The noise distribution on each side of the oscillator signal is subdivided into a large number of strips of width Δf located at the distance f_m away from the signal. It should be noted that, generally, the spectrum of the output signal consists of the amplitude and phase noise components. Hence, to measure the phase noise close to the carrier frequency, one needs to make sure that any contributions of parasitic amplitude modulation to the oscillator output noise spectrum are negligible compared with those from frequency modulation. The single-sideband phase noise $L(f_m)$ usually given logarithmically is defined as the ratio of signal power $P_{ss\Delta f}$ in one phase modulation sideband per bandwidth $\Delta f = 1$ Hz, at an offset f_m away from the carrier, to the total signal power P_s.

Despite some limitations of the linear Leeson model when a device is operated in a conduction angle large-signal mode and the output signal is not purely sinusoidal, which have an effect on the active device noise factor and low-frequency flicker noise up-conversion, such an approach gives a sense of the phase noise performance for oscillators with different resonant circuits. This applies especially, if the theoretical results can be supported by sufficiently accurate measurements of the oscillator resonant circuit loaded quality factor and simulations of the effective noise figure based on the modeled active device parametres and operation conditions.

In addition, such a simple model indicates the basic factors and provides the design rules which are necessary to follow in order to minimize the oscillator phase noise:

- choose the resonator with maximum unloaded quality factor Q_0 and optimize the loaded quality factor Q_L of the oscillator resonant circuit by proper load coupling;
- maximize the output power P_L delivered to the load by maximizing the RF voltage amplitude across the resonant circuit with limitations due to active device breakdown voltage and operation in the saturation mode;
- choose a device with the lowest noise figure F and corner frequency f_c for low-frequency flicker noise.

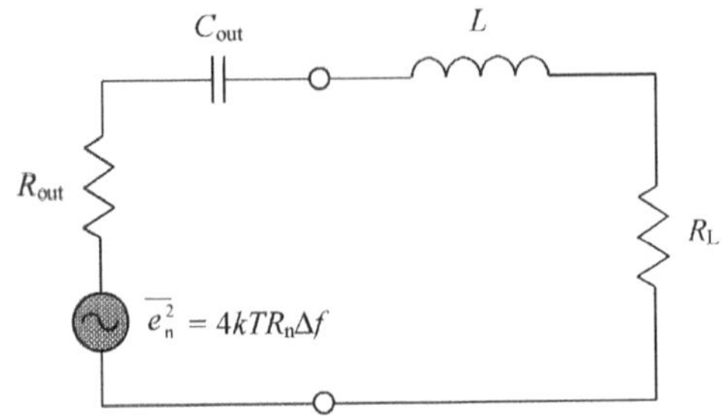

Figure 5.16 Simplified negative resistance oscillator noise model

5.4.2 Negative resistance oscillator

Now consider the equivalent circuit of a simple single-resonant negative resistance oscillator shown in Figure 5.16, where the available noise power is assumed to be totally from the active device. Here, R_n is the equivalent noise resistance associated with active device noise sources, the negative resistance R_{out} and the equivalent output capacitance C_{out} represent the device negative output impedance, L is the tank inductance and R_L is the load resistance. The derivation of the power spectral density will be based on the fact that the available noise power in the active device is amplified in a frequency-selective way, resulting at resonance in the output power P_L being dissipated in the load resistance R_L [60]. This will happen in a steady-state condition when the values of the negative resistance and the load resistance are close to each other.

Then, assuming that $R_L + \Delta R = -R_{out}$ and defining the magnitude of the mean-square noise current flowing into load from the mean-square noise voltage source, we can write

$$\overline{I_n^2} = \frac{\overline{e_n^2}}{(\Delta R)^2 + \left(\omega L - \dfrac{1}{\omega C_{out}}\right)^2} \tag{5.102}$$

Equation (5.102) can be rewritten in the common form

$$\overline{I_n^2} = \frac{1}{\left(\dfrac{\Delta R}{R_L}\right)^2 + Q_L^2\left(\dfrac{\omega}{\omega_0} - \dfrac{\omega_0}{\omega}\right)^2}\frac{\overline{e_n^2}}{R_L^2} \tag{5.103}$$

where ω_0 is the resonant frequency and Q_L is the oscillator loaded quality factor at the resonant frequency.

By normalizing to the power P_L dissipated in the load resistor R_L, Equation (5.102) can be rewritten through the spectral power densities as

$$S_\phi = \frac{S_{\Delta\phi}}{\left(\dfrac{\Delta R}{R_L}\right)^2 + Q_L^2\left(\dfrac{\omega}{\omega_0} - \dfrac{\omega_0}{\omega}\right)^2} \tag{5.104}$$

where $S_{\Delta\phi} = 4kTR_n/R_LP_L$, $S_\phi = \overline{i_n^2}R_L/P_L$ is the power spectral density of the noise current across the load resistor R_L and $\Delta f = 1$ Hz.

Since at small offset frequencies $\omega_m = \omega - \omega_0$ close to the resonant frequency,

$$\frac{\omega}{\omega_0} - \frac{\omega_0}{\omega} \cong \frac{2\omega_m}{\omega_0}$$

Equation (5.104) can be rewritten as

$$S_\phi = \frac{S_{\Delta\phi}}{\left(\frac{\Delta R}{R_L}\right)^2 + \left(\frac{2Q_L\omega_m}{\omega_0}\right)^2} \tag{5.105}$$

which is similar to the power spectral density at frequency offsets close to the resonant frequency for the parallel feedback oscillator. Equation (5.105) represents a Lorentz function corresponding to an exponential decay of the autocorrelation function in the time domain [61].

Since the total output power delivered to the load is equal to P_L,

$$\frac{1}{2\pi}\int_0^\infty S_\phi(\omega)\,d\omega = \left(\frac{\omega_0}{2Q_L}\right)^2 \frac{S_\theta}{2\Delta\omega_n} = 1 \tag{5.106}$$

where

$$\Delta\omega_n = \frac{\Delta R}{R_L}\frac{\omega_0}{2Q_L}$$

is the Lorentzian linewidth (half-width at half-maximum) which is an oscillator spectrum linewidth characterized by the natural phase fluctuations, due to the thermal and shot noises of the oscillator. However, in a common case, due to the variation of the oscillator resonant circuit parametres, flicker noise, pushing or pulling effects, the effective spectrum linewidth widens, especially close to the resonant frequency.

By using a widely used definition of the loaded quality factor of the passive resonator in the form

$$Q_L = \frac{\omega_0}{\Delta\omega_{3dB}} \tag{5.107}$$

where $\Delta\omega_{3dB}$ is the full linewidth at half-maximum level, one can write

$$2\Delta\omega_n = \frac{S_{\Delta\phi}}{4}\Delta\omega_{3dB}^2 = kT\frac{R_n}{R_L}\frac{\Delta\omega_{3dB}^2}{P_L} \tag{5.108}$$

As a result, there is a complete analogy between Lorentzian linewidth defined by Equation (5.108) and the expression for semiconductor laser homogeneous linewidth [62]. The characteristic energy kT corresponds to the photon energy, the oscillator output power P_L corresponds to the laser output power and the oscillator noise–gain ratio R_n/R_L corresponds to the inversion factor representing the ratio between spontaneous emission rate and the optical gain rate. The widening of the oscillator spectral line due to the low-frequency fluctuations is similar to an inhomogeneous laser line broadening due to the Doppler effect.

The normalized power spectral density can be expressed through the Lorentzian linewidth as

$$S_\phi(\omega_m) = \frac{2\Delta\omega_n}{\Delta\omega_n^2 + \omega_m^2} \cong \frac{2\Delta\omega_n}{\omega_m^2} \tag{5.109}$$

showing a simple linear relationship between Lorentzian linewidth and oscillator phase noise spectrum at offset frequencies $\omega_m \gg \Delta\omega_n$. Substituting Equation (5.108) into Equation (5.109) and taking into account that $F = R_n/R_L$ result in the single-sideband noise-to-carrier ratio

$$L(f_m) = \frac{kTF}{2P_L}\left(\frac{f_0}{Q_L f_m}\right)^2 \tag{5.110}$$

which is similar to the Edson noise formula [60, 63].

5.4.3 Colpitts oscillator

As an example, let us define the linear phase noise model for a popular Colpitts oscillator, the simplified circuit schematic of which is shown in Figure 5.17a. The power loss in the tank inductor L is included in the load resistance R_L. The transistor equivalent circuit with voltage and current noise sources in shown in Figure 5.17b, where the resonator noise is modeled by noise current i_{nR}. The noise voltage and current sources can be given through their mean-square

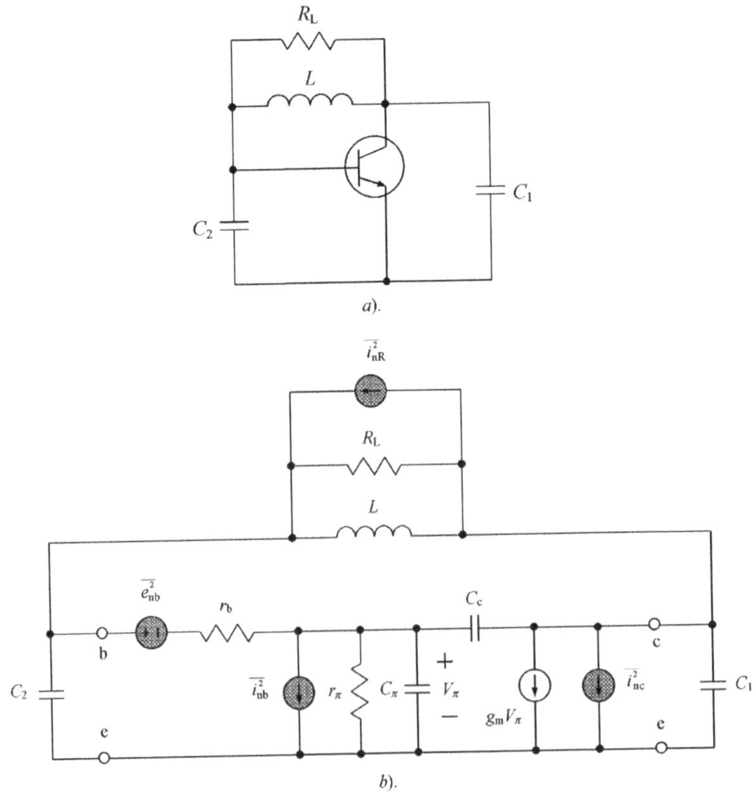

Figure 5.17 Equivalent circuits of the Colpitts oscillator

values as

$$\overline{i_{\mathrm{nR}}^2} = \frac{4kT\,\Delta f}{R_{\mathrm{L}}} \tag{5.111}$$

$$\overline{v_{\mathrm{nb}}^2} = 4kTr_{\mathrm{b}}\Delta f \tag{5.112}$$

$$\overline{i_{\mathrm{nb}}^2} = \frac{2qI_{\mathrm{c}}\Delta f}{\beta} \tag{5.113}$$

$$\overline{i_{\mathrm{nc}}^2} = 2qI_{\mathrm{c}}\Delta f \tag{5.114}$$

where the input noise current source $\overline{i_{\mathrm{nb}}^2}$ is related to shot noise at the emitter–base junction due to electrons which recombine with holes inside the neutral base and $\overline{i_{\mathrm{nc}}^2}$ represents the shot noise generated at the collector–base junction due to collector electrons. The mean-square values of these noise sources in a bandwidth Δf are given as

$$\overline{i_{\mathrm{nb}}^2} = 2qI_{\mathrm{b}}\Delta f = 2kTn_{\mathrm{b}}g_{\pi}\Delta f \tag{5.115}$$

$$\overline{i_{\mathrm{nc}}^2} = 2qI_{\mathrm{c}}\Delta f = 2kTn_{\mathrm{c}}g_{\mathrm{m}}\Delta f \tag{5.116}$$

where $g_{\pi} = 1/r_{\pi}$, I_{b} and I_{c} are the dc base and collector currents, n_{b} and n_{c} are the ideality factors of the emitter–base and collector–base junctions, respectively [64].

The admittance Y-parametres of the internal transistor excluding the base resistance r_{b} can be obtained as

$$[Y] = \begin{bmatrix} \dfrac{1}{r_{\pi}} + j\omega(C_{\pi} + C_{\mathrm{c}}) & -j\omega C_{\mathrm{c}} \\ g_{\mathrm{m}} - j\omega C_{\mathrm{c}} & j\omega C_{\mathrm{c}} \end{bmatrix} \tag{5.117}$$

where $g_{\mathrm{m}} = \beta/r_{\pi}$ is the device transconductance.

For a Colpitts oscillator with parallel feedback capacitors C_1 and C_2, it is convenient to represent all noise sources in a parallel configuration with input and output noise current sources. Figure 5.18 shows the transformation of the device noise model with the series thermal voltage noise source due to the base resistance into the equivalent device noise model with two parallel current noise sources at the input and output only, using the transmission $ABCD$-parametres. By using formulas of the transformation from the admittance Y-parametres to the transmission $ABCD$-parametres, one can write

$$\frac{1}{B} = -Y_{21} = -(g_{\mathrm{m}} - j\omega C_{\mathrm{c}}) \tag{5.118}$$

$$\frac{D}{B} = Y_{11} = \frac{1}{r_{\pi}} + j\omega(C_{\pi} + C_{\mathrm{c}}) \tag{5.119}$$

Let us simplify the analytical calculations by taking into account that, from Equations (5.118) and (5.119), the contribution of the collector capacitance C_{c} is not significant. As a result, the mean-square input current source $\overline{i_{\mathrm{ni}}^2}$ and output current source $\overline{i_{\mathrm{no}}^2}$ can be

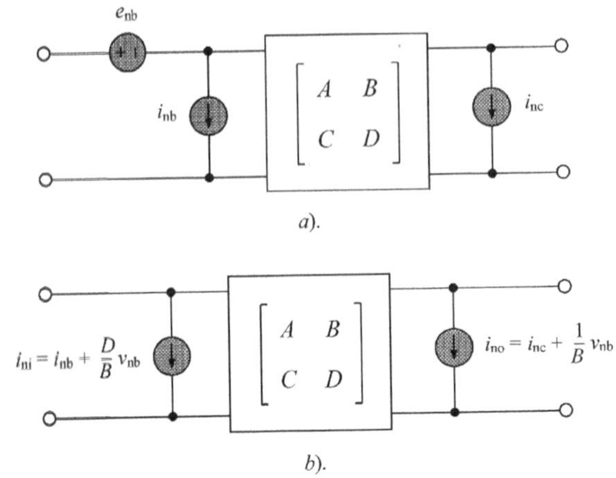

Figure 5.18 Transformation of base resistance thermal voltage noise

written as

$$\overline{i_{ni}^2} = \frac{2qI_c\Delta f}{\beta} + \frac{4kT\,\Delta f}{r_b}\left(\frac{r_b}{r_\pi}\right)^2\left(1+\beta^2\frac{f^2}{f_T^2}\right) \tag{5.120}$$

$$\overline{i_{n0}^2} = 2qI_c\Delta f + \frac{4kT\,\Delta f}{r_b}\left(\frac{\beta r_b}{r_\pi}\right)^2 \tag{5.121}$$

where f is the operating frequency, $f_T = g_m/2\pi C_\pi$ is the transition frequency, and it is assumed that the dc and small-signal values of the current gain β are equal.

To calculate the noise figure of the oscillator, the noise current sources $\overline{i_{ni}^2}$ and $\overline{i_{no}^2}$ should be transformed in parallel with $\overline{i_{nR}^2}$. By using Equations (5.115) and (5.116) for the case of ideal junctions when $n_b = n_c = 1$, the total equivalent noise source $\overline{i_{n\Sigma}^2}$ connected in parallel to the resonator and load can now be obtained as

$$\overline{i_{n\Sigma}^2} = \frac{4kT\,\Delta f}{R_L} + \frac{4kT\,\Delta f}{2r_\pi}\left[1+\frac{2r_b}{r_\pi}\left(1+\beta^2\frac{f^2}{f_T^2}\right)\right]\left(\frac{C_1}{C_1+C_2}\right)^2$$
$$+\frac{4kT\beta\Delta f}{2r_\pi}\left(1+\frac{2\beta r_b}{r_\pi}\right)\left(\frac{C_2}{C_1+C_2}\right)^2 \tag{5.122}$$

The noise figure of the oscillator F is defined as the ratio of the total noise power due to all noise current sources and the noise power from the loaded resonator due to the noise source $\overline{i_{nR}^2}$. As a result,

$$F = 1+\frac{R_L}{2r_\pi}\left[1+\frac{2r_b}{r_\pi}\left(1+\beta^2\frac{f^2}{f_T^2}\right)\right]\left(\frac{C_1}{C_1+C_2}\right)^2 + \frac{R_L}{2r_\pi}\left(1+\frac{2\beta r_b}{r_\pi}\right)\left(\frac{C_2}{C_1+C_2}\right)^2 \tag{5.123}$$

In most cases, the contribution of the collector shot noise dominates the contribution of the shot noise caused by the base current. Consequently, for $r_b \ll r_\pi$ and $f \ll f_T$, the expression for the oscillator noise figure is simplified to

$$F = 1 + \frac{R_L}{2r_\pi}\left(1 + \frac{2\beta r_b}{r_\pi}\right)\left(\frac{C_2}{C_1 + C_2}\right)^2 \tag{5.124}$$

Finally, the single sideband noise-to-carrier ratio at the output of the Colpitts oscillator in a linear consideration, using Equation (5.110), can be defined by

$$L(f_m) = \frac{kT}{2P_L}\left(\frac{f_0}{Q_L f_m}\right)^2\left[1 + \frac{R_L}{2r_\pi}\left(1 + \frac{2\beta r_b}{r_\pi}\right)\left(\frac{C_2}{C_1 + C_2}\right)^2\right] \tag{5.125}$$

5.5 OSCILLATOR NOISE SPECTRUM: NONLINEAR MODEL

5.5.1 Kurokawa approach

The one-port negative resistance oscillator can generally be represented by the circuit shown in Figure 5.19a. The active device represented by its output impedance $Z_{out} = R_{out} + jX_{out}$ and the load R_L are coupled through a multiple-resonant reciprocal circuit. Looking from the active device, the circuit can be equivalently expressed through series connection of the active device impedance Z_{out} and the load impedance Z_L shown in Figure 5.19b, where $e_n(t)$ is the noise voltage. A similar representation of an oscillator circuit can be obtained in terms of admittances. For a particular case of a single series resonant circuit, $R = R_L$.

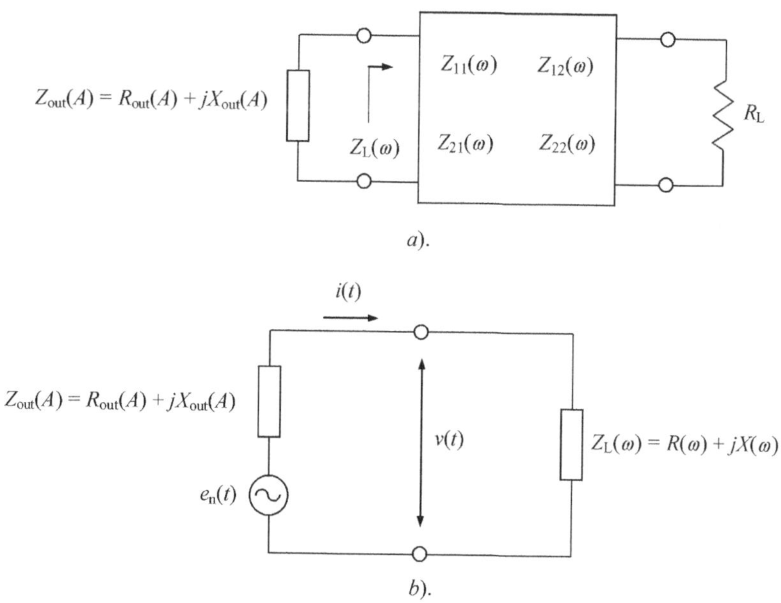

Figure 5.19 Equivalent negative resistance oscillator noise models

Consider the equivalent negative resistance oscillator circuit shown in Figure 5.19b, where the active device output impedance $Z_{\rm out}(A)$ with negative real part is a function of the oscillation amplitude and the load impedance $Z_{\rm L}(\omega)$ is a frequency-dependent function [65]. Let the current flowing through the active device be

$$i(t) = A(t)\cos[\omega t + \phi(t)] \tag{5.126}$$

where $A(t)$ and $\phi(t)$ are assumed to be slowly varying functions of time. The voltage drop across the active device is given by

$$v(t) = R_{\rm out} A\cos(\omega t + \phi) - X_{\rm out} A\sin(\omega t + \phi) \tag{5.127}$$

where $R_{\rm out}$ and $X_{\rm out}$ are functions of the current amplitude.

By calculating the first- and higher-order time derivatives of $i(t)$ and assuming $\mathrm{d}\phi/\mathrm{d}t \ll \omega_0$ and $(1/A)(\mathrm{d}A/\mathrm{d}t) \ll \omega$, the two equations for slowly varying amplitude and phase can be obtained in the form

$$R(\omega) + R_{\rm out}(A) + \frac{\partial R(\omega)}{\partial\omega}\frac{d\theta}{dt} + \frac{\partial X(\omega)}{\partial\omega}\frac{1}{A}\frac{dA}{dt} = \frac{1}{A}e_{\rm n}^{\rm c}(t) \tag{5.128}$$

$$-X(\omega) - X_{\rm out}(A) + \frac{\partial X(\omega)}{\partial\omega}\frac{d\theta}{dt} + \frac{\partial R(\omega)}{\partial\omega}\frac{1}{A}\frac{dA}{dt} = \frac{1}{A}e_{\rm n}^{\rm s}(t) \tag{5.129}$$

where

$$e_{\rm n}^{\rm c}(t) = \frac{2}{T_0}\int\limits_{t-T_0}^{t} e_{\rm n}(t)\cos(\omega t + \phi)\mathrm{d}t \tag{5.130}$$

$$e_{\rm n}^{\rm s}(t) = \frac{2}{T_0}\int\limits_{t-T_0}^{t} e_{\rm n}(t)\sin(\omega t + \phi)\mathrm{d}t \tag{5.131}$$

where T_0 is the oscillation period, $e_{\rm n}^{\rm c}(t)$ and $e_{\rm n}^{\rm s}(t)$ represent the cosine and sine components of the noise voltage $e_{\rm n}(t)$.

Equations (5.128) and (5.129) both contain the derivatives $\mathrm{d}\phi/\mathrm{d}t$ and $\mathrm{d}A/\mathrm{d}t$. However, by multiplying Equation (5.128) by $\partial X/\partial\omega$ and Equation (5.129) by $\partial R/\partial\omega$ and adding, as well as by multiplying Equation (5.128) by $\partial R/\partial\omega$ and Equation (5.129) by $-\partial X/\partial\omega$ and adding, we obtain the basic equations for the amplitude and phase of the oscillating current

$$[R(\omega) + R_{\rm out}(A)]\frac{\partial X}{\partial\omega} - [X(\omega) + X_{\rm out}(A)]\frac{\partial R}{\partial\omega} + \left|\frac{\partial Z_{\rm L}(\omega)}{\partial\omega}\right|^2 \frac{1}{A}\frac{\mathrm{d}A}{\mathrm{d}t}$$
$$= \frac{1}{A}\left[\frac{\partial X}{\partial\omega}e_{\rm n}^{\rm c}(t) + \frac{\partial R}{\partial\omega}e_{\rm n}^{\rm s}(t)\right] \tag{5.132}$$

$$[R(\omega) + R_{\rm out}(A)]\frac{\partial R_{\rm L}}{\partial\omega} + [X(\omega) + X_{\rm out}(A)]\frac{\partial X}{\partial\omega} + \left|\frac{\partial Z_{\rm L}(\omega)}{\partial\omega}\right|^2 \frac{\mathrm{d}\phi}{\mathrm{d}t}$$
$$= \frac{1}{A}\left[\frac{\partial R}{\partial\omega}e_{\rm n}^{\rm c}(t) - \frac{\partial X}{\partial\omega}e_{\rm n}^{\rm s}(t)\right] \tag{5.133}$$

For the deviation of the current amplitude A from its steady-state value A_0 by a small value ΔA, we can write

$$R(\omega_0) + R_{\text{out}}(A) = A_0 \frac{\partial R_{\text{out}}(A_0)}{\partial A} \frac{\Delta A}{A_0} \tag{5.134}$$

$$X(\omega_0) + X_{\text{out}}(A) = A_0 \frac{\partial X_{\text{out}}(A_0)}{\partial A} \frac{\Delta A}{A_0} \tag{5.135}$$

where all derivatives are evaluated in a steady-state mode characterized by the oscillation frequency ω_0 and oscillation amplitude A_0.

By substituting Equations (5.134) and (5.135) into Equations (5.132) and (5.133), respectively, and assuming the deviation of the phase ϕ from its initial value ϕ_0 by a small value $\Delta\phi$, the frequency spectra of ΔA and $\Delta\phi$ can be obtained as

$$|\Delta A|^2 = \frac{2\left|\dfrac{\partial Z_{\text{L}}(\omega_0)}{\partial \omega}\right|^2 |e_{\text{n}}|^2}{\omega^2\left|\dfrac{\partial Z_{\text{L}}(\omega_0)}{\partial \omega}\right|^4 + A_0^4\left[\dfrac{\partial R_{\text{out}}(A_0)}{\partial A}\dfrac{\partial X(\omega_0)}{\partial \omega} - \dfrac{\partial X_{\text{out}}(A_0)}{\partial A}\dfrac{\partial R(\omega_0)}{\partial \omega}\right]^2} \tag{5.136}$$

$$|\Delta\phi|^2 = \frac{2|e_{\text{n}}|^2}{\omega^2 A_0^2} \frac{\omega^2\left|\dfrac{\partial Z_{\text{L}}(\omega_0)}{\partial \omega}\right|^2 + A_0^2\left[\dfrac{\partial R_{\text{out}}(A_0)}{\partial A}\right]^2 + A_0^2\left[\dfrac{\partial X_{\text{out}}(A_0)}{\partial A}\right]^2}{\omega^2\left|\dfrac{\partial Z_{\text{L}}(\omega_0)}{\partial \omega}\right|^4 + A_0^4\left[\dfrac{\partial R_{\text{out}}(A_0)}{\partial A}\dfrac{\partial X(\omega_0)}{\partial \omega} - \dfrac{\partial X_{\text{out}}(A_0)}{\partial A}\dfrac{\partial R(\omega_0)}{\partial \omega}\right]^2} \tag{5.137}$$

By defining $e_{\text{r}}(t) = A_0\Delta R$ and $e_{\text{x}} = A_0\Delta X$, where ΔR and ΔX are the fluctuating parts of $R(\omega) + R_{\text{out}}(A)$ and $X(\omega) + X_{\text{out}}(A)$, respectively, the contribution from $1/f$ noise can also be introduced into Equations (5.128) and (5.129) [66]. In this case, their power spectral densities $|e_{\text{r}}(f)|^2$ and $|e_{\text{x}}(f)|^2$ are both proportional to $1/f$.

To analyse the stability conditions of the negative resistance oscillator, the right-hand sides of the denominators in Equations (5.136) and (5.137) can be rewritten as

$$\begin{aligned}\frac{\partial R_{\text{out}}}{\partial A}\frac{\partial X}{\partial \omega} - \frac{\partial X_{\text{out}}}{\partial A}\frac{\partial R}{\partial \omega} &= \frac{\partial R_{\text{out}}}{\partial A}\frac{\partial R}{\partial \omega}(\tan\Phi - \tan\phi) = \frac{\partial R_{\text{out}}}{\partial A}\frac{\partial R}{\partial \omega}\frac{\sin(\Phi - \phi)}{\cos\Phi\cos\phi} \\ &= \sqrt{\left(\frac{\partial R_{\text{out}}}{\partial A}\right)^2 + \left(\frac{\partial X_{\text{out}}}{\partial A}\right)^2}\left|\frac{\partial Z_{\text{L}}}{\partial \omega}\right| \sin(\Phi - \phi)\end{aligned} \tag{5.138}$$

where $\Phi = \tan^{-1}[\partial X(\omega_0)/\partial R(\omega_0)]$ is the load line angle and $\phi = \tan^{-1}[\partial X(A_0)/\partial R(A_0)]$ is the active device line angle at the steady-state point. This can be illustrated graphically on the complex plane with lines representing output device impedance $Z_{\text{out}}(A)$ and load impedance $Z_{\text{L}}(\omega)$, as shown in Figure 5.20. The arrows indicate the directions of increasing parametres A and ω. The intersection of these two curves gives a steady-state operation point with the oscillation amplitude A_0 and frequency ω_0.

From Equation (5.138) it follows that the stability condition is proportional to the sine of the angle between the device line and the load line, having maximum value when the two lines are perpendicular. This denotes a situation of maximum stability when the normalized dynamic

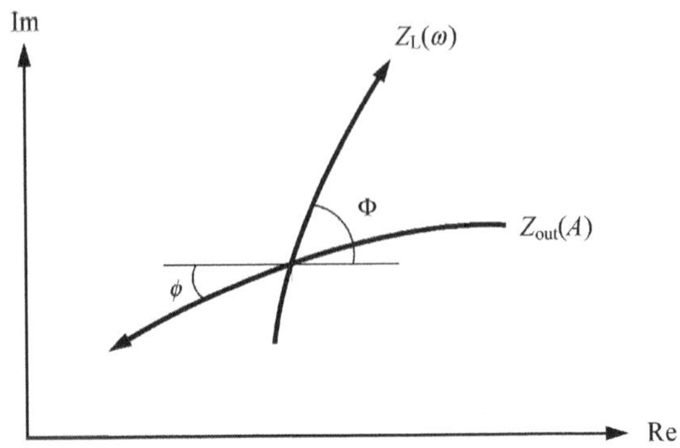

Figure 5.20 Device and load lines on the complex plane

stability factor

$$S = \frac{\dfrac{\partial R_{out}}{\partial A}\dfrac{\partial X}{\partial \omega} - \dfrac{\partial X_{out}}{\partial A}\dfrac{\partial R}{\partial \omega}}{\sqrt{\left(\dfrac{\partial R_{out}}{\partial A}\right)^2 + \left(\dfrac{\partial X_{out}}{\partial A}\right)^2}\sqrt{\left(\dfrac{\partial R}{\partial \omega}\right)^2 + \left(\dfrac{\partial X}{\partial \omega}\right)^2}} = \sin(\Phi - \phi) \tag{5.139}$$

has a maximum value of 1.0 which would characterize a free-running oscillator dynamically stabilized in an optimum way [67]. The normalized dynamic stability is a measure of the relative stability that appears in calculations of the oscillator amplitude and phase noise [65].

Equations (5.136) and (5.137) obtained by Kurokawa for a particular case of the negative resistance oscillator, where the device impedance is a function of only the oscillation amplitude and the load impedance depends only on the oscillation frequency, can be extended for the negative conductance oscillator. In a generalized case, the device and load immitances both depend on the oscillation frequency and amplitude. As a result, the double-sideband power spectral densities for the amplitude and phase fluctuations can be rewritten as

$$S_a\left(\frac{\Delta A}{A_0}, f_m\right) = \frac{S_{\Delta\phi}(f_m)}{p^2 + \left(\dfrac{2Q_L f_m}{f_0}\right)^2} \tag{5.140}$$

$$S_\phi(\Delta\theta, f_m) = S_{\Delta\phi}(f_m)\left(\frac{f_0}{2Q_L f_m}\right)^2\left[1 + \frac{q^2}{p^2 + \left(\dfrac{2Q_L f_m}{f_0}\right)^2}\right] \tag{5.141}$$

where

$$p = \frac{A_0 f_0}{2Q_L(\mathrm{Re}W_L)^2}\left(\frac{\partial \mathrm{Re}W}{\partial A}\frac{\partial \mathrm{Im}W}{\partial f} - \frac{\partial \mathrm{Im}W}{\partial A}\frac{\partial \mathrm{Re}W}{\partial f}\right)$$

$$q = \frac{A_0 f_0}{2Q_L(\mathrm{Re}W_L)^2}\left(\frac{\partial \mathrm{Re}W}{\partial A}\frac{\partial \mathrm{Re}W}{\partial f} + \frac{\partial \mathrm{Im}W}{\partial A}\frac{\partial \mathrm{Im}W}{\partial f}\right)$$

$$Q_L = \frac{f_0}{2\mathrm{Re}W_L}\sqrt{\left(\frac{\partial \mathrm{Re}W}{\partial f}\right)^2 + \left(\frac{\partial \mathrm{Im}W}{\partial f}\right)^2}$$

is the oscillator loaded quality factor, $S_{\Delta\phi}$ is the spectral power density of the noise voltage source e_n normalized to the output power delivered to the load P_L, $W = W_{out} + W_L$ is the overall circuit immitance, f_0 is the oscillation or carrier frequency, f_m is the offset frequency from the carrier, f_c is the corner frequency where the level of the low-frequency noise becomes equal to the level of the noise floor [68].

The parameter p is a function of the oscillator stability conditions, given in parentheses, characterizing the velocity of the establishment of the steady-state self-sustained oscillations under small perturbations. The parameter q illustrates the dependence of the oscillation frequency on the oscillation amplitude in a large-signal mode of operation. From Equations (5.140) and (5.141) it follows that the oscillation becomes very noisy as one approaches the boundary of the stable region, that is, as the parameter p becomes close to zero. In addition, the greater value of the parameter q, the higher the phase noise level is expected in the oscillator spectrum due to amplitude-to-phase conversion.

The linear dependences of the oscillator amplitude and phase noise on the noise power spectral density for voltage e_n in Equations (5.136) and (5.137) indicate that the specifics of the noise generation mechanism and its translation to the oscillator frequency spectrum are not considered. Therefore, the mixing effect from the nonlinear behaviour of the active device is not inherent in this theory and must be included explicitly in the evaluation of $e_n(t)$. For a general case of the active device described by a two-port network equivalent circuit, in order to evaluate the output port noise voltage generator, it is necessary to provide a transformation of the noise source from the input port to the output port of the device. An example of the procedure for the phase noise calculation of a microwave GaAs MESFET oscillator where the nonlinear transistor equivalent circuit contains two basic nonlinear elements (the device transconductance and the drain–source conductance) is presented in [68]. The noise analysis based on the extended MESFET model with dominant nonlinear elements, including also the gate–source capacitance, shows that the gate–source capacitance is responsible for the conversion of the low-frequency noise into the phase noise, whereas the amplitude noise is primarily determined by the nonlinear transconductance [57, 69]. Compared with the transconductance and gate–source capacitance, the nonlinear drain–source conductance is of minor importance for the noise up-conversion process. As a result, using an improved structure of the MESFET device with the substantially reduced value of $\partial\omega_0/\partial V_{gs}$, where V_{gs} is the gate–source voltage amplitude, can lead to a 15-dB reduction in the phase noise compared with a typical structure of the MESFET device [70].

The theoretical approaches to calculate the oscillator amplitude and phase noise spectra derived from the system of equations describing deterministic and stochastic behaviour based on numerical solutions are given in [71, 72]. These methods are neither limited to a certain circuit topology nor to certain types of oscillator noise sources. The general noise analysis of nonlinear microwave circuits by the piecewise harmonic balance technique is discussed elsewhere [73]. Here, the noise generation in autonomous systems such as free-running oscillators is shown to be originated by two different mechanisms: direct frequency modulation of the oscillator, dominant at low-frequency offsets, and conventional frequency conversion, dominant at high-frequency offsets. Noise analysis based on a frequency conversion theory alone are shown to be insufficient for predicting the near-carrier noise in general oscillators at very-low-frequency offsets, and should be complemented by modulation noise analysis. Being implemented in a simulator, the noise model including frequency modulation and frequency conversion mechanisms shows accurate correspondence for all regions when compared with the experimental data [74]. In this case, the phase noise spectral behaviour due to frequency modulation is more accurately predicted at smaller offsets, while the mixing data (noise in oscillator mixes the carrier and its harmonics to produce noise on either side of the carrier) give more accurate results at larger offsets corresponding to a noise floor.

5.5.2 Impulse response model

The behaviour of autonomous second-order weakly nonlinear oscillation systems with low damping factor close to linear conservative systems and small time-varying external force $f(t)$ can be described by

$$\frac{d^2x}{d\tau^2} + x = n(t) \tag{5.142}$$

where x is the time-dependent variable, voltage or current, and $\tau = \omega_0 t$ is the time normalized by the angular resonant frequency ω_0.

The phase plane method is one of the theoretical approaches that allows one to analyse qualitatively and quantitatively the dynamics of the oscillation systems described by second-order differential equations such as Equation (5.142) [75]. By setting the small external force equal to zero, the solution of the linear second-order differential equation takes the form

$$x = A\cos(\tau + \phi) = A\cos\psi \tag{5.143}$$

$$\frac{dx}{d\tau} = -A\sin(\tau + \phi) = -A\sin\psi \tag{5.144}$$

where A is the amplitude of the oscillations, ϕ is the phase of the oscillation. The phase portrait shown in Figure 5.21a, being similar to that of shown in Figure 3.12b (Chapter 3), represents the family of circular trajectories enclosing each other with radii $r = A$ (limit cycles) depending on the energy stored in the system.

Let us define the variations of the amplitude $A(t)$ and phase $\phi(t)$ under the effect of the external force applied to the oscillation system [76]. Assuming that the effect of the external force is small and these variations are slow, the amplitude and phase can be considered constant during a natural period of the oscillation. Then, Equation (5.142) can be rewritten in the form

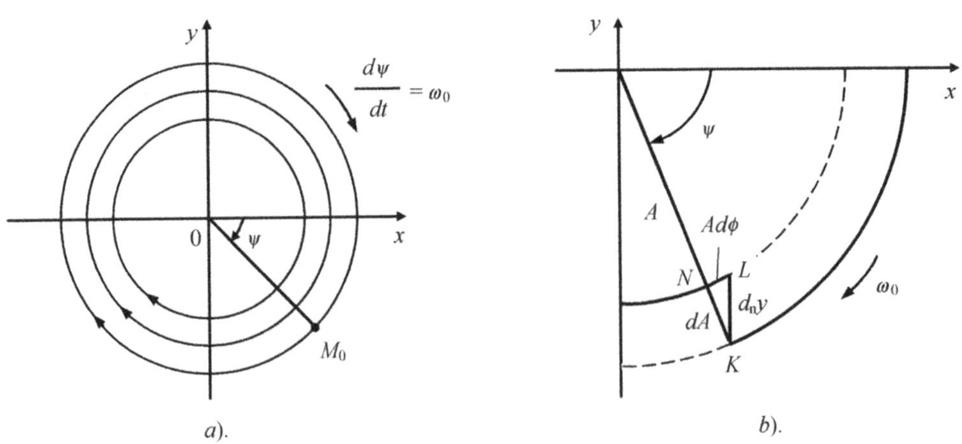

Figure 5.21 Phase portrait of second-order oscillation system and effect of injected impulse

of two first-order equations by

$$\frac{dx}{d\tau} = y \tag{5.145}$$

$$\frac{dy}{d\tau} = -x + n(t). \tag{5.146}$$

From Equations (5.145) and (5.146) it follows that the instantaneous change of the ordinate y by a value of $d_n y = n d\tau$ will occur, due to the small external force injected to the oscillation system. This corresponds to a step change of the representative point M_0 from the position K to the position L, resulting in the amplitude and phase changes, as shown in Figure 5.21b. These changes can be determined from consideration of a triangle *KLN* as

$$dA = -d_n y \sin\psi = -n \sin\psi \, d\tau \tag{5.147}$$

$$d\phi = -\frac{d_n y}{A} \cos\psi = -\frac{n}{A} \cos\psi \, d\tau \tag{5.148}$$

Thus, separate first-order differential equations for the time-varying amplitude and phase can be obtained from Equations (5.147) and (5.148) as

$$\frac{dA}{d\tau} = -n \sin\psi \tag{5.149}$$

$$\frac{d\phi}{d\tau} = -\frac{n}{A} \cos\psi \tag{5.150}$$

Since the right-hand sides of Equations (5.149) and (5.150) are small, time-averaged differential equations can be used instead of the differential equations for the instantaneous values of the amplitude and phase. Hence, the changes of the amplitude and phase for a time period $t \leq T$ are defined as

$$\Delta A(t) = -\int_0^{\omega_0 t} n(\tau) \sin\tau \, d\tau \tag{5.151}$$

$$\Delta\phi(t) = -\int_0^{\omega_0 t} \frac{n(\tau)}{A} \cos\tau \, d\tau \tag{5.152}$$

Figure 5.22a shows the equivalent circuit of the negative resistance oscillator with injected small perturbation current $i(t)$. In a steady-state oscillation mode, when the losses in the resonant circuit are compensated by the energy inserted into the circuit by the active device, the second-order differential equation of the oscillator is written as

$$\frac{d^2 v}{dt^2} + \omega_0^2 v = \frac{1}{C} \frac{di}{dt} \tag{5.153}$$

where $\omega_0 = 1/\sqrt{LC}$ is the resonant frequency and $v(t)$ is the voltage across the resonant circuit. If a current impulse $i(t)$ is injected, the amplitude and phase of the oscillator will have time-dependent responses. According to Equations (5.151) and (5.152), the resultant amplitude and phase changes have quadrature dependence with respect to each other. When an impulse is applied at the peak of the voltage across the capacitor, there will be a maximum amplitude deviation with no phase shift, as shown in Figure 5.22b. On the other hand, if the current

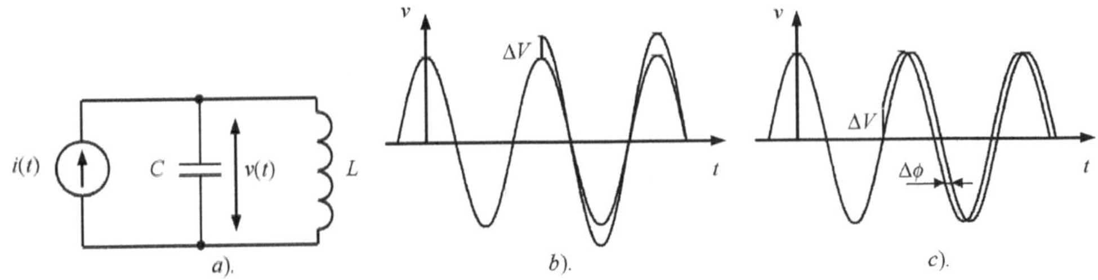

Figure 5.22 Second-order *LC* oscillator and effect of injected impulse

impulse is applied at the zero crossing, it will result in a maximum phase deviation with no amplitude response, as shown in Figure 5.22c.

Suppose that a perturbation current $i(t)$, injected into the oscillation circuit, is a periodic function that can generally be expanded into a Fourier series

$$i(t) = I_0 \cos \Delta\omega t + \sum_{k=1}^{\infty} \{I_{kc} \cos[(k\omega_0 + \Delta\omega)t] + I_{ks} \sin[(k\omega_0 + \Delta\omega)t]\} \tag{5.154}$$

where I_0 is the dc current, I_{kc} is the kth cosine current harmonic amplitude, I_{ks} is the kth sinusoidal current harmonic amplitude, and $\Delta\omega \ll \omega_0$. In this case, the small external force can be redefined as

$$n = \frac{1}{\omega_0 C}\frac{\mathrm{d}i}{\mathrm{d}\tau} = L\frac{\mathrm{d}i}{\mathrm{d}t}. \tag{5.155}$$

Consequently, substituting Equation (5.154) into Equations (5.151) and (5.152) and taking into account that $\omega_0 + \Delta\omega \cong \omega_0$ result in

$$\Delta V(t) = -\frac{I_{1s}}{2C}\frac{\cos(\Delta\omega t) - 1}{\Delta\omega} \tag{5.156}$$

$$\Delta\phi(t) = -\frac{I_{1c}}{2CV}\frac{\sin(\Delta\omega t)}{\Delta\omega} \tag{5.157}$$

where V is the voltage amplitude across the capacitor C. In this case, only the fundamental components in the injected current $i(t)$ can contribute to the amplitude (sine amplitude) and phase (cosine amplitude) fluctuations given by Equations (5.156) and (5.157), because for the dc and kth-order current components, the arguments for all their integrals in Equations (5.151) and (5.152) are significantly attenuated by the averaging over the integration period.

The output voltage of an ideal cosine oscillator with constant amplitude V and phase fluctuations $\Delta\phi$ can be written as

$$\begin{aligned} v(t) &= V \cos[\omega_0 t + \Delta\phi(t)] \\ &= V \cos[\Delta\phi(t)] \cos\omega_0 t - V \sin[\Delta\phi(t)] \sin\omega_0 t \end{aligned} \tag{5.158}$$

resulting in an output spectrum of the oscillator with sidebands close to the oscillation frequency ω_0. Since, for a narrowband phase modulation with small phase fluctuations, $\sin[\Delta\phi(t)] \cong \Delta\phi(t)$ and $\cos[\Delta\phi(t)] \cong 1$, the phase modulation spectrum given by Equation (5.158) can be rewritten using Equation (5.157) as

$$v(t) = V\cos\omega_0 t + \frac{I_{1c}}{4C\,\Delta\omega}\cos[(\omega_0 t - \Delta\omega)t] - \frac{I_{1c}}{4C\,\Delta\omega}\cos[(\omega_0 t + \Delta\omega)t] \quad (5.159)$$

which is similar to the single-tone amplitude modulation spectrum containing the spectral components corresponding to the carrier frequency ω_0 and two close sideband frequencies $\omega_0 - \Delta\omega$ and $\omega_0 + \Delta\omega$.

The injection of the current

$$i(t) = I_0 \cos\Delta\omega_0 t + \sum_{k=1}^{\infty}\{I_{kc}\cos[(k\omega_0 - \Delta\omega)t] + I_{ks}\sin[(k\omega_0 - \Delta\omega)t]\} \quad (5.160)$$

has similar effect, resulting in twice the noise power at the sidebands. Therefore, an injected total current $i(t)$ results in a pair of equal sidebands at $\omega_0 \pm \Delta\omega$ with a sideband power P_{sb} relative to the carrier power P_c given by

$$\frac{P_{sb}(\omega_0 \pm \Delta\omega)}{P_c(\omega_0)} = 2\left(\frac{I_{1c}}{4CV\,\Delta\omega}\right)^2 \quad (5.161)$$

Now let us assume that a stationary thermal noise current with a white power spectral density $\overline{i_n^2}$ is injected into the oscillator circuit close to carrier. Then, by making a replacement between the amplitude and root-mean-square current values when $I_{1c}^2/2 = \overline{i_n^2}$, the single-sideband power spectral density for the phase fluctuations at $\Delta\omega$ offset from the carrier ω_0 in $1/f^2$ region can be written using Equation (5.161) as

$$L(f_m) = \frac{\overline{i_n^2}}{4C^2V^2\Delta\omega^2} \quad (5.162)$$

Taking into account that $\overline{i_n^2} = 4FkT/R_L$ for $\Delta f = 1$ Hz, $Q_L = \omega_0 C R_L$ and $P_L = V^2/2R_L$, where R_L is the tank parallel or load resistance, Equation (5.162) can be rewritten as

$$L(f_m) = \frac{2FkT}{P_L}\left(\frac{\omega_0}{2Q_L\Delta\omega}\right)^2 \quad (5.163)$$

which is similar to Equation (5.110) for the negative resistance oscillator.

Consideration of both cosine current and cosine voltage across the device output terminals implies the device operation in a linear active region only when the establishment of the oscillations can be achieved by using a separate diode as a nonlinear element. Therefore, the mixing effect from the nonlinear behaviour of the active device is not taken into account. For a general case of the active device described by a two-port network equivalent circuit, in order to evaluate the output port noise voltage generator, it is necessary to provide a transformation of the noise source from the input port to the output port of the device.

As an example, consider an oscillator with a nonlinear output resistance dependent on the applied dc bias voltage and on the amplitude of the self-sustained oscillations. The basic oscillator circuit with the nonlinear negative output resistance R_{out}, capacitance C, inductance L, load resistance R_L and noise current $i_n(t)$ is shown in Figure 5.23a. The electrical behaviour of such an oscillator, in terms of voltage $v(t)$ across the capacitance, can be represented by a second-order nonlinear differential equation

$$LC\frac{d^2v}{dt^2} + \frac{L}{R_L}\frac{dv}{dt} + v + L\frac{di}{dt} = e_n(t) \quad (5.164)$$

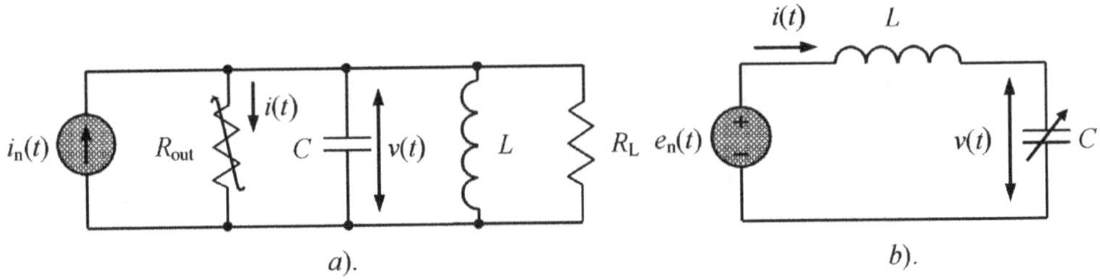

Figure 5.23 Second-order nonlinear oscillation systems

where

$$e_n(t) = L\frac{di_n(t)}{dt} \tag{5.165}$$

is the equivalent noise voltage and

$$i = I_0 + \frac{v}{R_{out}} + \sum_{k=2}^{\infty} G_k v^k \tag{5.166}$$

represents a power series expansion where I_0 is the dc current and G_k are the small coefficients.

In a steady-state operation mode, when the active device compensates for the losses in the load resistance, that is, $R_{out} + R_L = 0$, Equation (5.164) can be rewritten as

$$LC\frac{d^2v}{dt^2} + v + L\frac{d}{dt}\left(\sum_{k=2}^{\infty} G_k v^k\right) = e_n(t) \tag{5.167}$$

Seeking the general solution of the inhomogeneous differential equation as the superposition of the general solution of the homogeneous (noise-free) and specific solutions of Equation (5.167) as

$$v(t) = V(t)\cos[\omega_0 t + \phi(t)] + e_n(t) \tag{5.168}$$

and applying a van der Pol approach for the slowly time-varying amplitude $V(t)$ and phase $\phi(t)$, allow us to rewrite Equation (5.167) in the form

$$2\omega_0\frac{dV}{dt}\sin(\omega_0 t + \phi) + 2\omega_0 V\frac{d\phi}{dt}\cos(\omega_0 t + \phi) = \frac{1}{C}\frac{d}{dt}\left(\sum_{k=2}^{\infty} G_k v^k\right) \tag{5.169}$$

where $\omega_0 = 1/\sqrt{LC}$. It is assumed that $e_n(t)$ is a small slowly time-varying low-frequency noise voltage, for which

$$LC\frac{d^2e_n(t)}{dt^2} \ll e_n(t)$$

As a result, substituting Equation (5.168) into the right-hand side of Equation (5.169) and using trigonometric identities yields

$$\frac{d\phi}{dt} = 0 \tag{5.170}$$

which means that the nonlinear output resistance has no impact on the phase fluctuations. However, the amplitude fluctuations are not equal to zero because all factors on the right-hand side of Equation (5.169) have first-order sine components. Thus, the resistive type of nonlinearities alone would cause amplitude noise only, since the reactive elements determining the oscillation frequency remain constant. However, if the high-frequency noise current is injected close to the carrier frequency ω_0 (for example, at small offset $\Delta\omega$), it will cause phase fluctuations according to Equations (5.157) and (5.169).

Now consider a varactor-controlled oscillator with the varactor as a nonlinear element, whose capacitance depends not only on the applied dc bias voltage, but also on the amplitude of the self-sustained oscillations. The basic VCO circuit consists of the varactor with a nonlinear capacitance C, an inductance L, and a noise voltage $e_n(t)$, as shown in Figure 5.23b [77]. The voltage $e_n(t)$ can represent all the noise coming from both inside and outside the circuit, including any thermal noise from the resistors, flicker noise from the active device and noise from the power supply. The electrical behaviour of the oscillator can be described by

$$v + L\frac{di}{dt} = e_n(t) \tag{5.171}$$

$$i = \left(C + v\frac{dC}{dv}\right)\frac{dv}{dt} \tag{5.172}$$

where the nonlinear term $v\,dC/dv$ is included in Equation (5.172).

By expanding a nonlinear capacitance C into a power series

$$C = C_0 + \sum_{k=1}^{\infty} C_k v^k \tag{5.173}$$

with the small coefficients C_k, substituting Equation (5.172) into Equation (5.171) and applying an asymptotic perturbation procedure with decomposition of the perturbed and unperturbed equations, the first-order differential equation for phase fluctuations with the slowly time-varying noise voltage e_n can be derived as

$$\frac{d\phi}{dt} = -\frac{\omega_0}{C_0}\left[C_1 e_n + C_2\left(\frac{3}{4}V^2 + 3e_n^2\right) + C_3 e_n\left(3V^2 + 4e_n^2\right) + \ldots\right] \tag{5.174}$$

where V is the voltage amplitude across the varactor [77]. Note that nonlinear capacitance has no impact on the amplitude noise of the oscillator.

From Equation (5.174) it follows that

- the first order capacitance nonlinearity described by the coefficient C_1 contributes to the up-conversion of the low-frequency noise $e_n(t)$ to the sideband noise near carrier ω_0;
- the second-order nonlinearity described by the coefficient C_2 generates a phase noise, due to both amplitude-to-phase conversion and low-frequency noise up-conversion;
- the higher-order nonlinearities described by the coefficients C_k, $k = 3, 4, 5, \ldots$, cause a more complicated noise behaviour of the oscillator based on hybrid up-conversion and amplitude-to-phase conversion due to cross-terms of V and e_n.

In the case of a single-frequency *LC* oscillator, the main contributor to the phase noise is the nonlinear collector capacitance of the bipolar device or the gate–source capacitance of the FET device.

In a general case, the equivalent circuit of the active device is very complicated, including both nonlinear intrinsic and linear parasitic external elements. This means that it is difficult to

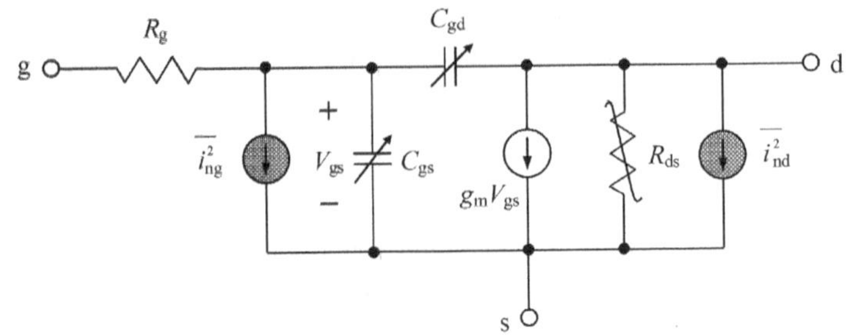

Figure 5.24 Equivalent circuit of MESFET device including noise sources

evaluate analytically the impact of each nonlinear element on the up-conversion mechanism. Moreover, the joint effect of different nonlinear circuit elements will result in both amplitude and phase fluctuations. Figure 5.24 shows the nonlinear MESFET equivalent circuit with input and output current noise sources. As can be determined from numerical calculations, the phase noise can be significantly reduced by linearizing both the transconductance g_m and the gate–source capacitance C_{gs}; in other words, both nonlinearities are important contributors to the phase noise [78].

The amplitude noise also depends on the capacitance and transconductance nonlinearities. However, the capacitance nonlinearity will not affect the output current if the series gate resistance R_g is set to zero. The nonlinearities of the gate–drain capacitance C_{gd} and drain–source resistance R_{ds} have negligible effect on the amplitude and phase noise.

Generally, the transition from soft start-up oscillation conditions to steady-state self-sustained oscillations is provided as a result of the degradation of the device transconductance in a large-signal mode, when the active device is operated in both pinch-off and active regions. As a result, for the cosine voltage across the resonant circuit, the output collector (or drain) current $i(t)$ represents a Fourier series expansion

$$i(t) = I_0 + \sum_{n=1}^{\infty} I_n \cos(n\omega_0 t) \tag{5.175}$$

where I_0 is the dc current and I_n is the amplitude of the nth harmonic component.

If the oscillation frequency is equal to the resonant circuit frequency, which means that the active device has no effect on the oscillation frequency, then the fundamental component of the collector voltage will be in phase with the fundamental component of the collector current. However, for all higher-order voltage harmonics, the impedance of the resonant circuit will be capacitive since the collector current harmonics are mostly flowing through the shunt capacitance. Therefore, for the Meissner oscillator circuit shown in Figure 5.25, the voltage at the input of the active device can be approximately represented as

$$v_{in}(t) = V_{in1} \cos \omega_0 t + \sum_{k=2}^{\infty} V_{ink} \cos\left(k\omega_0 t - \frac{\pi}{2}\right) \tag{5.176}$$

where $V_{ink} \ll V_{in1}$ for a high value of the oscillator loaded quality factor.

As a result, when the active device transfer characteristic is approximated by a polynomial, the presence of higher-order voltage harmonic contributes, first, to changes in the fundamental amplitude and, second, to the appearance of the phase shift between the fundamental voltage

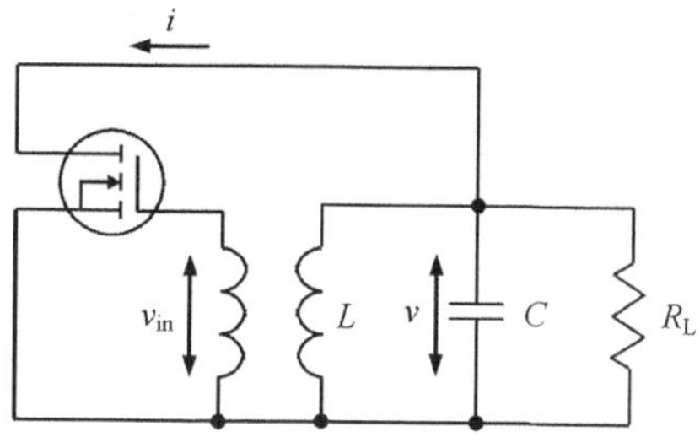

Figure 5.25 Schematic of parallel feedback oscillator

and the fundamental current given by Equations (2.17) in Chapter 2. In a general form, the frequency deviation $\Delta\omega$ caused by the presence of the second- and higher-order harmonic components of the voltage v on the resonant circuit can be obtained from

$$\frac{\Delta\omega}{\omega_0} = -\frac{1}{2}\sum_{k=2}^{\infty}(k^2 - 1)m_k^2 \tag{5.177}$$

where k is the order of the harmonic component and $m_k = V_k/V_1$ is the ratio of the harmonic voltage component to the fundamental voltage amplitude [79].

In view of the multiharmonic representation of the oscillator output spectrum due to the device and resonant circuit nonlinearities, the total phase fluctuations can be represented by a superposition integral as a result of each harmonic contribution. This is similar to the Fourier harmonic expansion in a frequency domain of the voltage waveform, when the phase trajectory on the phase plane is a result of the phase trajectories with different radii and velocities corresponding to the dc shift and harmonic amplitudes. In this case, Equation (5.152) can take a general form

$$\Delta\phi(t) = -\frac{1}{A}\int_0^{\omega_0 t} n(\tau)\,\Gamma(\tau)\,d\tau \tag{5.178}$$

where

$$\Gamma(t) = \frac{c_0}{2} + \sum_{n=1}^{\infty} c_n \cos(n\omega_0 t + \phi_n) \tag{5.179}$$

is a dimensionless periodic function characterizing the shape of the limit cycle or phase trajectory corresponding to the oscillation waveform and depending on the oscillator topology. It is called the impulse sensitivity function (ISF) for an approximate model for the oscillator phase behaviour [80] and serves a similar role as the perturbation projection vector (PPV) for the exact model [81]. The initial phase ϕ_n in Equation (5.179) is not important for random noise sources and can be neglected. For an ideal case of a purely sinusoidal oscillator, $c_1 = 1$ and $\Gamma(t) = \cos\omega_0 t$.

Now if any stationary noise current with a white power spectral density $\overline{i_n^2}/\Delta f$ is injected into the oscillator circuit close to any harmonic $n\omega_0 + \Delta\omega$ or $n\omega_0 - \Delta\omega$, it will result in a pair of equal sidebands at $\omega_0 \pm \Delta\omega$. Then, the total single-sideband power spectral density for the

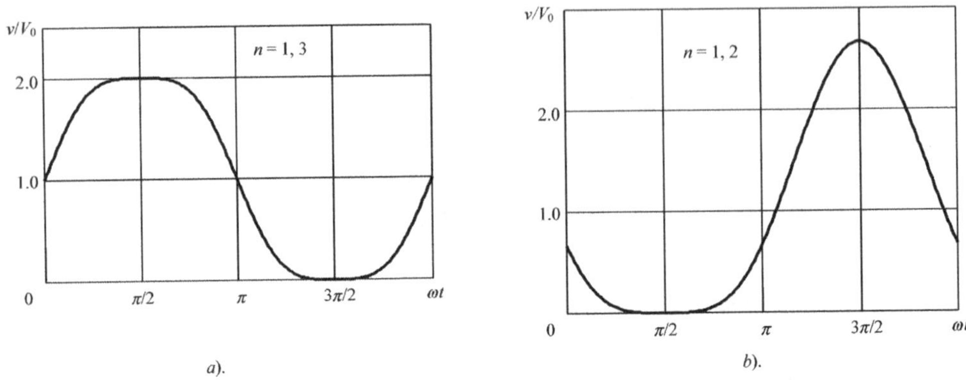

Figure 5.26 Voltage waveforms for *n* harmonic peaking

phase fluctuations in a bandwidth $\Delta f = 1$ Hz can be written, based on Equation (5.162), as

$$L(f_{\mathrm{m}}) = \frac{\overline{i_{\mathrm{n}}^2} \sum\limits_{\mathrm{n}=0}^{\infty} c_{\mathrm{n}}^2}{4C^2V^2\Delta\omega^2} = \frac{\overline{i_{\mathrm{n}}^2}\Gamma_{\mathrm{rms}}^2}{2C^2V^2\Delta\omega^2} \tag{5.180}$$

where Γ_{rms} is the root-mean-square value of $\Gamma(t)$ [82]. Thus, the total noise power near the carrier frequency of the oscillator is a result of the up-converted $1/f$ noise near dc, weighted by coefficient c_0, noise near the carrier weighted by coefficient c_1, and down-converted white noise near the second- and higher-order harmonics weighted by coefficients c_{n}, $n = 2, 3, \ldots$. The converted phase noise due to the conversion from one sideband to another can be of the order of 6 dB higher than the additive noise in the oscillator [83].

From Equation (5.180) it follows that the effect of the converted phase noise can be reduced by minimizing the dc coefficient c_0 and the higher-order harmonic coefficients c_{n} of the $\Gamma(t)$ approximating the cosine waveform of the injected node voltage. Figure 5.26 shows the voltage waveforms corresponding to (a) Class F with optimum flattened waveform consisting of the fundamental and third harmonics only and (b) inverse Class F consisting of the fundamental and second harmonics only [84]. To realize the symmetric flattened voltage waveform shown in Figure 5.26a, the ratio between the fundamental and third harmonic should be equal to $V_1/V_3 = 9$, while the ratio between the fundamental and second harmonic is equal to $V_1/V_2 = 4$ for the symmetric waveform close to the half-cosine shown in Figure 5.26b. Hence, the level of higher-order harmonics is significantly smaller for the symmetric flattened voltage waveforms. It should be noted that, in Class E operation with nonsymmetric voltage waveform, the effect of the second- and higher-order harmonics is significant, resulting in a high value of the voltage peak factor. The importance of the symmetry is also necessary to minimize the coefficient c_0 responsible for the low noise up-conversion and amplitude-to-phase conversion [80]. As is seen from Equation (5.174), the phase noise improvement can be achieved by reducing the effect of the device and circuit nonlinear capacitances. Due to the amplitude-to-phase conversion, the phase for each higher-order harmonic component changes with amplitude, resulting in a generally asymmetric voltage waveform.

Equation (5.178) describes an approximate phase noise behaviour, compared with the accurate equation [81], where the phase $\phi(t)$ also appears in its right-hand side. Such a simplified phase noise model is valid for the case of stationary noise sources such as white noise.

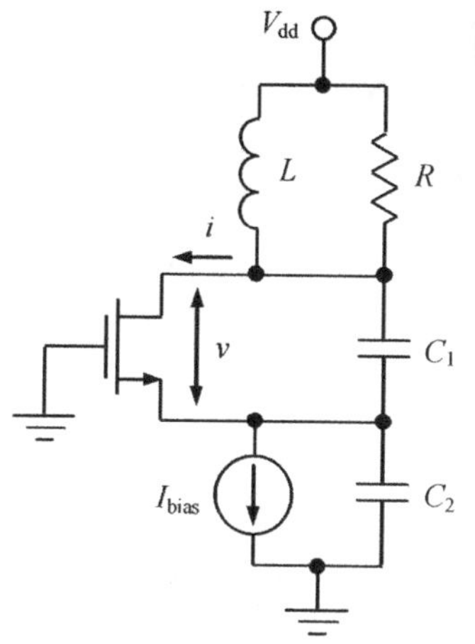

Figure 5.27 Simplified Colpitts oscillator schematic

However, when the noise sources are no longer stationary, it can be accurate only in the limits of an assumption of the small phase shifts for which cos $\Delta\phi$ is close to unity. This implies that the approximate model is not accurate enough to analyse neither the injection-locking phenomenon nor related issues such as behaviour of phase differences of coupled oscillators [85].

The response of the oscillation system to an impulsive noise can be provided by its direct measurement with a SPICE simulator, when an impulse is injected into the node of interest of the oscillator circuit and the oscillator is simulated for a few cycles afterwards [80]. By substituting the equivalent current noise source of each individual node in Equation (5.178), the phase contribution of each node can be calculated [86]. However, any further analytical simplification based on the orthogonal decomposition of noise into amplitude and phase components may not yield the correct result [87].

The noise sources in an oscillator generally cannot be modelled as purely stationary since the statistical properties of some of them may change with time in a periodic manner. Such types of noise sources are referred to as cyclostationary. If the thermal noise of the resistor has a stationary nature, then the collector (or drain) shot noise of the transistor is an example of cyclostationary noise due to the time-varying nature of the collector current. The most important issue is that the collector shot noise is dominant, compared with the noise from base resistance or tank losses, and can achieve about 70% of the total phase noise of the oscillator [86].

Figure 5.27 shows a simplified single-ended common gate CMOS Colpitts oscillator configuration where the required regeneration factor for the start-up oscillation conditions is chosen using a proper ratio of the feedback capacitances C_1 and C_2. The idealized voltage and current waveforms corresponding to the large-signal operation in idealized Class B with zero saturation voltage are shown in Figure 5.28.

Equation (5.175) for the drain time-varying current can be rewritten in the form

$$i(t) = I_{max}\left[\alpha_0 + \sum_{n=1}^{\infty}\alpha_n \cos(n\omega_0 t)\right] \tag{5.181}$$

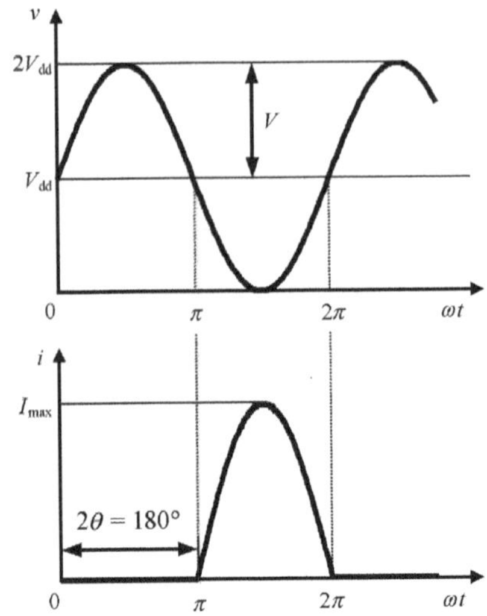

Figure 5.28 Ideal Class B drain voltage and current waveforms

where α_n is the ratio of the nth current harmonic amplitude to the peak output current I_{max}, expressed through half the conduction angle θ as

$$\alpha_n = \frac{I_n}{I_{max}} = \frac{\gamma_n(\theta)}{1 - \cos\theta} \tag{5.182}$$

where $\gamma_n(\theta)$ are the current coefficients introduced in Equations (1.25–1.27) in Chapter 1.

To account for the cyclostationary drain noise source as a result of total noise sources injected at frequencies $n\omega_0 \pm \Delta\omega$, Equation (5.180), in general form, can be rewritten as

$$L(f_m) = \frac{\overline{i_{nd}^2} \sum\limits_{n=0}^{\infty} (\alpha_n c_n)^2}{4C^2V^2\Delta\omega^2} \tag{5.183}$$

where $\overline{i_{nd}^2} = 2qI_{max}$ is the drain current noise power density in a frequency bandwidth $\Delta f =$ 1 Hz. The Fourier components for the current waveform close to half-cosinusoidal show that the drain shot noise is mixed mostly with the fundamental and second harmonics to contribute to the total phase noise of the oscillator.

To minimize the oscillator phase noise, it is very important to choose the optimum value of feedback ratio $k = C_2/C_1$ for the same total capacitance $C = C_1C_2/(C_1 + C_2)$. This is because different values of the conduction angle correspond to different harmonic contributions to the output spectrum. In the case of Class B with $\theta = 90°$, the third-, fifth- and higher-order harmonics can be eliminated since their current coefficients γ_n (for $n = 3, 5, \ldots$) become equal to zero. As a rule-of-thumb, the optimum feedback ratio for a Colpitts oscillator can be chosen to be approximately $k = 3.5 \div 4$ [80, 86].

5.6 LOADED QUALITY FACTOR

The loaded quality factor Q_L of the oscillator resonant circuit is an important factor of circuit behaviour, being a key parameter of the complete resonant system. Several definitions of Q_L are in general use. All definitions are valid when correctly applied. The classical definition applicable to any simple oscillatory system is

$$Q_L = 2\pi \frac{\text{total energy stored in the system}}{\text{energy lost per cycle from the system}} \tag{5.184}$$

where the numerator is taken as the average of the energy stored during the period of one full cycle [88].

In a steady-state operation mode, the energy loss is supplied by the external source, and Equation (5.184) is applicable, provided the denominator is interpreted as the energy supplied. For a negative resistance oscillator with a series LC resonant circuit, by applying this definition, we can write

$$Q_L = 2\pi \frac{LI^2}{R_L I^2/f} = \frac{\omega L}{R_L} \tag{5.185}$$

where I is the root-mean-square amplitude of the current flowing into the load resistance. This expression is the dual of the familiar $\omega C R_L$ of the parallel resonant circuit. In the oscillator with a multi-resonant circuit, each distinct mode of oscillations has a distinct value of Q_L. However, when coupling is present, there is an interchange of energy accompanied by beats which make it difficult or impossible to apply any concept of Q_L. Normal modes are free from this difficulty and therefore have uniquely defined values of Q_L.

Now let us understand an applicability of the conventional definition of the loaded quality factor obtained by Equation (5.184) to the feedback oscillator circuit shown in Figure 5.29a by following the derivation procedure given by Razavi [89]. The closed-loop voltage transfer function of such an oscillation system $T(j\omega) = V_{out}(j\omega)/V_{in}(j\omega)$ can be written as

$$T(j\omega) = \frac{H(j\omega)}{1 - H(j\omega)} \tag{5.186}$$

where $H(j\omega)$ is the open-loop voltage transfer function.

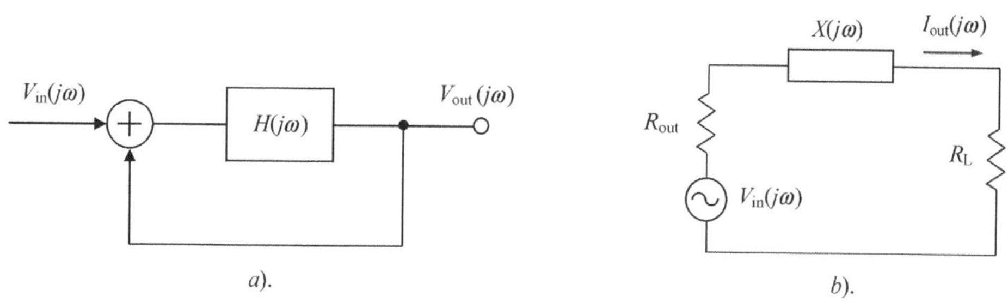

Figure 5.29 Equivalent models of oscillator circuit

For offset frequencies close to the carrier, $\omega_m = \omega - \omega_0$, the open-loop transfer function $H(j\omega)$ can be represented by a linear Taylor series expansion around ω_0 as

$$H(j\omega) \cong H(j\omega_0) + \frac{dH}{d\omega}\omega_m \tag{5.187}$$

the substitution of which into Equation (5.186) results in

$$T(j\omega) = \frac{H(j\omega_0) + \dfrac{dH}{d\omega}\omega_m}{1 - H(j\omega_0) - \dfrac{dH}{d\omega}\omega_m} \tag{5.188}$$

According to the steady-state oscillation condition known as a *Nyquist criterion*, if in a feedback system the open-loop gain has a total phase shift of 180° at some frequency ω_0, such a system will oscillate at that frequency provided that the open-loop gain is unity, i.e., $H(j\omega_0) = 1$ at the oscillation frequency ω_0. Since $|(dH/d\omega)\omega_m| \ll 1$ for most practical cases, Equation (5.188) reduces to

$$T(j\omega) \cong -\frac{1}{\dfrac{dH}{d\omega}\omega_m} \tag{5.189}$$

By representing the open-loop transfer function as $H(j\omega) = A(\omega)\exp[j\phi(\omega)]$ in differential form

$$\frac{dH}{d\omega} = \left(\frac{dA}{d\omega} + jA\frac{d\phi}{d\omega}\right)\exp(j\phi) \tag{5.190}$$

and taking into account that $A \approx 1$ for $\omega \approx \omega_0$, the oscillator noise power spectral density can be written as

$$|T(j\omega)|^2 = \frac{1}{\omega_m^2}\frac{1}{\left(\dfrac{dA}{d\omega}\right)^2 + \left(\dfrac{d\phi}{d\omega}\right)^2} \tag{5.191}$$

By introducing the definition of the oscillator loaded quality factor Q_L as

$$Q_L = \frac{\omega_0}{2}\sqrt{\left(\frac{dA}{d\omega}\right)^2 + \left(\frac{d\phi}{d\omega}\right)^2} \tag{5.192}$$

Equation (5.191) can be rewritten as

$$|T(j\omega)|^2 = \left(\frac{\omega_0}{2Q_L\omega_m}\right)^2 \tag{5.193}$$

representing a similar form of the Leeson model given by Equation (5.81).

Similarly, consider the negative resistance oscillator circuit shown in Figure 5.29b where $X = X_{out} + X_L$ is the total circuit reactance. The voltage transfer function of such an oscillation system $T(j\omega) = V_{out}(j\omega)/V_{in}(j\omega)$, where $V_{out}(j\omega) = I_{out}(j\omega)R_L$, can be written as

$$T(j\omega) = \frac{R_L}{Z(j\omega)} \tag{5.194}$$

where $Z(j\omega) = R(\omega) + jX(\omega)$, $R = R_{out} + R_L$. The steady-state oscillation conditions are established at $\omega = \omega_0$ when $X = 0$ and $Z(j\omega_0) = 0$.

The noise shaping function of such an oscillation system can be written as

$$|T(j\omega)|^2 = \frac{R_L^2}{|Z(j\omega)|^2} \tag{5.195}$$

For offset frequencies close to the carrier, $\omega_m = \omega - \omega_0$, the total circuit impedance $Z(j\omega)$ can be represented by a linear Taylor series expansion around ω_0 as

$$Z(j\omega) \cong Z(j\omega_0) + \frac{dZ}{d\omega}\omega_m \tag{5.196}$$

Since at the oscillation frequency $Z(j\omega_0) = 0$, substituting of Equation (5.196) into Equation (5.195) results in

$$|T(j\omega)|^2 = \frac{R_L^2}{\left|\dfrac{dZ}{d\omega}\omega_m\right|^2} \tag{5.197}$$

By representing the total circuit impedance $Z(j\omega)$ through its real and imaginary parts, Equation (5.197) can be rewritten as

$$|T(j\omega)|^2 = \frac{1}{\omega_m^2}\frac{R_L^2}{\left(\dfrac{dR}{d\omega}\right)^2 + \left(\dfrac{dX}{d\omega}\right)^2} \tag{5.198}$$

where both derivatives are evaluated at the oscillation frequency ω_0.

As a result, introducing the definition of the oscillator loaded quality factor Q_L as

$$Q_L = \frac{\omega_0}{2R_L}\sqrt{\left(\frac{dR}{d\omega}\right)^2 + \left(\frac{dX}{d\omega}\right)^2} \tag{5.199}$$

leads to exactly the same expression for the oscillator noise shaping function obtained by Equation (5.193). Similar expressions for the loaded quality factor of the negative resistance oscillator circuit are described elsewhere [68, 90].

The oscillator power spectral density can be generally rewritten using the noise shaping function $|T(j\omega)|^2$ as

$$S_\phi = S_{\Delta\phi}\,|T(j\omega)|^2 \tag{5.200}$$

where $T(j\omega)$ is the closed-loop transfer function of the oscillation system. Should $1/|T|^2$ not be parabolically shaped, it can be expanded to second order with respect to ω_m around the oscillation frequency ω_0. This corresponds to an oscillator power spectral density reduction of approximately −6 dB/octave, which is usually the case. Such an expansion results in another general definition of the oscillator loaded quality factor in the form

$$Q_L = \frac{\omega_0}{2}\sqrt{\frac{1}{2}\frac{d^2\left(1/|T|^2\right)}{d\omega^2}} \tag{5.201}$$

where the derivative is evaluated at the oscillation frequency ω_0 [61].

As a simple example, let us derive a relationship between the oscillator loaded quality factor Q_L and the phase slope of the circuit impedance $d\phi/d\omega$ for the case of a negative resistance

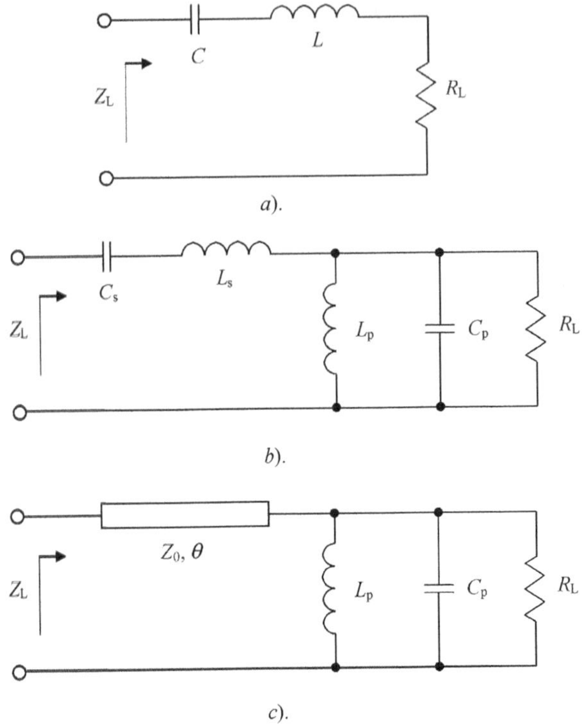

Figure 5.30 Negative resistance oscillators with different resonant circuits

oscillator with a series LC resonant circuit and frequency-independent load resistance R_L shown in Figure 5.30a. The loaded resonant circuit impedance Z_L and phase ϕ can be written, respectively, as

$$Z_L = R_L + j\frac{\omega^2 LC - 1}{\omega C} \tag{5.202}$$

$$\phi = \tan^{-1}\frac{\omega^2 LC - 1}{\omega C R_L} \tag{5.203}$$

The derivative of Equation (5.203) with respect to frequency ω results in

$$\frac{d\phi}{d\omega} = C R_L \frac{1 + \omega^2 LC}{(\omega C R_L)^2 + (\omega^2 LC - 1)^2} \tag{5.204}$$

At the resonant frequency $\omega_0 = 1/\sqrt{LC}$, Equation (5.204) can be rewritten as

$$\frac{d\phi}{d\omega} = \frac{2}{\omega_0^2 C R_L} \tag{5.205}$$

Since the loaded quality factor of this series single-resonant circuit at the resonant frequency is also can be defined as

$$Q_L = \frac{1}{\omega_0 C R_L} \tag{5.206}$$

then Equation (5.205) can finally be rewritten as

$$Q_L = \frac{\omega_0}{2} \frac{d\phi}{d\omega} \tag{5.207}$$

A similar expression can be applied to a feedback oscillator with parallel or series tuned nature of the feedback tank [91, 92]. Comparison of Equations (5.206) and (5.207) shows that in the case of a single-resonant circuit both definitions of Q_L are equivalent.

The double-resonant circuit consisting of a parallel resonant circuit connected in series to a series resonant circuit is shown in Figure 5.30b. Such a resonant circuit provides susceptance compensation when the susceptance slope of the series-tuned circuit compensates for the susceptance slope of the parallel-tuned circuit. At the resonant frequency ω_0 when

$$\omega_0^2 L_s C_s = \omega_0^2 L_p C_p = 1$$

these two basic definitions of the oscillator loaded quality factors can be represented by

$$Q_L = \frac{\omega_0 \left(L_s + C_p R_L^2\right)}{R_L} \tag{5.208}$$

$$\frac{\omega_0}{2} \frac{d\phi}{d\omega} = \frac{\omega_0 \left(C_p R_L^2 - L_s\right)}{R_L} \tag{5.209}$$

that are not identical. This means that the classical definition of Q_L based on average stored energy is not appropriate to characterize the dynamic behaviour of this circuit and correctly calculate the phase noise spectrum.

Figure 5.30c shows the oscillator resonant circuit with distributed parametres consisting of a transmission line of electrical length θ loaded by a parallel resonant circuit. In this case, when $Z_0 = R_L$, at the resonant frequency ω_0 corresponding to the electrical length θ_0, we can write

$$Q_L = \omega_0 R_L \left(C + \frac{\theta_0}{\omega_0 R_L}\right) \tag{5.210}$$

$$\frac{\omega_0}{2} \frac{d\phi}{d\omega} = \omega_0 C R_L \left|\cos(2\theta_0)\right| \tag{5.211}$$

that are also not identical [91]. Thus, the classical definition of Q_L based on average stored energy is not able to properly characterize the distributed resonant circuit having a periodic nature of its spectral performance.

5.7 AMPLITUDE-TO-PHASE CONVERSION

Substituting Equation (5.140) into Equation (5.141) allows us to rewrite the oscillator normalized double-sideband phase noise spectral power density in the form

$$S_\phi(\Delta\phi, f_m) = \left(\frac{f_0}{2Q_L f_m}\right)^2 \left[S_{\Delta\phi}(f_m) + q^2 S_a\left(\frac{\Delta A}{A_0}, f_m\right)\right] \tag{5.212}$$

which shows the direct relationship between the amplitude and phase noise in the oscillator. In this case, the coefficient q is responsible for the conversion of the amplitude noise into the phase fluctuations (AM/PM conversion). Equation (5.212) can be rewritten as

$$S_\phi(\Delta\phi, f_m) = S_\phi^0(\Delta\phi, f_m) + q^2 S_a^0\left(\frac{\Delta A}{A_0}, f_m\right) \tag{5.213}$$

where S_ϕ^0 is the phase noise of the isochronous oscillator when $q = 0$ and S_a^0 is the amplitude noise of isochronous oscillator [93]. It should be noted that $p = 1$ when the oscillator circuit is adjusted to maximum power [60, 65].

Equation (5.212) can be represented for a range of offset frequencies $f_m \ll f_0$ in the form

$$S_\phi(f_m) = S_{\Delta\phi}(f_m)\left[\left(\frac{f_0}{2Q_L f_m}\right)^2 + q_\phi^2\right] \tag{5.214}$$

where q_ϕ is the AM/PM conversion coefficient defined as

$$q_\phi = q\left(\frac{f_0}{2Q_L f_m}\right)^2 \tag{5.215}$$

Then, assuming that

$$S_{\Delta\phi} = \frac{4FkT}{P_L}\left(1 + \frac{f_c}{f_m}\right) \tag{5.216}$$

the single-sideband noise-to-carrier ratio can be rewritten as

$$L(f_m) = \frac{2FkT}{P_L}\left(1 + \frac{f_c}{f_m}\right)\left[\left(\frac{f_0}{2Q_L f_m}\right)^2 + q_\phi^2\right] \tag{5.217}$$

where, for the simplified case when the active immitances do not depend on the frequency, that is $\partial \mathrm{Re}W/\partial f = 0$, the AM/PM conversion coefficient q_ϕ can be defined by

$$q_\phi = \frac{A_0}{f_m}\frac{\mathrm{d}f}{\mathrm{d}A} \tag{5.218}$$

For a particular case of voltage-controlled oscillator (VCO), when it is assumed that the amplitude noise is mainly caused by the thermal noise of the varactor, the following expression can be written

$$L(f_m) = \frac{2FkT}{P_L}\left(1 + \frac{f_c}{f_m}\right)\left(\frac{f_0}{2Q_L f_m}\right)^2 + \frac{2kTR_e K_{VCO}^2}{f_m^2} \tag{5.219}$$

where R_e is the equivalent noise resistance of the varactor, K_{VCO} is the oscillator voltage gain in Hz/V [94, 95].

Consider a simple negative conductance oscillator with a parallel resonant circuit shown in Figure 5.31, where C_{v1} is the large-signal fundamentally averaged capacitance. To calculate the oscillator phase noise degradation, it is best to introduce a ratio of the large-signal single-sideband noise-to-carrier ratio $L_1(f_m)$ to the corresponding small-signal one $L(f_m)$ by using Equation (5.217) as

$$\frac{L_1(f_m)}{L(f_m)} = \frac{Q_L}{Q_{L1}} + q_\phi^2\left(\frac{2Q_L f_m}{f_0}\right)^2 \tag{5.220}$$

where $Q_{L1} = \omega C_{v1} R_L$ is the loaded quality factor of the circuit with the variable capacitance under large-signal operation.

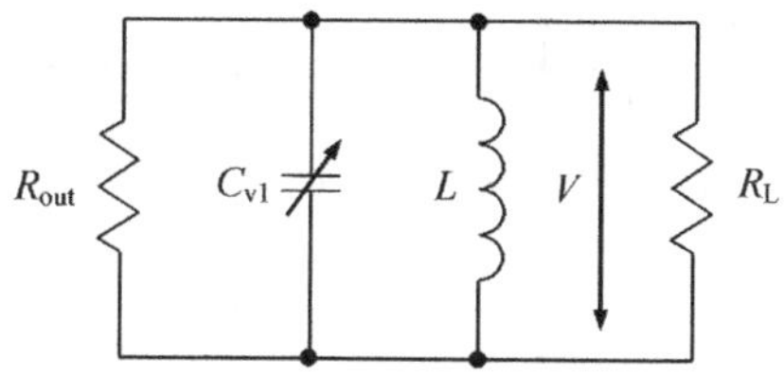

Figure 5.31 Negative conductance oscillator with a parallel resonant circuit

Let us assume that the tank capacitance represents the device abrupt junction collector capacitance or varactor capacitance which can generally be approximated under large-signal operation by Equation (6.13) given in Chapter 6 as $C_{\text{v1}} = C_{\text{v}}(1 + \alpha_{\text{v}} V^2)$, where C_{v} is the small-signal capacitance and V is the sinusoidal voltage amplitude across the load. Then,

$$q_{\phi} = \frac{V_0}{f_{\text{m}}}\frac{\text{d}f}{\text{d}V} \Big/ \left(1 + \frac{V_0}{f}\frac{\text{d}f}{\text{d}V} + \frac{V_0}{C_{\text{v1}}}\frac{\text{d}C_{\text{v1}}}{\text{d}V}\right) \tag{5.221}$$

Since

$$\frac{\text{d}f}{\text{d}V} = \frac{\text{d}f}{\text{d}C_{\text{v1}}}\frac{\text{d}C_{\text{v1}}}{\text{d}V}$$

and, for a single-resonant oscillator circuit,

$$\frac{\text{d}f}{\text{d}C_{\text{v1}}} = -\frac{f}{2C_{\text{v1}}} \tag{5.222}$$

then the AM/PM conversion coefficient as a function of the voltage amplitude across the varactor can be obtained as

$$q_{\phi} = -\frac{f}{f_{\text{m}}}\frac{\alpha_{\text{v}} V_0^2}{1 + 2\alpha_{\text{v}} V_0^2} \tag{5.223}$$

As a result, Equation (5.220) can be rewritten as

$$\frac{L_1(f_{\text{m}})}{L(f_{\text{m}})} = \frac{1}{1 + \alpha V_0^2} + \left(\frac{2Q_{\text{L}} f}{f_0}\right)^2 \left(\frac{\alpha_{\text{v}} V_0^2}{1 + 2\alpha_{\text{v}} V_0^2}\right)^2 \tag{5.224}$$

which allows the optimization of the oscillator phase noise performance with a nonlinear capacitance. For example, for a fixed Q_{L} and output voltage across the load, the optimum value of α_{v} can be calculated, minimizing the ratio of $L_1(f_{\text{m}})/L(f_{\text{m}})$. Similarly, for a fixed Q_{L} and α_{v}, the optimum oscillator output power can be defined.

5.8 OSCILLATOR PULLING FIGURE

The term ‘pulling’ is most widely used to express the frequency change produced in oscillators by a variation in the load impedance. In particular, the pulling figure of the microwave oscillator is defined as the maximum frequency excursion produced by a voltage standing wave ratio (VSWR) of 1.5, which is varied arbitrarily in phase angle [88]. Pulling is closely related to frequency stability and to the possibility of multiple frequencies. At the boundary between

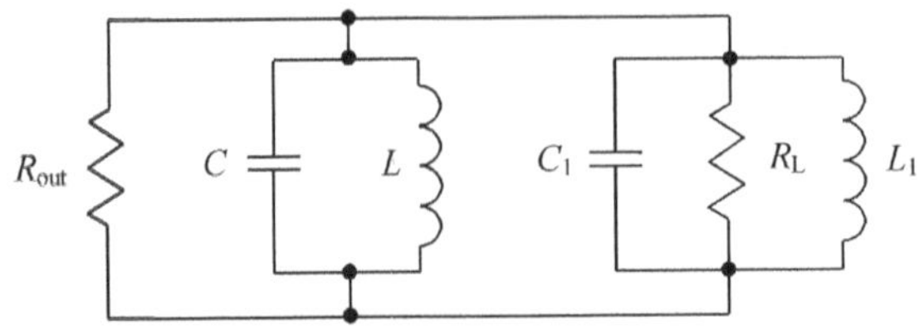

Figure 5.32 Pulling in a lumped resonant system

single and multiple responses, the frequency stability is zero and the pulling figure becomes infinite. This is the desired condition in locked oscillators, but otherwise very undesirable.

Consider the simple case of a resonant system represented by lumped elements, as shown in Figure 5.32. The resonant circuit consists of a parallel inductor L and a shunt capacitor C, while a parallel inductance L_1 and shunt capacitance C_1 are assumed to be part of the load. Assuming that the load is tuned to the operating frequency, we can write

$$\omega_1 = \frac{1}{\sqrt{L_1 C_1}} = \omega_0 = \frac{1}{\sqrt{LC}} \tag{5.225}$$

$$Q_1 = \omega_0 C_1 R_L = \frac{R_1}{\omega_0 L_1} \tag{5.226}$$

$$Q_L = \omega_0 C R_L = \frac{R_1}{\omega_0 L} \tag{5.227}$$

In this case, an increment dC_1 in the load capacitance produces a frequency increment $d\omega$ given by

$$\frac{d\omega}{\omega_0} = -\frac{dC_1}{2(C + C_1)} \tag{5.228}$$

Equation (5.228) can be rewritten in the form

$$\frac{d\omega}{\omega_0} = -\frac{Q_1}{2(Q_L + Q_1)} \frac{dC_1}{C_1} \tag{5.229}$$

Similarly, a variation in the load inductance L_1 produces a frequency deviation

$$\frac{d\omega}{\omega_0} = -\frac{Q_1}{2(Q_L + Q_1)} \frac{dL_1}{L_1} \tag{5.230}$$

From Equations (5.229) and (5.230) it follows that a small value of Q_1 and a large value of Q_L are desirable in order to minimize the oscillator pulling.

To derive the pulling figure by taking into account the nonlinearity of the oscillator output admittance, let us consider the steady-state oscillation condition for a negative conductance oscillator circuit

$$Y = Y_{out} + Y_0 = 0 \tag{5.231}$$

where Y is the overall circuit admittance, Y_{out} is the oscillator output admittance and Y_0 is the load admittance.

For a load perturbation of $\Delta Y_L = \Delta G_L + j\Delta B_L$ the oscillation conditions can be written as

$$Y + \Delta Y_L + \frac{\partial Y}{\partial \omega}\Delta\omega + \frac{\partial Y}{\partial V}\Delta V = 0 \tag{5.232}$$

where V is the voltage amplitude across the load [96].

From Equations (5.231) and (5.232), by separating real and imaginary parts of $Y = G + jB$,

$$\Delta G_L + \frac{\partial G}{\partial \omega}\Delta\omega + \frac{\partial G}{\partial V}\Delta V = 0 \tag{5.233}$$

$$\Delta B_L + \frac{\partial B}{\partial \omega}\Delta\omega + \frac{\partial B}{\partial V}\Delta V = 0 \tag{5.234}$$

Solving Equations (5.233) and (5.245) for $\Delta\omega$ and ΔV results in

$$\Delta\omega = \frac{\Delta G_L \dfrac{\partial B}{\partial V} - \Delta B_L \dfrac{\partial G}{\partial V}}{\dfrac{\partial G}{\partial V}\dfrac{\partial B}{\partial \omega} - \dfrac{\partial G}{\partial \omega}\dfrac{\partial B}{\partial V}} \tag{5.235}$$

$$\Delta V = \frac{\Delta B_L \dfrac{\partial G}{\partial \omega} - \Delta G_L \dfrac{\partial B}{\partial \omega}}{\dfrac{\partial G}{\partial V}\dfrac{\partial B}{\partial \omega} - \dfrac{\partial G}{\partial \omega}\dfrac{\partial B}{\partial V}} \tag{5.236}$$

By using a transmission line with the characteristic impedance $Z_0 = 1/Y_0$ and electrical length θ variable between 0 and $\lambda/2$, the transferred load admittance Y_L at the oscillator output plane can be written as

$$Y_L = Y_0 \frac{Y_0 + \Delta G + jY_0 \tan\theta}{Y_0 + j(Y_0 + \Delta G)\tan\theta} = Y_0 \frac{S(1 + \tan^2\theta)}{1 + S^2 \tan^2\theta} + jY_0 \frac{(1 - S^2)\tan\theta}{1 + S^2 \tan^2\theta} \tag{5.237}$$

where S is the VSWR of the perturbating admittance in the output line, given by

$$S = 1 + \frac{\Delta G}{Y_0} \tag{5.238}$$

From Figure 5.33 it can be seen that any reactive load perturbation of value $j\Delta B$ shown in Figure 5.33a can be represented by a nonreactive load perturbation of ΔG shown in Figure 5.33b by suitably selecting the reference plane in the output line.

As a result, a load perturbation $\Delta Y_L = Y_L - Y_0$ expressed separately by its real and imaginary parts can now be represented by

$$\Delta G_L = Y_0 \frac{(S - 1)(1 - S\tan^2\theta)}{1 + S^2 \tan^2\theta} \tag{5.239}$$

$$\Delta B_L = Y_0 \frac{(1 - S^2)\tan\theta}{1 + S^2 \tan^2\theta} \tag{5.240}$$

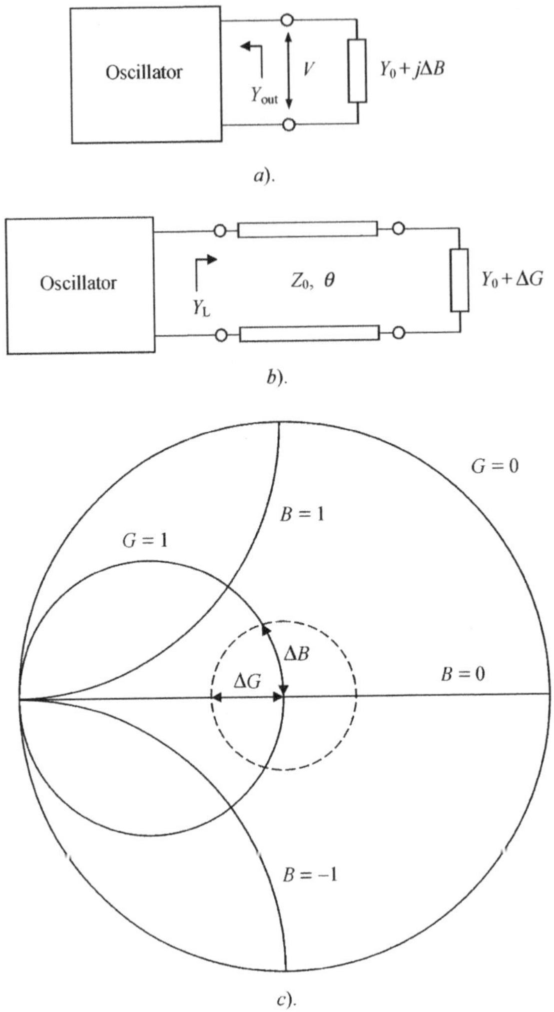

Figure 5.33 Oscillator with load variation

By substituting Equations (5.239) and (5.240) into Equation (5.235), the frequency variation $\Delta\omega$ can be obtained as

$$\Delta\omega = Y_0 \frac{\dfrac{\partial B}{\partial V}}{\dfrac{\partial G}{\partial V}\dfrac{\partial B}{\partial \omega} - \dfrac{\partial G}{\partial \omega}\dfrac{\partial B}{\partial V}} \frac{(S-1)(1 - S\tan^2\theta)}{1 + S^2\tan^2\theta} - Y_0 \frac{\dfrac{\partial G}{\partial V}}{\dfrac{\partial G}{\partial V}\dfrac{\partial B}{\partial \omega} - \dfrac{\partial G}{\partial \omega}\dfrac{\partial B}{\partial V}} \frac{(1 - S^2)\tan\theta}{1 + S^2\tan^2\theta} \tag{5.241}$$

The maximum frequency deviation $\Delta\omega_{max}$ as a function of θ can be found by equating the derivative $d(\Delta\omega)/d\theta$ to zero, resulting in a quadratic equation

$$S^2\tan^2\theta + 2S\alpha\tan\theta - 1 = 0 \tag{5.242}$$

where

$$\alpha = \frac{\partial B}{\partial V} \Big/ \frac{\partial G}{\partial V} \tag{5.243}$$

is the nonlinear oscillator parameter. Consequently, the two solutions corresponding to the extreme values of $\Delta\omega_1$ and $\Delta\omega_2$ can be calculated from

$$\Delta\omega_{1,2} = \frac{\pm(S-1)}{2S} \frac{\sqrt{\alpha^2+1}(2\alpha^2+S+1) \mp 2\alpha(1+\alpha^2)}{\alpha^2 \mp \alpha\sqrt{\alpha^2+1}+1} Y_0 K \tag{5.244}$$

where

$$K = \frac{\dfrac{\partial G}{\partial V}}{\dfrac{\partial G}{\partial V}\dfrac{\partial B}{\partial \omega} - \dfrac{\partial G}{\partial \omega}\dfrac{\partial B}{\partial V}} \tag{5.245}$$

The interesting feature resulting from Equation (5.244) is an asymmetry of the pulling range around the oscillation frequency as a function of the nonlinear oscillator parameter α. For example, for $\alpha = 0$, the pulling range is symmetric when $\Delta\omega_1$ and $\Delta\omega_2$ are equal. However, the pulling range becomes more and more asymmetric for increasing values of α.

The maximum total frequency variation $\Delta\omega_{\max} = \Delta\omega_1 - \Delta\omega_2$ can now be found to be

$$\Delta\omega_{\max} = Y_0 K \left(S - \frac{1}{S}\right)\sqrt{\alpha^2+1} \tag{5.246}$$

that explicitly represents the pulling figure of a nonlinear negative conductance oscillator.

Equation (5.246) can be rewritten through the reflection coefficient of the perturbed load Γ as

$$\Delta\omega_{\max} = 4Y_0 K \frac{|\Gamma|}{1-|\Gamma|^2}\sqrt{\alpha^2+1} \tag{5.247}$$

where

$$S = \frac{1+|\Gamma|}{1-|\Gamma|} \tag{5.248}$$

For a particular case when $\partial G/\partial\omega = 0$ and $|\Gamma|^2 \ll 1$, Equation (5.247) can be simplified to

$$\Delta\omega_{\max} = \frac{2\omega_0}{Q_L}\frac{S-1}{S+1}\sqrt{\alpha^2+1} \tag{5.249}$$

where

$$Q_L = \frac{\omega_0}{2Y_0}\frac{\partial B}{\partial \omega} \tag{5.250}$$

is the loaded quality factor of the oscillator resonant circuit [97].

REFERENCES

1. A. van der Ziel, *Noise*, Englewood Cliffs: Prentice-Hall (1954).
2. H. L. Krauss, C. W. Bostian and F. H. Raab, *Solid State Radio Engineering*, New York: John Wiley & Sons (1980).

3. B. L. Buckingham, *Noise in Electronic Devices and Systems*, New York: John Wiley & Sons (1983).
4. H. C. Montgomery, 'Transistor Noise in Circuit Applications', *Proc. IRE*, **40**, 1461–1471 (1952).
5. H. Rothe and W. Dahlke, 'Theory of Noise Fourpoles', *Proc. IRE*, **44**, 811–818 (1956).
6. A. G. T. Becking, H. Groendijk and K. S. Knol, 'The Noise Factor of Four-Terminal Networks', *Philips Res. Rep.*, **10**, 349–357 (1955).
7. IRE Subcommittee 7.9 on Noise (H. A. Haus, Chairman), 'Representation of Noise in Linear Twoports', *Proc. IRE*, **48**, 69–74 (1960).
8. H. Hillbrand and P. H. Russer, 'An Efficient Method for Computer Aided Noise Analysis of Linear Amplifier Networks', *IEEE Trans. Circuits and Systems*, **CAS-23**, 235–238 (1976).
9. F. N. Hooge, '$1/f$ Noise Sources', *IEEE Trans. Electron Devices*, **ED-41**, 1926–1934 (1996).
10. S. Mohammadi and D. Pavlidis, 'A Nonfundamental Theory of Low-Frequency Noise in Semiconductor Devices', *IEEE Trans. Electron Devices*, **ED-47**, 2009–2017 (2000).
11. X. N. Zhang, A. Van der Ziel, K. H. Duh and H. Morkoc, 'Burst and Low-Frequency Generation-Recombination Noise in Double-Heterojunction Bipolar Transistors', *IEEE Trans. Electron Device Lett.*, **5**, 277–279 (1984).
12. J. L. Plumb and E. R. Chenette, 'Flicker Noise in Transistors', *IEEE Trans. Electron Devices*, **ED-10**, 304–308 (1963).
13. R. C. Jaeger and A. J. Broderson, 'Low-Frequency Noise Sources in Bipolar Junction Transistors', *IEEE Trans. Electron Devices*, **ED-17**, 128–134 (1970).
14. T. G. M. Kleinpenning, 'Low-Frequency Noise in Modern Bipolar Transistors: Impact of Intrinsic Transistor and Parasitic Series Resistances', *IEEE Trans. Electron Devices*, **ED-41**, 1981–1991 (1994).
15. G. Niu, J. Tang, Z. Feng, A. Joseph and D. L. Harame, 'SiGe HBT Scaling Implications on $1/f$ Noise and Oscillator Phase Noise', *2004 IEEE RFIC Symp. Dig.*, 299–302 (2004).
16. O. Jantsch, 'Flicker ($1/f$) Noise Generated by a Random Walk of Electrons in Interfaces', *IEEE Trans. Electron Devices*, **ED-34**, 1100–1115 (1987).
17. H. Mikoshiba, '$1/f$ Noise in *n*-Channel Silicon-Gate MOS Transistors', *IEEE Trans. Electron Devices*, **ED-29**, 965–969 (1982).
18. K. K. Hung, P. K. Ko, C. Hu and Y. C. Cheng, 'A Unified Model for the Flicker Noise in Metal-Oxide-Semiconductor Field-Effect Transistors', *IEEE Trans. Electron Devices*, **ED-37**, 654–664 (1990).
19. F.-C. Hou, G. Bosman and M. E. Law, 'Simulation of Oxide Trapping Noise in Submicron *n*-Channel MOSFETs', *IEEE Trans. Electron Devices*, **ED-50**, 846–852 (2003).
20. G. V. Kuznetsova, I. Balaz, V. N. Kuleshov and B. E. Leshukov, 'Experimental Investigation of Low-Frequency Noise in Field-Effect Transistors (in Russian)', *Poluprovodnikovaya Elektronika v Tekhnike Svyazi*, **26**, 56–62 (1986).
21. L. K. J. Vandamme, X. Li and D. Rigaud, '$1/f$ Noise in MOS Devices, Mobility or Number Fluctuations?' *IEEE Trans. Electron Devices*, **ED-41**, 1936–1944 (1994).
22. C. F. Hiatt, A. van der Ziel and K. M. van Viet, 'Generation–Recombination Noise Produced in the Channel of JFETs', *IEEE Trans. Electron Devices*, **ED-22**, 614–616 (1975).
23. F. M. Klaassen and J. R. Robinson, 'Anomalous Noise Behaviour of the Junction Gate Field Effect Transistor at Low Temperatures', *IEEE Trans. Electron Devices*, **ED-17**, 852–857 (1970).
24. A. van der Ziel, 'Flicker Noise in Electronic Devices', *Advances in Electronics and Electron Physics*, **49**, 225–297 (1979).
25. R. A. Pucel, 'The GaAs FET Oscillator — Its Signal and Noise Performance', *Proc. 40th Annual Frequency Control Symp.*, 385–391 (1986).
26. M. Chertouk and A. Chovet, 'Origins and Characterization of Low-Frequency Noise in GaAs MESFET's Grown on InP Substrates', *IEEE Trans. Electron Devices*, **ED-43**, 123–129 (1996).
27. C. Enz, 'An MOS Transistor Model for RFIC Design Valid in All Regions of Operation', *IEEE Trans. Microwave Theory Tech.*, **MTT-50**, 342–359 (2002).

28. G. Massobrio and P. Antognetti, *Semiconductor Device Modeling with SPICE*, New York: McGraw-Hill (1993).
29. S. Tedja, J. van der Spiegel and H. H. Williams, 'Analytical and Experimental Studies of Thermal Noise in MOSFETs', *IEEE Trans. Electron Devices*, **ED-41**, 2069–2075 (1994).
30. S. Asgaran, M. J. Deen and C.-H. Chen, 'Analytical Modeling of MOSFET's Channel Noise and Noise Parametres', *IEEE Trans. Electron Devices*, **ED-51**, 2109–2114 (2004).
31. H. Statz, H. A. Haus and R. A. Pucel, 'Noise Characteristics of Gallium Arsenide Field-Effect Transistors', *IEEE Trans. Electron Devices*, **ED-21**, 549–562 (1974).
32. W. Baechtold, 'Noise behaviour of GaAs Field-Effect Transistors with Short Gate Lengths', *IEEE Trans. Electron Devices*, **ED-19**, 674–680 (1972).
33. A. van der Ziel, 'Thermal Noise in the Hot Electron Regime in FET's', *IEEE Trans. Electron Devices*, **ED-18**, 977 (1971).
34. R. A. Pucel, D. F. Masse and C. F. Krumm, 'Noise Performance of Gallium Arsenide Filed-Effect Transistors', *IEEE Trans. Solid-State Circuits*, **SC-11**, 243–255 (1976).
35. A. van der Ziel, 'Thermal Noise in Field Effect Transistor', *Proc. IRE*, **50**, 1808–1812 (1962).
36. A. van der Ziel, 'Gate Noise in Field Effect Transistors at Moderately High Frequencies', *Proc. IRE*, **51**, 461–467 (1963).
37. V. Rizzoli, F. Mastri and C. Cecchetti, 'Computer-Aided Noise Analysis of MESFET and HEMT Mixers', *IEEE Trans. Microwave Theory Tech.*, **MTT-37**, 1401–1410 (1989).
38. A. Cappy, 'Noise Modeling and Measurement Techniques', *IEEE Trans. Microwave Theory Tech.*, **MTT-36**, 1–9 (1988).
39. P. Heymann and H. Prinzler, 'Improved Noise Model for MESFETs and HEMTs in Lower Gigahertz Frequency Range', *Electronics Lett.*, **28**, 611–612 (1992).
40. D.-S. Shin, J. B. Lee, H. S. Min, J.-E. Oh, Y.-J. Park, W. Jung and D. S. Ma, 'Analytical Noise Model with the Influence of Shot Noise Induced by the Gate Leakage Current for Submicrometer Gate-Length High-Electron Mobility Transistors', *IEEE Trans. Electron Devices*, **ED-44**, 1883–1887 (1997).
41. J. Gao, C. L. Law, H. Wang, S. Aditya and G. Boeck, 'A New Method for pHEMT Noise-Parameter Determination Based on 50-Ω Noise Measurement System', *IEEE Trans. Microwave Theory Tech.*, **MTT-51**, 2079–2089 (2003).
42. H. Fukui, 'Design of Microwave GaAs MESFET's for Broad-Band Low-Noise Amplifiers', *IEEE Trans. Microwave Theory Tech.*, **MTT-27**, 643–650 (1979).
43. H. Fukui, 'Addendum to Design of Microwave GaAs MESFET's for Broad-Band Low-Noise Amplifiers', *IEEE Trans. Microwave Theory Tech.*, **MTT-29**, 1119 (1981).
44. H. Fukui, 'Optimal Noise Figure of Microwave GaAs MESFET's', *IEEE Trans. Electron Devices*, **ED-26**, 1032–1037 (1979).
45. E. G. Nielsen, 'Behavior of Noise Figure in Junction Transistors', *Proc. IRE*, **45**, 957–963 (1957).
46. H. Fukui, 'The Noise Performance of Microwave Devices', *IEEE Trans. Electron Devices*, **ED-13**, 329–341 (1966).
47. L. Escotte, J.-P. Roux, R. Plana, J. Graffeuil and A. Gruhle, 'Noise Modeling of Microwave Heterojunction Bipolar Transistors', *IEEE Trans. Electron Devices*, **ED-42**, 883–889 (1995).
48. M. Rudolph, R. Doerner, L. Klapproth and P. Heymann, 'An HBT Noise Model Valid Up to Transit Frequency', *IEEE Electron Device Lett.*, **20**, 24–26 (1999).
49. G. Niu, J. D. Cressler, S. Zhang, W. E. Ansley, C. S. Webster and D. L. Harame, 'A Unified Approach to RF and Microwave Noise Parameter Modeling in Bipolar Transistors', *IEEE Trans. Electron Devices*, **ED-48**, 2568–2574 (2001).
50. J. Gao, X. Li, H. Wang and G. Boeck, 'Microwave Noise Modeling for InP-InGaAs HBTs', *IEEE Trans. Microwave Theory Tech.*, **MTT-52**, 1264–1272 (2004).
51. D. B. Leeson, 'A Simple Model of Feedback Oscillator Noise Spectrum', *Proc. IEEE*, **54**, 329–330 (1966).

52. D. Scherer, 'Today's Lesson — Learn about Low-Noise Design', *Microwaves*, **18**, 116–122 (1979).
53. T. E. Parker, 'Characteristics and Sources of Phase Noise in Stable Oscillators', *Proc. 41st Annual Frequency Control Symp.*, 99–110 (1987).
54. D. J. Healy III, 'Flicker of Frequency and Phase and White Frequency and Phase Fluctuations in Frequency Sources', *Proc. 26th Annual Frequency Control Symp.*, 43–49 (1972).
55. J. K. A. Everard, 'A Review of Low Noise Oscillator: Theory and Design', *Proc. IEEE Int. Frequency Control Symp.*, 909–918 (1997).
56. T. E. Parker, 'Current Developments in SAW Oscillator Stability', *Proc. 31st Annual Frequency Control Symp.*, 359–364 (1977).
57. H. J. Siweris and B Schiek, 'Analysis of Noise Upconversion in Microwave FET Oscillators', *IEEE Trans. Microwave Theory Tech.*, **MTT-33**, 223–242 (1985).
58. G. Sauvage, 'Phase Noise in Oscillators: A Mathematical Analysis of Leeson's Model', *IEEE Trans. Instrum. Meas.*, **IM-26**, 408–410 (1977).
59. T. Hajder, 'Higher Order Loops Improve Phase Noise of Feedback Oscillators', *Applied Microwave & Wireless*, **14**, 24–31 (2002).
60. S. Hamilton, 'FM and AM Noise in Microwave Oscillators', *Microwave J.*, **21**, 105–109 (1978).
61. F. Herzel, 'An Analytical Model for the Power Spectral Density of a Voltage-Controlled Oscillator and Its Analogy to the Laser Linewidth Theory', *IEEE Trans. Circuits and Systems—I: Fundamental Theory Appl.*, **CAS-I-45**, 904–908 (1998).
62. F. Herzel, K. Hennenberger and W. Vogel, 'The Semiconductor Laser Linewidth: A Green's Function Approach', *IEEE J. Quantum Electron.*, **QE-29**, 2891–2897 (1993).
63. W. A. Edson, 'Noise in Oscillators', *Proc. IRE*, **48**, 1454–1466 (1960).
64. A. van der Ziel, 'Noise in Solid-State Devices and Lasers', *Proc. IEEE*, **58**, 1178–1206 (1970).
65. K. Kurokawa, 'Some Basic Characteristics of Broadband Negative Resistance Oscillator Circuits', *Bell Sys. Tech. J.*, **48**, 1937–1955 (1969).
66. K. Kurokawa, 'Noise in Synchronized Oscillators', *IEEE Trans. Microwave Theory Tech.*, **MTT-16**, 234–240 (1968).
67. G. H. B. Hansson and K. I. Lundstrom, 'Stability Criteria for Phase-Locked Oscillators', *IEEE Trans. Microwave Theory Tech.*, **MTT-20**, 641–645 (1972).
68. B. T. Debney and J. S. Joshi, 'A Theory of Noise in GaAs FET Microwave Oscillators and Its Experimental Verification', *IEEE Trans. Electron Devices*, **ED-30**, 769–775 (1983).
69. K. Hosoya, S. Tanaka and K. Honjo, 'Theoretical Analysis of Relationships between Resonator Coupling Coefficient and Phase Noise in Microwave Negative-Resistance Oscillators', *IEICE Trans. Electron.*, **E87-C**, 2132–2142 (2004).
70. H. Rohdin, C.-Y. Su and C. Stolte, 'A Study of the Relation between Device Low-Frequency Noise and Oscillator Phase Noise for GaAs MESFETs', 1984 *IEEE MTT-S Int. Microwave Symp. Dig.*, 267–271 (1989).
71. W. Anzill and P. Russer, 'A General Method to Simulate Noise in Oscillators Based on Frequency Domain Techniques', *IEEE Trans. Microwave Theory Tech.*, **MTT-41**, 2256–2263 (1993).
72. F. X. Kaertner, 'Analysis of White and $f^{-\alpha}$ Noise in Oscillations', *Int. J. Circuit Theory Appl.*, **18**, 485–519 (1990).
73. V. Rizzoli, F. Mastri and D. Masotti, 'General Noise Analysis of Nonlinear Microwave Circuits by the Piecewise Harmonic-Balance Technique', *IEEE Trans. Microwave Theory Tech.*, **MTT-42**, 807–819 (1994).
74. R. Poore, 'Accurate Simulation of Mixer Noise and Oscillator Phase Noise in Large RFICs', *Proc. 1997 Asia-Pacific Microwave Conf.*, **1**, 357–360 (1997).
75. A. A. Andronov, A. A. Vitt and S. E. Khaikin, *Theory of Oscillations*, New York: Dover Publications (1987).
76. K. A. Samoylo, *Method of Analysis of Second-Order Oscillation Systems* (in Russian), Moskva: Sov. Radio, 1976.

77. T. Ohira, 'Higher-Order Analysis on Phase Noise Generation in Varactor-Tuned Oscillators — Baseband Noise Upconversion in GaAs MESFET Oscillators', *IEICE Trans. Electron.*, **E76-C**, 1851–1854 (1993).
78. S. Lee and K. J. Webb, 'The Influence of Transistor Nonlinearities on Noise Properties', *IEEE Trans. Microwave Theory Tech.*, **MTT-53**, 1314–1321 (2005).
79. J. Groszkowski, 'The Interdependence of Frequency Variation and Harmonic Content, and the Problem of Constant-Frequency Oscillators', *Proc. IRE*, **21**, 958–981 (1933).
80. A. Hajimiri and T. H. Lee, 'A General Theory of Phase Noise in Electrical Oscillators', *IEEE J. Solid-State Circuits*, **SC-33**, 179–194 (1998).
81. A. Demir, A. Mehrotra and J. Roychowdhury, 'Phase Noise in Oscillators: A Unifying Theory and Numerical Methods for Characterization', *IEEE Trans. Circuits and Systems—I: Fundamental Theory Appl.*, **CAS-I-47**, 655–673 (2000).
82. T. H. Lee and A. Hajimiri, 'Oscillator Phase Noise: A Tutorial', *IEEE J. Solid-State Circuits*, **SC-35**, 326–336 (2000).
83. J.-C. Nallatamby, M. Prigent and J. Obregon, 'On the Role of the Additive and Converted Noise in the Generation of Phase Noise in Nonlinear Oscillators', *IEEE Trans. Microwave Theory Tech.*, **MTT-53**, 901–906 (2005).
84. A. Grebennikov, *RF and Microwave Power Amplifier Design*, New York: McGraw-Hill (2004).
85. P. Vanassche, G. Gielen and W. Sansen, 'On the Difference between Two Widely Publicized Methods for Analyzing Oscillator Phase Behavior', *Proc. IEEE/ACM Int. Conf. Computer Aided Design*, 229–233 (2002).
86. M. A. Margarit, J. L. Tham, R. G. Meyer and M. J. Deen, 'A Low-Noise, Low-Power VCO with Automatic Amplitude Control for Wireless Applications', *IEEE J. Solid State Circuits*, **SC-34**, 761–771 (1999).
87. G. J. Coram, 'A Simple 2-D Oscillator to Determine the Correct Decomposition of Perturbations into Amplitude and Phase Noise', *IEEE Trans. Circuits and Systems—I: Fundamental Theory Appl.*, **CAS-I-48**, 896–898 (2001).
88. W. E. Edson, *Vacuum-Tube Oscillators*, New York: John Wiley & Sons (1953).
89. B. Razavi, 'A Study of Phase Noise in CMOS Oscillators', *IEEE J. Solid-State Circuits*, **SC-31**, 331–343 (1996).
90. M. Ohtomo, 'Experimental Evaluation of Noise Parametres in Gunn and Avalanche Oscillators', *IEEE Trans. Microwave Theory Tech.*, **MTT-20**, 425–437 (1972).
91. J.-C. Nallatamby, M. Prigent, M. Camiade and J. J. Obregon, 'Extension of the Leeson Formula to Phase Noise Calculations in Transistor Oscillators with Complex Tanks', *IEEE Trans. Microwave Theory Tech.*, **MTT-51**, 690–696 (2003).
92. J.-C. Nallatamby, M. Prigent, M. Camiade and J. J. Obregon, 'Phase Noise in Oscillators — Leeson Formula Revisited', *IEEE Trans. Microwave Theory Tech.*, **MTT-51**, 1386–1394 (2003).
93. A. N. Malakhov, *Fluctuations in Self-Oscillation Systems* (in Russian), Moskva: Nauka (1968).
94. U. L. Rohde and F. Hagemeyer, 'Feedback Technique Improves Oscillator Phase Noise', *Microwaves & RF*, **37**, 61–70 (1998).
95. U. L. Rohde, 'A Novel RFIC for UHF Oscillators', 2000 *IEEE RFIC Symp. Dig.*, 53–56 (2000).
96. J. J. Obregon and A. P. S. Khanna, 'Exact Derivation of the Nonlinear Negative-Resistance Oscillator Pulling Figure', *IEEE Trans. Microwave Theory Tech.*, **MTT-30**, 1109–1111 (1982).
97. G. S. Hobson, 'Measurement of External *Q*-Factor of Microwave Oscillators Using Frequency Pulling or Frequency Locking', *Electronics Lett.*, **8**, 191–193 (1973).

6

Varactor and oscillator frequency tuning

Voltage-controlled oscillators (VCOs) are key components in many applications, especially in wireless communication systems, measurement equipment, or military applications. A growing market of wireless applications requires highly integrated circuit solutions, where both high-performance transistors and passive elements with high quality factors can be used. Since the VCO performance is a significant aspect of the overall effectiveness of the system, SiGe technology with high-resistive substrates and thick top metallizations provides excellent prospects for further widening of the VCO applications. This chapter discusses the varactor modelling issues, varactor nonlinearity and its effect to frequency modulation, and resonant circuit techniques to improve VCO tuning linearity using lumped and transmission line elements. Various practical examples of VCO implementation techniques based on using different types of the active devices, circuit schematic approaches and hybrid or monolithic integrated circuit technologies are shown and described.

6.1 VARACTOR MODELLING

The term 'varactor' comes from the phrase 'variable reactor' and means a device whose reactance can be varied in a controlled manner, in this case, with dc bias voltage [1]. The discovery of the capacitance of a rectifying contact extends back to 1929 when the first comprehensive investigation of this phenomenon was made [2]. In 1949, Shockley published his classic paper on the theory of *p–n* junctions where not only the expressions for the current–voltage relationships in a diode are given, but the capacitance effect is considered as well [3]. The possibility of tuning diodes with exponents greater than one-half was firstly described in 1958 [4]. The term 'hyperabrupt' was coined for such sensitive devices.

A simplified varactor equivalent circuit is shown in Figure 6.1, where C_v is the variable depletion layer capacitance, C_p is the package capacitance, R_s is the series contact and bulk resistance, L_s is the series inductance incorporating package inductance, *VD* is the diode junction [5, 6]. The diode is necessary to take into account because of the rectifying effect during a positive voltage swing. Series resistance R_s is a function of applied voltage and operating frequency, although in most practical cases it can be considered constant. Such a model neglects some parasitic linear components, which should be taken into account for microwave applications including the distributed line package model and some capacitances due to the ground proximity. However, for most high-frequency applications, up to 2.5 GHz,

RF and Microwave Transistor Oscillator Design A. Grebennikov

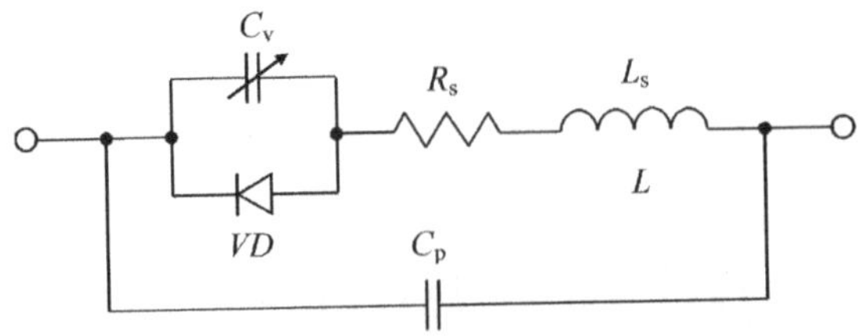

Figure 6.1 Simplified varactor equivalent circuit

these parasitics would not be significant unless higher-order harmonics due to the varactor nonlinearity affect VCO performance.

The varactor junction capacitance C_v as a function of reverse dc bias voltage V_v can be expressed by

$$C_v(V_v) = C_{v0}\left(1 + \frac{V_v}{\varphi}\right)^{-\gamma} \tag{6.1}$$

where

$$\gamma = -\frac{dC_v}{C_v} \bigg/ \frac{dV_v}{(V_v + \varphi)}$$

is the varactor junction sensitivity ($\gamma = 0.5$ for abrupt varactors, $1 \leq \gamma \leq 2$ for hyperabrupt varactors), φ is the contact potential. Its values depend on a doping profile of varactor; $C_v(0)$ is the varactor junction capacitance at $V_v = 0$. The voltage V_v is positive since it is assumed a reverse connection of the varactor.

When the junction is reverse-biased, a large electric field exists in the depletion region. As the bias voltage is increased, this field increases to a sufficiently high value where the thermally generated carriers traversing the depletion layer will generate additional hole–electron pairs by collision. These hole–electron pairs will in turn generate additional pairs, thus causing a multiplication effect or avalanche of carriers. The breakdown voltage at which avalanche occurs is the upper limit for the varactor voltage-control range and determines the minimum varactor junction capacitance C_{vmin}. The reverse breakdown voltage is relatively insensitive to temperature variations.

The doping concentration level in the varactor depletion region defines the difference between an abrupt junction varactor and a hyperabrupt junction one. For (a) abrupt junctions, the doping density is constant across the depletion region; for (b) hyperabrupt junctions, the doping density is a nonlinear function, as shown in Figure 6.2 [7]. In the latter case, ion implantation or nonlinear epitaxial growth techniques accomplish this. As a result, as the reverse voltage is increased, the higher doping density contributes to a greater capacitance change in the hyperabrupt varactor than in the abrupt varactor with a constant doping density. However, the averaged doping concentration of the undepleted epitaxial region for the abrupt varactor is higher approximately by a factor of two, which gives the higher value of the series resistance R_s for the hyperabrupt varactor.

The quality factor of a varactor Q_v (taking into account that the varactor junction capacitance C_v is substantially higher than the package capacitance C_p) is defined by

$$Q_v(V_v, \omega) = \frac{1}{\omega R_s C_v(V_v)} \tag{6.2}$$

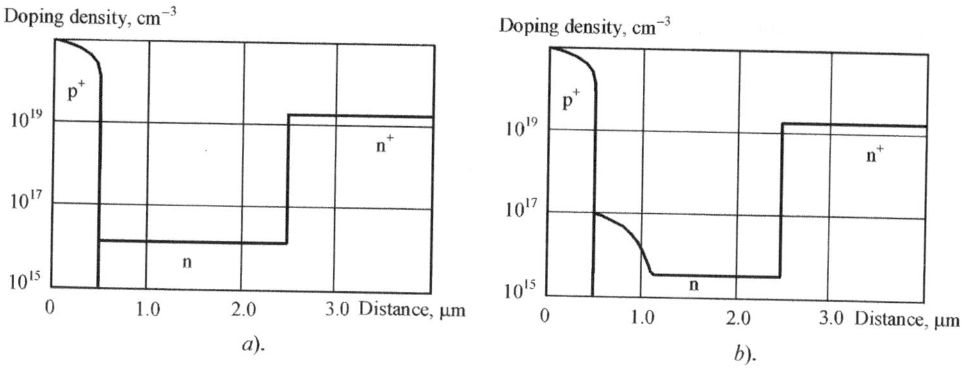

Figure 6.2 Doping density for (a) abrupt and (b) hyperabrupt varactors

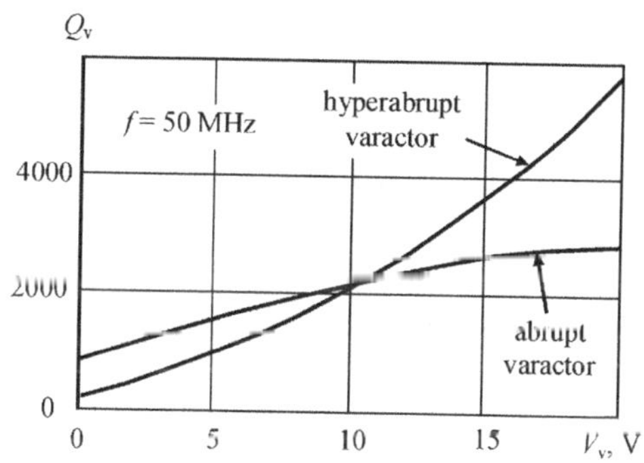

Figure 6.3 Varactor quality factor versus bias voltage

being a function of operating frequency and applied voltage. Since with the increase of R_s the varactor quality factor decreases, at low reverse bias voltage Q_v of the abrupt varactor is higher than that of the hyperabrupt one. However, at higher reverse voltages the quality factor of the hyperabrupt varactor becomes higher due to the more rapid decrease in the hyperabrupt varactor capacitance. As shown in Figure 6.3, usually over the linear tuning range for reverse-bias voltage in the range 1–10 V, the Q_v for the hyperabrupt varactor is lower. As a result, the output power of such a VCO with hyperabrupt varactor should be lower due to higher power losses in the varactor.

To approximate the experimental voltage–capacitance dependence, it is necessary to determine three unknown parameters C_{v0}, φ and γ from Equation (6.1). In this case, these parameters do not reflect the precise physical or electronic varactor properties because they are only the empirical variables under optimization procedure. A very convenient and simple analytical approach to calculate C_{v0}, φ and γ is described elsewhere [8]. The slope of the voltage–capacitance characteristic is defined as follows:

$$S_v = \frac{dC_v}{dV_v} = -\gamma \frac{C_{v0}}{\varphi} \left(1 + \frac{V_v}{\varphi}\right)^{-\gamma-1} \tag{6.3}$$

Under the condition of zero bias voltage when $V_v = 0$

$$S_{v0} = -\gamma \frac{C_{v0}}{\varphi} \tag{6.4}$$

Substituting Equation (6.4) into Equation (6.3) yields

$$S_v \left(1 + \frac{V_v}{\varphi}\right)^{-\gamma} = \frac{S_v}{S_{v0}} \left(1 + \frac{V_v}{\varphi}\right) \tag{6.5}$$

which allows one to obtain the final linear equation by substituting Equation (6.1)

$$C_v = -\frac{\varphi S_v}{\gamma} \left(1 + \frac{V_v}{\varphi}\right) \tag{6.6}$$

The parameters φ and γ can be easily defined by a least-squares method using the well-known algorithm for calculating coefficients a and b for a linear approximation function

$$y = ax + b \tag{6.7}$$

where $y = C_v/S_v$, $x = V_v$, $a = -1/\gamma$ and $b = -\varphi/\gamma$.

Figure 6.4 shows the analytical (solid line) and experimental (dotted line) voltage–capacitance dependencies for a silicon hyperabrupt varactor. The maximum error between empirical and experimental capacitance values is about 12% near reverse-bias voltage of about 7 V. As a result, the analytical voltage–capacitance dependence of such a hyperabrupt varactor can be written as

$$C_v(V) = 23 \left(1 + \frac{V_v}{2.5}\right)^{-1.05} \tag{6.8}$$

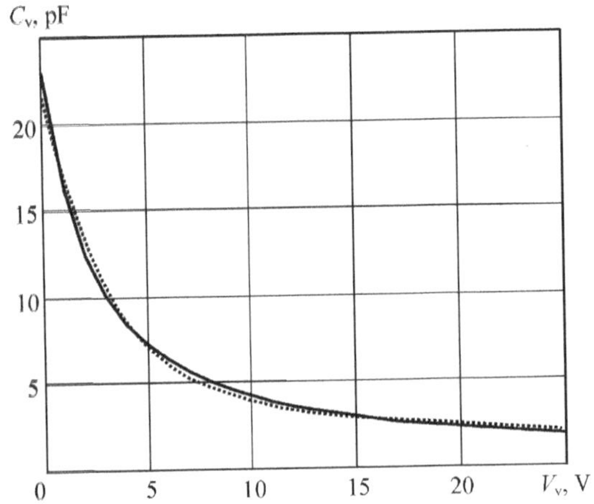

Figure 6.4 Silicon hyperabrupt varactor analytical (solid line) and experimental (dotted line) voltage–capacitance dependencies

6.2 VARACTOR NONLINEARITY

Under small-signal operation when the amplitude of the RF signal on the varactor is small, the device capacitance can be considered as a function of dc bias only since the varactor capacitance practically does not change. However, under large-signal operation when the amplitude of the RF signal can be of the same magnitude as the dc bias voltage, the instantaneous varactor capacitance is a function of both dc voltage and RF signal amplitude.

To analyse the effect of the varactor nonlinearity under large-signal operation, the following model of the varactor transferred-charge characteristic can be defined with initial condition of $q_{\mathrm{v}} = 0$ for $C_{\mathrm{v}} = 0$ in the vicinity of a bias point V_{v}:

$$q_{\mathrm{v}} = \int C_{\mathrm{v}} \mathrm{d}V_{\mathrm{v}} = \frac{\varphi C_{\mathrm{v0}}}{1-\gamma}\left(1+\frac{V_{\mathrm{v}}}{\varphi}\right)^{1-\gamma} \qquad \text{for } \gamma \neq 1 \tag{6.9}$$

$$q_{\mathrm{v}} = \int C_{\mathrm{v}} \mathrm{d}V_{\mathrm{v}} = \varphi C_{\mathrm{v0}} \ln\left(1+\frac{V_{\mathrm{v}}}{\varphi}\right) \qquad \text{for } \gamma = 1 \tag{6.10}$$

In this case, the total signal on the varactor should represent a sum of dc bias and ac harmonic components. Assume that the ac voltage on the varactor is limited to the fundamental frequency only when

$$v_{\mathrm{v}} = V_{\mathrm{v}} + V\cos\omega t \tag{6.11}$$

Substituting Equation (6.11) into the varactor transferred-charge characteristic given by Equation (6.9), applying a power series expansion with respect to the incremental time-varying voltage $V\cos\omega t$, taking into account only the first three terms and equating the fundamental charge component terms result in

$$\begin{aligned} q_{\mathrm{v}} &= Q_{\mathrm{v}} + Q\cos\omega t \\ &= \frac{C_{\mathrm{v0}}\varphi}{1-\gamma}\left(1+\frac{V_{\mathrm{v}}}{\varphi}\right)^{1-\gamma}\left[1-\frac{\gamma(1-\gamma)}{\varphi^2}\left(1+\frac{V_{\mathrm{v}}}{\varphi}\right)^{-2}\frac{V^2}{4}\right] \\ &\quad + C_{\mathrm{v0}}\left(1+\frac{V_{\mathrm{v}}}{\varphi}\right)^{1-\gamma}\left[1+\frac{\gamma(1+\gamma)}{\varphi^2}\left(1+\frac{V_{\mathrm{v}}}{\varphi}\right)^{-2}\frac{V^2}{8}\right]V\cos\omega t. \end{aligned} \tag{6.12}$$

Thus, the equivalent fundamentally averaged capacitance C_{v1} defined as $C_{\mathrm{v1}} = Q/V$ can be obtained as

$$C_{\mathrm{v1}} = C_{\mathrm{v0}}\left(1+\frac{V_{\mathrm{v}}}{\varphi}\right)^{-\gamma}(1+\alpha_{\mathrm{v}}V^2) \tag{6.13}$$

where

$$\alpha_{\mathrm{v}} = \frac{\gamma(1+\gamma)}{8\varphi^2}\left(1+\frac{V_{\mathrm{v}}}{\varphi}\right)^{-2}$$

It should be noted that Equation (6.13) gives the value of the equivalent linearized capacitance C_{v1} when the amplitude of the fundamental current has the same value as in the circuit with such a nonlinear capacitance under purely cosinusoidal voltage. Another result will be obtained when calculating the fundamental voltage amplitude across the varactor under consideration of purely cosinusoidal current.

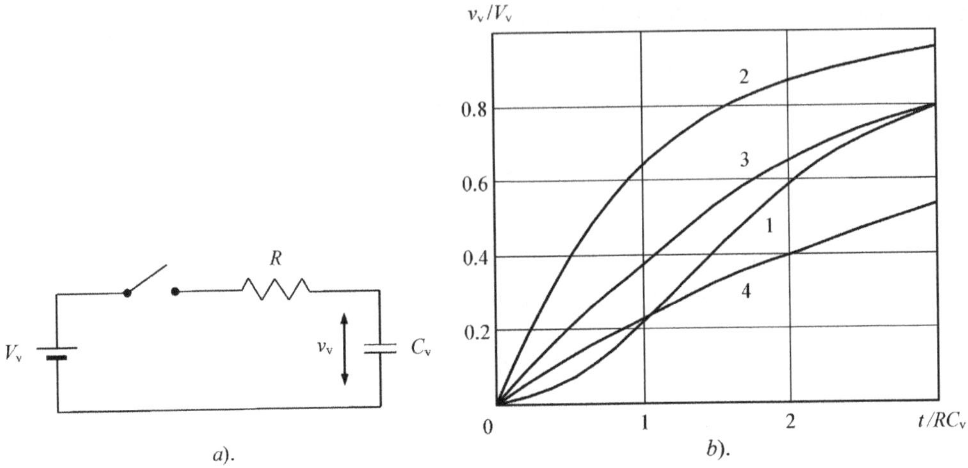

Figure 6.5 Varactor capacitance charging circuit and transient response

The deviation of $C_{v1}(V)/C_v(V = 0)$ from unity does not exceed 15% for the abrupt varactors when the varactor reverse bias $V_v = 1$ V and the fundamental amplitude $V \leq 1$ V. Therefore, to avoid the rectifying effect, which contributes to an increase of the series varactor resistance and noise level, a minimum varactor bias in the range 1–2 V is required. In this case, to evaluate the low-power VCO frequency tuning properties, it is sufficient to use a small-signal varactor model.

Since the varactor capacitance is nonlinear, it is necessary to have a quantitative knowledge of the transient characteristics of the varactor. To obtain the analytical expressions for the transient responses, it is advisable to rewrite Equation (6.1) in the form

$$C_v(v_v) = C_V \left(\frac{\varphi + V_v}{\varphi + v_v} \right)^{\gamma} \tag{6.14}$$

where $C_V = C_v$ is the capacitance at $v_v = V_v$.

Now consider the circuit shown in Figure 6.5a which corresponds to the charging process when the switch is turned on, with the following assumptions:

- the change in the varactor series bulk resistance with voltage is small compared with the total series resistance;
- the junction capacitance follows the junction voltage in a very short time compared with the transient response time of the diode;
- the value of the bias voltage V_v is much larger than that of the contact potential φ;
- the varactor junction has an abrupt characteristic with sensitivity $\gamma = 0.5$.

In this case, the current i in the circuit can be written, using Equation (6.14), as

$$i = C_v(v_v)\frac{dv_v}{dt} = C_V\sqrt{\frac{V_v dv_v}{v_v dt}} \tag{6.15}$$

On the other hand, from Kirchhoffs voltage law it follows that

$$i = \frac{V_v - v_V}{R} \tag{6.16}$$

Substituting Equation (6.16) into Equation (6.15) results in a nonlinear first-order differential equation describing the charging process, given by

$$\frac{dv_V}{(V_v - v_V)\sqrt{v_V}} = \frac{dt}{RC_V\sqrt{V_v}} \tag{6.17}$$

Integrating both parts of Equation (6.17) by taking into account zero initial conditions yields the varactor transient response in the form [9, 10]

$$\frac{v_V}{V_v} = \tanh^2 \frac{t}{2RC_V} \tag{6.18}$$

Figure 6.5b shows the normalized transient response of the abrupt varactor (curve 1). Here, the charging curves for the linear capacitance C are plotted for comparison (curve 2: $C = C_V$; curve 3: $C = 2C_V$; curve 4: $C = 4C_V$). From Figure 6.5b it follows that the abrupt varactor junction capacitance is charging faster than the linear capacitance, having values of $C > 2C_V$, but slower than the linear capacitance $C = C_V$. Unlike the linear capacitance case, the varactor capacitance transients exhibit a lower slope at small values of the reverse-bias voltage and steeper slope at its larger values. The value of the slope at its steepest point decreases with increasing varactor junction sensitivity γ [11].

For the circuit shown in Figure 6.6a, which corresponds to the discharging process when the switch is turned on, the current i as a function of time can be written, using Equation (6.16), as

$$i = \frac{v_V}{R}. \tag{6.19}$$

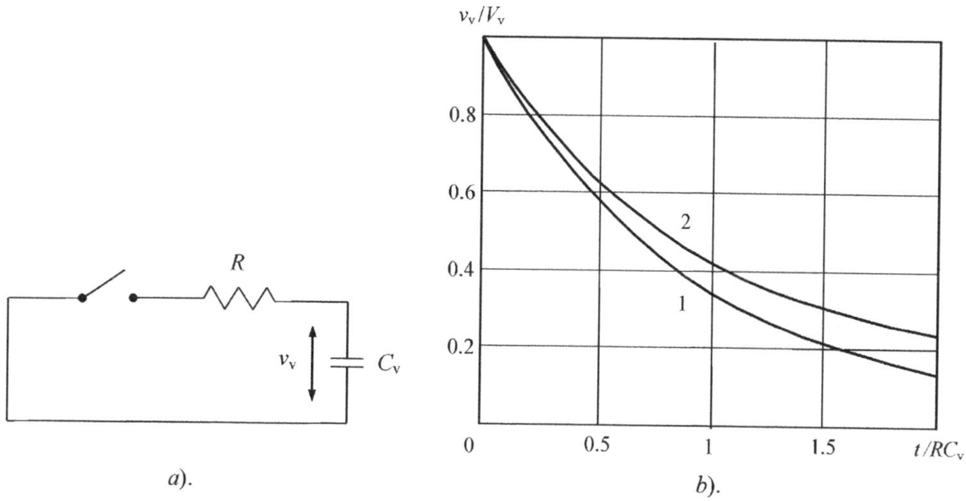

Figure 6.6 Varactor capacitance discharging circuit and transient response

Substituting Equation (6.19) into Equation (6.15) results in a nonlinear first-order differential equation describing the discharging process given by

$$\frac{dv_V}{v_V\sqrt{v_V}} = -\frac{dt}{RC_V\sqrt{V_v}} \tag{6.20}$$

Integrating both sides of Equation (6.20) by taking into account the initial conditions of $v_V = V_v$ for $t = 0$ yields the varactor transient response in the form [9]

$$\frac{v_V}{V_v} = \left(1 + \frac{t}{2RC_V}\right)^{-2} \tag{6.21}$$

Figure 6.6b shows the normalized transient response of the abrupt varactor (curve 1) where the discharging curve for the linear capacitance (curve 2) is plotted for comparison. From Figure 6.6b it follows that the discharging time for the abrupt varactor junction capacitance is faster than the discharging time of the linear capacitance.

6.3 FREQUENCY MODULATION

A varactor whose capacitance is changed in accordance to an applied reverse-bias voltage is used as a frequency tuning element in the oscillators. By providing a proper reverse-bias voltage and low-frequency modulating signal to the varactor, being a part of the oscillator resonant circuit, the formation of the oscillator output high-frequency signal with frequency modulation is achieved. To evaluate the quality of such a frequency-modulated transmitting signal, it is very important to know the behaviour of the varactor voltage–capacitance characteristic and its influence on the oscillator static modulation characteristic, representing a dependence of the oscillation frequency on slowly varying varactor bias voltage. Together with the dynamic modulation characteristic, parasitic amplitude modulation or dependence of the modulation index on the amplitude of the modulated signal, the oscillator static modulation characteristic is the most important oscillator characteristic, based on which the oscillator frequency-modulated output signal with minimal nonlinear distortions can be provided.

To investigate the nonlinearity of the static modulation characteristic, let us represent the voltage across the varactor v_v as a sum of dc bias voltage V_v and low-frequency ac component v_Ω in the form

$$v_v = V_v + v_\Omega \tag{6.22}$$

It is convenient to represent the static modulation characteristic as a dependence of the relative frequency deviation $y = (\omega - \omega_0)/\omega_0 = \Delta\omega/\omega_0$ on the normalized ac low-frequency voltage $\varepsilon = v_\Omega/V_v$. Generally, the dependence $y(\varepsilon)$ is nonlinear, and it can be written as an nth-order power series expansion

$$y = S_1\varepsilon + S_2\varepsilon^2 + S_3\varepsilon^3 + \ldots + S_n\varepsilon^n \tag{6.23}$$

where S_1 is the transconductance of the static modulation characteristic, the coefficients $S_2, S_3, \ldots, S_n$ characterize its curvature, and

$$S_n = \left.\frac{1}{n!}\frac{d^n y(\varepsilon)}{d\varepsilon^n}\right|_{\varepsilon=0} \tag{6.24}$$

In a common case, the capacitance of the varactor (or any variable capacitor) as a function of the voltage v_v can be written as

$$C_v = C_V(V_v) + \Delta C_v(v_\Omega) \tag{6.25}$$

where C_V is the average capacitance at $v_v = V_v$ and ΔC_v is the variation of the capacitance C_v under the modulating signal. In order to define the dependence $\Delta\omega(v_\Omega)$, it is necessary to calculate the functions $\Delta\omega(\Delta C)$, where ΔC is the variation of the overall capacitance of the oscillator resonant circuit C_0, $\Delta C(\Delta C_v)$ and $\Delta C_v(v_\Omega)$ in succession. Generally, the procedure of analytical calculation of these dependencies is complicated. Therefore, it is best to introduce the following power series expansions [11]:

$$y(x) = a_{11}x + a_{12}x^2 + a_{13}x^3 + \ldots + a_{1n}x^n \tag{6.26}$$

$$x(\xi) = a_{21}\xi + a_{22}\xi^2 + a_{23}\xi^3 + \ldots + a_{2n}\xi^n \tag{6.27}$$

$$\xi(\varepsilon) = a_{31}\varepsilon + a_{32}\varepsilon^2 + a_{33}\varepsilon^3 + \ldots + a_{3n}\varepsilon^n \tag{6.28}$$

where $x = \Delta C/C_0$ and $\xi = \Delta C_v/C_V$.

The function $y(x)$ depends on the number of resonant circuits in the oscillator. For a single resonant circuit oscillator,

$$\Delta\omega = \frac{1}{\sqrt{L(C + \Delta C)}} - \frac{1}{\sqrt{LC}}$$

where L is the resonant circuit inductance. As a result, from Equation (6.26) it follows that

$$y = \frac{1}{\sqrt{1+x}} - 1 = -\frac{x}{2} + \frac{3x^2}{8} - \frac{5x^3}{16} + \ldots. \tag{6.29}$$

with the first three coefficients

$$a_{11} = -\frac{1}{2} \qquad a_{12} = \frac{3}{8} \qquad a_{13} = -\frac{5}{16} \tag{6.30}$$

The function $x(\xi)$ depends on the particular type of the varactor connection to the oscillator resonant circuit, whereas the function $\xi(\varepsilon)$ depends on the voltage–capacitance characteristic of the particular varactor. This means that the coefficients of the power series given by Equations (6.27) and (6.28) can be determined only for particular cases of the oscillator configuration and varactor used.

The expressions for the first two coefficients S_1 and S_2 in Equation (6.23) can be obtained by substituting one series into the other twice

$$S_1 = a_{11}a_{21}a_{31} \tag{6.31}$$

$$S_2 = a_{11}\left(a_{21}a_{32} + a_{22}a_{31}^2\right) + a_{12}(a_{21}a_{31})^2 \tag{6.32}$$

Figure 6.7a shows the general case of the partial varactor connection to the oscillator resonant circuit. When $C_1 \gg C_2 + C_V$, it can represent a parallel resonant circuit with variable capacitance C_v, as shown in Figure 6.7b. When $C_2 \ll C_V$, the oscillator can be represented by a series resonant circuit with variable capacitance C_v, as shown in Figure 6.7c.

For a parallel resonant circuit with $C_0 = C_2 + C_v$ when $\varepsilon = 0$, $C_v = C_V$,

$$x = \frac{\Delta C}{C_0} = \frac{C_2 + C_V + C_v}{C_2 + C_V} - 1 = \frac{C_V}{C_2 + C_V}\frac{\Delta C_v}{C_V} = p\xi \tag{6.33}$$

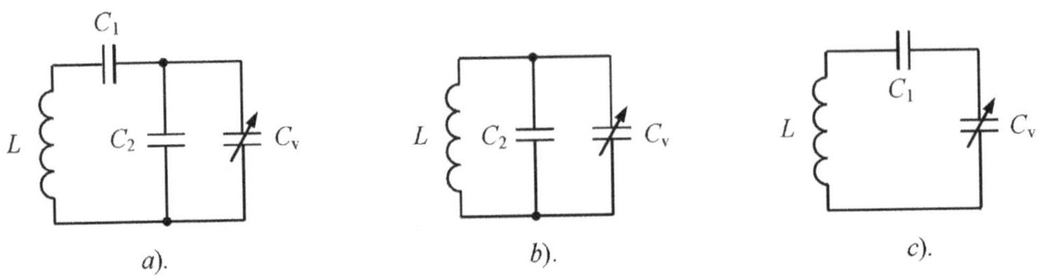

Figure 6.7 Varactor connection to a resonant circuit

where $p = C_V/(C_2 + C_V)$ is the coefficient of partial varactor connection into the oscillator resonant circuit. Consequently, here the power series $x(\xi)$ has the coefficients $a_{21} = p$, $a_{22} = a_{23} = \ldots = a_{2n} = 0$.

For a series resonant circuit with $C_0 = C_v C_1/(C_v + C_1)$,

$$x = \frac{\Delta C}{C_0} = \frac{C_1 C_V (1+\xi)(C_1 + C_V)}{[C_1 + C_V(1+\xi)]\, C_1 C_V} - 1 = \frac{p\xi}{1 + (1-p)\xi}$$
$$= p\xi - p(1-p)\xi^2 + \ldots \tag{6.34}$$

where $p = C_1/(C_1 + C_V)$ is the coefficient of partial varactor connection into the oscillator resonant circuit. Consequently, here the power series $x(\xi)$ has the first two coefficients $a_{21} = p$ and $a_{22} = -p(1-p)$. It should be noted that, for a series resonant circuit, the coefficient p has an effect on the curvature of the static modulation characteristic.

The resonant circuit, which is shown in Figure 6.7a, represents a combination of the series and parallel resonant circuits. The relative deviation ξ_1 of the capacitance $C'_v = C_2 + C_v$ can be defined from Equation (6.33) as

$$\xi_1 = \frac{\Delta C'_v}{C'_V} = p_1 \xi \tag{6.35}$$

where $C'_V = C_2 + C_V$ and $p_1 = C_V/(C_2 + C_V)$. The capacitance $C'_v = C'_V + \Delta C'_v$ is the variable capacitance for the circuit consisting of the capacitance C_1 and inductance L. Therefore, using Equation (6.34) results in

$$x(\xi) = p_2 \xi_1 - p_2(1 - p_2)\xi_1^2 + \ldots \tag{6.36}$$

where $p_2 = C_1/(C_1 + C'_V) = C_1/(C_1 + C_2 + C_V)$. Substituting Equation (6.35) into Equation (6.36) allows us to determine the first two coefficients as

$$a_{21} = p_1 p_2 \qquad a_{22} = -p_1^2 p_2 (1 - p_2) \tag{6.37}$$

As a result, by choosing the coefficients p_1 and p_2 with proper values of C_1, C_2 and C_V, the behaviour of the static modulation characteristic can be optimized for linear frequency tuning.

Generally, the voltage across the varactor υ_v consists of dc bias voltage V_v, low-frequency ac component with amplitude V_Ω and high-frequency ac component with amplitude V, which can be written as

$$\upsilon_v = V_v + V_\Omega \cos \Omega t + V \cos \omega t \tag{6.38}$$

as shown in Figure 6.8. The bias voltage $V_v = V_{v0}$ is normally chosen at the centre point of

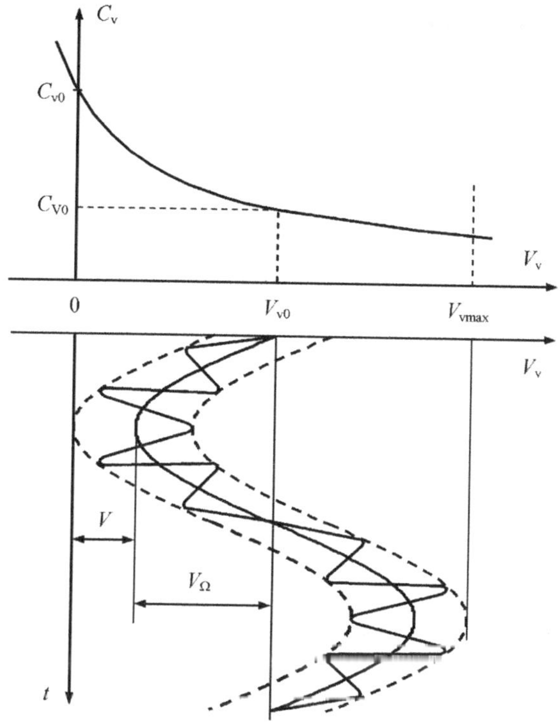

Figure 6.8 Varactor capacitance and frequency modulation

the voltage–capacitance characteristic to maximize the frequency tuning bandwidth without performance degradation due to the forward conduction of the active device or its breakdown [12]. Under these assumptions, the borders of voltage variations across the varactor junction should satisfy the condition

$$V_\Omega + V = 0.5V_{vmax} \tag{6.39}$$

To obtain an equation for the static modulation characteristic of the bipolar common base oscillator shown in Figure 6.9, it is necessary to use the varactor–capacitance dependence given by Equation (6.14). Assuming $V_v = V_{v0} + \Delta V_v$, then we can find the dependence of the relative capacitance variation $\xi = \Delta C_v / C_{V0}$ on the normalized voltage $\varepsilon = v_\Omega/(\varphi + V_{v0})$ as

$$\xi = \left(\frac{\varphi + V_{v0}}{\varphi + v_v + \Delta V_v} \right)^\gamma - 1 = \frac{1}{(1+\varepsilon)^\gamma} - 1 \tag{6.40}$$

Expanding Equation (6.40) into an nth-order power series given by Equation (6.28) for the case of $\gamma = 0.5$ yields

$$\xi = -\frac{\varepsilon}{2} + \frac{3\varepsilon^2}{8} - \ldots. \tag{6.41}$$

which enables us to write the coefficients $a_{31} = -1/2$ and $a_{32} = 3/8$.

Thus, by setting the correspondence between the parameters of the resonant circuit of the bipolar oscillator shown in Figure 6.9 and the parameters of the series–parallel resonant

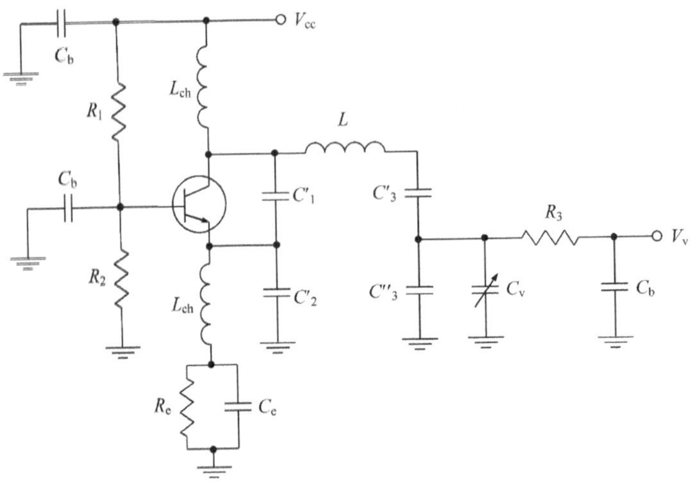

Figure 6.9 Schematic of a bipolar varactor-tuned oscillator

circuit shown in Figure 6.7a as $C_V = C_{V0}$, $C_2 = C_3''$ and $1/C_1 = 1/C_1' + 1/C_2' + 1/C_3'$, the coefficients of the static modulation characteristic, taking into account Equations (6.30), (6.37) and (6.41), can be obtained from Equations (6.31) and (6.32) as

$$S_1 = \frac{p_1 p_2}{4} \tag{6.42}$$

$$S_2 = S_1 \left[\frac{3}{4} - \frac{p_1}{2} \left(1 - \frac{p_2}{4} \right) \right] \tag{6.43}$$

The maximum frequency tuning bandwidth y_{max} can be found under assumptions of constant voltage amplitude across the resonant circuit V_0, constant value of the coefficient p_2 and constant slope of the static modulation characteristic when

$$y = \frac{\Delta\omega}{\omega_0} = S_1 \varepsilon \tag{6.44}$$

for different varactor bias conditions. As a result, using Equations (6.39) and (6.42) yields

$$y_{max} = \frac{p_1 p_2}{2} \frac{0.5 V_{vmax} - p_2 V_0}{2\varphi + V_{vmax}} \tag{6.45}$$

where $p_1 = C_{V0}/(C_2 + C_{V0})$, $p_2 = C_{V0}/(C_1 + C_2 + C_{V0})$ and $V_0 = V/p_2$.

From Equation (6.45) it follows that the tuning bandwidth initially increases with increase of p_2, but then decreases. Hence, the maximum tuning bandwidth can be chosen for optimum value of coefficient p_2, i.e., optimum connection of the varactor to the resonant circuit.

6.4 ANTI-SERIES VARACTOR PAIR

Since the voltage–capacitance characteristic of the varactor is nonlinear, resulting in the nonlinear distortion of the voltage-controlled oscillator, the level of which depends on the dc bias voltage on the varactor, it is important to analyse the anti-series varactor pair where the

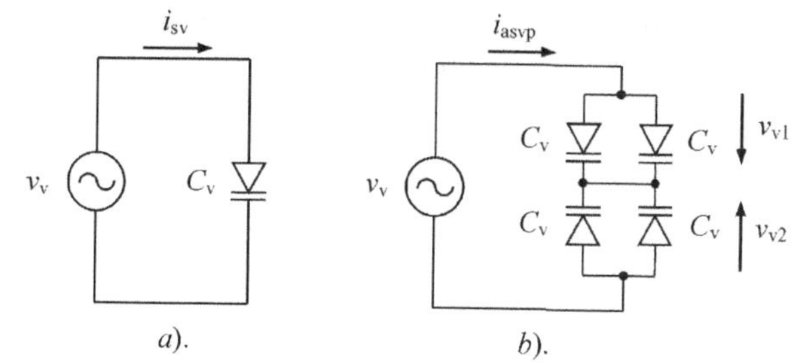

Figure 6.10 Single varactor and anti-series varactor pair

varactors are connected back-to-back. Figure 6.10 shows (a) the single varactor with capacitance C_v and (b) two anti-series varactor pairs with overall capacitance C_v connected to the harmonic voltage source v_v. Based on these simplified schematics, an analytical approach can be used to demonstrate better harmonic suppression when using the anti-series varactor pair.

Generally, the nonlinear varactor junction voltage–capacitance characteristic can be approximated by an nth-order polynomial form where the polynomial is expanded around the dc bias point. Then, the current i_{sv} flowing into the single varactor, as shown in Figure 6.10a, can be written as

$$i_{sv} = C_v(v_v)\frac{dv_v}{dt} = \left(C_{v0} + C_{v1}v_v + C_{v2}v_v^2\right)\frac{dv_v}{dt} \tag{6.46}$$

which is a series form of the incremental varactor capacitance limited to the second order, assuming that higher-order terms are small enough to neglect.

Also assume that voltage v_v across the varactor is cosinusoidal given by Equation (6.11) with $V_v = 0$. By substituting it into Equation (6.46) and applying trigonometric formulas for squares and products of cosines, we can obtain for a single varactor

$$i_{sv} = -\left(\omega C_{v0}V + \frac{\omega C_{v2}V^3}{4}\right)\sin\omega t - \frac{\omega C_{v1}V^2}{2}\sin 2\omega t - \frac{\omega C_{v2}V^3}{4}\sin 3\omega t \tag{6.47}$$

From Equation (6.47) it follows that current i_{sv} contains the fundamental, second and third harmonics, and the level of the fundamental and third harmonics depends strongly on the coefficient C_{v2}.

To analyse the nonlinearity caused by the anti-series varactor pair, let us apply the same analytical approach to the two anti-series varactor pairs shown in Figure 6.10b [13]. Here, the current i_{asvp} flowing into the anti-series varactor pairs can be written as

$$\begin{aligned} i_{asvp} &= 2\left(C_{v0} + C_{v1}v_{v1} + C_{v2}v_{v1}^2\right)\frac{dv_{v1}}{dt} \\ &= -2\left(C_{v0} + C_{v1}v_{v2} + C_{v2}v_{v2}^2\right)\frac{dv_{v2}}{dt} \end{aligned} \tag{6.48}$$

where v_{v1} and v_{v2} are the voltages across the upper and lower varactors respectively, and

$$v_{v1} - v_{v2} = v_v = V\cos\omega t \tag{6.49}$$

By substituting Equation (6.49) into Equation (6.48) and performing some calculations, the final equation for i_{asvp} can be written in the form

$$i_{\mathrm{asvp}} = -\omega V \left[C_{\mathrm{v0}} + \frac{V^2}{16}\left(C_{\mathrm{v2}} - \frac{C_{\mathrm{v1}}^2}{2C_{\mathrm{v0}}} \right) \right] \sin \omega t + \frac{\omega V^3}{16}\left(\frac{3C_{\mathrm{v1}}^2}{2C_{\mathrm{v0}}} - C_{\mathrm{v2}} \right) \sin 3\omega t \tag{6.50}$$

From Equation (6.50) it follows that the second harmonic component is cancelled for the symmetric anti-series varactor pair. This is explained by the fact that the overall varactor capacitance C_{v} becomes an even function of voltage v_{v} around dc bias voltage due to symmetrical anti-series connection of the varactors. In addition, the condition

$$\frac{3C_{\mathrm{v1}}^2}{2C_{\mathrm{v0}}} = C_{\mathrm{v2}} \tag{6.51}$$

can provide zero amplitude for the third harmonic component. For the case when $C_{\mathrm{v1}} = 0$, using an anti-series varactor pair enables a reduction of the third harmonic component by 12 dB compared with a single varactor case. Numerical analysis using a fourth-order polynomial to approximate the varactor capacitance shows that the least nonlinear distortion can be provided by abrupt varactor capacitances having junction sensitivity $\gamma = 0.5$ [14].

Let us now analyse in detail the oscillator resonant frequency dependence on the voltage on the varactor under large-signal operation for the case of an anti-series abrupt varactor pair [15]. Figure 6.11a shows the resonant circuit of the oscillator with two identical back-to-back abrupt varactors with $\gamma = 0.5$, where V_{v} is the dc bias voltage, v_1 and v_2 are the RF voltage components applied to the varactors, v is the RF voltage component on the resonant circuit, q_1 and q_2 are the RF charge components flowing into the varactors, and i is the current in the resonant circuit. Assume that the choke inductor L_{ch} is lossless and has infinite impedance for the current in the resonant circuit. By using Equation (6.9), the following model of the varactor transferred-charge characteristic under large-signal operation can be used:

$$q_{\mathrm{v}} = 2\sqrt{\varphi} C_{\mathrm{v0}} \sqrt{\varphi + v_{\mathrm{v}}} \tag{6.52}$$

where v_{v} is the voltage on the varactor.

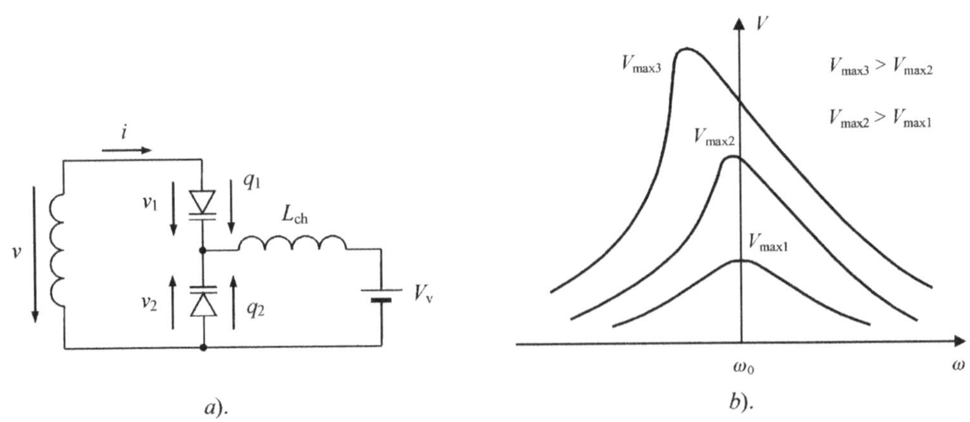

Figure 6.11 Resonant circuit with anti-series varactor pair

The relationships between dc and RF voltage components on the varactors can be written as

$$v = v_1 - v_2 \qquad v_{v1} = V_v - v_1 \qquad v_{v2} = V_v - v_2 \tag{6.53}$$

where v_{v1} and v_{v2} are the voltages on the varactors.

The charges q_{v1} and q_{v2} can be obtained from Equations (6.52) and (6.53) as

$$q_{v1} = 2\sqrt{\varphi}C_{v0}\sqrt{\varphi + V_v - v_1} \tag{6.54}$$

$$q_{v2} = 2\sqrt{\varphi}C_{v0}\sqrt{\varphi + V_v - v_2} \tag{6.55}$$

which can also be represented by the dc and RF components in the form

$$q_{v1} = Q_{v1} - q_1 \qquad q_{v2} = Q_{v2} - q_2 \tag{6.56}$$

where $q_1 = -q_2 = q$, since the current i is flowing only in the resonant circuit and the varactors are identical.

As a result,

$$v_1 = -\frac{(Q_{v1} - q)^2}{4\varphi C_{v0}^2} + V_v + \varphi \tag{6.57}$$

$$v_2 = -\frac{(Q_{v2} + q)^2}{4\varphi C_{v0}^2} + V_v + \varphi \tag{6.58}$$

$$v = -\frac{1}{4\varphi C_{v0}^2}\left(Q_{v1}^2 - Q_{v2}^2\right) + \frac{1}{2\varphi C_{v0}^2}(Q_{v1} + Q_{v2})q \tag{6.59}$$

According to Equation (6.53), the dc charge component Q_v in Equation (6.59) is equal to zero, i.e., $Q_{v1} = Q_{v2} = Q_v$. Consequently,

$$v = -\frac{Q_v}{\varphi C_{v0}^2}q \tag{6.60}$$

which shows the linear relationship between the RF components of the charge and voltage on the varactors for the case of the anti-series abrupt varactor pair.

However, the cancellation of all nonlinear effects for such a resonant circuit with an anti-series abrupt varactor pair does not occur because dc charge component Q_v at a fixed value of bias voltage V_v is also a function of RF voltage amplitude. For example, for a cosinusoidal charge of

$$q = Q\cos\omega t \tag{6.61}$$

where Q is the amplitude of the RF charge component, from Equations (6.57) and (6.58) it follows that

$$v_1 = -\left[\frac{2Q_v^2 + Q^2}{8\varphi C_{v0}^2} - (V_v + \varphi)\right] + \frac{Q_v Q}{2\varphi C_{v0}^2}\cos\omega t - \frac{Q^2}{8\varphi C_{v0}^2}\cos 2\omega t \tag{6.62}$$

$$v_2 = -\left[\frac{2Q_v^2 + Q^2}{8\varphi C_{v0}^2} - (V_v + \varphi)\right] - \frac{Q_v Q}{2\varphi C_{v0}^2}\cos\omega t - \frac{Q^2}{8\varphi C_{v0}^2}\cos 2\omega t \tag{6.63}$$

Equations (6.62) and (6.63) show that, for a purely sinusoidal current $i = dq/dt = -I \sin \omega t$ in the resonant circuit, the junction capacitance of the abrupt varactor with $\gamma = 0.5$ contributes to the appearance of the second voltage harmonic. Since the RF voltage components v_1 and v_2 defined by Equations (6.62) and (6.63) must not contain the dc components, equating their dc components to zero yields

$$Q_\mathrm{v} = \sqrt{2\varphi}C_{\mathrm{v}0}\sqrt{\varphi + V_\mathrm{v}}\sqrt{1 - \frac{Q^2}{8\varphi C_{\mathrm{v}0}^2 (\varphi + V_\mathrm{v})}} \tag{6.64}$$

which illustrates the fact that the dc charge component Q_v depends on both the dc voltage V_v and the RF charge component Q.

In this case, the equivalent fundamentally averaged total capacitance $C_{\mathrm{v}1}$ defined as $C_{\mathrm{v}1} = Q/V_1$, where V_1 is the fundamental voltage amplitude, can be obtained as

$$C_{\mathrm{v}1} = \frac{\varphi C_{\mathrm{v}0}^2}{Q_\mathrm{v}} \tag{6.65}$$

which depends not only on the external dc bias voltage V_v, but on the RF charge component Q as well.

For a series connection of two identical abrupt varactors, the RF voltage across them can be written as

$$v = \frac{Q_\mathrm{v} Q}{\varphi C_{\mathrm{v}0}^2} \cos \omega t - \frac{Q^2}{4\varphi C_{\mathrm{v}0}^2} \cos 2\omega t \tag{6.66}$$

which shows the presence of the second voltage harmonic, unlike the case of the anti-series varactor pair. However, the dc voltage components for both varactor configurations as well as the equivalent average capacitances $C_{\mathrm{v}1}$ are the same.

By substituting Equation (6.64) into Equation (6.65) and using Equation (6.1), the equivalent average capacitance $C_{\mathrm{v}1}$ of the anti-series varactor pair can be obtained from

$$C_{\mathrm{v}1} = \frac{1}{2} \frac{C_\mathrm{v}}{\sqrt{1 - \frac{1}{8}\left(\frac{1}{\varphi + V_\mathrm{v}}\right)^2 \left(\frac{V}{2}\right)^2}} \tag{6.67}$$

where $V = I/\omega C_\mathrm{v}$ is the voltage amplitude across the linear capacitor C_v [16].

Figure 6.11b shows the resonant curves of the oscillator resonant circuit with anti-series varactor pair, where V is the voltage amplitude on the circuit, having its maximum value $V_{\max}$ at the resonant frequency ω_0 for the linear case. For low voltage amplitudes, the equivalent average capacitance $C_{\mathrm{v}1}$ is close to its small-signal value C_v, and the maximum voltage amplitude $V_{\max} = V_{\max 1}$ almost corresponds to the resonant frequency ω_0. However, for higher voltage amplitudes ($V_{\max 2} > V_{\max 1}$), the equivalent average capacitance $C_{\mathrm{v}1}$ increases according to Equation (6.67), resulting in a decrease of the resonant circuit frequency. Under larger voltage amplitudes ($V_{\max 3} > V_{\max 2}$), the resonant frequency continues to deviate from its value ω_0 for a linear resonant circuit. In some cases, for sufficiently large voltage amplitudes, the slope of the resonant curve becomes negative, resulting in a hysteresis region characterized by frequency jumping. Thus, an analysis of the varactor nonlinearity shows that, even when using identical anti-series varactor pair, the deviations of the amplitude of the oscillations can affect the frequency stability.

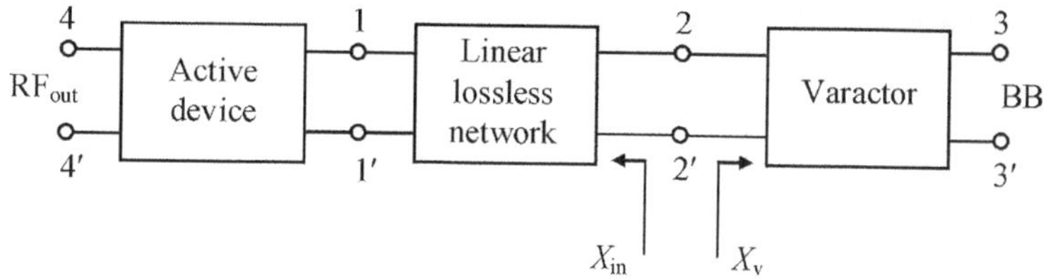

Figure 6.12 Block diagram of VCO with a linear circuit

6.5 TUNING LINEARITY

6.5.1 VCOs with lumped elements

To analyse the tuning linearity of the VCO circuit, a general approach, describing the oscillation circuit in terms of natural frequencies of a lossless two-port network when one of its ports is short-circuited, can be used [17]. Figure 6.12 shows the block diagram of the general VCO equivalent circuit, where a linear lossless network can incorporate one or several resonant circuits, including active device reactive elements, the load is connected to the port 1–1′, the baseband modulation signal is brought to the varactor using port 3–3′. Such a block representation enables the description of the VCO modulated curve in terms of poles and zeros, irrespective of any particular circuit diagram.

The oscillation frequency can be found form the phase balance condition

$$X_{in}(\omega_0) = -X_v(\omega_0) \tag{6.68}$$

where $X_{in}(\omega_0)$ is the reactance seen by the varactor and $X_v(\omega_0)$ is the reactance of the varactor, and ω_0 is the oscillation frequency. The reactance seen by the active device is equal to zero at the oscillation frequency.

Generally, the reactance X_{in} can be expressed in terms of poles and zeros. For instance, for the case shown in Figure 6.13 where the poles occur at the origin and ω_2 while the zeros occur

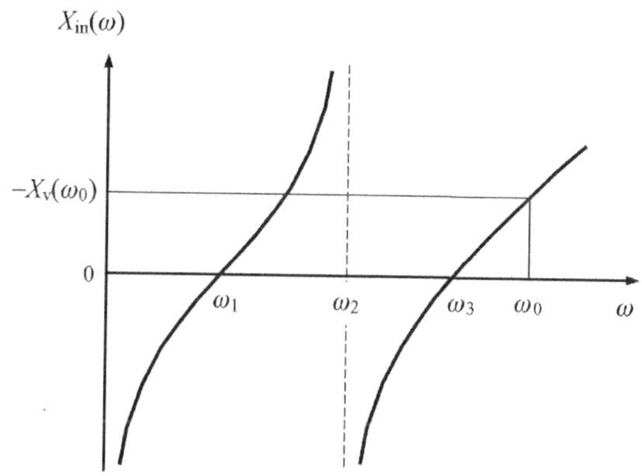

Figure 6.13 Reactance of a lossless network

at ω_1 and ω_3, the reactance X_{in} is written as

$$X_{in}(\omega) = K\frac{(\omega^2 - \omega_1^2)(\omega^2 - \omega_3^2)}{\omega(\omega^2 - \omega_2^2)} \tag{6.69}$$

The reactance due to the varactor junction capacitance can be written, using Equation (6.1) in the form

$$X_v = -\frac{(v_v + \varphi)^\gamma}{\omega C_{vo}\varphi^\gamma} \tag{6.70}$$

where the voltage v_v on the varactor consists of the dc bias voltage V_v and of the modulation voltage v_m.

The modulation voltage can be represented in the normalized form

$$v = \frac{v_m}{V_v + \varphi}. \tag{6.71}$$

For the initial conditions of $\omega = \omega_0$, v is equal to zero. This leads to

$$1 + v = \left[\frac{(\omega^2 - \omega_1^2)(\omega_0^2 - \omega_2^2)(\omega^2 - \omega_3^2)}{(\omega_0^2 - \omega_1^2)(\omega^2 - \omega_2^2)(\omega_0^2 - \omega_3^2)}\right]^{1/\gamma} \tag{6.72}$$

Introducing the frequency normalization in the form

$$\frac{\omega_i}{\omega_0} = \Omega_i \qquad \text{for } i = 1, 2, 3 \tag{6.73}$$

and taking into account that, in the vicinity of the operating point

$$\frac{\omega}{\omega_0} = 1 + \eta \tag{6.74}$$

where $\eta = (\omega - \omega_0)/\omega_0$ is a small number, the modulation curve can be rewritten as

$$1 + v = \left[\frac{(1+\eta)^2 - \Omega_1^2}{1 - \Omega_1^2}\,\frac{1 - \Omega_2^2}{(1+\eta)^2 - \Omega_2^2}\,\frac{(1+\eta)^2 - \Omega_3^2}{1 - \Omega_3^2}\right]^{1/\gamma} \tag{6.75}$$

When the linear lossless circuit between the varactor and the active device has more than three finite natural frequencies, Equation (6.75) can be appropriately expanded to include any additional zeros and poles.

The simplest two-port network between the varactor and the active device is a single series resonant circuit shown in Figure 6.14a. The modulation curve for such a resonant circuit can

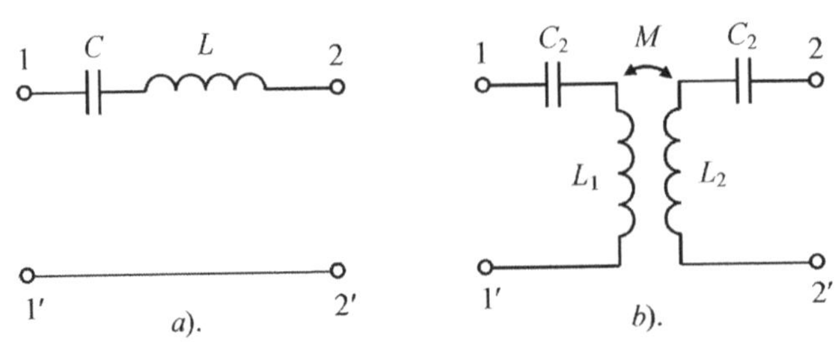

Figure 6.14 Schematics of two-port networks

be obtained from Equation (6.75) as

$$1 + v = \left[1 + A_1\eta + \frac{A_1}{2}\eta^2\right]^{1/\gamma} \tag{6.76}$$

where

$$A_1 = \frac{2}{1 - \Omega_1^2}$$

Since η is a small number, it is convenient to use the binomial expansion to obtain the first two terms of the Taylor series for voltage v

$$v = C_1\eta + C_2\eta^2 \tag{6.77}$$

where

$$C_1 = \frac{A_1}{\gamma}$$

$$C_2 = \frac{A_1}{2\gamma}\left[1 + \left(\frac{1}{\gamma} - 1\right)A_1\right]$$

The slope coefficient C_1 is the inverse of the normalized circuit sensitivity S_n equal to

$$S_n = \left.\frac{d\eta}{dv}\right|_{\omega_0} = S\frac{V_v + \varphi}{f_0} \tag{6.78}$$

where S is the slope of the modulation curve. As a result, for the series resonant circuit shown in Figure 6.14a, the normalized sensitivity must satisfy the condition

$$S_n = \frac{\gamma}{2}\left(1 - \Omega_1^2\right) \tag{6.79}$$

The modulation curve is linearized by setting $C_2 = 0$, resulting in

$$\gamma = \frac{2}{1 + \Omega_1^2} \tag{6.80}$$

From Equation (6.80) it follows that varactor junction sensitivity γ must be larger than unity because $\Omega_1 < 1$. Consequently, the only possibility of linearizing the modulation curve for a series resonant circuit is to use a hyperabrupt varactor. However, such a resonant circuit for a fixed operating point does not provide enough flexibility to accommodate small variations in γ.

The inductively coupled pair of lumped resonant circuits shown in Figure 6.14b offers more degrees of freedom to linearize the modulation curve, having the position of natural frequencies ω_1, ω_2, and ω_3 (when port 2–2′ is short-circuited) as shown in Figure 6.13. To achieve resonance with the varactor capacitance, reactance X_{in} must be positive with oscillation frequency ω_0 located above the zero ω_3.

The linearity of the modulation curve of the resonant circuit with two coupled resonators can be defined by expanding Equation (6.75) in a three-term Taylor series

$$v = C_1\eta + C_2\eta^2 + C_3\eta^3 \tag{6.81}$$

where

$$C_1 = \frac{1}{\gamma}(A_1 - A_2 + A_3)$$

$$C_2 = \frac{1}{2}\left[C_1 + C_1^2 - \frac{1}{\gamma}\left(A_1^2 - A_2^2 + A_3^2\right)\right]$$

$$C_3 = -\frac{C_1^3}{3} - \frac{C_1^2}{2} - \frac{C_1}{2} + C_2\left(\frac{1}{2} + C_1\right) + \frac{1}{3\gamma}\left(A_1^3 - A_2^3 + A_3^3\right)$$

where

$$A_i = \frac{2}{1 - \Omega_i^2} \qquad \text{for } i = 1, 2, 3$$

Since the normalized sensitivity S_{n0} at the centre frequency ω_0 can be written as $S_{n0} = 1/C_1$, and the linearity of the modulation curve requires $C_2 = C_3 = 0$, then the three nonlinear equations to be solved can be obtained as

$$A_1 - A_2 + A_3 = \frac{\gamma}{S_{n0}} \tag{6.82}$$

$$A_1^2 - A_2^2 + A_3^2 = \frac{\gamma}{S_{n0}}\left(1 + \frac{\gamma}{S_{n0}}\right) \tag{6.83}$$

$$A_1^3 - A_2^3 + A_3^3 = \frac{\gamma}{S_{n0}}\left[\frac{3}{2} + \frac{1}{S_{n0}}\left(\frac{3}{2} + \frac{1}{S_{n0}}\right)\right] \tag{6.84}$$

which should be solved numerically to find the optimum position of natural frequencies forcing elimination of both the second and third coefficients in Equation (6.81).

The calculated values Ω_1, Ω_2, and Ω_3 are substituted in Equation (6.76), and the relative frequency η is gradually varied so that normalized modulation curve $v(\eta)$ is evaluated [17]. The higher the chosen value of S_{n0}, the wider bandwidth within which the deviation from linearity stays within prescribed limits. For example, by using an abrupt varactor with $\gamma = 0.5$, for 1% deviation from linearity, the sensitivity $S_{n0} = 0.02$ results in the relative bandwidth of 0.0214, while the relative bandwidth is 0.0319 for $S_{n0} = 0.03$. To provide oscillation stability, the following approximate condition should be satisfied:

$$\Omega_2^2 \geq \Omega_r^2\left(1 + \frac{kQ_v\Omega_r}{2\Omega_v^2}\right) \tag{6.85}$$

where $\Omega_r = \omega_r/\omega_0$ is the normalized resonant frequency associated with the varactor side,

$$\omega_r = \sqrt{\frac{C_2 + C_v}{L_2 C_2 C_v}}$$

$\Omega_v = \omega_v/\omega_0$ is the normalized resonant frequency associated with the varactor only,

$$\omega_v = \frac{1}{\sqrt{L_2 C_v}}$$

$k^2 = M^2/L_1 L_2$ is the inductive coupling coefficient, and Q_v is the varactor quality factor.

Another approach to linearize the oscillator tuning performance is based on consideration of the resonant condition of $X_{in}(\omega) - 1/\omega C_v = 0$, defining the proper circuit reactance, choosing

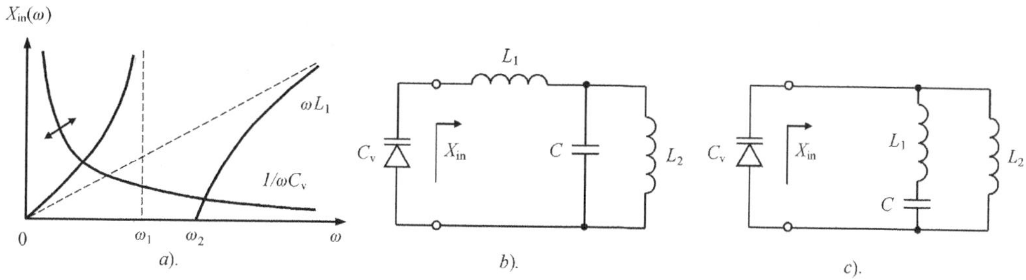

Figure 6.15 Double-resonant lossless circuits and their reactance

its structure depending on the varactor sensitivity with parametric optimization of resonant circuit elements [18]. By using Equation (6.1), the resonant condition can be rewritten as

$$[a\omega X_{\text{in}}(\omega)]^{1/\gamma} = V_{\text{v}} \tag{6.86}$$

where it is assumed that $V_{\text{v}} \gg \varphi$, and $a = \varphi^{\gamma} C_{\text{v0}}$.

To obtain linear dependence $\omega(V_{\text{v}})$, it is necessary to provide the inductive reactance X_{in} of the two-port network connected to the varactor, which can be represented as a function of frequency in the form

$$X_{\text{in}}(\omega) = b\frac{(\omega - \omega')^{\gamma}}{\omega} \tag{6.87}$$

where b and ω' are the constant coefficients, and $\omega > \omega'$. The main feature of such a reactance function is that, under the condition $\gamma > 0$, the derivative $\mathrm{d}X_{\text{in}}(\omega)/\mathrm{d}\omega$ reduces with increase of ω. Similar frequency behaviour shown in Figure 6.15a can be realized by a double-resonant circuit at frequencies $\omega > \omega_2$, where ω_2 is the series resonant frequency for the circuit shown in Figure 6.15b. Varactor C_{v} and capacitor C can be interchanged, resulting in the double-resonant circuit shown in Figure 6.15c.

The resonant condition for the circuit shown in Figure 6.15b can be obtained as

$$\frac{1}{C_{\text{v}}(\omega)} = \omega L_1 + \frac{\omega L_2}{1 - \omega^2 L_2 C} \tag{6.88}$$

By using the analytical varactor voltage–capacitance dependence, Equation (6.88) can be rewritten in the normalized form

$$\left(\frac{V_{\text{v}}}{V_{\text{vmin}}}\right)^{\gamma} = \Omega^2 + \frac{k_1 \Omega^2}{1 - k_1 k_2 \Omega^2} \tag{6.89}$$

where V_{vmin} is the varactor bias voltage corresponding to its maximum capacitance C_{vmax}, $\Omega = \omega/\omega_0$ is the normalized frequency, $\omega_0 = 1/\sqrt{L_1 C_{\text{vmax}}}$ is the minimum natural frequency of the circuit consisting of the varactor capacitance C_{v} and inductance L_1, $k_1 = L_2/L_1$, and $k_2 = C/C_{\text{vmax}}$.

If the voltage–capacitance characteristic of the varactor is determined experimentally and represented by the discrete sequence of the values of $C_{\text{v}n}$ ($n = 1, 2, \ldots, N$) corresponding to proper bias voltage intervals, Equation (6.88) can be also rewritten in normalized form

$$\frac{C_{\text{vmax}}}{C_{\text{v}n}} = \Omega^2 + \frac{k_1 \Omega^2}{1 - k_1 k_2 \Omega^2} \tag{6.90}$$

Table 6.1 Optimum parameters for varactor tuning linearity of a double-resonant circuit [18]

Varactor type	$k_{1\text{opt}}$	$k_{2\text{opt}}$	$K_f(\%)$	$\Delta(\%)$
$\gamma = 0.5$	0.0745	6.944	27	0.54
$\gamma = 1.05$	0.0678	6.99	15	1.48

with the same parameters on its right-hand side as those in Equation (6.89). The solution of these two equations gives the family of normalized tuning curves $\Omega(V_v/V_{vmin})$ depending on both the varactor sensitivity γ and parameters k_1 and k_2.

To determine the optimum values of $k_{1\text{opt}}$ and $k_{2\text{opt}}$ providing the required linearity of the tuning curve $\Omega(V_v/V_{vmin})$, initially the tuning curve obtained from Equation (6.89) is discretized by n points. Then, the minimax method is used to minimize the deviation of the discretized tuning curve $\Omega(n)$ from the linear one whose minimum and maximum values connect two boundary points of the tuning curve $\Omega(1)$ and $\Omega(N)$. The results of the parametrical optimization for the double-resonant circuit shown in Figure 6.15b with abrupt and hyperabrupt varactors are given in Table 6.1, where $K_f = \omega_{max}/\omega_{min}$ is the frequency tuning ratio, Δ is the linearity of the tuning curve. The approximated voltage–dependence characteristic of a silicon hyperabrupt varactor is shown in Figure 6.4.

The nonlinearity of the VCO tuning curve can be improved by using a series resonant circuit connected in parallel to a varactor [19]. Figure 6.16a shows the equivalent VCO schematic where C_{out} and L_{out} are the equivalent output capacitance and inductance of the active device represented by a negative resistance one-port network, C_v and L_v are the varactor capacitance and inductance, and C_0 and L_0 are the capacitance and inductance of the series resonant circuit. The resonant frequency of such a VCO circuit can be calculated from

$$f = \frac{1}{2\pi}\sqrt{\frac{a_1 + \sqrt{a_1^2 - 4(1 + a_3)a_2}}{2a_2}} \tag{6.91}$$

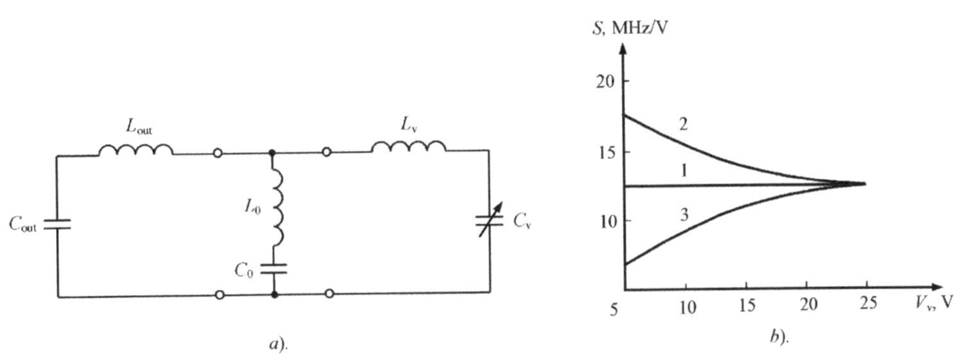

Figure 6.16 VCO resonant circuit with additional series circuit

where

$$a_1 = L_0C_0 + L_{out}C_{out} + a_3\left[L_vC_v + \frac{C_0C_{out}(L_0 + L_{out})}{C_0 + C_{out}}\right]$$

$$a_2 = L_0L_{out}C_0C_{out} + L_{out}C_{out} + a_3\frac{C_0C_{out}C_v(L_0 + L_{out})L_v}{C_0 + C_{out}}$$

$$a_3 = \frac{C_0 + C_{out}}{C_v}.$$

Figure 6.16b shows the dependence of the tuning curve slope $S = df/dV_v$ of the VCO resonant circuit with an abrupt varactor, $L_v = 1$ nH, $C_{out} = 0.575$ pF and $L_{out} = 2.17$ nH for different values of L_0 and C_0. The minimum bandwidth frequency of 4.2 GHz is realized at the varactor bias voltage of 5 V. From Figure 6.16b it follows that there is an optimum set of the parameters $L_0 = 1.75$ nH and $C_0 = 1.5$ pF (curve 1), at which the slope of the tuning curve S is constant over the entire tuning bandwidth. A negative slope of the tuning curve is realized at smaller values of the inductance L_0 (curve 2, $L_0 = 1$ nH, $C_0 = 1.5$ pF), whereas smaller values of the capacitance C_0 create a positive slope of the tuning curve (curve 3, $L_0 = 1.75$ nH, $C_0 = 1$ pF).

In addition, for each value C_0 there is an optimum value L_{0opt}, at which the slope of the tuning curve is approximately constant, and it can be obtained from the empirical approximated dependence of

$$L_{0opt} = \frac{L_{out}L_v}{L_{out} + L_v}\left(1 + \frac{C_{out}}{C_0}C_{out} + \frac{C_{v0} + C_{vmin}}{2C_0}\right) \tag{6.92}$$

where C_{v0} is the varactor capacitance at $V_v = 0$ and maximum bandwidth frequency corresponds to the minimum varactor capacitance C_{vmin}. It should be noted that inclusion of the additional series resonant circuit leads to reduction of the VCO resonant frequency. In order to keep the same resonant frequency, it is necessary to slightly change the output reactances of the active device C_{out} and L_{out} by changing the parameters of the feedback elements connected to the device terminals.

The increase in a VCO linear frequency tuning bandwidth can be achieved by using some combination of the capacitors and varactors used in the oscillator resonant circuit [20]. Figure 6.17 shows the schematic of a common gate MESFET oscillator with two varactor pairs, anti-series and parallel. The capacitors C_3 and C_4 provide the required feedback conditions corresponding to a Colpitts-type oscillator. The frequency-determining circuit includes the series capacitor C_1, variable capacitor C_2 and shunt inductor L. Such a tuning resonant circuit can provide almost constant slope of the frequency tuning curve with variations of about 15% over varactor bias range 3–8 V. When capacitor C_2 is varied from its maximum to its minimum value, the slope of the oscillator tuning curve can be changed from 1.5 to 1.1 MHz/V.

6.5.2 VCOs with transmission lines

For VCOs based on the microstrip technique, the simple approach to linearize the oscillator tuning curve is to use sufficiently low quality factor of the oscillator resonant circuit and a separate modulation linear circuit [21]. Figure 6.18a shows a schematic of a 1.8-GHz microstrip common collector VCO with linear modulation circuit, where the feedback capacitor C is selected to provide negative resistance in as large a frequency bandwidth as possible. The

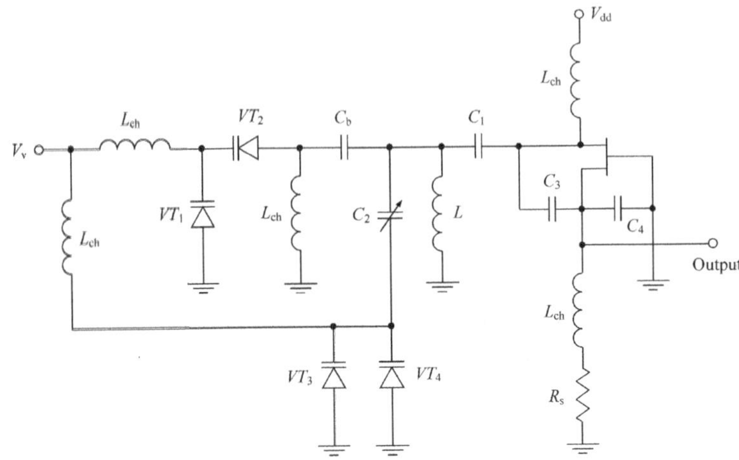

Figure 6.17 Schematic of a MESFET varactor-tuned oscillator [20]

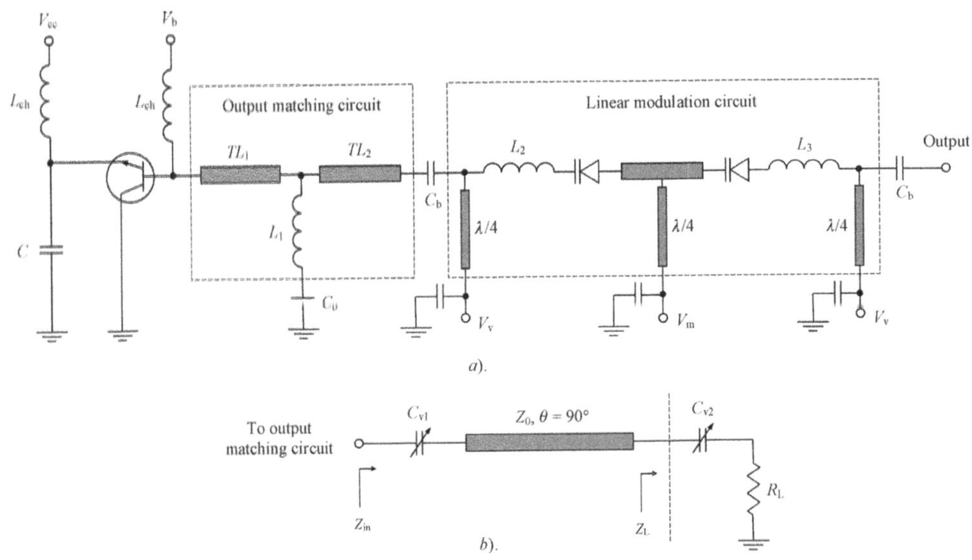

Figure 6.18 Schematic of 1.8 GHz bipolar VCO with linear modulation curve (© 1978 IEEE)

oscillator loaded quality factor was measured as 6, resulting in a modulation sensitivity of 16.5 MHz/V, or peak-to-peak deviation of 33 MHz for a modulation input of ± 1 V. The output matching circuit was designed to place the oscillator at its centre frequency.

The linear modulation circuit is composed of two varactors separated by a quarter-wave transmission line and connected in series with oscillator output port. To present the correct impedance for the oscillator, each of the varactors is resonated with corresponding series inductance. The values of the varactors must be selected in such a way that when the modulation voltage is applied they will present a reactive change large enough to pull the oscillation frequency. Figure 6.18b shows the simplified schematic of the linear modulation circuit which is resonant at some frequency where $\mathrm{Im}Z_{\mathrm{in}} = 0$. Hence, the equation to define the resonant frequency can be written as

$$\frac{1}{\omega C_{\mathrm{v2}}} - Z_0 \frac{\left(1 - \frac{1}{Z_0\omega C_{\mathrm{v1}}}\cot\theta\right)\left(\cot\theta + \frac{1}{Z_0\omega C_1}\right) - \left(\frac{R_{\mathrm{L}}}{Z_0}\right)^2 \cot\theta}{\left(\cot\theta + \frac{1}{Z_0\omega C_{\mathrm{v1}}}\right)^2 + \left(\frac{R_{\mathrm{L}}}{Z_0}\right)^2} = 0 \tag{6.93}$$

At operating frequencies within $\pm\Delta\omega$ of the centre bandwidth frequency ω_0, it is possible to make the approximation

$$\cot\theta = \cot\left(\frac{\pi}{2}\frac{\omega}{\omega_0}\right) = \cot\left(\frac{\pi}{2} \pm \frac{\pi}{2}\frac{\Delta\omega}{\omega_0}\right) \cong \mp\frac{\pi}{2}\frac{\Delta\omega}{\omega_0} \tag{6.94}$$

Then, by substituting Equation (6.94) and Equation (6.1) for each varactor voltage–capacitance characteristic with capacitances $C_{\mathrm{v01}} = C_{\mathrm{v02}}$ and voltages $v_1 = V_{\mathrm{v}} + v_{\mathrm{m}}$ and $v_2 = V_{\mathrm{v}} - v_{\mathrm{m}}$, where v_{m} is the modulation voltage, into Equation (6.93), we can find numerically a solution for $\pm\Delta\omega$ as a function of v_{m}. The linear modulation curve can be obtained in the vicinity of the centre bandwidth frequency. For example, for varactor sensitivity $\gamma = 1/3$ and reverse bias voltage $V_{\mathrm{v}} = 4$ V, linearity of the modulation curve less than 1% for a deviation of ± 20 MHz was achieved with no external frequency compensation [21].

Using a transmission line with low characteristic impedance between the resonant circuit and the varactor can also linearize the frequency tuning characteristic of the microwave VCO providing less than 1% deviation from linear curve over ± 3.5 MHz from the carrier frequency ranging from 2.3 to 2.43 GHz with modulation slope of approximately 2.9 MHz/V [22]. The effective varactor sensitivity can be increased by using an external inductor connected in parallel to the varactor composing an additional parallel resonant circuit. Figure 6.19 shows the schematic of a 9.5 GHz MESFET transmission-line oscillator with the resonator between the active device and the varactor [23]. In order to provide the high quality factor of the oscillator resonant circuit, the resonator is connected to the active device and the varactor through small capacitances. The loaded Q-factor was estimated as large as 50 or more, resulting in a phase noise level better than -100 dB/Hz at 100 kHz offset across the control voltage with capacitors

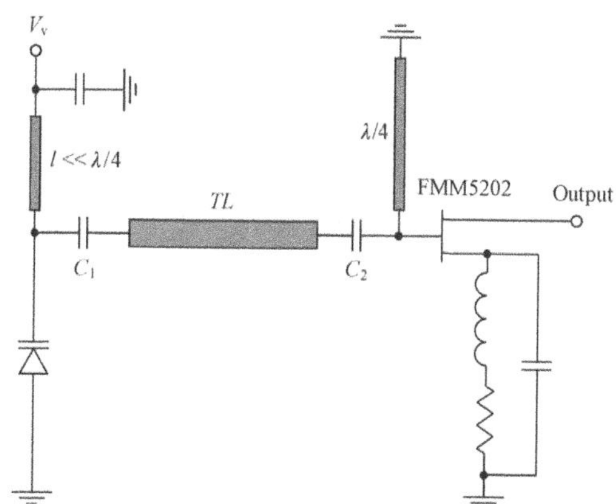

Figure 6.19 Basic schematic of a MESFET transmission-line VCO (permission by IEICE)

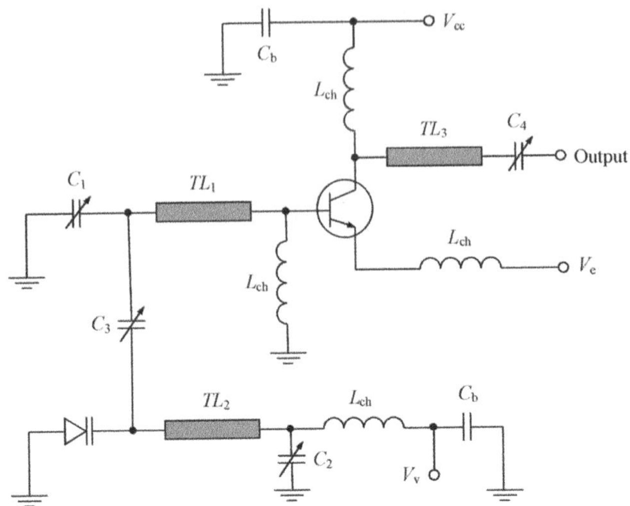

Figure 6.20 Schematic of a bipolar VCO with linear frequency tuning [25]

$C_1 = 0.34$ pF and $C_2 = 0.26$ pF. To linearize the oscillator tuning curve, the short-length bias feed microstrip line with high characteristic impedance is used in parallel to a hyperabrupt varactor MA46452-134, contributing to an increase of the nominal varactor sensitivity from $\gamma = 0.75$ to its equivalent value $\gamma = 1.5$ in a wide control range. As a result, a constant tuning sensitivity was measured as 61 MHz/V with associated modulation linearity as good as 1.2% between 5 and 7 V of control bias voltage V_v. For a miniature integrated monolithic application using similar VCO structure, matching the capacitance slope of the varactor to the reactance slope of the remaining network resulted in a slope variation of less than 1.6 to 1 over tuning bandwidth from 6.4 to 7.3 GHz [24].

Figure 6.20 shows a microstrip common base oscillator intended for frequency modulation where the linearity of the modulating circuit is adjusted by using a microstrip line as a secondary resonator TL_2 having an electrical length of $\lambda/4$ in conjunction with a variable capacitor C_2 [25]. The varactor is connected to one end of this resonator, which end is also connected to the main resonator TL_1 through a variable capacitor C_3. The oscillation frequency is mainly determined by a microstrip main resonator TL_1 which is connected to the base terminal of the active device and terminated by variable capacitor C_1. The modulation circuit consists of secondary microstrip resonator TL_2, terminated on one side by a varactor and on the other side by capacitor C_2 which can also be implemented as a varactor. The advantage of such a circuit configuration is that it provides independence in adjustment of operating frequency and tuning linearity.

6.6 REACTANCE COMPENSATION TECHNIQUE

For the first time, to increase the frequency tuning bandwidth of a negative resistance oscillator, a reactance compensation technique using a single-resonant circuit was applied to a varactor-tuned diode oscillator [26]. To describe such a reactance compensation technique, consider the simplified equivalent load network with the series L_0C_0 and shunt L_1C_1 resonant circuits

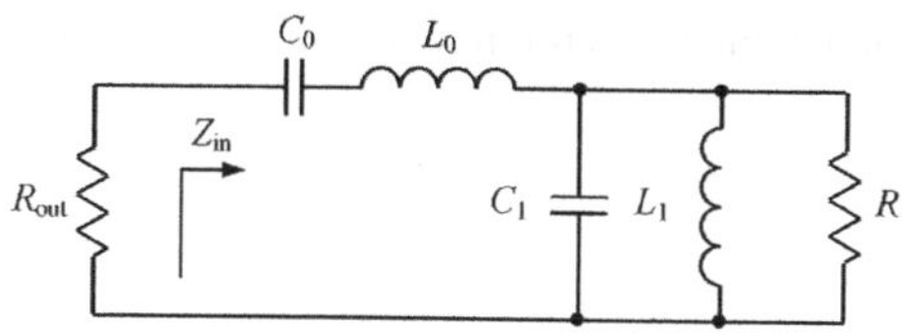

Figure 6.21 Single reactance compensation circuit

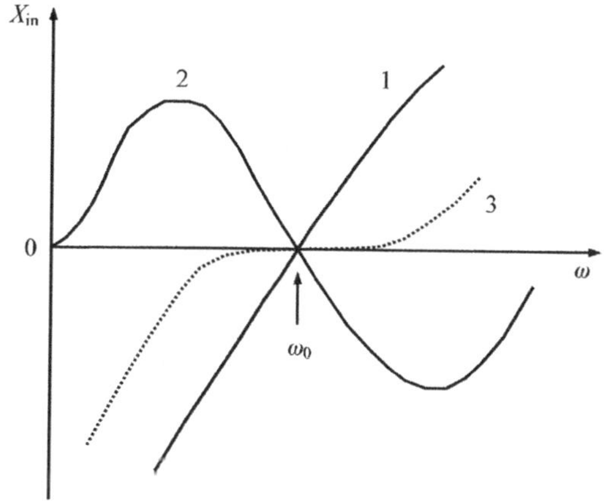

Figure 6.22 Reactance compensation principle

both tuned to the centre bandwidth frequency, as shown in Figure 6.21, where R_{out} is the oscillator output negative resistance. The reactances of the series and shunt resonant circuits vary with frequency as shown in Figure 6.22 by curve 1 and curve 2, respectively. Near the resonant frequency ω_0, the reactance of the series circuit has positive slope, whereas the slope of the shunt circuit reactance is negative, thus reducing the overall reactance slope of the load network. By proper choice of inductance L_1 and capacitance C_1, the overall reactance slope can be reduced to zero, as shown by curve 3 (dotted line).

The load network input impedance Z_{in} can be written as

$$Z_{in} = \left(j\omega L_0 + \frac{1}{j\omega C_0} + \frac{R}{1 + j\omega' C_1 R} \right) \tag{6.95}$$

where $\omega' = \omega \left(1 - \frac{\omega_0^2}{\omega^2} \right)$, $\omega_0 = 1/\sqrt{L_1 C_1}$ is the shunt circuit resonant frequency.

The frequency bandwidth will be maximized if, at resonant frequency ω_0,

$$\left. \frac{\mathrm{d}X_{in}(\omega)}{\mathrm{d}\omega} \right|_{\omega=\omega_0} = 0 \tag{6.96}$$

where $X_{in} = \mathrm{Im}Z_{in}$ is the load network reactance. Consequently, an additional equation can be obtained

$$L_0 + \frac{1}{\omega_0^2 C_0} - 2C_1 R = 0 \tag{6.97}$$

As a result, the shunt capacitance C_1 and inductance L_1 can be calculated from

$$C_1 = \frac{L_0}{R^2} \tag{6.98}$$

$$L_1 = R^2 C_0. \tag{6.99}$$

Such a reactance compensation technique can be used as a technique to linearize the frequency tuning characteristic of the negative resistance oscillator [27]. For this purpose, let us rewrite Equations (6.98) and (6.99) as

$$C_1 = \frac{L_0}{AR^2} \tag{6.100}$$

$$L_1 = AR^2 C_0 \tag{6.101}$$

where A is a constant. The case of $A = 1$ corresponds to a complete compensated overall reactance with its zero slope. However, for certain values of A, a reasonably high degree of linearity can be obtained over tuning ranges comparable to or more than the overall tuning bandwidth of the uncompensated VCO. For example, for values of A from 1.4 to 1.6, all of the tuning bandwidths are wider, having linearity in the range 3.8–8.4% or maximum slope ratio of 1.5. Minimum slope ratio of 1.08 corresponds to the optimum case of $A = 1.55$, providing a linear tuning bandwidth of 1.7 GHz for an X-band negative resistance oscillator.

Further widening of the oscillator frequency bandwidth can be achieved using the double reactance compensation circuit shown in Figure 6.23a, where L_1C_1 and L_2C_2 are the shunt and series resonant circuits, respectively [28]. Similarly, the double susceptance compensation circuit shown in Figure 6.23b can be used to widen the frequency tuning bandwidth having two series and one shunt resonant circuits, where $G = 1/R$ and $G_{out} = 1/R_{out}$. To determine the oscillator circuit parameters, a system of two equations should be solved by making the

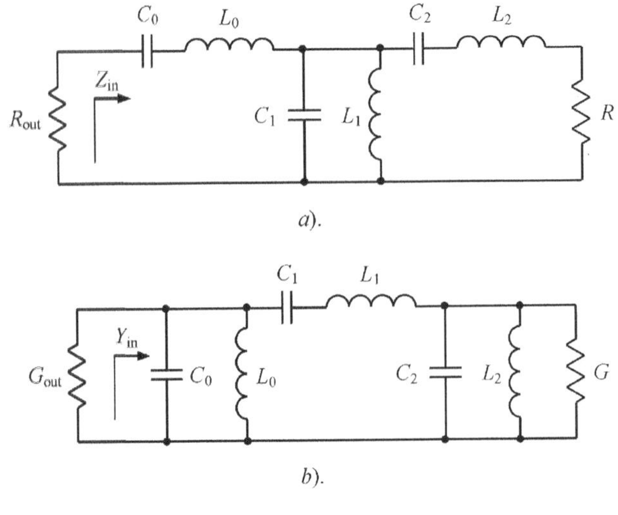

Figure 6.23 Double reactance and susceptance compensation circuits

first- and third-order derivatives zero at the centre bandwidth frequency ω_0 according to

$$\left.\frac{dX_{in}}{d\omega}\right|_{\omega=\omega_0} = \left.\frac{d^3X_{in}}{d\omega^3}\right|_{\omega=\omega_0} = 0 \qquad (6.102)$$

since the second derivative cannot provide an appropriate analytical expression.

The input reactance of the oscillator circuit with double reactance compensation can be expressed as

$$X_{in}(\omega') = \omega' L_0 + \omega' \frac{L_2\left[1-(\omega')^2 L_2 C_1\right] - C_1 R^2}{\left[1-(\omega')^2 L_2 C_1\right]^2 + (\omega' C_1 R)^2} \qquad (6.103)$$

resulting into the two following equations by applying zero derivative conditions given by Equation (6.102):

$$L_0 + L_2 - C_1 R^2 = 0 \qquad (6.104)$$

$$L_2^2 + (C_1 R^2 - 2L_2)(L_2 - C_1 R^2) = 0 \qquad (6.105)$$

Thus, the parameters of the series and shunt resonant circuits corresponding both to reactance and susceptance compensation techniques can be calculated from

$$L_1 = C_0 \frac{2R^2}{\sqrt{5}+1} \qquad C_1 = L_0 \frac{\sqrt{5}+1}{2R^2} \qquad (6.106)$$

$$L_2 = L_0 \frac{\sqrt{5}-1}{2} \qquad C_2 = C_0 \frac{2}{\sqrt{5}-1} \qquad (6.107)$$

Table 6.2 shows the frequency tuning bandwidths of the diode negative resistance oscillator for various types of reactance compensation and different load resistances. The series variable capacitance C_0 varying from 0.12 to 0.22 pF was used. As a result, for the circuit having a 50-Ω load, theoretically an improvement of 4% in tuning range can be achieved using double reactance compensation whereas, for the circuit operating into 100-Ω load, the increase in tuning range is 17%.

Using triple reactance compensation with two shunt and one series resonant circuits, each tuned to the fundamental frequency, enables one to further increase the linear tuning range. For example, a linear frequency bandwidth of 3 GHz for an X-band VCO with slope ratio 1.3 instead of the linear frequency bandwidth of 2.4 GHz with slope ratio of 1.5 can be obtained [29].

Table 6.2 Tuning bandwidths for various types of reactance compensation [28]

Compensation type	Load resistance (Ω)	Low bandwidth frequency (GHz)	High bandwidth frequency (GHz)	Bandwidth (GHz)
None	50	8.1	11.0	2.9
Single	50	7.0	12.2	5.2
Double	50	6.9	12.3	5.4
Single	100	6.2	13.4	7.2
Double	100	6.8	14.2	8.4

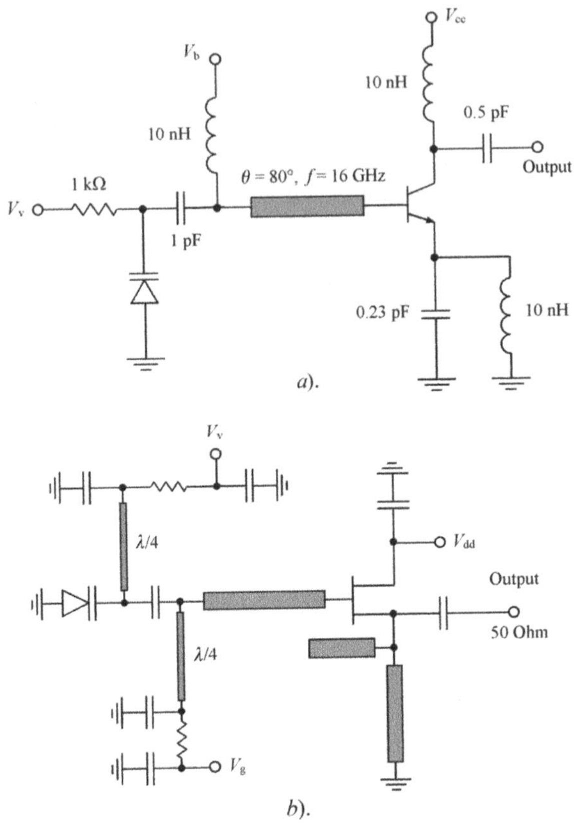

Figure 6.24 Circuit schematics of microstrip VCOs [30, 32]

6.7 PRACTICAL VCO SCHEMATICS

6.7.1 VCO implementation techniques

Depending on the operating frequency and application requirements, there is a variety of VCO implementation techniques based on using different types of active devices, circuit schematic approaches and hybrid or monolithic integrated circuit technologies. For example, their low-cost implementation and low-phase-noise performance are required in wireless communication systems. At microwave frequencies, most of these VCOs use MESFET or heterojunction bipolar devices (HBT) in order to produce lower phase noise than VCOs based on high electron mobility transistors (HEMT). However, high-performance HEMT oscillators are essential if they are to be integrated together with amplifiers and mixers for single-chip receivers or transmitters using the same technology. Figure 6.24a shows the 15-GHz monolithic common base AlGaAs/GaAs HBT VCO with capacitive feedback in the emitter circuit where the double collector layer was used for both active device and varactor [30]. The measured transition frequency f_T and maximum oscillation frequency f_{max} were about 85 and 55 GHz, respectively. The value of f_{max} was lower in the high current density region because of the high base contact resistivity realized. For a varactor capacitance ratio of about 1.72 when the tuning voltage is changed from 0 to 4 V, a frequency tuning range of about 600 MHz with output power more than −4 dBm was obtained. The phase noise level measured at 15.6 GHz was of −85 and −110 dBc/Hz at offset frequencies of 100 kHz and 1 MHz, respectively. The oscillation frequency changes almost linearly as tuning voltage varies between 0 and 2 V. Using SiGe HBT technology providing $f_T = 200$ GHz enables oscillator operation at very high frequencies. For example, based on similar common base oscillator topology with the series connected coplanar transmission line and varactor, the frequency tuning from 95.2 to 98.4 GHz with maximum phase noise of −85 dBc/Hz at 1 MHz offset was realized [31].

The circuit schematic of a fully monolithic microstrip VCO using a combined pseudomorphic HEMT (pHEMT) device and Schottky varactor diode process is shown in Figure 6.24b [32]. To simulate the circuit performance using commercially available circuit simulators, the varactor diode with zero voltage capacitance $C_{v0} = 0.505$ pF and series resistance $R_s = 1.3\ \Omega$ was modelled as an abrupt Schottky diode with $\gamma = 0.5$ and $\varphi = 0.8V$. The Curtice cubic MESFET model was used for the HEMT device to describe its nonlinear behaviour [33]. The VCO circuits with 0.25 μm HEMT devices connected in a common drain configuration were optimized for maximum and flat output power throughout the frequency tuning bandwidth. Open-circuited and short-circuited stubs connected to the source terminal provide output matching as well as second harmonic suppression. As a result, the VCO with a single quarter-wave microstrip line enables varactor tuning in the frequency bandwidth 21.2–22.3 GHz with output power of about 15 dBm and variation of the phase noise at 100 kHz offset from −68.8 to −76.1 dBc/Hz. By using a 0.15 μm *p*HEMT process with $f_T = 110$ GHz and $f_{max} = 180$ GHz, the varactor frequency tuning of 3.8 GHz with output power more than 10 dBm and phase noise −68 dBc/Hz at 100 kHz offset and −102 dBc/Hz at 1 MHz offset at centre bandwidth frequency of 28.3 GHz was achieved [34].

However, using the open-ended coupled quarter-wave microstrip lines, as shown in Figure 6.25a, has contributed to a slightly higher frequency bandwidth of 20.4–22.0 GHz with output power 12 dBm and variation of the phase noise at 100 kHz offset from −71.3 to −80.3 dBc/Hz, which is lower by 3 dB on average than that of the VCO with a single quarter-wave microstrip line [32]. Further increase in a VCO frequency tuning range can be achieved by using the resonator in the form of short-circuited coupled microstrip lines having inductive reactances at resonant frequencies, as shown in Figure 6.25b [35]. For a resonator designed with characteristic impedance 70 Ω and coupling coefficient 8 dB at a frequency of 5 GHz, the width of the microstrip lines is about 0.02 inches and the spacing is about 0.0038 inches when fabricated on a 0.02-inch-thick substrate of dielectric constant equal to 2.94. As a result, frequency tuning from 4.9 to 6.3 GHz with phase noise less than −105 dBc/Hz at 100 kHz offset from the carrier was achieved.

For a portable wireless communication system, the power consumption of its components should be minimized. Therefore, the design of a highly efficient and high-performance VCO in such applications is of significant importance. Figure 6.26 shows the circuit schematic of a Class E common gate MESFET VCO MMIC [36]. To provide the close approximation to a Class E operation mode, the series inductance, including bondwire inductance L_{bond} and output equivalent capacitance C_{out}, should be chosen according to Equations (4.101) and (4.102) respectively, corresponding to idealized Class E load network parameters for a load resistance $R = 50\ \Omega$. It should be noted that a finite value of the choke inductance can affect the exact values of the load network parameters as well as output performance [37]. To make the active device unstable with negative resistance seen from the gate terminal, the feedback capacitance of 1.3 pF is connected to the source terminal. To minimize MMIC chip size using

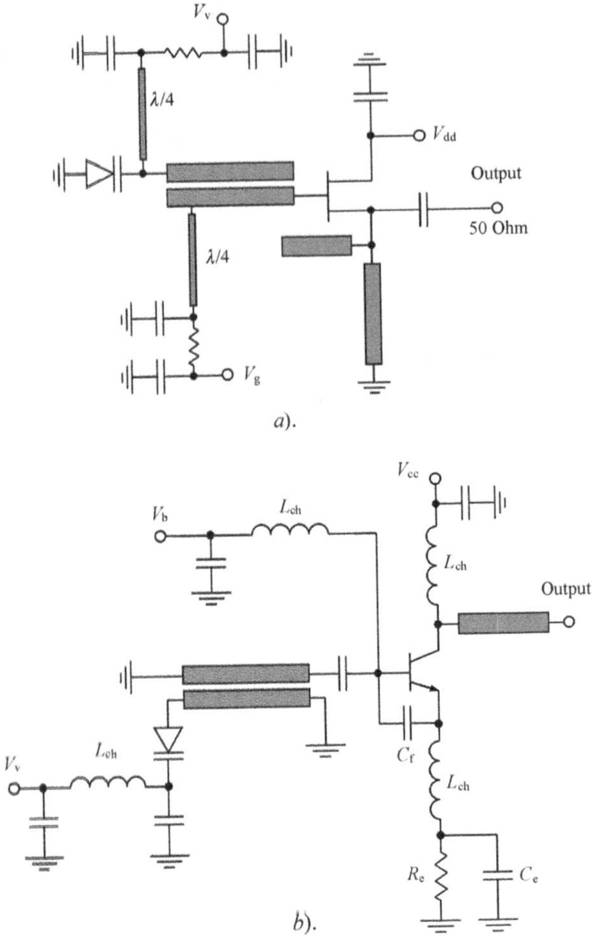

Figure 6.25 Circuit schematics of VCOs with coupled transmission lines

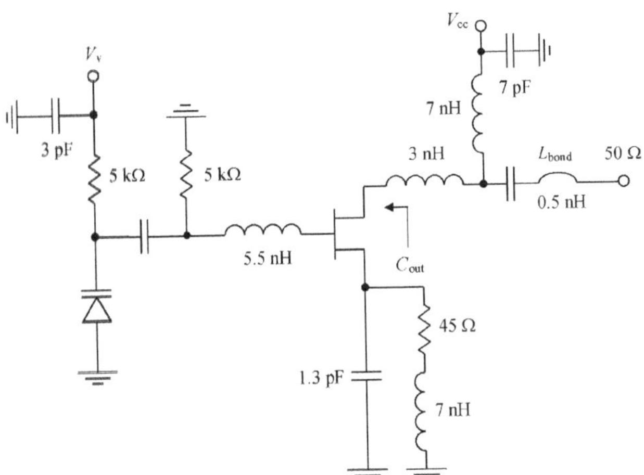

Figure 6.26 Circuit schematic of a Class E MESFET VCO (© 2001 IEEE)

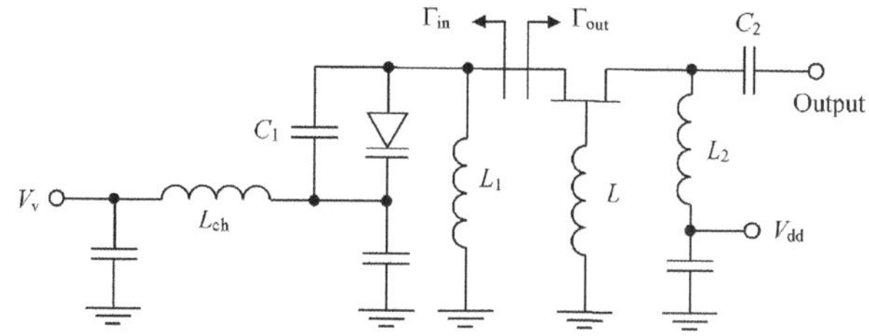

Figure 6.27 Circuit schematic of a common gate MESFET VCO

a 0.6-μm standard GaAs MESFET process, a relatively small dc-feed inductance of 7 nH and no additional harmonic filtering at the output is used. The *LC* resonator at the gate consists of an inductor of 5.5 nH with a quality factor of 20 at 4 GHz and a varactor implemented using a depletion diode-connected FET. The periphery of the varactor is 1000 μm to provide a good trade-off between size, quality factor and VCO tuning range. As a result, an efficiency of 43% with output power of 6.5 dBm and tuning range of 150 MHz was achieved at centre frequency 4.4 GHz and supply voltage 1.8 V. At a supply voltage of 1.2 V, a phase noise of −132 dBc/Hz at 5 MHz offset and suppression of the second harmonic of −16 dBc were measured.

The common gate VCO configuration is usually used to generate strong negative resistance over a wide frequency range. However, due to the series resistance of the varactor, the phase noise of such a VCO is relatively higher than that of a single frequency oscillator. Nevertheless, the phase noise can be reduced by placing the varactor into the source rather than into the gate circuit. Figure 6.27 shows the circuit schematic of a common gate MESFET VCO MMIC where an output L_2C_2 matching network is incorporated in the form of a high-pass section to eliminate low-frequency parasitic oscillations [38]. Based on an analysis of the trajectories of the reflection coefficient lines of the resonator Γ_{in} and the device Γ_{out} as a function of drain supply voltage, the oscillator loaded quality factor Q_L was maximized. As the drain voltage increases, the angle between these trajectories approaches 90° where Q_L becomes maximal [39]. Increasing the drain voltage also reduces the phase noise by extending the depletion region in the device channel to the drain side, thus reducing the sensitivity of the oscillator to the gate–source voltage. Being implemented in a commercial 0.6-μm GaAs MESFET process, such a VCO demonstrates the phase noise of −91 dBc/Hz at 100 kHz offset with output power of 11.5 dBm and frequency tuning of 500 MHz around the centre frequency of 11.5 GHz.

Figure 6.28 shows the basic circuit schematic of the VCO using the two source-coupled GaAs JFET devices CFY30 and the varactor BBY52 [40]. The shunt capacitor C_d connected in parallel to the resistor with a value of 470 Ω is necessary to prevent parasitic oscillations at high frequencies and its value is about twice the value of the feedback capacitor C_f. For frequencies higher than 2 GHz, the input capacitive reactance of the common drain transistor should be taken into account. Therefore, the tank circuit consists of a series connection of two inductors L_1 and L_2. The varactor diode is biased via a potentiometer through a 1-MΩ resistor, so that tuning voltage can be varied between zero and the supply voltage values. The value of the capacitor C determines the degree of coupling between the varactor and resonant circuit, i.e., the frequency tuning range. Its higher value will allow a larger tuning variation, but will pull down the centre bandwidth frequency. For lower-frequency oscillators, to increase the tuning bandwidth, it is necessary to connect two or four varactors in parallel. Some typical element values for the specific frequency bandwidths are given in Table 6.3.

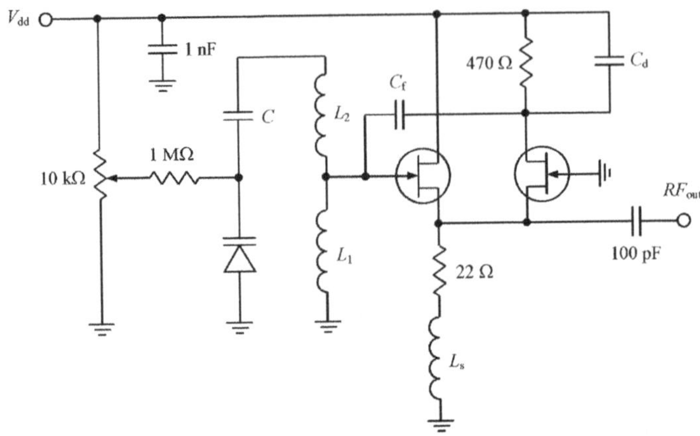

Figure 6.28 Circuit schematic of a source-coupled JFET VCO [40]

Figure 6.29a shows the circuit schematic of the bipolar VCO structure with common collector and common emitter transistor configurations intended for use in commercial applications in the 900 MHz frequency band [41]. The VCO circuit was built on a multilayer printed circuit board (PCB) consisting of four equally spaced metallized layers where the circuit elements are mounted on the top metallized layer, while the second and the fourth metallized layers serve as ground. This enables one to locate the bias circuit transmission lines TL_1 and TL_2 and resonant circuit transmission line TL_3 in the third metallized layer using stripline technology. One of the key features of such a VCO is the sharing by two transistors of the same dc current, which minimizes its power consumption. The feedback emitter capacitance C_e is necessary to make the bottom transistor in the common collector configuration unstable, resulting in a negative resistance seen from the resonant circuit.

To calculate the input impedance Z_{in} between the base and emitter terminals of this transistor seen from the resonant circuit, it is advisable to transform the simplified bipolar transistor equivalent circuit shown in Figure 6.29b to the equivalent circuit shown in Figure 6.29c. Subscript 2 is added to the equivalent circuit elements corresponding to the top transistor in a common base configuration. As a result, with capacitance C_3 set to zero, the input impedance

Table 6.3 Typical element values for source-coupled JFET VCOs [40]

Frequency	C(pF)	L_1(nH)	L_2(nH)	L_s(nH)	C_f(pF)	C_d(pF)	V_{dd}(V)	P_{out}(dBm)
3.2 GHz	2.2	1.35	2.7	8.2	0.25		3.0	6.4
2.45 GHz	3.3	2.7	2.7	8.2	0.25		2.5	6.7
920 GHz	1.8	10		18	0.5		2.0	9.9
870 GHz	2.7	10		18	0.5		2.0	10.1
430 GHz	3.3	33		39	1.0	2.2	2.0	10.7
230 GHz	6.8	68		68	2.2	5.6	2.0	9.8
145 GHz	12	136		136	4.7	10	2.0	10.3
50 GHz	56	470		470	22	47	2.0	10.0
20 GHz	82	940		940	56	100	2.0	8.7

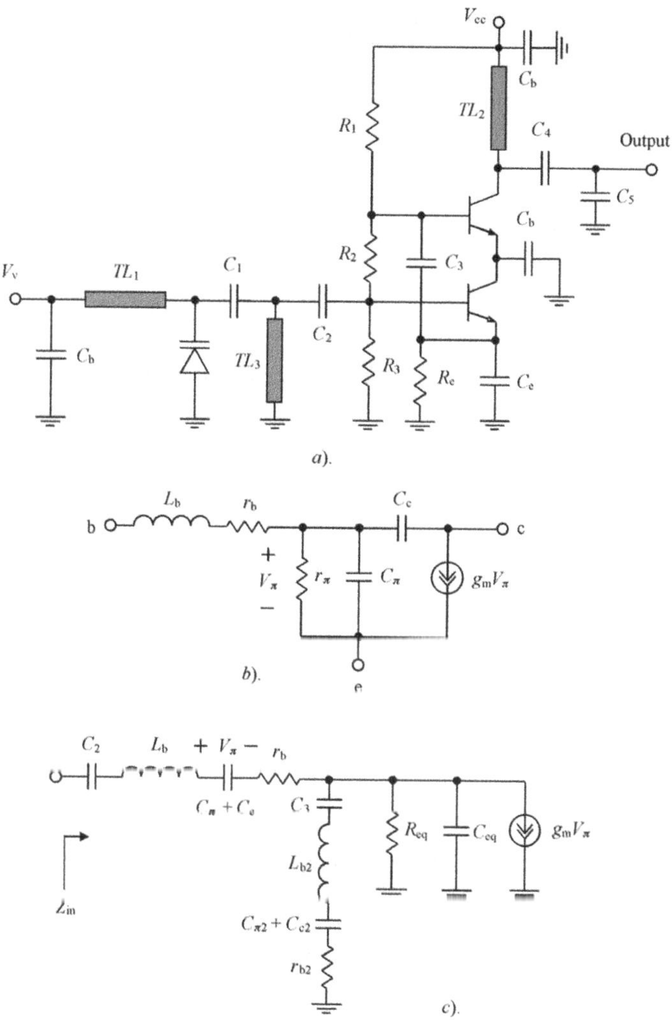

Figure 6.29 Circuit schematics of a dual transistor bipolar VCO (© 1998 IEEE)

Z_{in} can be written as

$$Z_{in} \approx r_b + j\omega L_b + \frac{C_c + C_2}{j\omega C_c C_2} + \left(1 + \frac{g_m}{j\omega C_\pi}\right)\frac{R_{eq}}{1 + j\omega C_{eq} R_{eq}} \tag{6.108}$$

where $R_{eq} = R_e/(1 + \omega_T C_e R_e)$, $C_{eq} = C_e + C_c$ and $\omega_T = g_m/C_\pi$ [41]. In this case, the practical maximum oscillation frequency f_{pmax}, including some margin required to cancel out the resistance from the resonant circuit, can be introduced in the form

$$f_{pmax} = \frac{f_{max}}{\sqrt{2}}\sqrt{\frac{R_e}{R_e + \dfrac{1}{\omega_T C_c}}} \tag{6.109}$$

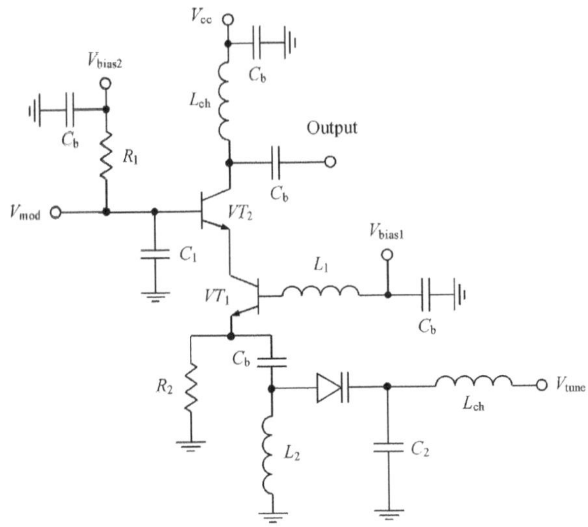

Figure 6.30 Circuit schematic of a cascode bipolar VCO [42]

from which it follows that, to maximize the oscillation frequency for the particular bipolar transistor, it is necessary to choose a sufficiently high value of the emitter resistance R_e.

To provide higher isolation of the load from the VCO resonant circuit and separate modulation and oscillation paths, a cascode VCO configuration, shown in Figure 6.30, can be used [42]. The negative resistance oscillation conditions for a common emitter transistor VT_1 are provided by using the feedback inductance L_1. The tank circuit in the emitter circuit of the transistor VT_1, which includes the parallel inductance L_2, varactor and shunt capacitance C_2, determines the resonant frequency of the oscillator. The modulating signal is applied to the base of the common base transistor VT_2 where a value of the capacitance C_1 is chosen to have high reactance at the modulating frequencies and low reactance at the oscillation frequency. The modulating signal voltage is in turn also applied to the collector–base abrupt junction capacitance of the transistor VT_1 through the base–emitter junction of the transistor VT_2, thus altering this capacitance, resulting in the frequency-modulated signal at the oscillator output.

6.7.2 Differential VCOs

The differential VCOs represent a particular realization of *LC* oscillators with a negative differential resistance based on a differential amplifier schematic, in which two cross-coupled transistors have a common emitter or source circuit and are biased by a tail current source. Such a VCO differential architecture inherently provides low even-order harmonics and has proved to be very useful in low-power and low-noise applications and very convenient for RFIC integration [43, 44]. The main operating principles can be conveniently described by using the general circuit schematic with one output terminal shown in Figure 6.31a. The resonant circuit is connected to the base terminal of the grounded collector transistor. The common base configuration is provided for the second device. The nonlinearity of the device capacitances can be compensated. The collector junction capacitances C_{c1} and C_{c2} are connected in parallel when the voltages across them have opposite phases. At the same time, the emitter–base capacitances

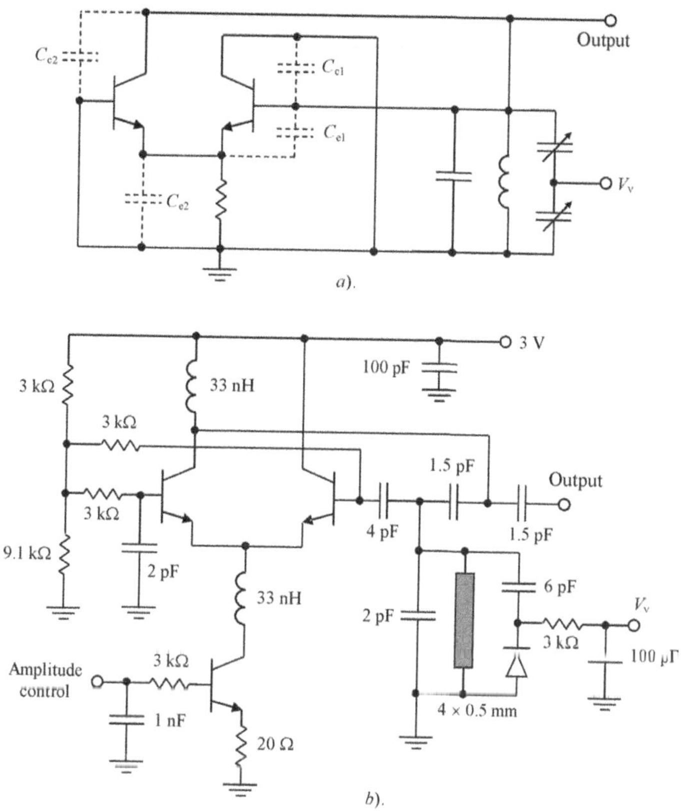

Figure 6.31 Circuit schematics of differential VCOs [43, 44]

C_{e1} and C_{e2} are in series with back-to-back connection, which allows cancellation of even current harmonic components. For differential configuration, the parallel resonant circuit is loaded by the input base-to-base impedance of two back-to-back base–emitter junctions, which is two times higher than the value for one transistor. To minimize its influence on the oscillation frequency, it is best to choose devices with maximum transition frequency.

Figure 6.31b shows a practical example of a VCO circuit designed for wireless application using a printed circuit board (PCB) technique [44]. The feedback capacitances in the base circuits of both bipolar transistors NE68519 are chosen to provide minimum flatness of output power over wide tuning bandwidth. The inductances of 33 nH are used as RF chokes. The 20-Ω resistance in the emitter circuit provides the low-frequency feedback that contributes to improve the VCO spectrum performance. Series resistances of 51 Ω can be connected between the collectors of each transistor and voltage supply to stabilize operation mode. By varying the bias voltage of the varactor SMV1493-079 from 0 to 3 V, a frequency tuning range from 910 to 985 MHz with phase noise level of −93 dBc/Hz at 100 kHz offset and dc supply current of 8 mA was achieved.

To integrate differential VCOs fully into RF transceivers when it is necessary to provide both high-performance transistors and high-quality passive components, using SiGe BiCMOS technology is very attractive with its high-resistive substrates and thick top metallizations.

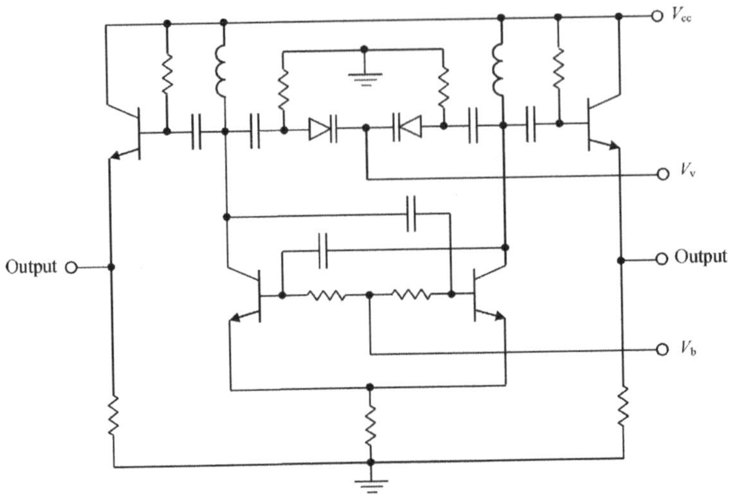

Figure 6.32 Schematic diagram of SiGe BiCMOS VCO for wireless application

Figure 6.32 shows the schematic diagram of such a VCO intended for wireless RF application where negative feedback is generated by a capacitive cross-coupling of the collector and base terminals of a differential pair [45]. To improve the pulling performance of the VCO and the appropriate matching with load impedance, emitter followers are used in the feedback path. The *LC* parallel resonator at the collectors determines the tuning frequency bandwidth. The inductor geometry should be designed to achieve as high *Q*-factor as possible, which can be increased by using a multimetal layer and removal of highly doped channel stop layer from under the inductor [46]. It is also very important to optimize the overall size of the inductor as a compromise between its quality factor and resonant frequency. For an inductor with a diameter of 213 μm, a quality factor of 10 is obtained at 2 GHz. For large a 320-μm inductor, the quality factor can be increased up to 11 [45]. For the varactor diodes, a typical quality factor of 20–40 for capacitance values of 1–2 pF is usually obtained. The base–emitter junction of the transistor can be used as a varactor diode for VCO frequency tuning. As a result, a tuning range from 1.9 to 2.6 GHz for a varactor bias voltage range from 0 to 4 V with tuning sensitivities 150–200 MHz/V and phase noise of −128.5 dBc/Hz at 3 MHz offset and dc current of 5.6 mA is obtained at a supply voltage of 3.6 V.

The tank inductor can be implemented differentially, resulting in an area saving compared with two single-ended inductors and a higher quality factor due to mutual coupling between two inductors [47]. As an alternative, the modified classical Armstrong oscillator topology can be used for fully monolithic and differential realization. Figure 6.33 shows the circuit schematic of such a high-performance differential VCO where the integrated differential transformer used instead of two single-ended inductors provides smaller overall physical size, higher inductance quality factor and higher self-resonant frequency [48]. The base bias of both transistors is fed from the centre tap of a secondary winding of the transformer, while the power supply is coupled to the centre tap of its primary winding. By using emitter degeneration resistors, it is possible to improve the pulling figure by about a factor of two, but at the expense of about 4 dB phase noise degradation at 3 MHz offset. As a result, when implemented in a 0.35-μm BiCMOS process, the oscillator demonstrates a frequency tuning range of 1 GHz, pulling

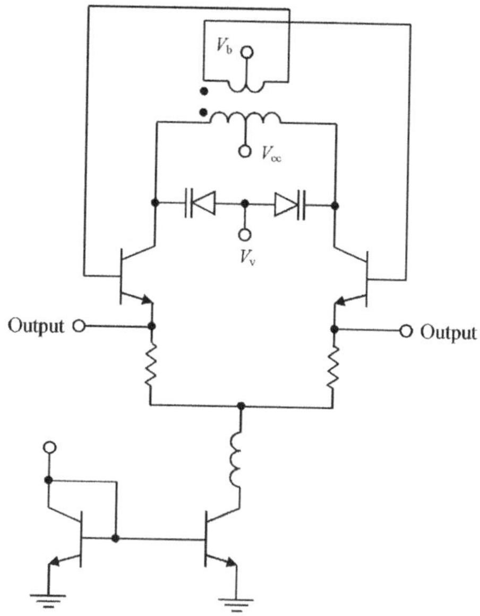

Figure 6.33 Circuit schematic of symmetrical Armstrong VCO

figure of 5.4 MHz and phase noise of −125 dBc/Hz at 3 MHz offset from carrier of 4.7 GHz, delivers single-ended output power of −6.5 dBm and consumes dc current of less than 4 mA from a supply voltage of 2.5 V.

Figure 6.34 shows the circuit schematic of a cross-coupled bipolar VCO with differentially tuned varactor structure, providing better common-node noise rejection due to more symmetric inherent structure and higher quality factor due to lower substrate effects [49]. Compared with the similar VCO using an antiparallel varactor structure, such a differential VCO implemented in a 0.5-μm BiCMOS process demonstrates better phase noise of about 5 dB at 1 MHz offset. The quality factor of a tank spiral inductor was 14 at 4.3 GHz. A frequency tuning bandwidth from 4.251 to 4.428 GHz with phase noise of −119 dBc/Hz at 1 MHz offset was obtained at a supply voltage of 2.5 V.

Millimetre-wave VCOs based on a differential topology can also be fully integrated into a SiGe bipolar technology. Figure 6.35 shows the circuit schematic of such a fully integrated millimetre-wave VCO where all inductances are realized by short microstrip lines [50]. The microstrip lines L_{e2} are necessary to compensate the negative effect of the output equivalent capacitance of the current source transistor on the frequency tuning range. In addition, their inductances together with this capacitance form low-pass filters, which decouple high-frequency phase noise generated by the current source transistor, mostly due to the second harmonic. The microstrip lines L_{e1} increase the oscillator loaded quality factor compared with the case of the pure varactors. The increased quality factor is a result of the smaller varactor capacitance value due to nonzero series inductance contributing to a steeper phase characteristic near the resonant frequency.

To implement the varactors, transistor base–collector junctions were used, modified by an additional implantation step that results in a wider tuning capacitance ratio and higher quality factor. To adjust the VCO oscillation frequency for the different frequency range, the

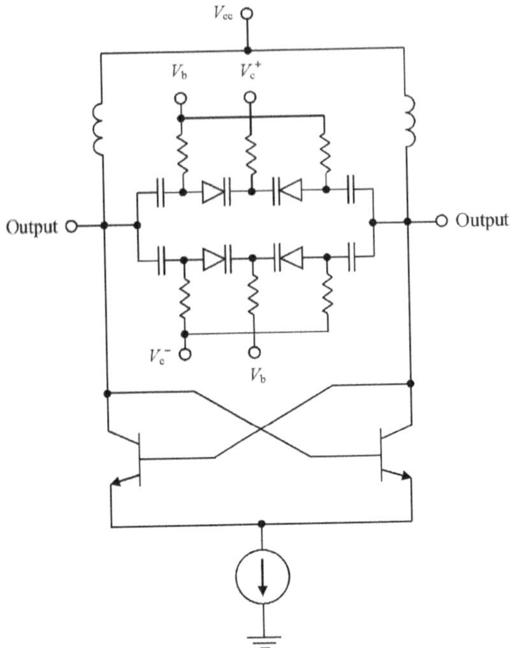

Figure 6.34 Circuit schematic of a cross-coupled VCO with differential varactor structure

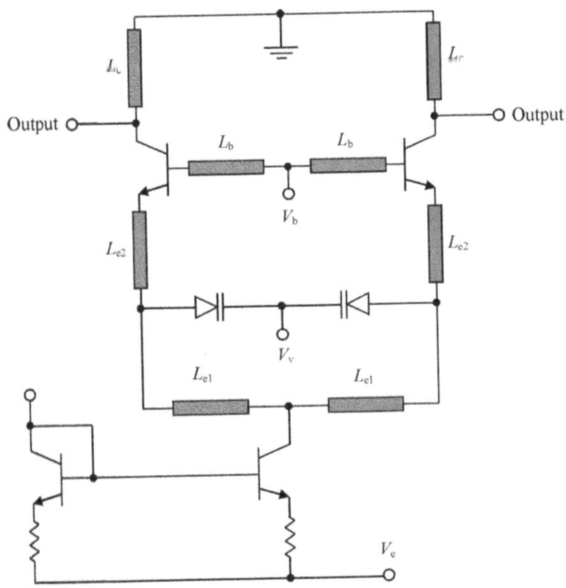

Figure 6.35 Circuit schematic of a millimetre-wave differential VCO

transmission lines are partly realized in the form of symmetric differential microstrip lines, each of which represents the two complementary microstrip lines with shorting bars between the lines. Using the effect of a virtual ground, the effective lengths of the microstrip lines can be easily increased by cutting some or all of these shorting bars in the upper metallization layer. The lengths of the differential lines and the distance between shorting bars are chosen so as to provide continuous adjustment in a frequency range 28–50 GHz. Such a millimetre-wave differential VCO fabricated in a commercial 0.35-μm SiGe bipolar technology with $f_T = 70$ GHz and $f_{max} = 75$ GHz provides a wide frequency tuning range from 36 to 46.9 GHz by varying the varactor bias from −3 to +3 V with phase noise level between −107 and −110 dBc/Hz at 1 MHz offset. The corresponding differential and single-ended output powers are 6.5 and 3.5 dBm, respectively, with total power consumption of 280 mW at a supply voltage of −5.5 V.

In modern wireless transceiver architectures, it is often necessary for the local oscillator to produce two tones with a quadrature phase relationship. This can be done by coupling two oscillators such that their outputs are locked in quadrature. Several techniques for obtaining quadrature relationship are known, based on parallel or series coupling schemes using the fundamental-frequency or second-harmonic coupling. For the differential oscillators, it is preferable to use second-harmonic coupling due to the second harmonic circulating at common node points which are ideal to inject a signal at twice the natural oscillator frequency. Figure 6.36 shows the quadrature VCO circuit schematic using standard cross-coupled differential transistor pairs and differential tank inductors with inductance values of 2 nH and quality factors of 27 at 6 GHz [51]. The varactors with a capacitance tuning ratio of 3.3 from 0 to 5 V have a minimum value of 0.52 pF at a reverse-bias voltage of 5 V. The two differential oscillators are mutually injection locked at nodes A and B using capacitive coupling through their current

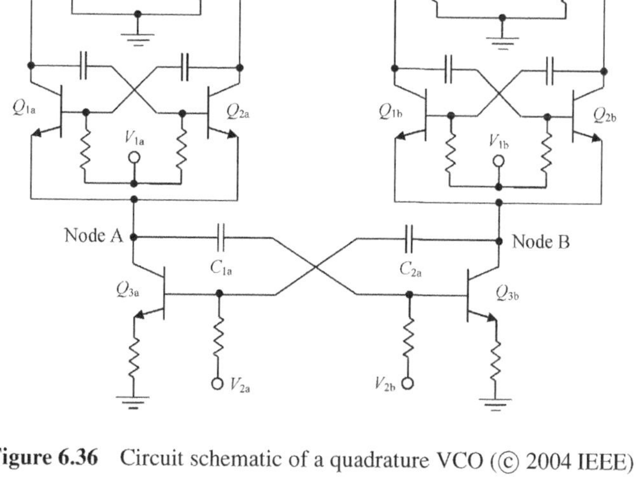

Figure 6.36 Circuit schematic of a quadrature VCO (© 2004 IEEE)

sources. The current source transistors are grounded through degeneration resistors. The use of these resistors enables one to decrease the noise contributed by the current source transistors and reduce the transistor effective transconductances to prevent parasitic oscillations at node A and node B by making the loop gain due to cross-coupled capacitances less than unity.

When the tail current source transistors are capacitively cross-coupled, a bilateral coupling network is formed that forces the second harmonics at node A and node B to be 180° out of phase, resulting in a quadrature 90° phase shift between the fundamentals of the two oscillators. It should be noted that perfect quadrature operation can be achieved only for two absolutely identical differential oscillators. The quadrature oscillator was implemented into SiGe HBT technology with $f_T = 80$ GHz and $f_{max} = 90$ GHz. A frequency tuning range of 24% with maximum frequency 5.92 GHz and average output power −5.3 dBm was obtained. Based on measurements of the single differential and quadrature VCOs, it was found that the superharmonic coupling does not affect the output power or the tuning range. The phase noise was measured to be −105.8 dBc/Hz at 1 MHz offset, which is 3 dB lower than that of the single differential VCO due to mutual injection locking.

6.7.3 Push–push VCOs

The simplified lumped push–push VCO configuration with a series resonant circuit is shown in Figure 6.37a where the load is connected to the centre point of the oscillator symmetry.

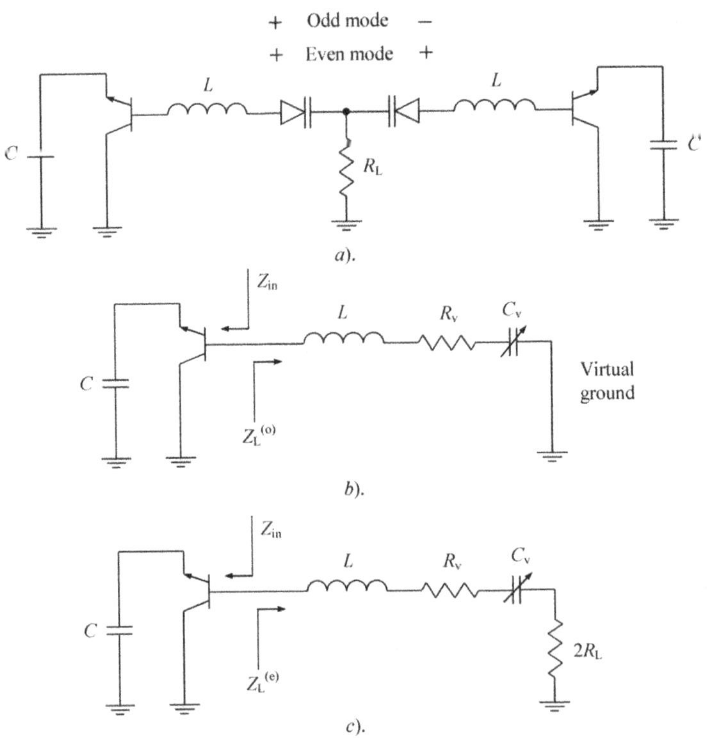

Figure 6.37 Operation modes of a bipolar push–push VCO

Push–push VCO topology creates an opportunity to provide frequency tuning at twice the operating frequencies when each half-circuit oscillates at one-half of the desired output frequency. This means that a push–push VCO must operate in only the odd mode when the second harmonic currents from both transistors flow into the load in the same direction during both half-periods or 180° phases of the active device. Since in the odd mode of operation the centre point of the oscillator resonant circuit becomes a virtual ground, the circuit can be represented as a series combination of transistor in a common collector configuration, a tank inductance L and a varactor with capacitance C_v and series resistance R_v, as shown in Figure 6.37b. For the even mode of operation, the transistors are in phase at the fundamental and each of them sees the load of $2R_L$, as shown in Figure 6.37c. To provide stable operation in the odd mode and to inhibit the oscillations in the even mode, the start-up amplitude oscillation conditions can be written as

$$\mathrm{Re}Z_{in} + R_v < 0 \tag{6.110}$$

$$\mathrm{Re}Z_{in} + R_v + 2R_L > 0 \tag{6.111}$$

when the even harmonics dissipate at the load R_L. These conditions are usually easily met in practice.

Two push–push VCOs using HBT devices with maximum oscillation frequency $f_{max} >$ 50 GHz and GaAs abrupt varactors with $C_{v0} = 0.5$ pF can cover a frequency range of 25–42 GHz [52]. Each HBT device was biased with a single current source on its emitter, while both the base and the collector terminals were dc grounded. The first VCO is tuned from 25 to 31.1 GHz. The second has a tuning bandwidth from 33 to 42.3 GHz. By using a combination of a low-pass filter for fundamental and a high-pass filter for second harmonic, the fundamental component was suppressed by more than 20 dB across the entire tuning bandwidths for both VCOs. By using InP HBT technology with $f_T = 75$ GHz and $f_{max} = 200$ GHz, a maximum frequency of 107.37 GHz with tuning bandwidth of 2.6%, output power of 0.92 dBm and phase noise of −88 dBc/Hz at 1 MHz offset was achieved [53]. To improve the phase noise characteristic of a push–push VCO with a series tank circuit, it is necessary to maximize the inductance and minimize the value of the varactor capacitance. However, the oscillator quality factor can be easily improved by using an oscillator configuration with a parallel tank circuit with a small inductance and a large capacitance, which also works as a high-pass filter, providing more efficient coupling of second harmonics [54]. In this case, the two parallel resonant circuits are arranged between the bases of the two transistors, each of which includes a short-circuited microstrip line connected in parallel with a varactor.

Generally, push–push VCO architecture provides a better phase noise level in comparison with a fundamental-frequency VCO developed under the same conditions for the same output frequency and using the same active device [55]. A simplified analysis based on the pushing factor showing the oscillator frequency sensitivity to the small dc voltage perturbations indicates ideally a 9 dB phase noise improvement using a push–push VCO topology. This is because the evaluated pushing factor at twice higher frequency contributes to 12 dB phase noise worsening. The circuit schematic of the push–push VCO with two half-frequency oscillators, each of which was designed to operate at 1 GHz, is shown in Figure 6.38. It was found that the phase noise around 2 GHz under push–push operation is exactly 3 dB worse than that around 1 GHz of the single oscillator. The practical design of the X-band push–push VCO with parallel coupled resonant circuits using InGaP/GaAs HBT technology results in the frequency tuning from 8.2 to 8.6 GHz with output power of about 8 dBm and phase noise of −83.2 dBc/Hz at

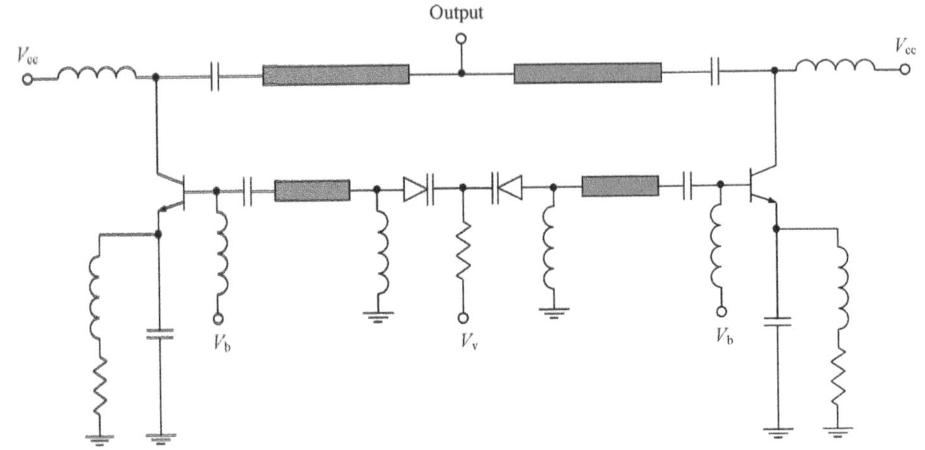

Figure 6.38 Bipolar push–push VCO configuration

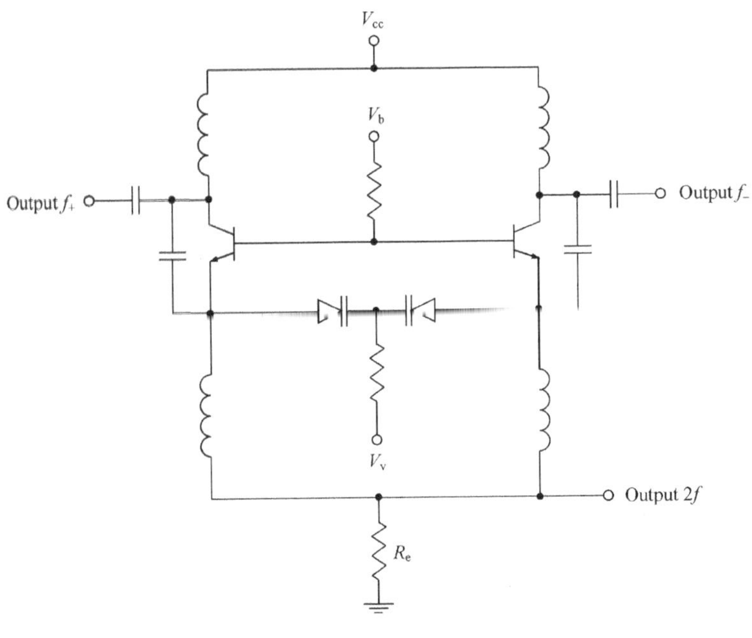

Figure 6.39 Bipolar balanced VCO schematic

100 kHz offset from maximum bandwidth frequency [56]. The 2 GHz push–push monolithic VCO fabricated using 2-μm InGaP/GaAS HBT technology with a tuning range of 280 MHz, optimized for minimum noise level based on the advanced noise model of the HBT device, achieved −118 dBc/Hz at 100 kHz offset from centre bandwidth frequency [57].

A VCO based on the balanced topology is required to provide differential output for direct compatibility with prescaler and double-balanced mixer inputs [58]. Figure 6.39 shows a typical circuit schematic of the balanced VCO using the common base Colpitts configuration, taking into account the effect of a virtual grounding at the points of oscillator symmetry. The signals at the collectors are differential, 180° out of phase for identical oscillator sides. If the value of the emitter tail resistor R_e is nonzero, the behaviour of such a balanced oscillator is similar to the conventional cross-coupled differential VCO when the second-harmonic signals are flowing into this resistor in phase. As a result, the second-harmonic signal can be taken from this oscillator together with two differential fundamental ones. However, due to emitter and collector capacitive feedback, the balanced VCO can oscillate in a lower tail current and have larger amplitude for the same tail current [59]. This can be simply explained by the fact that the presence of the two feedback capacitances in a Colpitts configuration gives the possibility of optimizing the oscillator feedback ratio to realize a significant margin of regeneration factor for the oscillation build-up and optimum steady-state conditions corresponding to the maximum available output power. In addition, due to a shunting effect of the collector feedback capacitances, the contribution of the collector current noise in a balanced VCO is decreased, resulting in a lower phase noise compared with a cross-coupled differential VCO.

Two varactor pairs with differential tuning can be placed between the two emitters instead of a single varactor pair [60]. In this case, common-mode noise voltages result in changing varactor capacitances in opposite directions, thus leaving the total capacitance of the parallel connection unchanged. This increases the frequency stability, improving the VCO phase noise level by about 3 dB compared with the case of only two varactors. Such a balanced VCO implemented in a 0.25-μm SiGe BiCMOS process with $f_T = 70$ GHz and $f_{max} = 90$ GHz provides frequency tuning from 20.8 to 21.86 GHz for the fundamental with minimum output power of −7 dBm and power consumption of about 130 mW at a supply voltage of 3.2 V. The output power at the second-harmonic output was about −17 dBm with suppression of the fundamental of 20 dB relative to the second-harmonic level. The minimum measured phase noise was −113 dBc/Hz for 21.5 GHz output and −107 dBm for 43 GHz output at 1 MHz offset.

Figure 6.40 shows a balanced frequency doubler based on a cross-coupled differential oscillator with two back-to-back varactors located between the collector terminals [61]. The

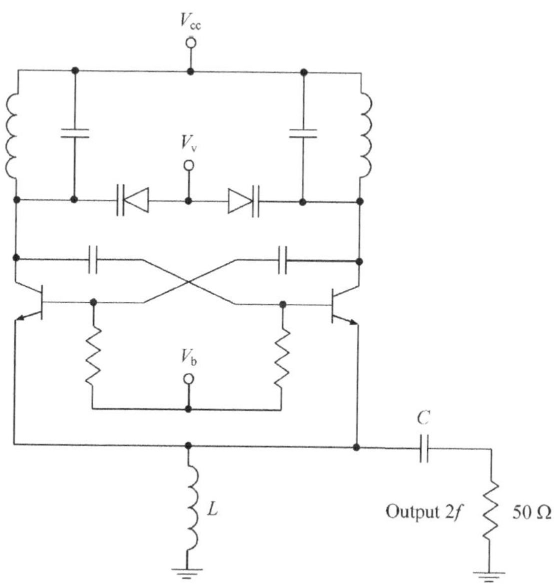

Figure 6.40 Bipolar push–push cross-coupled VCO schematic

cross-coupling capacitors are used to decouple the collector and base dc biasing. A high-pass *LC* matching circuit is used in the emitter circuit to provide optimum output matching to the load resistance of 50 Ω for second-harmonic signal. A commercial InGaP/GaAs HBT technology with $f_T = 60$ GHz and $f_{max} = 110$ GHz was used for the VCO implementation. By varying the varactor bias voltage from 0 to 2.5 V, frequency tuning from 21.17 to 21.02 GHz with variations of the output power from 1.7 to −4.5 dBm and phase noise from −110 to −105.3 dBc/Hz at 1 MHz offset, suppression of the fundamental of about −13 dBc and power consumption of 120 mW was obtained.

REFERENCES

1. M. H. Norwood and E. Shatz, 'Voltage Variable Capacitor Tuning: A Review', *Proc. IEEE*, **56**, 788–798 (1968).
2. W. Schottky and W. Deutschmann, 'Zum Mechanismus der Richtwirkung in Kupferoxydulgleichrichten', *Phys. Z.*, **30**, 839–846 (1929).
3. W. Shockley, 'The Theory of *p–n* Junctions in Semiconductors', *Bell Sys. Tech. J.*, **28**, 435–450 (1949).
4. M. E. McMahon and G. F. Straube, 'Voltage-Sensitive Semiconductor Capacitor', *IRE WESCON Conv. Rec.*, pt. **3**, 72–82 (1958).
5. P. Penfield and R. P. Rafuse, *Varactor Applications*, Cambridge: The M.I.T Press (1962).
6. Application Note 1004, Varactor SPICE Models for RF VCO Applications, Alpha Industries (1998).
7. R. Beach, 'High Speed Linear Oscillators', *Microwave J.*, **21**, 59–65 (1978).
8. R. B. Smith, B. Bramer and D. Y. Croft, 'Determination of the Constants of the Capacitance/Voltage Law of a Varactor Diode', *Electronics Lett.*, **4**, 2740–275 (1968).
9. I. M. Chudakov, *Frequency Modulation Using p–n Junction Capacitances* (in Russian), Moskva: Svyaz (1968).
10. D. Schulz, 'Transient Response of Variable Capacitance Diodes', *IEEE Trans. Component Parts*, **CP-7**, 49–53 (1960).
11. M. V. Blagoveshchensky and G. M. Utkin, *Radio Transmitting Devices* (in Russian), Moskva: Radio i Svyaz (1982).
12. R. Fekete, 'Varactors in Voltage Tuning Applications', *Microwave J.*, **7**, 53–61 (1964).
13. Q. Han, K. Inagaki, K. Iigusa, R. Schlub, T. Ohira and M. Akaike, 'Harmonic Distortion Suppression Technique for Varactor-Loaded Parasitic Radiator Antennas', *IEICE Trans. Electron.*, **E85-C**, 2015–2021 (2002).
14. M. Akaike, T. Ohira, K. Inagaki and Q. Han, 'An Analysis of Nonlinear Terms in Capacitance-Voltage Characteristic for Anti-Series-Connected Varactor-Diode Pair', *Int. J. RF and Microwave Computer-Aided Eng.*, **14**, 274–282 (2004).
15. V. N. Kuleshov and M. P. Savchenko, 'Non-isochronous oscillator with anti-series varactor pair (in Russian)', *Radiotekhnika*, **41**, 38–40 (1986).
16. V. K. Labutin, *Frequency-Selective Circuits with Electronic Tuning* (in Russian), Moskva-Leningrad: Energiya (1966).
17. D. Kajfez and E. J. Hwan, 'Linearity Limits of the Varactor-Controlled Osc-Mod Circuits', *IEEE Trans. Microwave Theory Tech.*, **MTT-33**, 620–625 (1985).
18. L. Y. Rotkov, S. M. Ryzhakov and E. V. Silin, 'Optimization of Parameters of Resonant Circuit with Varactor Frequency Tuning (in Russian)', *Izvestiya Vuzov USSR, Ser. Radioelektronika*, **33**, 97–100 (1990).
19. V. Y. Myakinkov, 'Method of Linearity Improvement of Oscillator Tuning Characteristic (in Russian)', *Elektronnaya Tekhnika, Ser. Microwave Engineering*, no. **2(462)**, 36–39 (1994).
20. G. N. Mears, Voltage Controlled Oscillator, U.S. Patent 4,450,416 (May 1984).

21. G. F. Bock and B. L. Walsh, 'An RF Linear Modulation Circuit', 1978 *IEEE MTT-S Int. Microwave Symp. Dig.*, 3150–317 (1978).
22. E. Marazzi and V. Rizzoli, 'Computer-Aided Design of Highly Linear, High-Power Varactor-Tuned Frequency Modulators', 1978 *IEEE MTT-S Int. Microwave Symp. Dig.*, 88–90 (1978).
23. T. Tokumitsu, O. Baba and K. Kajii, 'Very Linear and Low-Noise Ka/Ku-Band Voltage Controlled Oscillators', *IEICE Trans. Electron.*, **E85-C**, 2008–2014 (2002).
24. K. J. Anderson and D. L. Allen, 'A Miniature Integrated Monolithic VCO Module', *1988 IEEE MTT-S Int. Microwave Symp. Dig.*, 495–498 (1988).
25. A. F. Schneiter and J. L. Fikart, 'Circuit for Linearizing Frequency Modulated Oscillators on Microstrip', U.S. Patent 4,375,621 (March 1983).
26. C. S. Aitchison and R. V. Gelsthorpe, 'A Circuit Technique for Broadbanding the Electronic Tuning Range of Gunn Oscillators', *IEEE J. Solid-State Circuits*, **SC-12**, 21–28 (1977).
27. W. H. Lockyear, 'Linearize VCOs by Reactance Compensation', *Microwave J.*, **23**, 60–68 (1980).
28. R. V. Gelsthorpe and C. S. Aitchison, 'Analytical Evaluation of the Components Necessary for Double Reactance Compensation of an Oscillator', *Electronics Lett.*, **12**, 485–486 (1976).
29. D. F. Luan, G. Y. Zhou and S. Qin, '*X*-Band Hybrid-Integrated VCOs with High Tuning Linearity', *Proc. 14th Europ. Microwave Conf.*, 290–295 (1984).
30. Y. Yamauchi, H. Kamitsuna, M. Nakatsugawa, H. Ito, M. Muraguchi and K. Osafune, 'A 15-GHz Monolithic Low-Phase-Noise VCO Using AlGaAs/GaAs HBT Technology', *IEEE J. Solid-State Circuits*, **SC-27**, 1444–1447 (1992).
31. W. Perndl, H. Knapp, K. Auginger, T. F. Meister, W. Simburger and A. L. Scholtz, 'Voltage-Controlled Oscillators up to 98 GHz in SiGe Bipolar Technology', *IEEE J. Solid State Circuits*, **SC-39**, 1773–1777 (2004).
32. O. Sevimli, J. W. Archer and G. J. Griffiths, 'GaAs HEMT Monolithic Voltage-Controlled Oscillators at 20 and 30 GHz Incorporating Schottky-Varactor Frequency Tuning', *IEEE Trans. Microwave Theory Tech.*, **MTT-46**, 1572–1576 (1998).
33. W. R. Curtice and M. Ettenberg, 'A Nonlinear GaAs FET Model for Use in the Design of Output Circuits for Power Amplifiers', *IEEE Trans. Microwave Theory Tech.*, **MTT-33**, 1383–1394 (1985).
34. B. Piernas, K. Nishikawa, T. Nakagawa and K. Araki, 'A Compact and Low-Phase-Noise *Ka*-Band pHEMT-Based VCO', *IEEE Trans. Microwave Theory Tech.*, **MTT-51**, 778–783 (2003).
35. J. L. Merenda, Varactor Tuned Strip Line Resonator and VCO Using Same, U.S. Patent 5,942,950 (August 1999).
36. F. Ellinger, U. Lott and W. Bachtold, 'Design of a Low-Supply-Voltage High-Efficiency Class-E Voltage-Controlled MMIC Oscillator at *C*-Band', *IEEE Trans. Microwave Theory Tech.*, **MTT-49**, 203–206 (2001).
37. R. E. Zulinski and J. W. Steadman, 'Class E Power Amplifiers and Frequency Multipliers with Finite DC-Feed Inductance', *IEEE Trans. Circuits and Systems*, **CAS-34**, 1074–1087 (1987).
38. C.-H. Lee, S. Han, B. Matinpour and J. Laskar, 'A Low Phase Noise X-band MMIC GaAs MESFET VCO', *IEEE Microwave and Guided Wave Lett.*, **10**, 325–327 (2000).
39. M. J. Howes and D. V. Morgan, *Microwave Devices*, New York: John Wiley & Sons (1976).
40. B. Koster, P. Waldow and I. Wolff, 'A Unique, Low-Voltage, Source-Coupled J-FET VCO', *RF Design*, 58–68 (April 2001).
41. K.-W. Yeom, 'An Investigation of the High-Frequency Limit of a Miniaturized Commercial Voltage-Controlled Oscillators Used in 900-MHz-Band Mobile-Communication Handset', *IEEE Trans. Microwave Theory Tech.*, **MTT-46**, 1165–1168 (1998).
42. M. J. Barclay, Cascode VCO Frequency Modulator, U.S. Patent 5,502,420 (March 1996).
43. Application Note 1012, VCO Designs for Wireless Handset and CATV Set-Top Applications, Alpha Industries (1999).
44. Application Note 1013, A Differential VCO Design for GSM Handset Applications, Alpha Industries (1999).

45. B. H. Klepser and J. Kucera, 'A Fully Integrated SiGe Bipolar 2.4 GHz Bluetooth Voltage-Controlled Oscillator', *2000 IEEE RFIC Symp. Dig.*, 61–64 (2000).
46. H. Jacobsson, S. Gevorgian, M. Mokhtari, B. Hansson, C. Hedenas, T. Lewin, W. Rabe and A. Schuppen, 'Low Phase Noise, Low Power IC VCOs for 5-8 GHz Wireless Applications', *2000 IEEE MTT-S Int. Microwave Symp. Dig.*, 723–726 (2000).
47. T. M. Hancock, I. Gresham and G. M. Rebeiz, 'Compact Low Phase-Noise 23 GHz VCO Fabricated in a Commercial SiGe Bipolar Process', *Proc. 33rd Europ. Microwave Conf.*, 575–578 (2003).
48. N. T. Tchamov, T. Niemi and N. Mikkola, 'High-Performance Differential VCO Based on Armstrong Oscillator Topology', *IEEE J. Solid-State Circuits*, **SC-36**, 139–141 (2001).
49. H. Moon, S. Kang, Y. T. Kim and K. Lee, 'A Fully Differential LC-VCO Using a New Varactor Control Structure', *IEEE Microwave and Wireless Components Lett.*, **14**, 410–412 (2004).
50. H. Li and H.-M. Rein, 'Millimeter-Wave VCOs with Wide Tuning Range and Low Phase Noise, Fully Integrated in a SiGe Bipolar Production Technology', *IEEE J. Solid-State Circuits*, **SC-38**, 184–191 (2003).
51. T. M. Hancock and G. M. Rebeiz, 'A Novel Superharmonic Coupling Topology for Quadrature Oscillator Design at 6 GHz', *2004 IEEE RFIC Symp. Dig.*, pp 285–288 (2004).
52. D. M. Smith, J. C. Canyon and D. L. Tait, '25-42 GHz Heterojunction Bipolar Transistor Low Phase Noise Push-Push VCOs', *1989 IEEE MTT-S Int. Microwave Symp. Dig.*, 725–728 (1989).
53. D. M. Smith and B. R Allen, Voltage Controlled Push-Push Oscillator with Parallel Resonant Tank Circuits, U.S. Patent 5,231,361 (July 1993).
54. K. W. Kobayashi, A. K. Oki, L. T. Tran, J. C. Cowles, A. Gutierrez-Aitken, F. Yamada, T. R. Block and D. C. Streit, 'A 108-GHz InP-HBT Monolithic Push-Push VCO with Low Phase Noise and Wide Tuning Bandwidth', *IEEE J. Solid-State Circuits*, **SC-34**, 1225–1231 (1999).
55. R. Cignani, C. Florian, A. Costantini, G. Vannini, F. Filicori and L. Manfredi, 'Circuit Architectures for Low-Phase-Noise Oscillators', *Proc. 10th Europ. GaAs Symp.*, 77–80 (2002).
56. C. Florian, M. Pirazzini, R. Cignani, G. Vannini, G. Favre and F. Filicori, 'Push–Push X Band GaInP/GaAs VCO with a Fully Monolithic Microstrip Resonator', *2004 IEEE MTT-S Int. Microwave Symp. Dig.*, **3**, 1999–2001 (2004).
57. A. Sion, P. Cortese, C. Gourdon, M. Camiade, J. C. Nallatamby, M. Prigent and J. Obregon, 'Low Phase Noise 2 GHz HBT Push-Push VCO Based on an Advanced Low Frequency Noise Model', *Proc. 12th Europ. GAAS Symp.*, 143–146 (2004).
58. L. Dauphinee, M. Copeland and P. Schvan, 'A Balanced 1.5GHz Voltage Controlled Oscillator with an Integrated LC Resonator', *1997 IEEE Int. Solid-State Circuits Conf. Dig.*, 390–391, 491.
59. D. Baek, S. Ko, J.-G. Kim, D.-W. Kim and S. Hong, '*Ku*-band InGaP-GaAs HBT MMIC VCOs with Balanced and Differential Topologies', *IEEE Trans. Microwave Theory Tech.*, **MTT-52**, 1353–1359 (2004).
60. M. Bao, Y. Li and H. Jacobsson, 'A 21.5/43-GHz Dual-Frequency Balanced Colpitts VCO in SiGe Technology', *IEEE J. Solid-State Circuits*, **SC-39**, 1352–1355 (2004).
61. S.-H. Sim, S. Ko and S. Hong, 'A K-Band Push-Push VCO MMIC Using Embedded Doubling Mechanism', *Proc. 12th Europ. GaAS Symp.*, 139–142 (2004).

7

CMOS voltage-controlled oscillators

The rapid growth of new-generation wireless communication systems has created a strong demand for designing single-chip radio transceivers in a fully monolithic CMOS process with extremely small size due to better integration, low cost and low operating voltage. To increase the integration level, all passive components must be integrated monolithically into a single chip. In this case, the elements of a resonant LC-circuit of the voltage-controlled oscillator (VCO) as a core part of the synthesizers should feature high quality factors over the frequency tuning range. This chapter discusses the technological aspects to realize MOS varactors and spiral inductors, basic concepts of circuit design and implementation issues, oscillator phase noise and effect of low-frequency flicker noise. Also included are various practical examples of differential, complementary and quadrature CMOS VCOs using different process technologies.

7.1 MOS VARACTOR

It is well known that an MOS transistor can be used as a variable capacitance due to the dependence of its total charge (representing the charge on the gate, the effective interface charge and the charge in the semiconductor under the oxide) on the applied voltage between bulk and gate terminals [1]. In the case of a pMOS varactor with the source, drain and bulk terminals together connected, an inversion channel with mobile holes will be realized for bulk–gate voltages V_{bg} greater than the threshold voltage of the MOS transistor V_{th}. Further increase in V_{bg} results in a strong inversion of the MOS transistor operation. On the other hand, for voltages V_{bg} lower than V_{th}, an MOS transistor operates in the accumulation region, where the voltage at the interface between gate oxide and semiconductor is positive and high enough to allow electrons to move freely.

Figure 7.1a shows the cross-section of the MOS transistor where the movement behaviour of the majority-charge carriers in the inversion, depletion and accumulation regions are also shown [2]. A metal oxide structure is build on top of a lightly doped n-well diffusion layer with the gate and the two n^+ contacts inside the n-well. The device bias-dependent capacitance C can be modelled as oxide capacitance C_{ox} in series with the parallel connection of the capacitance C_b owing to the depletion region charge and C_i owing to the inversion layer charge at the gate–oxide interface. If C_b or C_i dominates, the MOS transistor is operated in a depletion or strong inversion region, respectively. Otherwise, if neither capacitance dominates, the MOS transistor is operated in a weak inversion region. The overall behaviour of the MOS capacitance C versus

RF and Microwave Transistor Oscillator Design A. Grebennikov

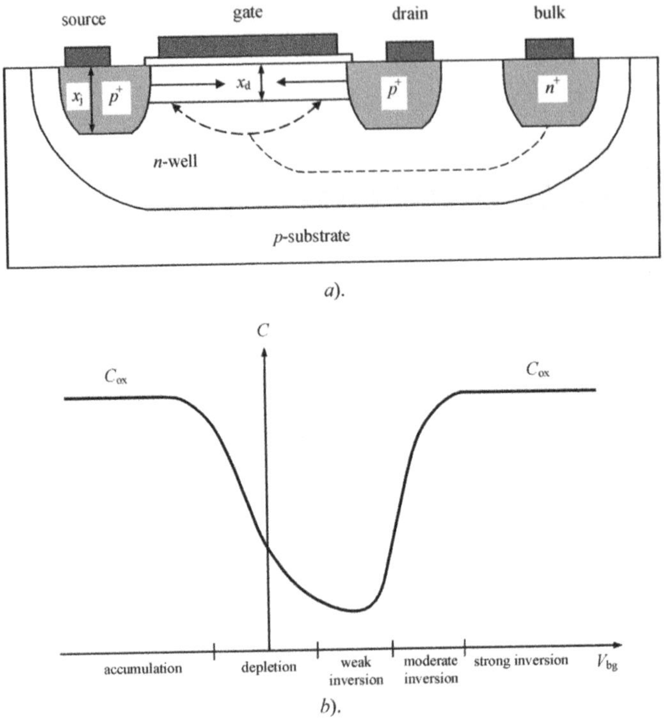

Figure 7.1 MOS varactor and its voltage–capacitance dependence

bulk–gate voltage V_{bg} is qualitatively shown in Figure 7.1b, the charge carrier movement in strong and moderate inversion regions is indicated by solid lines shown in Figure 7.1a.

When V_{bg} approaches V_{th} in the moderate inversion region, the concentration of holes at the oxide interface decreases steadily, but C_i continues to be much larger than C_b. However, when the MOS transistor enters the weak inversion region, modulation of the depletion region becomes of the same importance as hole injection when $C_i \approx C_b$, being the dominant effect in the depletion region when $C_i \ll C_b$. Thus, both in the strong inversion and accumulation regions, the overall device capacitance C_v approaches the oxide capacitance C_{ox}. Physically, an abundance of holes exists at high V_{bg} immediately below the oxide and provides the bottom plate of the oxide capacitor, just as abundance of electrons provided that plate in the case of accumulation. In this case, the parasitic resistance is associated with the resistive losses of electrons moving from the bulk contact to the interface between the bulk and depletion region, as shown in Figure 7.1a by dashed lines. Such a parasitic resistance can be reduced by using scaled technology with shorter device gate length L [3]. With the technology scaling, the oxide thickness is also reduced with corresponding increase in the oxide capacitance. Ideally, this should result not only in a better quality factor of a MOS varactor, but also in a wider tuning range since a minimum depletion capacitance increases at a lower rate.

To obtain monotonic dependence for C, it is necessary to provide the device operation without entering the accumulation region for a very wide range of gate voltage values. This can be accomplished by removing the connection between drain and source with bulk, by connecting the bulk terminal to the supply voltage as the highest dc voltage in the circuit.

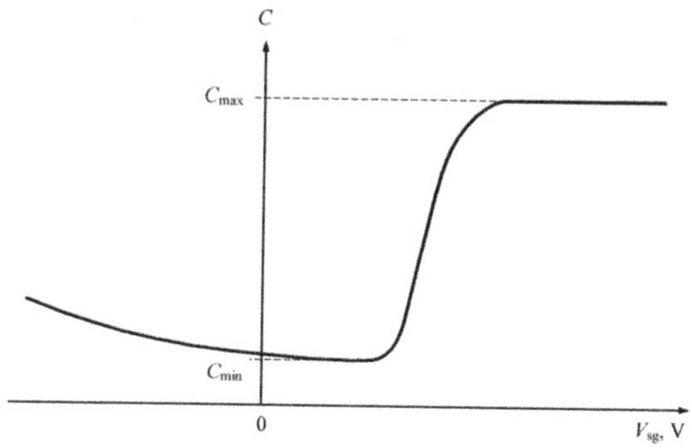

Figure 7.2 Inversion-mode MOS varactor tuning curve

Figure 7.2 shows the voltage–capacitance dependence of such an inversion-mode MOS varactor operating in the strong, moderate, or weak inversion regions only. However, a more attractive approach is to use the *p*MOS device in the depletion and accumulation regions only, resulting in a wider tuning range and better quality factor due to lower parasitic resistance since the electrons have mobility approximately three times higher than holes [4].

The voltage–capacitance dependence of such an accumulation-mode MOS varactor operating in the depletion and accumulation regions only is shown in Figure 7.3a. To realize an accumulation-mode MOS varactor, the formation of the strong, moderate and weak inversion regions must be inhibited; this requires the suppression of any injection of holes in the MOS

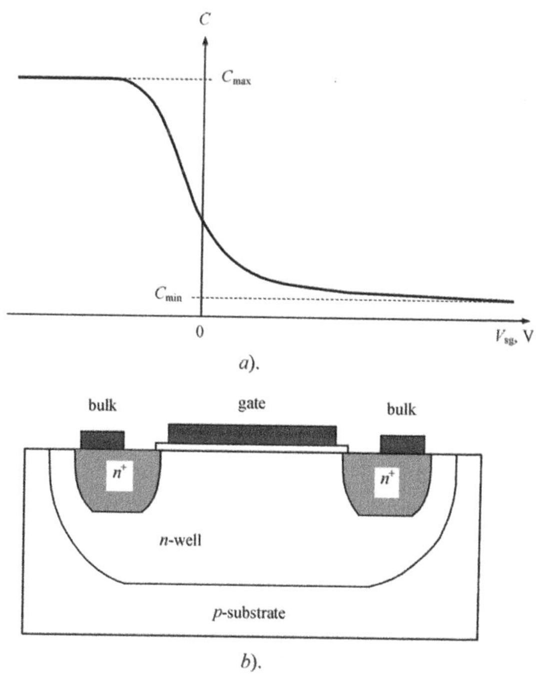

Figure 7.3 Accumulation-mode varactor and its tuning curve

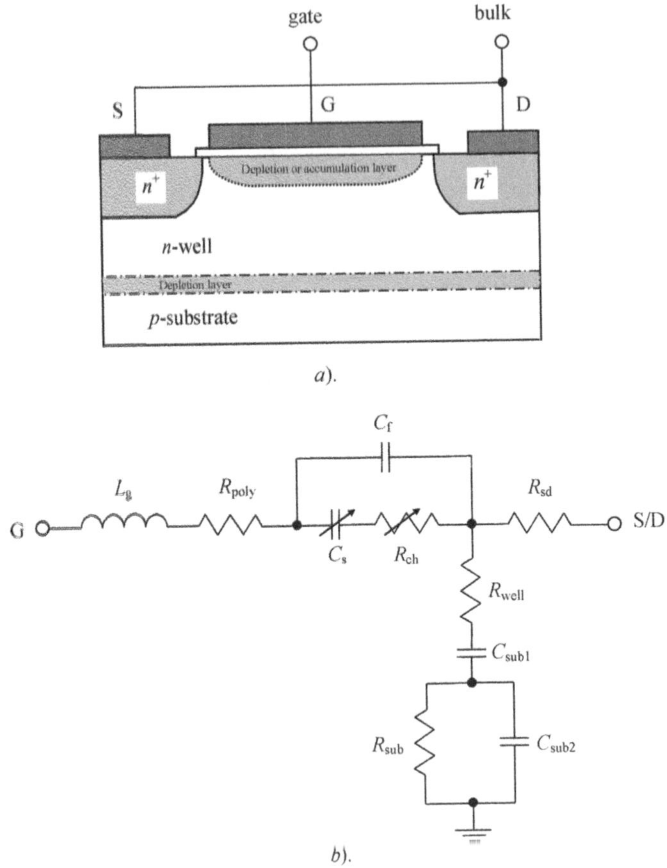

Figure 7.4 MOS varactor equivalent circuit

channel. This can be accomplished by removing the source and drain diffusion p^+-doped layers and implementing the bulk n^+-doped contacts instead of their minimizing the parasitic-well resistance, as shown in Figure 7.3b.

Since the physical models described the behaviour of a MOS device in the accumulation and depletion regions are different, it normally comprises separate models for these regions the integration of which into a common circuit simulator such as SPICE is complicated. Figure 7.4a shows the cross-section of an accumulation-mode MOS varactor where bulk represents shorted n-well contacts. Its single equivalent circuit is shown in Figure 7.4b where C_f represents the fringing capacitance mainly associated with the sidewall of the gate, L_g and R_{poly} are the parasitic inductance and resistance of the gate electrode, respectively [5]. The resistances R_{well}, R_{sub} and capacitances C_{sub1}, C_{sub2} are the substrate-related components. The resistance R_{sd} represents the source/drain regions. The channel resistance R_{ch} is the only bias-dependent resistance. It can be modeled as $R_{ch} = R_s + R_{acc}//R_p$ where R_s is the bias-independent n-well resistance between n^+ contact and accumulation or depletion regions underneath the gate, R_{acc} is the bias-dependent resistance of the accumulation layer, and R_p is the effective resistance along the edge of the depletion region.

In the accumulation region, R_{acc} becomes much smaller than R_p and R_{ch} is approximately equal to $R_s + R_{acc}$. In the depletion region, R_{acc} can be considered infinite, and R_{ch} approaches a constant value of $R_s + R_p$. The gate bias dependence model of R_{acc} based on the measurement results can be empirically given by

$$R_{acc}(V_{gb}) = \frac{1}{K_{acc}} \frac{1}{V_{gb} - V_{dep}} \qquad \text{for } V_{gb} \leq V_{dep} \tag{7.1}$$

$$R_{acc}(V_{gb}) = \infty \qquad \text{elsewhere} \tag{7.2}$$

where K_{acc} is the fitting parameter related to the device geometry and mobility of electrons in the accumulation region, and V_{dep} is the fitting parameter related to the flatband voltage in the depletion region.

The gate bias dependence of the variable capacitance C_s can be conveniently modelled with the hyperbolic tangent function [6, 7]. A hyperbolic tangent function can be used for each of two parts of the dependence $C(V_{gb})$ shown in Figure 7.1b where the gate bias-dependent capacitance C can be approximated by the following function:

$$C = C_{min} + C_b \left\{ 1 + \tanh \left[\frac{S}{C_b} (V_{gs} - V_b) \right] \right\} \tag{7.3}$$

where $C_b = (C_{max} - C_{min})/2$, C_{max} is maximum capacitance value, C_{min} is minimum capacitance value, $S = (S_1, S_2)$ is the slope of $C(V_{gb})$ at the bend point $V_{gb} = V_b = (V_{b1}, V_{b2})$ for each curve shown in Figures 7.2 and 7.3a [6].

An accurate model of the MOS varactor capacitance C based on the description of its bias-dependent behaviour in two regions separately, available in the SPICE simulator and valid under different bias conditions for a frequency range up to 10 GHz, is given by

$$C_v(V_{gb}) = C_j \left(1 - \frac{V_{gb}}{V_j} \right)^{M_j} \qquad \text{for } V_{gb} \leq F_c V_j \tag{7.4}$$

$$C_v(V_{gb}) = C_j \frac{1 - F_c(1 + M_j) + M_j \dfrac{V_{gb}}{V_j}}{(1 - F_c)^{-1-M_j}} \qquad \text{for } V_{gb} > F_c V_j \tag{7.5}$$

where C_j, V_j, F_c and M_j are the model fitting parameters [8].

It should be noted that, due to the higher average doping beneath the gate and enhanced parasitic interconnect capacitance in the device structure representing the more parallel connected segments for the same gate area and smaller gate length L, the minimum capacitance will increase, resulting in a lower capacitance ratio [9]. However, the minimum gate length devices have the highest minimum quality factors since the polysilicon gate resistance at lower L dominates the overall parasitic resistance R which is several magnitudes smaller than the n-well resistance at large L. As a result, there is trade-off between the quality factor and capacitance tuning ratio which can been achieved, for example, for medium 0.65 μm gate length MOS varactors.

To minimize parasitic capacitances which are mainly overlap and fringing capacitances between poly gate and n^+ diffusion contacts, the shallow trench isolation can be inserted separating the active area beneath the thin gate oxide and n-well, as shown in Figure 7.5a [10]. In this case, the overlap capacitance is significantly lower than for a conventional case, because the shallow trench isolation is much thicker than the gate oxide. The tuning voltage is applied to

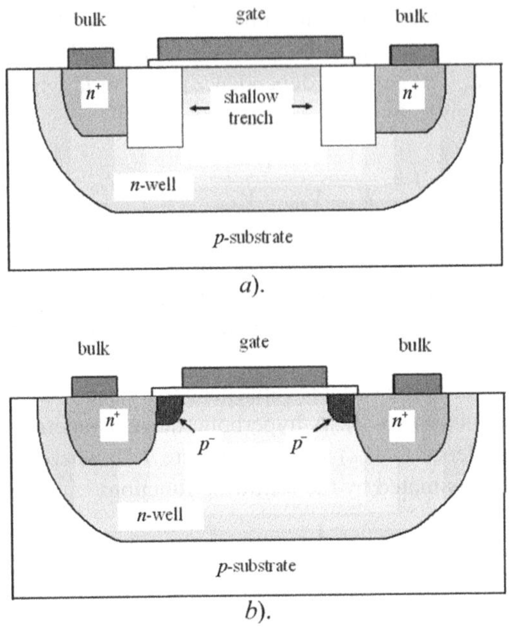

Figure 7.5 Accumulation-mode MOS varactor structures

the bulk n^+ contacts, the p^- substrate is grounded, and both dc voltage and RF signal appear at the gate. The varactor linear tuning range can be substantially increased by forming the lightly doped p^- diffusion regions between each bulk n^+ contact and lower-plate region inside the n well, as shown in Figure 7.5b [11]. These additionally incorporated diffusion regions increase the size of the depletion region resulting from a negative gate voltage which, in turn, increases the tuning range of varactor. For example, when ground is applied to bulk n^+ region, the varactor capacitance linearly increases from 6.8 to 15.5 fF for a gate voltage variation from -0.19 to $+0.5$ V, compared with a tuning range of the same varactor structure without a p^- layer of only 9.0–15 fF. However, the quality factor of varactors with these novel structures is lower compared with conventional ones.

Figure 7.6 shows the three dimensional physical structure of the MOS varactor where n^+ and p^+ diffusion regions are located in the same n-well of a standard CMOS process, separated by a small portion of n-well region to prevent early Zener breakdown [12]. With the p^+ terminal open, the device behaves like a two-terminal MOS varactor and its capacitance is varied by changing the voltage between gate and bulk terminals. For positive applied voltages, the device operates in the accumulation mode when the capacitance is maximal. However, for a reversed applied voltage, a depletion region is formed underneath the gate. As the negative voltage is increased, the depletion region widens and the capacitance is reduced, achieving a minimum value when $V_{gb} = V_{th}$. To further reduce the capacitance, the minority-carriers should be removed from the inversion layer, which can be done by biasing the p^+ contact at negative voltage with respect to bulk terminal. As a result, the capacitance tuning range of about 3.1 is achieved for the 0.35-μm CMOS process. Also, the difference in the gate doping type of the MOS varactor affects the capacitance tuning range and quality factor [13]. The multifinger varactor structure with mixed n^+ and p^+ doping demonstrates smoother capacitance and

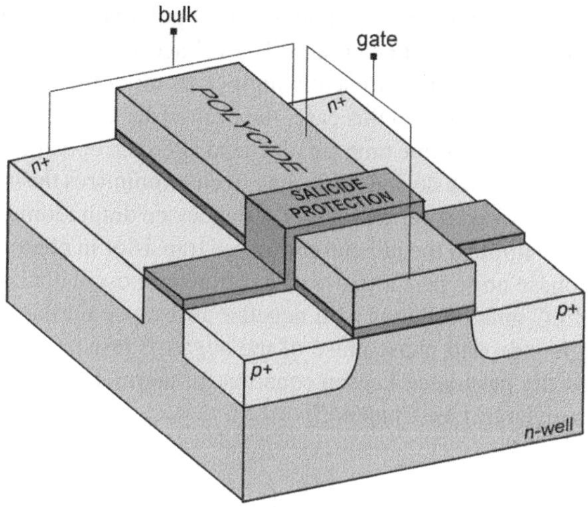

Figure 7.6 Three-dimensional physical structure of an accumulation-mode MOS varactor

resistance behaviour because of two depletion–accumulation transitions. Its quality factor and tuning range are always within the values corresponding to varactors with entirely n^+ or p^+ gate doping type.

7.2 PHASE NOISE

The differential cross-coupled *LC*-tank oscillator, the simplified circuit schematic of which is shown in Figure 7.7, is a preferred basic choice of oscillator in CMOS technology due to its good noise performance, ease of implementation and differential operation. The electrical

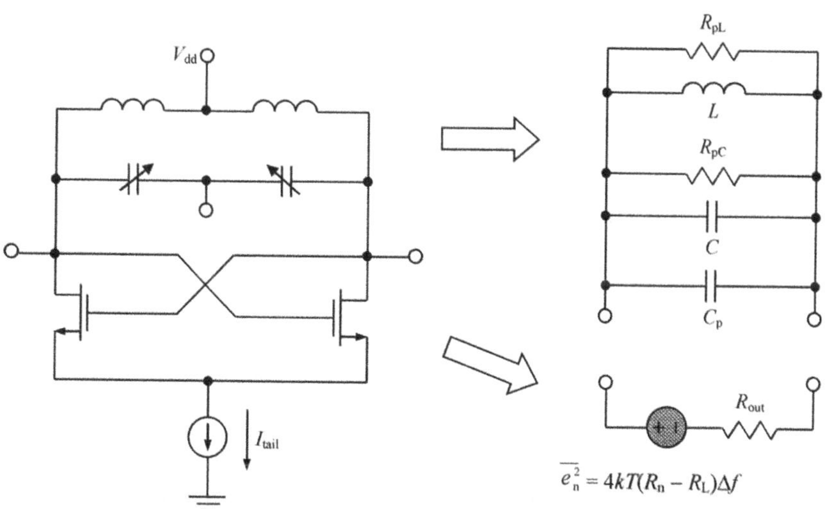

Figure 7.7 Equivalent representation of a differential cross-coupled CMOS VCO

behaviour of a differential oscillator is based on the push–pull operation principle of two out-of-phase identical oscillators with virtual ground. The voltage across the tank and at the inputs of the transistors is sinusoidal due to sufficiently high Q-factor of the parallel resonant circuit providing the harmonic attenuation. In this case, the even-order current harmonics flowing in the tank cancel each other. At the same time, an idealized alternate operational condition with a conduction angle of 180° for each device (50% duty cycle) minimizes the odd-harmonic level, thus resulting in half-cosinusoidal current pulses at the device drain terminals. The dc current and even harmonics flow through the tail current source transistor in phase. Consequently, the determination of the phase noise performance of a differential oscillator can be based on that of for a single-ended LC-tank oscillator with negative resistance during half a period of the oscillations. The single-sideband phase noise of the negative resistance oscillator written in the form of an empirically postulated Leeson equation for thermally induced phase noise of a feedback oscillator is given by (see Chapter 5)

$$L(f_m) = \frac{kTF}{2P_L}\left(\frac{f_0}{Q_L f_m}\right)^2 \tag{7.6}$$

where a cross-coupled differential transistor pair provides a negative resistance which is connected to an LC-tank, the output is assumed to be buffered to decouple an external load from the oscillator, and variable capacitors are composed of a pair of pMOS varactors. In Equation (7.6), P_L represents power dissipated in the load resistance R_L, Q_L is the quality factor of the resonant circuit and $F = R_n/R_L$ is the noise factor.

Now consider the constituent elements of the actual oscillator circuit separately [14]. By taking into account losses in reactive elements of the resonant circuit, the quality factor of such a resonator Q_L can be obtained through the inductor quality factor $Q_{ind} = R_{pL}/\omega L$ and capacitor quality factor $Q_{cap} = \omega C R_{pC}$ by

$$\frac{1}{Q_L} = \frac{1}{Q_{ind}} + \frac{1}{Q_{cap}}\frac{C}{C + C_p} \tag{7.7}$$

where C_p is a sum of the device output capacitance, loading capacitance due to the buffer stage and parasitic implementation capacitance.

Next, in order to calculate the noise factor F, the differential transistor pair is transformed into a noisy negative resistance with equivalent voltage noise power $\overline{e_n^2}$ shown in Figure 7.7, since the total noise consisting of the source noise and load noise must be equal to $4kTR_n$. In a steady-state operation mode, when the device is simply represented by an ideal voltage-controlled current source with transconductance g_m, the negative resistance of the differential pair can be written as

$$R_{out} = -\frac{2}{g_m} \tag{7.8}$$

It should be noted that, in a linear phase noise model, g_m in Equation (7.8) represents the small-signal device transconductance. However, under large-signal operation conditions using a piecewise-linear approximation of the device transfer characteristic, the value of g_m should be multiplied by the coefficient $\gamma_1(\theta)$ for a fundamental current component, which is a function of a half the conduction angle θ of the drain (or collector) current (see Chapter 1).

Using Equation (7.8) allows the transformation of the noise voltage source into a noise current with noise power in the form

$$\overline{i_n^2} = 4kT\frac{R_n - R_L}{R_{out}^2}\Delta f = kT(R_n - R_L)g_m^2\Delta f \tag{7.9}$$

On the other hand, the thermal noise in the channel of a MOS transistor is modelled as

$$\overline{i_{nd}^2} = \gamma 4kT g_m \Delta f \tag{7.10}$$

where $\gamma = 2/3$ for a long-channel device [1]. Assuming that the noise currents of both transistors are uncorrelated, the resulting current noise power of the differential pair is given by

$$\overline{i_n^2} = \frac{\overline{i_{nd}^2}}{2} = \gamma 2kT g_m \Delta f \tag{7.11}$$

From Equations (7.9) and (7.11) we can find that

$$R_n - R_L = \frac{2\gamma}{g_m} = R_{out}\gamma \tag{7.12}$$

which results in a final expression for the noise factor in the form

$$F = \frac{R_n}{R_L} = 1 + \gamma \tag{7.13}$$

where it assumed that $R_{out} \approx R_L$ at the oscillation frequency in a steady-state mode.

Generally, such a linear model neglects a contribution of the high-frequency multiplicative noise, resulting from the device nonlinear transfer characteristic, and low-frequency multiplicative noise, which takes place as a result of the frequency modulation of the tail current noise components [15]. However, for nMOS transistors with gate width up to 60 μm, these nonlinear contributions do not much exceed the value of 3 dB [14]. Thus, if the effect of high-frequency multiplicative noise is taken into account by including a factor of two, the final linear model for fast estimate of the phase noise in a differential CMOS oscillator can be obtained from Equation (7.6) as

$$L(f_m) = (1 + \gamma)\frac{kT}{P_L}\left(\frac{f_0}{Q_L f_m}\right)^2 \tag{7.14}$$

A figure-of-merit (FOM) for oscillators can be defined as a normalized phase noise spectrum given by

$$\text{FOM} = L(f_m)\left(\frac{f_m}{f_0}\right)^2 \frac{P_{tot}}{P_{ref}} \tag{7.15}$$

where P_{tot} is the total power consumption and $P_{ref} = 1$ mW. Substituting Equation (7.14) into Equation (7.15) yields

$$\text{FOM} \approx (1 + \gamma)\frac{kT}{P_{ref}}\frac{1}{\eta Q_L^2} \tag{7.16}$$

showing that FOM is essentially the reciprocal squared quality factor of the resonator divided by the oscillator efficiency $\eta = P_L/P_{tot}$.

Despite the heuristic nature of the Leeson noise equation and its analytical simplicity, it still continues to be very popular and can provide a sufficiently accurate description of the noise behaviour of any type of *LC* oscillator, in particularly differential CMOS oscillator, if the initially unspecified noise factor F is properly expressed through the transistor and resonant circuit parameters. For example, good agreement between predicted and measured data was achieved when the noise factor is defined for thermally induced phase noise arising from the resonator, differential pair and tail bias current [16]. In this case, for a resonator whose equivalent circuit is shown in Figure 7.7, by taking into account the modulation and mixing processes due to a piecewise-linear approximation of the device transfer characteristic, the oscillator noise factor can be described by

$$F = 1 + \gamma \frac{4}{\pi} \frac{I_{\rm tail}}{I_1} + \frac{4}{9} \gamma g_{\rm mbias} R_{\rm L} \tag{7.17}$$

where $g_{\rm mbias}$ is the transconductance of a tail transistor and I_1 is the fundamental current component flowing into the load resistor $R_{\rm L}$. Unlike the linear model for the noise factor given by Equation (7.13), the second term in Equation (7.17) is a result of differential pair noise contribution based on an assumption that the random pulse-width modulated at the switch output may be decomposed into the original periodic square wave in the absence of noise, superimposed with pulses of constant height, but random width. However, it is unlikely to expect a square waveform of the drain current with infinite transition from pinch-off mode to saturation mode of the device operation when the driving voltage at the device input is sinusoidal. Moreover, the presence of the high-Q parallel resonant circuit cannot provide high-impedance conditions for even current harmonics to minimize their effect on the drain current waveform, since ideally the square waveform is a combination of an infinite number of the fundamental-frequency and odd-harmonic components, properly phased. On the contrary, applying the sinusoidal voltage drive to the device, the transfer characteristic of which can be represented by a piecewise-linear approximation with sufficient accuracy, results in half-sinusoidal pulses for a 50% duty cycle with the conduction angle $2\theta = 180°$. In this case, the major noise contribution to the oscillator output spectrum is a result of the phase fluctuations due to the additive white noise at frequency offsets close to the carrier, similar to a narrowband phase modulation. The ratio of the tail bias current $I_{\rm tail}$ and fundamental component I_1 given in Equation (7.17) as a function of the half conduction angle θ can be written as

$$\frac{I_{\rm tail}}{I_1} = \frac{\gamma_0(\theta)}{\gamma_1(\theta)} = \frac{2}{\pi} \tag{7.18}$$

It follows from Equation (7.17) that the noise factor remains almost constant for the same conduction angle, slightly reducing with increase of the tail current since normally the transconductance reduces at high-current conditions. Consequently, phase noise performance can be improved by increasing the oscillation amplitude across the tank or power dissipated in the load resistance, as follows from Equation (7.6).

Figure 7.8 demonstrates the typical phase noise behaviour of the differential CMOS oscillator as a function of the tail bias current, whose increase results in an appropriate increase of the voltage amplitude across the tank, because, for the same conduction angle, the fundamental current flowing through the tank increases as well. This process continues in an active (or linear) device operation mode. However, beyond a critical value of the tail bias current, corresponding to a saturation mode of the device operation, the oscillation amplitude is limited by the supply voltage and can increase only slightly within the overall voltage waveform reached with harmonics. Such an operation mode is characterized by a strong device nonlinearity, resulting in

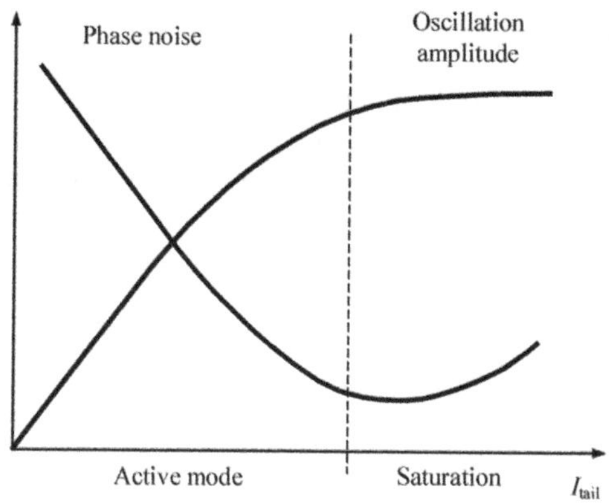

Figure 7.8 Phase noise and oscillation amplitude versus tail bias current

a significantly reduced value of its drain–source resistance, thus shunting the resonant circuit and reducing its loaded quality factor that, in turn, contributes to an increase of the harmonic content in oscillator output spectrum. As a result, the oscillator noise factor becomes higher at large values of the bias tail current. Therefore, the minimum phase noise can be achieved with maximum achievable oscillation amplitude in the active mode, just prior to saturation.

An analytical consideration of the cross-coupled CMOS oscillator without a tail current source based on the solution of the second order differential equations, taking into account the nonlinear device transfer characteristic, is presented in [17]. Assuming nearly sinusoidal oscillator behaviour, using a third-order polynomial to approximate the device transfer function and applying the averaging technique, the amplitude and phase noise properties were analysed, and a single sideband noise expression in terms of a random diffusion process was derived. It was found that increasing the device width is a more power-efficient approach to improve the phase noise rather than increasing the supply voltage. For instance, a fully integrated VCO fabricated using a 0.18-μm CMOS technology improves phase noise by 63% at 1 MHz offset from the carrier of 1 GHz with the increased device width from 25 to 50 μm at a supply voltage of 1 V. However, boosting V_{dd} from 1 to 1.5 V, while maintaining a width of 25 μm, achieves a similar phase noise improvement for a 125% power increase. Both effects applied simultaneously can provide more than 10 dB improvement in phase noise performance. However, using a tail current source, which is necessary to reduce sensitivity to power supply noise and variations of the bias current over process and temperature, results in a significantly increased phase noise level. In this case, the high-frequency bias current noise is a dominant contributor to the total phase noise at the oscillator output. To minimize phase noise, the width of the current source transistor must be minimized. For instance, reducing device width from 200 to 10 μm for a fixed bias current in a 0.18-μm SiGe BiCMOS technology reduces the bias noise contribution by 20 dB, making it negligible compared with the noise of the switching transistors [18]. At the same time, such a decrease of the device width results in an increase in phase noise contribution from the switching transistors by approximately 5 dB. In addition, the phase noise measurements show that the drain current thermal noise contribution of the switching transistors for the same transconductance g_m is about 4 dB less in the *p*MOS VCO than the *n*MOS VCO.

7.3 FLICKER NOISE

Flicker noise in CMOS transistors is dominant up to relatively high frequencies of several tens or hundreds of kilohertz. The physical origin of low frequency $1/f$ noise in CMOS transistors is similar to that of any type of MOSFET device based on two dominant processes: random fluctuation of the carriers in the channel due to fluctuations in the surface potential caused by trapping and releasing of the carriers by traps located near the Si–SiO_2 interface, and mobility fluctuations due to carrier interactions with lattice fluctuations [1]. However, depending on the type of CMOS transistor, one effect can prevail over the other. For example, flicker noise in n-channel devices is mostly attributed to carrier number fluctuations, while flicker noise in p-channel devices is often attributed to mobility fluctuations.

At strong inversion in the saturation region, the dependence of the equivalent gate voltage noise power $\overline{e_{\text{ng}}^2}$ can be modelled empirically as

$$\overline{e_{\text{ng}}^2} = \frac{K}{C_{\text{ox}}^2 LW} \frac{\Delta f}{f} \tag{7.19}$$

where $K = q^2 N_{\text{ot}}$ [19]. The effective density of oxide traps N_{ot} is a process-dependent parameter and is lower for pMOS devices. The bias voltages, where N_{ot} and K are nearly constant, can be experimentally determined from the measured dc transfer characteristics, and correspond to bias voltages where a nearly constant channel mobility is observed. Since the equivalent noise voltage appears in series with the gate, one can determine its effect on the drain current by multiplying by the device transconductance g_{m} and applying an appropriate expression for g_{m} at saturation, given by

$$g_{\text{m}} = \sqrt{\frac{2C_{\text{ox}}\mu W I_{\text{d}}}{L_{\text{eff}}}} \tag{7.20}$$

where μ is the carrier mobility, I_{d} is the fundamental drain current amplitude, and L_{eff} is the effective channel length. As a result, such a model will correspond to the SPICE drain current noise power model obtained by

$$\overline{i_{\text{nd}}^2} = \frac{K_{\text{F}} I_{\text{d}}}{C_{\text{ox}} L_{\text{eff}}^2} \frac{\Delta f}{f} \tag{7.21}$$

where the empirical flicker noise coefficient K_{F} can be calculated from $K_{\text{F}} = q^2 N_{\text{ot}} \mu$.

At weak inversion in the subthreshold region, the dependence of the equivalent gate voltage noise power $\overline{e_{\text{ng}}^2}$ can be well modelled by

$$\overline{e_{\text{ng}}^2} = \left(\frac{C_{\text{inv}}}{C_{\text{ox}}}\right)^2 \frac{1}{n^2} \frac{K}{C_{\text{ox}}^2 LW} \frac{\Delta f}{f} \tag{7.22}$$

where $n = (C_{\text{ox}} + C_{\text{d}})/C_{\text{ox}}$, C_{d} and C_{inv} are the depletion and inversion capacitances, respectively [19]. In this region, the gate noise power is lower because of the capacitive divider ratio $(C_{\text{ox}}/C_{\text{inv}})^2(1/n)^2$ and because $K = q^2 N_{\text{ot}}$ is lower at lower gate voltages.

By applying an appropriate expression for g_{m} at weak inversion given by

$$g_{\text{m}} = \frac{qI_{\text{d}}}{kT} \frac{C_{\text{ox}}}{C_{\text{ox}} + C_{\text{d}}} \tag{7.23}$$

the drain current noise power model will correspond to the SPICE model

$$\overline{i^2_{\mathrm{nd}}} = \frac{K_{\mathrm{F}} I_{\mathrm{d}}^2}{C_{\mathrm{ox}} L W} \frac{\Delta f}{f} \tag{7.24}$$

where the empirical flicker noise coefficient K_{F} can be calculated from

$$K_{\mathrm{F}} = \frac{C_{\mathrm{inv}}^2}{C_{\mathrm{ox}}^3} \frac{1}{n^4} \left(\frac{q}{kT}\right)^2 q^2 N_{\mathrm{ot}} \tag{7.25}$$

It was observed that pMOS transistors have significantly lower $1/f$ noise than nMOS transistors of the same size and fabricated with the same CMOS process (by one order of magnitude or more). This is because, when an n^+-polysilicon gate layer is used for both the nMOS and pMOS devices, nMOS transistors have a surface channel while pMOS transistors have a buried channel [20]. Consequently, the carriers flowing across the channel of the nMOS device are closer to the Si–SiO_2 interface, having higher probability of trapping and releasing by oxide traps. However, in the 0.25-μm CMOS process, to deal with the short-channel effects and high off current, p^+-poly gates are used for pMOS transistors while n^+-poly gates are used for nMOS transistors, thus making the surface channel for both types of transistor. Nevertheless, the measurement results show about an order of magnitude lower $1/f$ noise for pMOS transistors because of the difference in tunnelling coefficients, resulting from the differences in effective masses in SiO_2 and barrier heights of holes and electrons for tunnelling into SiO_2. Also, the effective mass of a hole in the oxide is 10–20 times heavier than that of an electron. To compare the phase noise of nMOS and pMOS VCOs, the same technology, integrated inductors and currents were used. As a result, a 5.4-GHz VCO using all surface channel pMOS transistors demonstrates lower phase noise by about 14 dB at 50 kHz offset from the carrier compared with its counterpart using all channel nMOS transistors.

The experimental results show that the flicker noise in nMOS transistors has no gate bias dependence, when the gate bias is varied from subthreshold to strong inversion and the measurement temperature is changed from room temperature to 5 K [21]. This can suggest that flicker noise in nMOS devices follows carrier density fluctuations. On the contrary, the flicker noise demonstrates the strong gate bias and temperature dependences in pMOS transistors that can be explained by mobility fluctuations rather than carrier density fluctuations. The p-channel noise amplitude increases with the gate bias and approaches the n-channel noise amplitude at high voltages. This means that, despite the fact that pMOS transistors generally suffer from mobility fluctuations, the carrier density fluctuation phenomenon becomes dominant under high-field conditions.

In a practical CMOS VCO design, the frequency tuning curve deviates from a required linear one due to the nonideal behaviour of the MOS varactor voltage–capacitance (C–V) dependence at high oscillation amplitudes resulting in AM-to-PM conversion when the oscillation frequency becomes dependent on the voltage amplitude. In a differential VCO with tail current source, the varactor bias conditions are also affected by the bias current variations. In this case, the low-frequency noise associated with the bias current provided by the tail current source transistor modulates the common-mode node voltage and thus the varactor bias voltage, resulting in an additional parasitic phase modulation. Generally, the tail current source together with its bias circuit is considered a main contributor of flicker noise in differential cross-coupled LC oscillators. Figure 7.9 shows the differential double cross-coupled VCO schematic and demonstrates the effect of low-frequency noise contribution in the VCO phase noise [22].

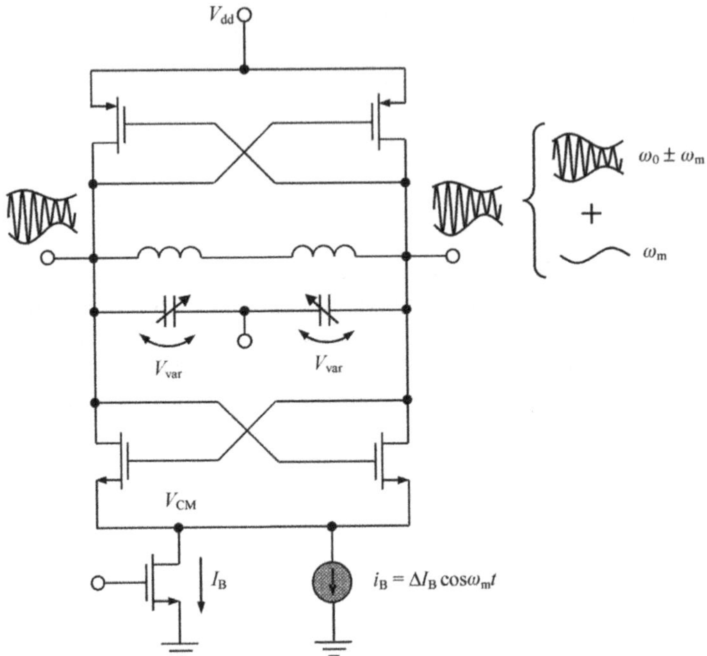

Figure 7.9 Differential cross-coupled CMOS VCO and effect of flicker noise

The total sensitivity of the oscillation frequency ω_0 to variations in the bias current I_B can be written as

$$\frac{d\omega_0}{dI_B} = \left(\frac{\partial\omega_0}{\partial I_B}\right]_{CM} + \left(\frac{\partial\omega_0}{\partial I_B}\right]_{AM} \tag{7.26}$$

where

$$\left(\frac{\partial\omega_0}{\partial I_B}\right]_{CM} = \left(\frac{\partial\omega_0}{\partial V_{var}}\right)\left(\frac{\partial V_{var}}{\partial I_B}\right) \tag{7.27}$$

is the contribution to the frequency deviation of variations of the common-mode node voltage V_{CM} to variations of the bias voltage across the varactor V_{var}, and

$$\left(\frac{\partial\omega_0}{\partial I_B}\right]_{AM} = \left(\frac{\partial\omega_0}{\partial V}\right)\left(\frac{\partial V}{\partial I_B}\right) \tag{7.28}$$

is the contribution to the frequency deviation of variations of the tail current I_B to variations of the oscillation amplitude V.

It follows from Equation (7.28) that the term $\partial V/\partial I_B$ can be significantly minimized by providing the device operation in a saturation mode when the oscillation amplitude is limited by the supply voltage V_{dd}. To eliminate the effect of the common mode on the phase noise, it is necessary to equate the term $\partial V_{var}/\partial I_B$ in Equation (7.27) to zero, which can be done by removing a cross-coupled pair of *p*MOS transistors. This allows connecting the common node point between two inductors directly to the voltage supply with no modulation of the varactor bias point induced by the bias current noise. However, there are several drawbacks in such an approach [22]. In this case, the varactors can be biased only at negative bias voltage values, thus reducing the maximum achievable tuning range, and the output voltage can reach considerably higher values, thus reducing the operation reliability. In addition, the differential VCO topology without a current source is exposed to a higher sensitivity of the oscillation frequency to the supply voltage variations (pushing factor).

7.4 TANK INDUCTOR

The key element of any oscillator is the tank inductor, the quality factor of which can significantly affect the oscillator phase noise performance. In contrast to the capacitors, high-quality inductors are not readily available in a standard CMOS technology. Therefore, it is necessary to use special techniques improving the inductor electrical performance. By using a standard CMOS technology with only two metal layers and a heavily doped substrate, the spiral inductor will have a large series resistance compared with three or four metal layer technologies, and the substrate losses become a very important factor due to the relatively low resistivity of silicon. The layout of a basic square spiral inductor is shown in Figure 7.10a. If the technology allows 45° routing, an octagonal shape can be used, as shown in Figure 7.10b. A major source of substrate losses is capacitive coupling when current is flowing not only through the metal strip, but also through the silicon substrate. Another important source of substrate losses is inductive coupling when, due to the planar inductor structure, the magnetic field penetrates deeply into the silicon substrate, inducing current loops and related losses. However, the latter effects are particularly important for large-area inductors and can be overcome by using silicon micromachining techniques [23].

The monolithic passive inductor normally represents a microstrip transmission-line structure consisting of a top metal strip above a conducting (or ground) plane, with the substrate and intermetal oxide layers sandwiched between the two conductor layers. The ratio of the series inductance to shunt capacitance per unit length defines the characteristic impedance of the microstrip transmission line. Maximizing the characteristic impedance using a narrow metal strip will reduce the length of line that is required to realize a given inductance. However, this also reduces the current density capability of an inductor. Among the different inductor

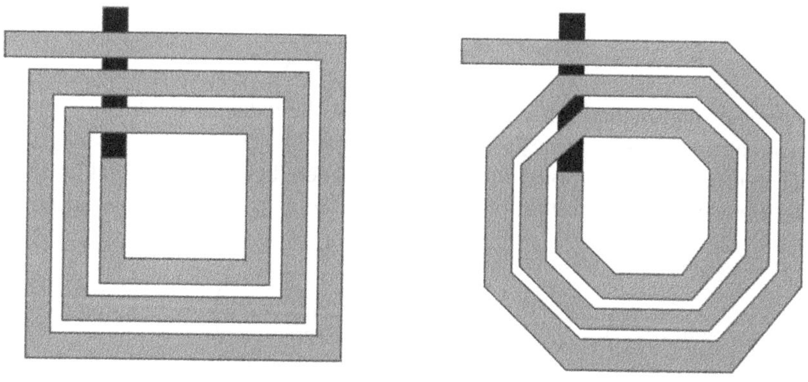

Figure 7.10 Spiral inductor layouts

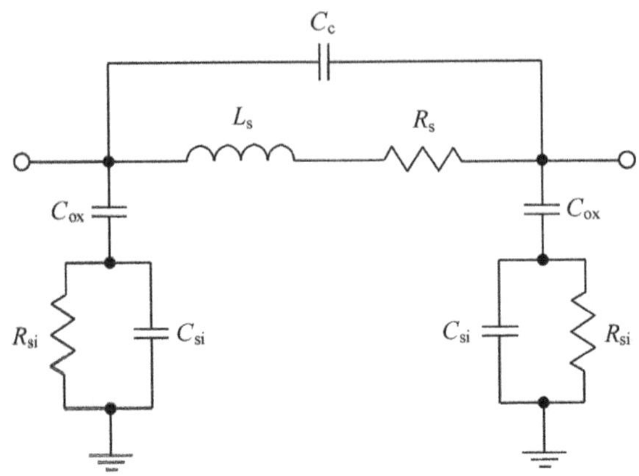

Figure 7.11 Equivalent circuit of a square spiral inductor

geometries, the circular spiral inductor structure can achieve up to 10% improvement in an inductor quality factor, but is more difficult to analyse.

The simplified compact equivalent circuit for the spiral microstrip inductor is shown in Figure 7.11, where L_s models the self and mutual inductances, R_s is the series coil resistance, C_{ox} is the parasitic oxide capacitance from the metal layer to the substrate, R_{si} is the resistance of the conductive silicon substrate, C_{si} is the silicon substrate parasitic capacitance and C_c is the parasitic coupling capacitance [24]. If the spiral inductor is treated as a lossless transmission line with total length much shorter than a quarter-wavelength, the capacitance C_{ox} is approximately equal to one-half the input capacitance of the open-circuited line. The parasitic silicon substrate capacitance C_{si} is sufficiently small and in most cases it can be neglected.

Such a model shows an accurate agreement between simulated and measured data within 10% across a variety of inductor geometries and substrate dopings up to 20 GHz [25]. At frequencies well below the inductor self-resonant frequency, the coupling capacitance C_c between metal segments due to fringing fields in both the dielectric and air regions can also be neglected since the relative dielectric constant of the oxide is small enough [26].

In this case, if one side of the inductor is grounded, the self-resonant frequency of the spiral inductor can approximately be calculated from

$$\omega_{SRF} = \frac{1}{\sqrt{L_s C_{ox}}} \sqrt{\frac{L_s - R_s^2 C_{ox}}{L_s - R_{si}^2 C_{ox}}} \tag{7.29}$$

At frequencies higher than self-resonant frequency, the inductor exhibits a capacitive behaviour. The self-resonant frequency ω_{SRF} is limited mainly by the parasitic oxide capacitance C_{ox}, which is inversely proportional to the oxide thickness between the metal layer and substrate. The frequency at which the inductor quality factor Q, is maximal can be obtained as

$$\omega_Q = \frac{1}{\sqrt{L_s C_{ox}}} \sqrt{\frac{R_s}{2R_{si}}} \left(\sqrt{1 + \frac{4R_{si}}{3R_s}} - 1 \right)^{0.5} \tag{7.30}$$

The inductor metal conductor series resistance R_s can be easily calculated at low frequencies as the product of the sheet resistance and the number of squares of the metal trace. However, at high frequencies, the skin effect and other magnetic field effects will cause a nonuniform current distribution in the inductor profile. In this case, a simple increase in the diameter of the inductor metal turn does not necessarily reduce the inductor series resistance correspondingly. For example, for the same inductance value, the difference in resistance between the two inductors, when one of which has a metal strip of twice the width, is only a factor of 1.35 [27]. Moreover, at very high frequencies, the largest contribution to the series resistance does not come from the longer outer turns, but from the inner turns. This phenomenon is a result of the generation of circular eddy currents in the inner conductors, whose direction is such that they oppose the original change in magnetic field. On the inner side of the inner turn, coil current and eddy current flow in the same direction so the current density is larger than average. On the outer side, both currents cancel and the current density is smaller than average. As a result, the current in the inner turn is pushed to the inside of the conductor.

Using three metal layers makes it possible to fabricate an inductor with larger inductance, whose value per unit area rises almost quadratically with the number of layers due to solenoid-like properties. Inductors with larger values are normally fabricated as aluminium spirals with many turns. As the inductance of the spiral is made larger, the parasitic capacitance to substrate increases, resulting in a progressively lower self-resonant frequency, according to Equation (7.29). For a standard silicon substrate, aluminium inductors are only as large as 10 nH with self-resonant frequency of 2 GHz [24]. However, such a silicon substrate problem can be overcome if the area underneath the inductor is made to appear locally insulating. This can be provided simply by selectively etching out the silicon, leaving the inductor encased in a suspended oxide layer attached at all four corners. As a result, the self-resonant frequency of a 100 nH square spiral inductor with a 20 turns and 4-μm-wide lines of the second aluminium metal layer separated by 4-μm spaces increases from 800 MHz to 3 GHz after removal of the underlying substrate [26].

Removing the silicon substrate leads not only to an increase in the self-resonant frequency, but to an increase in the inductor quality factor Q as well. For example, the maximum value of Q-factor changes from 3.5 at 1.25 GHz to about 20 at 5 GHz before and after etching of the silicon substrate [23]. Also, at higher frequencies there is an optimum metal strip width, which minimizes series resistance and maximizes the Q-factor, resulting from the effect of two main contributions into the inductor series resistance as sheet resistance, which is inversely proportional to the metal strip width, and magnetic induced losses, which depend directly on the time derivative of the magnetic field flux through the metal strip. Better results can be obtained if a different strip width is used for each turn of the inductor coil [23, 29, 30]. Narrow strips optimize losses in the inner turns, where the magnetic field reach its maximum value, while wide strips optimize the outer turns, where ohmic losses are predominant. As a result, values of Q-factor of about 17 were obtained for a 34-nH inductor at 1.5 GHz [23]. Moreover, the Q-factor for a 20-nH inductor working at 3.5 GHz after its layout optimization can reach a value of 40, which is about 1.6 times that of the best tank inductor with constant strip width.

An alternative solution to increase the inductor Q-factor is to use a pattern floating shield placed between the inductor and the substrate [31], [32]. The solid shield inserted between the inductor and the substrate has a serious drawback in disturbing the magnetic field, creating an induced loop current flowing in the opposite direction and, as a result, in reducing the overall inductance. The shield is patterned with slots orthogonal to the spiral and constructed from metal, silicided polysilicon, or low-resistivity buried layers close to the silicon surface [33]. The floating shield fingers lie directly underneath the inductor winding to block the capacitively coupled currents from entering the silicon substrate, thus reducing substrate dissipation. The

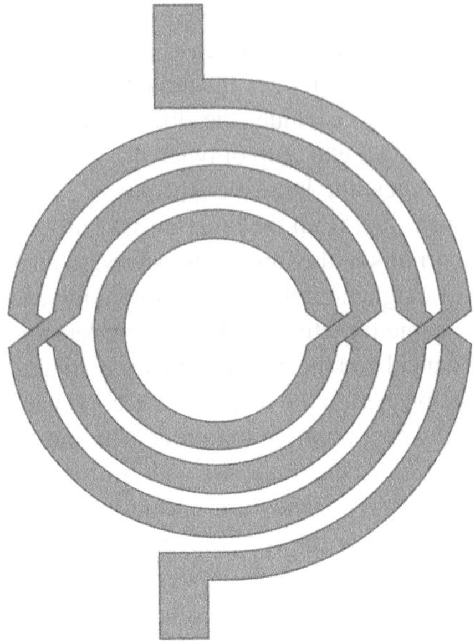

Figure 7.12 Balanced inductor layout

floating shield is most effective when the shield strips have much lower inductance than the multi-turn top-metal inductor winding and when it is made of metal (aluminium or copper) instead of a more resistive material (polysilicon).

For a differential CMOS VCO architecture, the tank circuit must exhibit a differential behaviour. Thus, either the tank inductor should have a balanced structure, or it is necessary to use two standard inductors symmetrically located. The latter case results in a larger area and correspondingly more substrate losses. A balanced inductor, the layout of which is shown in Figure 7.12, can occupy smaller area with less metal length and also exhibits more coupling between its turns [34]. In order to reduce the resistive losses of the metal tracks, two improvements can be made: connection in parallel with longitudinal vias (except in the underpass) of the two available metal layers to diminish the series coil resistance; and the design of a hollow spiral to avoid the high resistance due to circular eddy currents. As a result, when implemented in a standard two-metal layer 0.8-μm CMOS process, such a balanced inductor with radius of 105 μm and 3.5 turns provides an inductance of about 2.6 nH and Q-factor of about 7 between 1.2 and 2.9 GHz [34].

In order to realize a wideband frequency tuning of the VCO, it is necessary for the tank inductor to be sufficiently small to minimize the parasitic series resistance and maximize a minimum value of the MOS varactor and have low parasitic oxide capacitance. Therefore, a single-loop (horseshoe) inductor can provide enough inductance with the capability of wideband tuning for the tank resonator operating above 2 GHz. Compared to the multi-turn spiral inductance, the horseshoe inductor has worse magnetic coupling that generally leads to lower inductance per area and lower quality factor. Nevertheless, the absence of interwinding capacitance and reduced substrate capacitance, usual for such an inductor, contributes to a wider frequency tuning range. Such an approach to inductor implementation was applied to the design of 3.0–5.6 GHz CMOS VCO for wireless LAN applications using a 0.13-μm process [35]. In this case, the inductor Q-factor is relatively high over the entire tuning range. The series parasitic resistance is minimized due to wider strip width at lower bandwidth frequencies. Reduction of the entire inductor metal area decreases the power loss to the substrate at higher frequencies. Finally, the horseshoe inductor is perfectly symmetric without the need for an underpass. Being fabricated in a standard CMOS technology using the top two copper layers of 1.2 μm thickness and having a diameter of 460 μm, such an inductor has an inductance value of about 0.85 nH with the quality factor above 20 between 3.0 and 5.6 GHz and self-resonant frequency greater than 10 GHz.

7.5 CIRCUIT DESIGN CONCEPTS AND TECHNIQUE

7.5.1 Device operation modes

To better understand the oscillator operation and design principles, it is necessary to start with the basic physics and electrical behaviour of the MOS transistor as a main element of the oscillator, the typical output current–voltage (I–V) characteristic of which is shown in Figure 7.13a. Initially, consider the situation for a fixed gate bias V_{gs} and an increasing

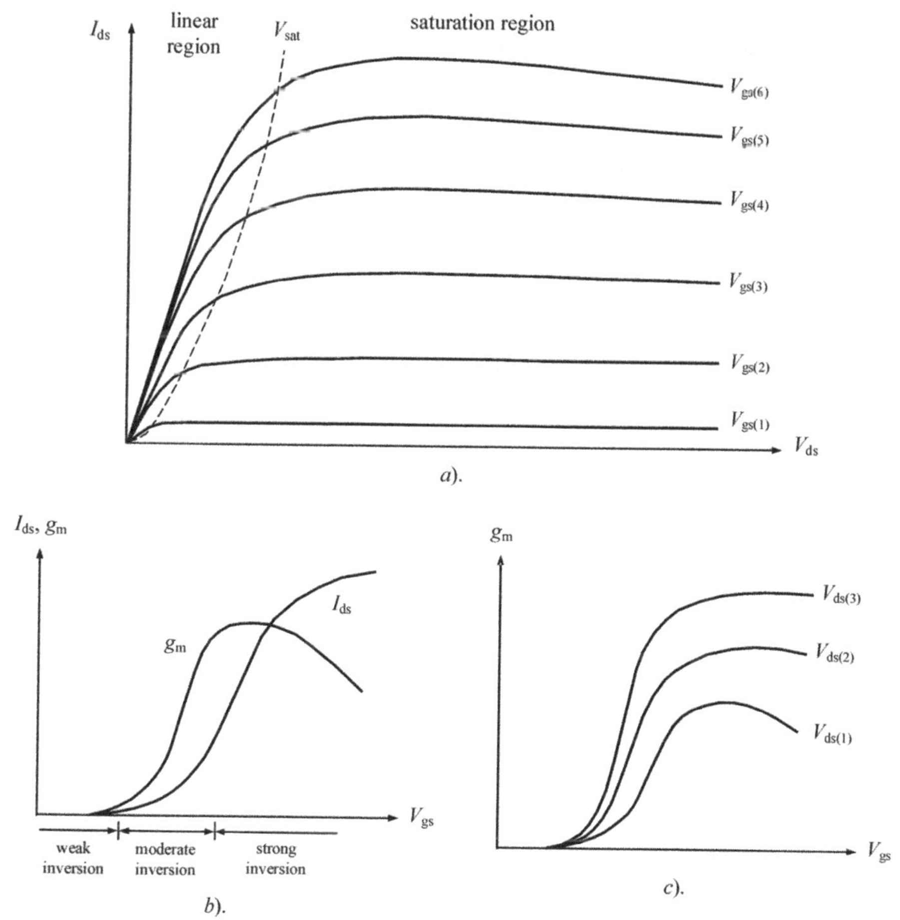

Figure 7.13 Drain current and device transconductance versus gate–source voltage

drain–source voltage V_{ds}. As seen for low drain bias voltages V_{ds}, the drain current I_d increases almost linearly with the drain bias. This region of the device output dc characteristic is therefore called the *linear region*. In this case, there is an inversion layer connecting the source and drain, which behaves as an ideal ohmic resistor. Then, at larger drain bias voltages, the dependence of the drain current on the gate bias voltage decreases. Finally, at some voltage, known as the saturation voltage V_{sat}, the drain current no longer increases with increasing drain–source voltage. This region is called the *saturation region*, due to the effect of the dc drain current saturation at high drain bias voltages for a certain gate–source bias voltage V_{gs}. At $V_{ds} = V_{sat}$, the channel is pinched off and the inversion layer is no longer affected by the drain voltages. At drain–source voltages below V_{sat}, the current continues to flow because there is no barrier to transfer carriers travelling down the channel toward the drain. As they arrive at the edge of the pinched-off region, they are pulled across it by the field directed from the drain toward the source. If the drain bias is increased further, any additional voltage is dropped across the depleted, high-field region near the drain electrode, and the point at which the channel is entirely depleted moves slightly toward the source [36]. Note that at very high gate bias voltages (voltages $V_{gs(5)}$ and $V_{gs(6)}$ in Figure 7.13a), due to a self-heating effect in the high dissipated power region, the slope of the output I–V curve will be negative, which means a degradation of the device transconductance [37].

The transfer characteristic of the MOS transistor behaves differently in different regions of the device operation. For example, the drain current in a weak-inversion region is mainly dominated by the diffusion component that increases exponentially with the gate voltage [1]. On the other hand, in the strong-inversion region when the gate-source voltage V_{gs} is greater than the threshold voltage V_{th}, the drain current is proportional to the square of $(V_{gs} - V_{th})$. However, with a further increase of V_{gs}, the sensitivity of the drain current to gate–source voltage increase becomes smaller, as shown in Figure 7.13b. Therefore, to link all regions of the device operation with the analytically derived closed-form expression, it is necessary to use transcendental functions [37, 38]. As seen, the transconductance g_m derived by differentiating I_{ds} reaches a maximum and then decreases with increasing V_{gs} due to the influence of V_{gs} on the effective carrier mobility and the increasing effect of the series parasitic resistances with increasing I_{ds}. For lower drain–sources voltages, $V_{ds(1)} < V_{ds(2)} < V_{ds(3)}$, the maximum value of the transconductance g_m is reduced, with corresponding significant degradation of the transconductance at large gate bias voltages, as shown in Figure 7.13c [39]. This means that the closer the drain bias voltage is to the saturation region, the faster the reduction in ΔI_{ds} for the same ΔV_{gs} at large gate–source bias voltages V_{gs}.

It is necessary, however, to distinguish the device dc operation regions from its operation regions under RF signal. Generally, the large-signal transistor behaviour is divided into three operation regions, as shown in Figure 7.14. The region where the input driving voltage v_{gs} is less than threshold voltage V_{th}, that is $v_{gs} \leq V_{th}$, is called the *pinch-off or cutoff region*. The term *pinch-off* is normally used for FET device. The operation region, where $v_{gs} > V_{th}$, is called the *active or linear region*, where the transistor can be considered as a voltage-controlled current source, responding linearly to the gate voltage drive. Finally, the transistor is roughly equivalent to a resistance r_{sat} operating in the *saturation region*. Unlike a dc current saturation region, the RF saturation region corresponds to voltage saturation, since there is a particular saturation voltage V_{sat}, corresponding to a fixed load resistance and defined by the intersection point between the load line and the linear part of the output I–V curves. The saturation or on-resistance r_{sat} is defined as a slope of the linear part of output I–V curves, that is, $r_{sat} = dv_{ds}/di_{ds}$.

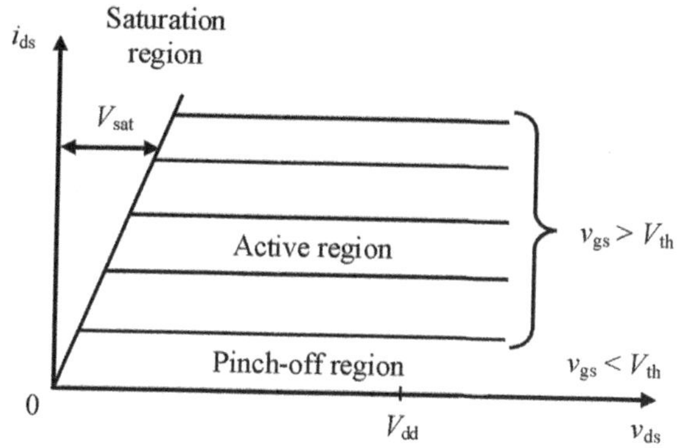

Figure 7.14 Idealized transistor output I–V characteristic and regions of RF operation

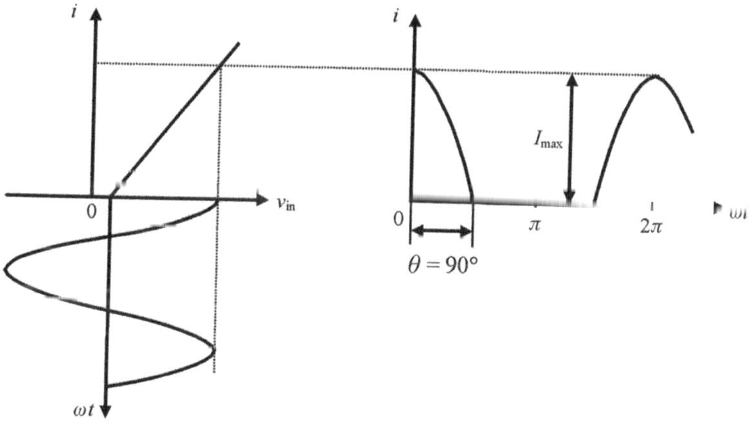

Figure 7.15 Output current waveform for device operation in active and pinch-off regions

In a Class B operation mode with a piecewise-linear approximation of the transistor transfer characteristic, an active device is operated both in the active and pinch-off regions. The magnitude of the output current exceeds a zero value during only half an entire signal natural period, representing a half-cosine waveform, as shown in Figure 7.15. In this case, because the parallel resonant LC-circuit has a high quality factor, ideally only the fundamental frequency signal is flowing into the load, whereas the higher-order harmonic components are short-circuited.

Analytically such an operation can be written as

$$i = \begin{cases} I_q + I\cos\omega t & -\theta \le \omega t < \theta \\ 0 & \theta \le \omega t < 2\pi - \theta \end{cases} \tag{7.31}$$

where I_q is the quiescent current, I is the output current amplitude and the θ is half the conduction angle, indicating the part of the RF current cycle for which device conduction occurs, and determines the moment when output current i takes a zero value.

At this moment,

$$i = I_q + I\cos\theta = 0 \tag{7.32}$$

and θ can be calculated from

$$\cos\theta = -\frac{I_q}{I} \tag{7.33}$$

Consequently, in a common case,

$$i = I(\cos\omega t - \cos\theta). \tag{7.34}$$

When $\omega t = 0$, the output collector current has a maximum amplitude of

$$i = I_{max} = I(1 - \cos\theta). \tag{7.35}$$

From Equation (7.33), the basic definitions are derived as

- when $\theta > 90°$, $\cos\theta < 0$ and $I_q > 0$, corresponding to Class AB operation;
- when $\theta = 90°$, $\cos\theta = 0$ and $I_q = 0$, corresponding to Class B operation;
- when $\theta < 90°$, $\cos\theta > 0$ and $I_q < 0$, corresponding to Class C operation.

As a result, the periodic half-cosine output current i can be represented as a Fourier series expansion

$$i = I_0 + I_1\cos\omega t + I_2\cos 2\omega t + I_3\cos 3\omega t + \ldots \tag{7.36}$$

where the dc, fundamental and higher-order harmonic components can be obtained from Equations (1.25–1.27) given in Chapter 1 in a common form

$$I_n = I\gamma_n(\theta) \tag{7.37}$$

where $\gamma_n(\theta)$ are the current coefficients for dc, fundamental and nth-harmonic components expressed through half the conduction angle θ.

It should be noted that, in an ideal Class B with $\theta = 90°$, the current coefficients for the third- and higher-order odd harmonics are equal to zero. Consequently, a half-cosine waveform consists of the dc and even-order harmonics only. In a real situation, when the transistor transfer characteristic has a quadratic dependence at its initial part, the output current waveform slightly deviates from an ideal waveform.

In an active region, the cosine voltage amplitude across the tank resistance R_L, connected in parallel to the tank, can be written using Equation (7.37) as

$$V = I_1 R_L = \frac{\gamma_1(\theta)}{\gamma_0(\theta)} I_0 R_L \tag{7.38}$$

which is a linear function of the dc current I_0 for a fixed conduction angle and tank resistance. However, for a varying conduction angle, defined by bias conditions and voltage drop across the current source, the voltage amplitude is a function of θ. For example, the voltage amplitude will be higher by $\pi/2$ in Class B with $\theta = 90°$, compared with an idealized nonharmonic condition of Class A with $\theta = 180°$. Using the tank loaded factor $Q_L = \omega_0 C R_L$, where $\omega_0 = 1/\sqrt{LC}$ is the oscillation frequency, Equation (7.38) can be rewritten as

$$V = \frac{\gamma_1(\theta)}{\gamma_0(\theta)} \frac{Q_L}{\omega_0 C} I_0 \tag{7.39}$$

showing the voltage amplitude reduction with the degradation of the loaded quality factor. This happens when voltage amplitude V approaches the saturation region with increasing shunting

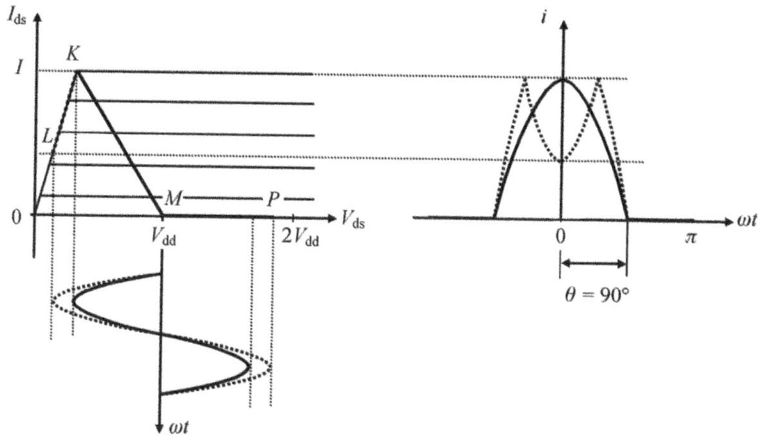

Figure 7.16 Collector voltage and current waveforms for device operation in saturation, active and pinch-off regions

effect of the drain–source resistance and reduces down to a value of r_{sat}. In this case, the voltage waveform across the tank resistance cannot be considered to be purely cosinusoidal because of the significant harmonic contribution. Consequently, any consideration of the voltage amplitude as constant in a saturation region, introduced in [40], can be applied only as a first-order approximation, since the contribution of the fundamental voltage spectral component will depend on the overall voltage waveform, which will be different for different load and bias conditions.

Now let us consider a Class B operation with increased amplitude of the cosinusoidal voltage across the tank using the transistor output I–V curves. In this case, it follows from Figure 7.16 that an active device is operated in saturation, active and pinch-off regions in succession, and the load line follows a broken line *LKMP* with three linear sections (*LK*, *KM*, and *MP*) [37]. The new section *KL* corresponds to the saturation region, resulting in the half-cosine current waveform with a depression in the top part. With further increase of the voltage amplitude, the output current pulse can be split into two symmetric pulses containing a significant level of the higher-order harmonic components. Similar simulated drain waveforms of a 1.8 GHz differential oscillator, based on a 0.18-μm CMOS process and operating in the pinch-off, active and saturation regions, are given elsewhere [41].

7.5.2 Start-up and steady-state conditions

The determination of the start-up and steady-state oscillation conditions is very often based upon a loop or nodal analysis of the circuit. However, the oscillator analysis using matrix techniques brings out the similarities between several types of oscillators and results in one group of equations, which can be used to analyse the different oscillator configurations [42]. In this case, a two-port network can represent both the active device and feedback element. Depending on the oscillator configuration, in the form of a parallel feedback or a negative resistance (conductance), an oscillator with parallel or series feedback using of admittance Y-or impedance Z-parameters can be respectively modelled.

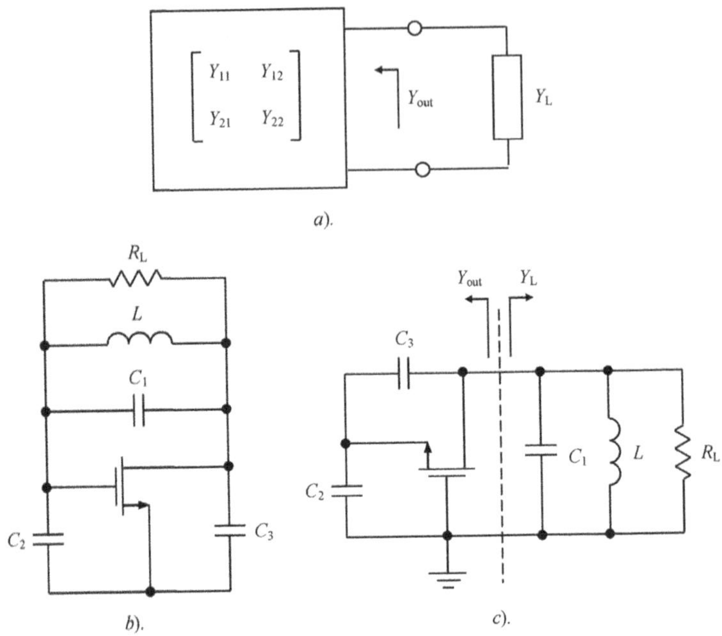

Figure 7.17 Basic CMOS oscillator schematics

The steady-state oscillation condition for a single-frequency negative conductance oscillator, the basic representation of which is shown in Figure 7.17a, can generally be expressed through Y-parameters as

$$Y_{out} + Y_L = 0 \tag{7.40}$$

where

$$Y_{out} = Y_{22} - \frac{Y_{12}Y_{21}}{Y_{11}} \tag{7.41}$$

is the output admittance expressed through the Y-parameters of a loaded two-port network, which includes the device and feedback elements.

Figure 7.17b shows the generic schematic of the modified Colpitts oscillator with a parallel resonant circuit, which is called a *Seiler oscillator* [43]. Here, C_2 and C_3 represent feedback capacitors and R_L is the load resistor, which generally can include any losses in the tank inductor L. Such an oscillator configuration is useful for wideband frequency tuning when the capacitance C_1 in the tank circuit is variable. Note that grounding of any terminal of the oscillator circuit does not change its electrical performance, provided there are no changes in the connection of the feedback elements and load to the active device. To analyse the oscillator start-up and steady-state conditions, it is convenient to consider a common gate configuration of this circuit, since the load resistance is connected between the drain and gate terminals. In this case, the common gate admittance matrix $[Y]_{CG}$ for a two-port network expressed through the common source Y-parameters can be written as

$$[Y]_{CG} = \begin{bmatrix} Y_{11} + Y_{12} + Y_{21} + Y_{22} & -(Y_{12} + Y_{22}) \\ -(Y_{21} + Y_{22}) & Y_{22} \end{bmatrix} \tag{7.42}$$

where Y_{ij} $(i, j = 1, 2)$ are the common source Y-parameters [37].

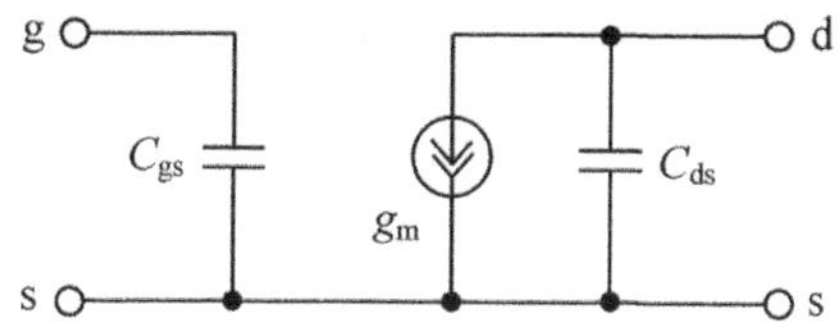

Figure 7.18 Simplified MOSFET equivalent circuit

Then, a steady-state oscillation condition for the oscillator circuit shown in Figure 7.17c can be rewritten through the device common source Y-parameters and feedback admittances $Y_2 = j\omega C_2$ and $Y_3 = j\omega C_3$ as

$$Y_{out} + Y_L = 0 \tag{7.43}$$

where

$$Y_{out} = Y_{22} + Y_3 - \frac{(Y_{12} + Y_{22} + Y_3)(Y_{21} + Y_{22} + Y_3)}{Y_{11} + Y_{12} + Y_{21} + Y_{22} + Y_2 + Y_3} \tag{7.44}$$

$$Y_L = Y_1 = \frac{1}{R_L} + j\omega C_1 + \frac{1}{j\omega L} \tag{7.45}$$

Separate equations for real and imaginary parts of the output and load admittances of a negative conductance oscillator can be obtained from Equation (7.43) as

$$\mathrm{Re}Y_{out} + \mathrm{Re}Y_L = 0 \tag{7.46}$$

$$\mathrm{Im}Y_{out} + \mathrm{Im}Y_L = 0 \tag{7.47}$$

Similarly, the start-up conditions for a negative conductance oscillator are written as

$$\mathrm{Re}Y_{out} + \mathrm{Re}Y_L < 0 \tag{7.48}$$

$$\mathrm{Im}Y_{out} + \mathrm{Im}Y_L = 0 \tag{7.49}$$

To obtain the explicit analytical relationships between the active device and resonant circuit parameters, consider the simplified high-frequency intrinsic MOSFET equivalent circuit shown in Figure 7.18, where C_{gs} is the gate-source capacitance, g_m is the device transconductance and C_{ds} is the drain–source capacitance. The admittance Y-parameters of the equivalent circuit are

$$\begin{matrix} Y_{11} = j\omega C_{gs} & Y_{12} = 0 \\ Y_{21} = g_m & Y_{22} = j\omega C_{ds} \end{matrix} \tag{7.50}$$

Substituting the device Y-parameters into Equation (7.44) allows the real and imaginary part of the output admittance to be represented through the elements of the device parameters as

$$\mathrm{Re}Y_{out} = -\frac{g_m\omega^2(C_{gs} + C_2)(C_{ds} + C_3)}{g_m^2 + \omega^2(C_{gs} + C_{ds} + C_2 + C_3)^2} \tag{7.51}$$

$$\mathrm{Im}Y_{out} = \omega\frac{\omega^2(C_{gs} + C_2)(C_{ds} + C_3)(C_{gs} + C_{ds} + C_2 + C_3)}{g_m^2 + \omega^2(C_{gs} + C_{ds} + C_2 + C_3)^2} \tag{7.52}$$

From Equations (7.51) and (7.52) it follows that, in a common gate configuration, the real part of the output admittance is negative and the imaginary part of the output admittance has a capacitive reactance. At frequencies $\omega \ll \omega_T$, where $\omega_T = 2\pi f_T$, $f_T = g_m/2\pi C_{gs}$ is the device transition frequency, Equation (7.51) can be simplified to

$$\mathrm{Re}Y_{out} \cong -g_m \left(\frac{\omega}{\omega_T}\right)^2 \frac{(C_{gs}+C_2)(C_{ds}+C_3)}{C_{gs}^2} \tag{7.53}$$

Substituting Equations (7.53) and (7.52) into Equations (7.48) and (7.49), representing the oscillator start-up conditions, yields

$$g_m > \frac{1}{R_L}\left(\frac{\omega_T}{\omega}\right)^2 \frac{C_{gs}^2}{(C_{gs}+C_2)(C_{ds}+C_3)} \tag{7.54}$$

$$\omega^2 = \frac{1}{L}\frac{1}{C_1 + \dfrac{C_{gs}+C_{ds}+C_2+C_3}{g_m R_L}} \tag{7.55}$$

From Equation (7.54) it follows that the build-up of the self-oscillations will be more easily provided at lower frequencies, lower ratio of C_{gs}/C_{ds} and smaller losses in the resonant circuit. In addition, the regeneration factor or start-up margin can be improved by selecting the proper values of the external feedback capacitances C_2 and C_3 connected between the gate–source and drain–source terminals in parallel to the gate–source and drain–source capacitances C_{gs} and C_{ds}, respectively. It follows from Equation (7.55) that the oscillation frequency is a function of not only the external reactive elements, but also the device parameters.

In a steady-state mode, the device small-signal transconductance should be considered as fundamentally averaged over the oscillation period during large-signal operation. By using a piecewise-linear approximation, its large-signal definition can be written as $g_m\gamma_1(\theta)$. Thus, assuming constant capacitances C_{gs} and C_{ds} whose values are given by the operating dc bias point, the oscillator steady-state conditions can be rewritten, for zero external feedback capacitances C_2 and C_3, as

$$g_m = \frac{1}{\gamma_1(\theta)R_L}\left(\frac{\omega_T}{\omega}\right)^2 \frac{C_{gs}}{C_{ds}} \tag{7.56}$$

$$\omega^2 = \frac{1}{L}\frac{1}{C_1 + \dfrac{C_{gs}+C_{ds}}{g_m\gamma_1(\theta)R_L}} \tag{7.57}$$

where g_m and ω_T are the small-signal transconductance and angular transition frequency, respectively. Typically, the value of the gate–source capacitance C_{gs} is equal to oxide capacitance at low and high bias voltages, and is reduced by approximately two to three times in the region near the threshold voltage. The drain–source capacitance C_{ds} can be considered as the junction capacitance, for which the maximum large-signal value deviates from the small-signal value by not more than by 10–20% [37]. Consequently, as a first-order approximation, the capacitances C_{gs} and C_{ds} can be modelled as the fixed capacitances measured at the quiescent dc bias voltage.

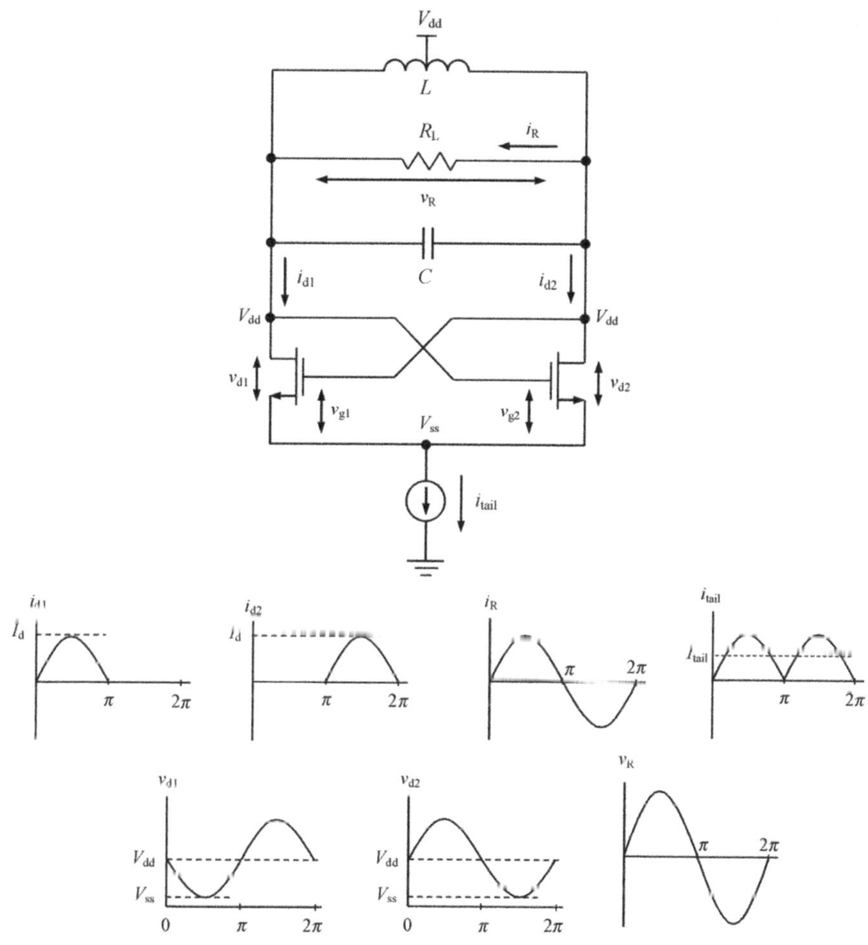

Figure 7.19 Differential cross-coupled *LC* oscillator: circuit schematic and operation principle

7.5.3 Differential cross-coupled oscillators

It is most convenient to describe the operational principle of the differential cross-coupled *LC* oscillator by the example of an ideal class B operation with a piecewise-linear approximation of each transistor transfer characteristic, which means that each transistor conducts exactly half a cycle with zero quiescent current. The simplified equivalent circuit of the differential cross-coupled *LC* oscillator with a tail current source is shown in Figure 7.19.

If the quality factor of the tank circuit is assumed sufficiently high to provide the sinusoidal voltages applied to the gate–source terminals of the transistors and across the tank resistor R_L, the drain current of each transistor can be represented in the following half-sinusoidal form:

for first transistor

$$i_{d1} = \begin{cases} +I_d \sin \omega t & 0 \leq \omega t < \pi \\ 0 & \pi \leq \omega t < 2\pi \end{cases} \tag{7.58}$$

for second transistor

$$i_{d2} = \begin{cases} 0 & 0 \le \omega t < \pi \\ -I_d \sin \omega t & \pi \le \omega t < 2\pi \end{cases} \tag{7.59}$$

Since, for an idealized piecewise-linear approximation, the third- and higher-order odd harmonics of the drain currents are equal to zero, the total current i_R flowing across the tank resistor R_L is a difference of the two out-of-phase drain currents,

$$i_R = i_{d1} - i_{d2} = I_d \sin \omega t \tag{7.60}$$

representing a purely sinusoidal waveform.

The current flowing into the tail current source through the centre point of the circuit is the sum of the drain currents:

$$i_{tail} = i_{d1} + i_{d2} = I_d \left| \sin \omega t \right| \tag{7.61}$$

containing the dc and even-order harmonic components.

Ideally, even-order harmonics are cancelled out and should not appear at the resistor. In practice, the second harmonic is suppressed by approximately 20 dB or more below the fundamental. It is necessary to connect a bypass capacitor to the centre point of the circuit in order to exclude power losses due to even-order harmonics. The current i_L produces a sinusoidal voltage across the resistor R_L equal to

$$v_R = I_d R_L \sin \omega t = V_R \sin \omega t \tag{7.62}$$

The dc component I_{tail} of the total drain current i_{tail} can be defined by integration over the oscillation period as

$$I_{tail} = \frac{1}{2\pi} \int_0^{2\pi} i_{tail}(\omega t)\, d(\omega t) = \frac{2}{\pi} I_d \tag{7.63}$$

7.5.4 Wideband tuning techniques

The VCO performance in terms of phase noise and tuning range determines the basic characteristics of a whole transceiver. However, a limited frequency-tuning range is usually a serious problem for VCO fully based on CMOS technology. There are four basic candidate single differential topologies, shown in Figure 7.20 [44]. Figure 7.20a shows an nMOS differential VCO topology with tail current source, while Figure 7.20b shows an nMOS differential topology with top current source. The pMOS differential topologies with tail and top current sources are shown Figure 7.20c and d, respectively. As regards tuning range capability, topologies shown in Figure 7.20b and c are preferable. This is because the anode varactor bias voltage is fixed to V_{dd} for an nMOS oscillator with tail current source, and the anode varactor bias voltage is fixed to zero for a pMOS oscillator with top current source. Since using pMOS devices results in a lower phase noise due to inherently smaller low-frequency $1/f$ noise, the pMOS differential topology with tail current source shown in Figure 7.20c represents the best choice for low-noise wideband tuning. Being implemented in a 0.25-μm standard CMOS process, such a differential VCO with a Q-factor of the tank inductor of 7.5 provides the phase noise of −109 dBc/Hz and −123 dBc/Hz at 100 kHz and 500 kHz offset from carrier

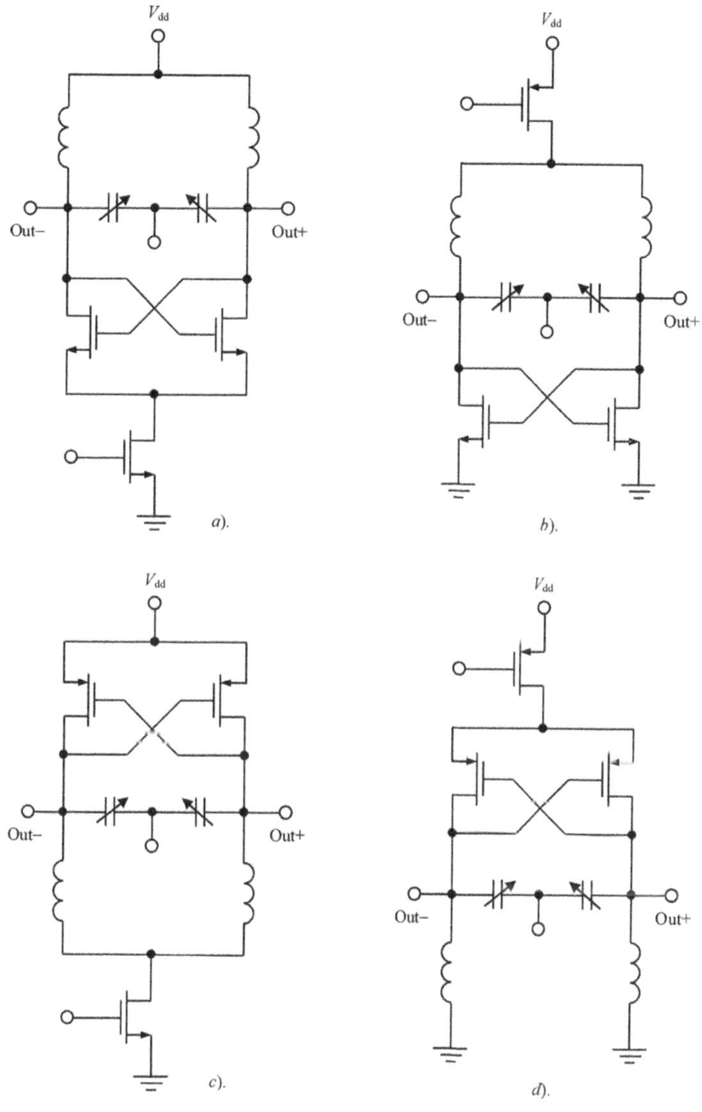

Figure 7.20 Single differential topologies of CMOS VCO

frequency of 1.3 GHz, respectively. The tuning range was 13.3% for $V_{dd} = 1.4$ V and 20.1% for $V_{dd} = 2.0$ V.

Using a proper configuration of the accumulation MOS varactors can significantly improve the tuning range. Such a varactor with high quality factor fabricated using the 0.13-μm CMOS silicon-on-insulator (SOI) technology demonstrates the capacitance ratio of 5 with $\pm$ 1 V voltage tuning that provides over 50% frequency-tuning range [45]. However, a high capacitance ratio implies high varactor tuning sensitivity or K_{VCO}, which makes the oscillator phase noise worse. A band switching solution can reduce the varactor tuning sensitivity, but requires extra control circuitry. A simple and effective solution can be provided by differential varactor tuning to avoid the effect of high K_{VCO}.

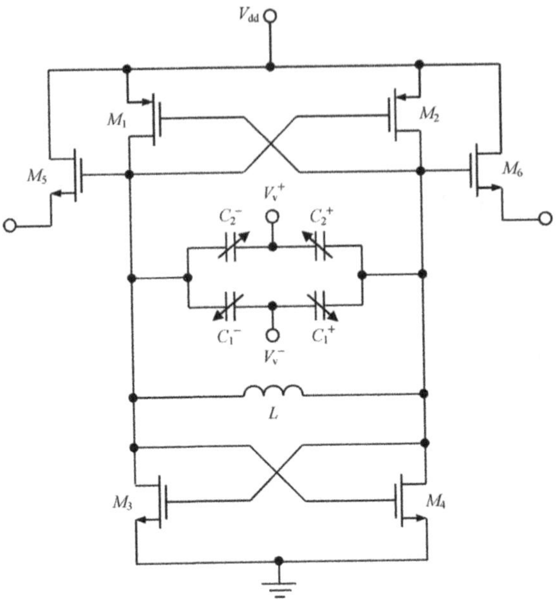

Figure 7.21 Schematic of a differentially tuned VCO

The differentially tuned varactor topology consists of two varactor pairs, as shown in Figure 7.21 [46]. One pair is excited by RF signal at the cathode (C_1^+ and C_1^-) and the other pair is excited by RF signal at the anode (C_2^+ and C_2^-). The buried oxide of SOI makes accumulation MOS varactors with higher Q-factor and more symmetric than those in bulk technology. As a result, the varactor capacitance ratio can be achieved as high as 6 with ± 1 V voltage tuning. At supply voltage of 1.4 V, the best phase noise of -122 dBc/Hz at 1 MHz offset was measured at the lower bandwidth frequency of 3.8 GHz. At the upper bandwidth frequency of 5.4 GHz, the measured phase noise was -112 dBc/Hz at 1 MHz offset. Due to common-mode noise rejection, the differential-tuning topology improves the phase noise by approximately 9 dB. To minimize the phase noise variations over frequency range and to make the entire phase-locked loop circuit more stable, it is important to provide an equalization of K_{VCO} over the frequency-tuning range. It can be done by connecting in parallel several back-to-back varactor pairs and biasing them at different voltages [47].

However, the band-switching technique has proved to be a successful way to increase tuning range. The generic binary-weighted band-switching LC tank configuration of nth size is shown in Figure 7.22 [48]. The MOS switch in a unit branch of the array contributes a parasitic capacitance C_{d} that is mainly composed of its drain-to-bulk junction and drain-to-gate overlap capacitances. In this case, by neglecting parasitic capacitances, the frequency tuning ratio $K_{\mathrm{f}} = f_{\max}/f_{\min}$ can be written as

$$K_{\mathrm{f}} = \sqrt{\frac{\dfrac{kK_{\mathrm{c}}}{K_{\mathrm{c}}-1}(\beta-1)+(2^n-1)\beta}{\dfrac{k}{K_{\mathrm{c}}-1}(\beta-1)+(2^n-1)}} \tag{7.64}$$

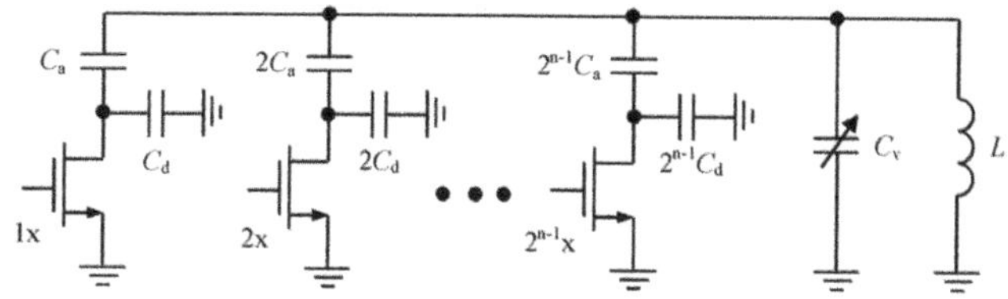

Figure 7.22 Generic binary-weighted band-switching *LC* tank configuration

where $K_c = C_{vmax}/C_{vmin}$ is the varactor capacitance tuning ratio, C_{vmax} is the maximum varactor capacitance, C_{vmin} is the minimum varactor capacitance, k is the overlap safety margin factor, being greater than unity, and $\beta = 1 + C_a/C_d$. For example, in order to achieve more than a 2:1 tuning range, the coefficient β should be more than 4 for $k = 1.1$, $n = 4$ and K_c greater than 2.5. As regards array size n (i.e., the number of bits controlling the binary-weighted array), adding more bits is beneficial to the tuning range, but only to a certain degree. Beyond a certain point, the minimum fixed capacitance in the design prevents any further improvement. When implemented in a 0.18-μm bulk CMOS technology with a device gate width of 32 μm, a very wide frequency range from 1.14 to 2.46 GHz was achieved with a maximum tuning sensitivity of 270 MHz/V. The frequency range is split into 16 sub-bands by means of a 4-bit binary weighted array of switched capacitors of 5 pF each and small accumulation MOS varactors with $K_c = 3.2$ covering each sub-band. Typical measured phase noise was −123.5 dBc/Hz at 600 kHz offset from 1.8 GHz for a core power consumption of 4.8 mW from a 1.5 V dc power supply.

A high tuning range can also realized by varying the tank inductance. Figure 7.23 shows the schematic of a CMOS VCO with self-tuned inductors [49]. For a high control voltage V_{ctr}, the transistors M_1 and M_2 are switched on most of the time, resulting in a low inductance, provided by the parallel inductors L_1 and L_2, and high oscillation frequency. Conversely,

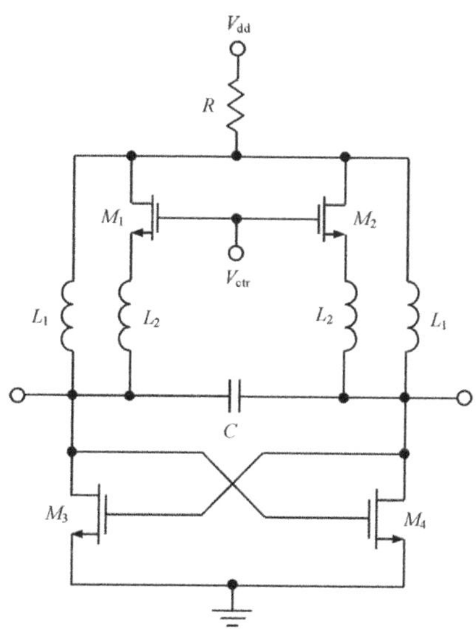

Figure 7.23 Schematics of a CMOS VCO with self-tuned inductors

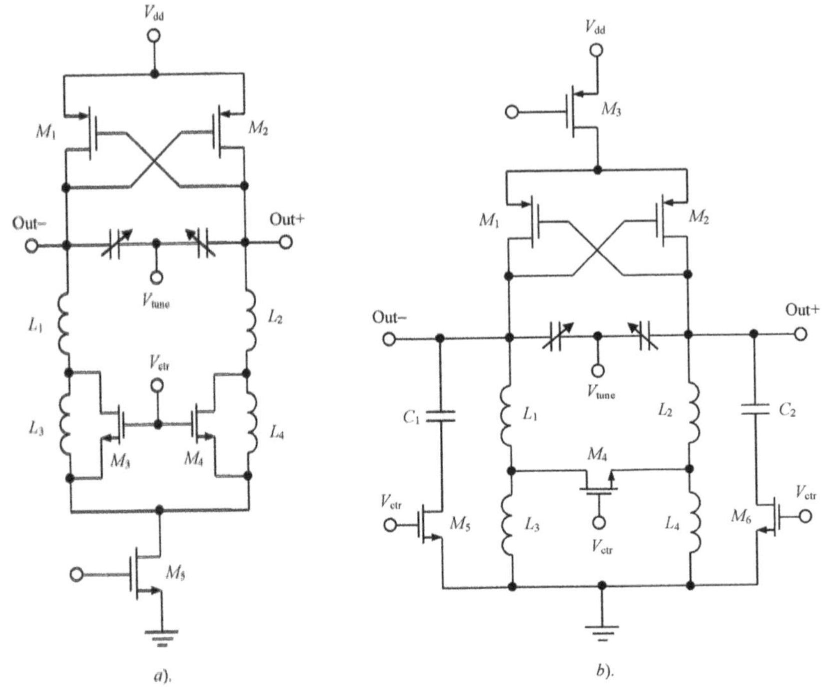

Figure 7.24 Schematics of multiband VCOs

for a low control voltage V_{ctr}, these transistors are open and the inner inductors L_2 are not active. Therefore, the resulting inductance is high, provided by the inductance L_1 only, and the oscillation frequency is low. The boundary frequencies are related to each other according to

$$f_{max} = \sqrt{1 + \frac{L_1}{L_2}} f_{min} \tag{7.65}$$

For a ratio $L_1/L_2 = 2$, a frequency tuning range of 1.34–2.14 GHz by varying a control voltage from 0 to 3 V was achieved.

To improve the poor phase noise performance demonstrated by a VCO circuit with self-tuned inductors only, the concept of switched resonators can be used. Figure 7.24a shows the differential VCO schematic with varactors and switched inductors [50]. The inductances are changed by turning transistors M_3 and M_4 on and off. When the transistors are off, each inductance is approximately equal to a sum of inductances L_1 and L_2. However, when these transistors are switched on, the inductors L_3 and L_4 are shunted by low transistor on-resistances. Since on-resistances of the nMOS transistors M_3 and M_4 are connected in series with the inductors L_1 and L_2, thus reducing the inductor quality factor, the size of these devices is increased. The large size of the tail current source pMOS transistor M_5 also reduces low-frequency $1/f$ noise. The VCO is coarsely tuned by switched inductors and finely tuned using MOS varactors. When the switching transistors M_3 and M_4 are turned off with low control voltage V_{ctr}, the output frequency is varied from 667 to 942 MHz by varying the varactor tuning voltage V_{tune} from 0 to 1.5 V. When the switching transistors are turned on with high control voltage V_{ctr}, the varactors tune VCO from 813 to 1156 MHz. At a supply voltage of 1.5 V, nearly constant phase noise between −123 and −124 dBc/Hz at 600 kHz offset was observed over the entire frequency tuning range.

The multiband top-biased differential CMOS VCO schematic with switched inductors and capacitors is shown in Figure 7.24b [51]. The differential inductors L_1 to L_4 and nMOS transistor M_4 form the switched inductor part, while capacitors C_1 and C_2 in combination with nMOS transistors M_5 and M_6 form the switched capacitor part of a tank circuit. The switched components coarsely tune VCO to four frequency bands: 2.40–2.44 GHz, 2.47–2.52 GHz, 4.65–4.80 GHz and 4.92–5.12 GHz. Being implemented in a 0.18-μm CMOS process, such an all-pMOS multiband VCO demonstrates phase noise of −134 dBc/Hz at 1 MHz offset from 2.4 GHz and −124 dBc/Hz at 1 MHz offset from 4.7 GHz with power consumption of 6 and 4.5 mW at a supply voltage of 1 V, respectively.

7.5.5 Quadrature VCOs

Recently, due to the need for accurate quadrature signals in modern image-rejection wireless transceivers, the quadrature frequency generation technique has attracted a lot of attention. Generally, three options can be used to generate quadrature signals: combination of the VCO, polyphase filters and output buffer (or limiter) stages; a VCO operating at double the desired frequency followed by a frequency divider performed either in the digital or analogue form, and coupling of the two symmetric cross-coupled differential VCOs to each other. Since both the first and second approaches introduce an excessive power consumption, although the second approach can provide the smallest area and be beneficial to avoid pulling effect, the third approach outperforms others in terms of both the power consumption and the phase noise performance (provided that either good quality on-chip inductors are available or off-chip inductors are allowed).

The block schematic of a quadrature VCO consisting of two identical differential VCOs is shown in Figure 7.25a. The combination of the direct and inverted connection forces the two VCOs to oscillate in quadrature. The original circuit implementation of this design principle, based on the two identical differential cross-coupled tail-biased nMOS VCOs, with the coupling transistors M_{cpl} connected in parallel to the switching transistors M_{sw}, is shown in Figure 7.25b [52]. However, despite being based on LC-tank VCOs, such a parallel quadrature CMOS VCO has a poor phase noise performance that essentially limits its application. In this design, the gate widths of both coupling and switching transistors are equal.

Figure 7.26 shows an alternative approach to design quadrature VCOs, in which the coupling transistor M_{cpl} is connected in series with the switching transistor M_{sw}, forming a cascode-type connection that enables the phase noise performance to be significantly improved [53]. There are two circuit configurations corresponding to this approach: a top-series quadrature VCO where M_{cpl} is placed on the top of M_{sw}, shown in Figure 7.26a and a bottom-series quadrature VCO where M_{sw} is placed on the top of M_{cpl}, shown in Figure 7.26b. To compare both phase noise and phase error of the parallel and series quadrature VCOs, it was convenient to introduce the parameter $\alpha = W_{cpl}/W_{sw}$, where W_{cpl} and W_{sw} are the gate widths of the coupling and switching transistors, respectively. It is assumed that the gate lengths of both transistors are equal. The simulations show that reduction in α results in higher phase error, but lower phase noise for a parallel quadrature VCO. This means that it is easy to improve the phase noise performance of such a quadrature VCO at the expense of its phase error. Unlike

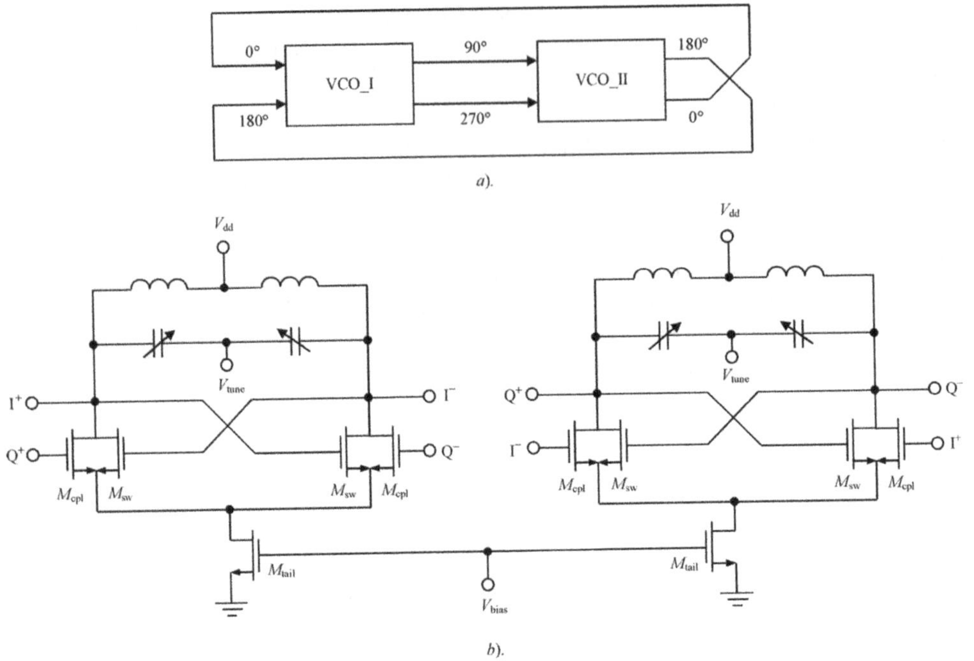

Figure 7.25 Basic schematics of a quadrature CMOS VCO

a parallel quadrature VCO, the phase error in series quadrature VCO is almost independent of α for all reasonable values. As a result, if the phase errors of both parallel and top-series quadrature VCOs are approximately the same at $\alpha = 1/2$, then the phase noise of a top-series quadrature VCO is lower than the phase noise of a parallel quadrature VCO by 8–11 dB over wide range of the offset frequencies from 10 kHz to 1 MHz [53]. The measurement results for such a quadrature VCO implemented in a 0.35-μm standard CMOS process show a tuning frequency range from 1.64 to 1.97 GHz, a phase noise of −140 dBc/Hz or less at 3 MHz offset and a maximum phase error of 0.25°, for a current consumption of 25 mA at a supply voltage of 2 V. The comparison of the phase noise between the top-series and bottom-series quadrature VCOs gives the same values at small offsets up to 10 kHz, but better phase noise by 6 dB at 3 MHz offset for a bottom-series quadrature VCO [54]. The measurement results for a bottom-series quadrature VCO implemented into the same CMOS process show the worst-case phase noise of −140 dBc/Hz at 3 MHz offset over a tuning range from 1.91 to 2.27 GHz, for a current consumption of 16 mA at a supply voltage of 1.3 V.

Figure 7.27 shows the schematic of the complementary cross-coupled differential quadrature VCO fully fabricated using a 0.25-μm standard CMOS technology [55]. The complementary structure compared with its all-*n*MOS counterpart offers higher transconductance for a given current, better phase noise at all bias points and lower dc voltage drop across the channel [41]. The frequency tuning was provided by using of a parallel connection of *n*MOS and *p*MOS varactors with separate tuning voltage inputs. Such a connection enables in first-order a differentially tuned VCO with equal, but opposite sign K_{VCO} to *n*MOS and *p*MOS tuning input. A quadrature phase accuracy of approximately 3° was measured. Such a sufficiently high phase error is a result of a reduced coupling between two VCOs with $\alpha = 1/3$ when the better

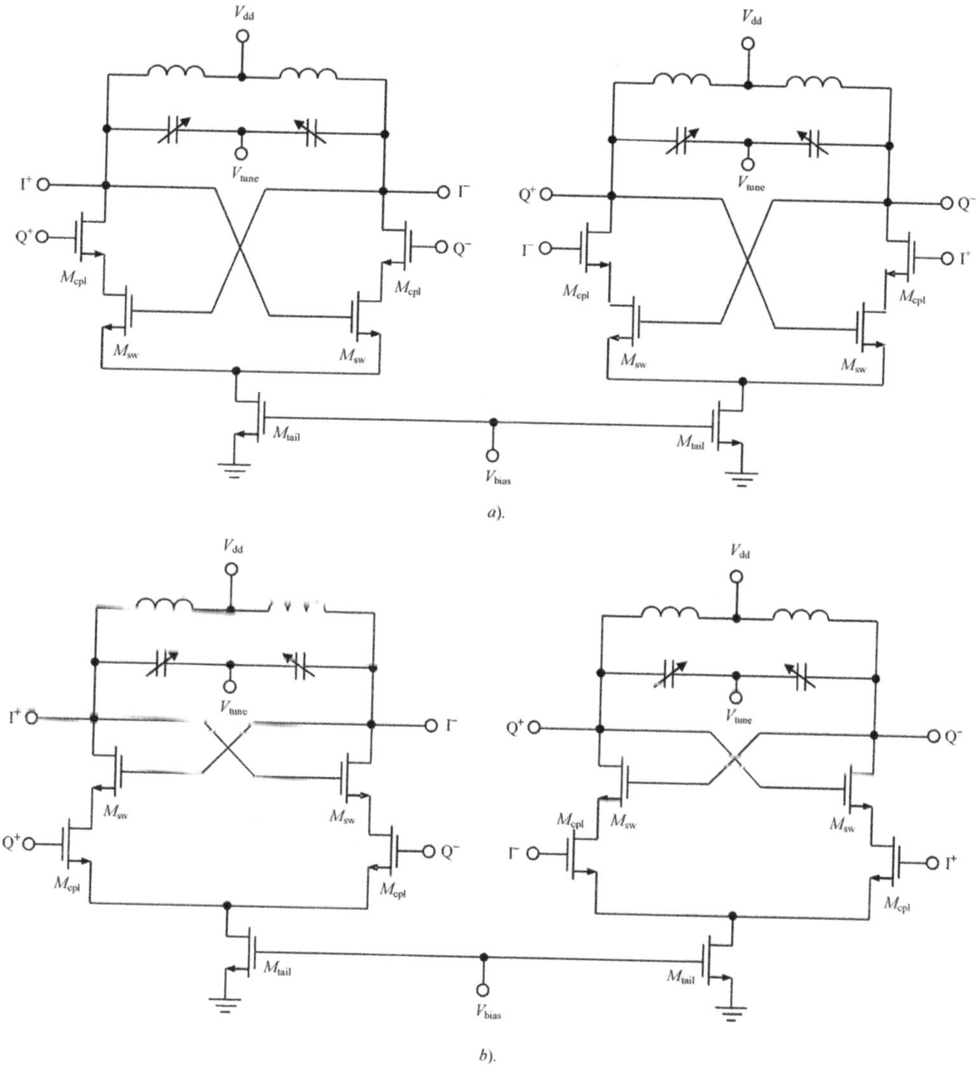

Figure 7.26 Top-series and bottom-series quadrature CMOS VCO schematics

noise performance is achieved. At a supply voltage of 2.5 V and a total power consumption of 20 mW, such a complementary quadrature VCO demonstrates a worst-case phase noise of −143 dBc/Hz at 3 MHz offset over the tuning range 1.71–1.91 GHz [55].

7.6 IMPLEMENTATION TECHNOLOGY ISSUES

With the continual progress in CMOS technology providing a reduction of the device channel length to deep submicrometre values, it has become feasible to integrate the baseband and signal-processing blocks together with the RF transceiver on a single chip using CMOS

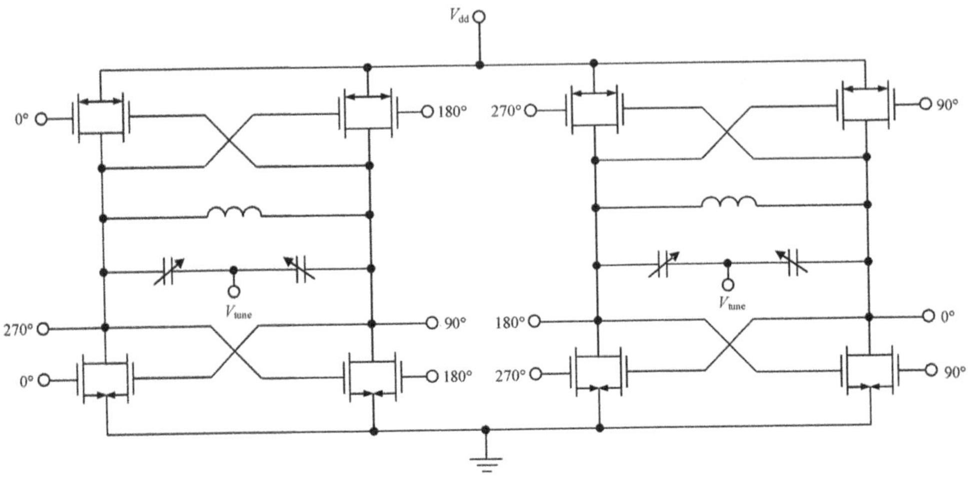

Figure 7.27 Complementary cross-coupled quadrature CMOS VCO schematic

process exclusively. However, the performance of the CMOS devices with short channel can significantly be affected by such a phenomenon as hot-carrier degradation. The high-energy carriers known as hot carriers, when travelling from the source toward the drain in a strong electric field, can damage the device structure by introducing interface traps and oxide-trapped charges, resulting in degradation of the device electrical parameters such as threshold voltage V_{th} and transconductance g_m. Under hot-carrier stress conditions, the region of maximal interface damage generation is shifted toward drain. Experimental verification based on the LC-tank cross-coupled differential top-biased CMOS oscillator using 0.18-μm process technology shows the degradation of the current–voltage (I–V) characteristics of the nMOS devices, resulting in the smaller oscillation amplitude [56]. The hot-carrier stress contributes to an increase of the output conductance of the device in the saturation region. For long stress times beyond 500 s, the output conductance G_{ds} can increase by more than three times, whereas the transconductance g_m can reduce by more than 30%. Also, hot-carrier stress affects the values of the device capacitances, resulting in higher oscillation frequency due to the changes in the gate–drain, gate–source and drain–substrate junction capacitances, reducing the total capacitance of the resonant circuit.

Since $1/f$ noise and hot-carrier degradation are assumed to have the same physical origin produced by traps at the interface between oxide and silicon, it is important to understand the hot-carrier effect on the oscillator phase noise performance. Generally, the noise-related hot-carrier behaviours of pMOS and nMOS transistors are different [57]. For an nMOS device, the $1/f$ noise increases strongly, especially after 10^5 s stress time, and this increase is much stronger than the decrease in the drain current. The $1/f$ noise and drain current in the saturation region degrade almost linearly as a function the channel length. The shorter channel, the more degradation in drain current and $1/f$ noise occurs due to shift of the threshold voltage and damage induced by hot carriers, respectively. A degradation of the oscillator phase noise by 4 dB at stress time 500 s is observed at 100 kHz offset from the carrier [56]. Similar to the nMOS devices, the interface traps induced by stress are responsible for $1/f$ noise in pMOS transistors. However, the overall situation is different since, in the latter case, $1/f$ noise degradation effects caused by interface traps are masked by electron trapping, which is dominated at low stresses whose trap charges do not generate $1/f$ noise [57]. But after long stress times, the behaviour of a pMOS transistor becomes very similar to its nMOS counterpart when the noise increases strongly with decreasing drain voltage. The drain current shows a similar behaviour and also degrades in the saturation mode as a function of channel length. Thus, a more reliable operation condition can be provided by choosing the device with longer channel length, by appropriate choice of the transistor operation region, or by using the pMOS device as being generally more resistant to hot-carrier effects.

The soft breakdown phenomenon for ultra-thin gate oxide layer of less than 5 nm corresponds to an anomalous increase of the stress-induced leakage current and the occurrence of fluctuations in drain current. The soft breakdown model can consist of an oxide breakdown without strong thermal effect that leads to an irreversible failure of the device [58]. It is assumed that multiple tunnelling via electron traps generated in the ultra-thin gate oxide layer is the physical mechanism of the electron transport in a localized small area of the MOS capacitor after soft breakdown. The model implies a generation of electron traps in the SiO_2 layer over the whole capacitor area and the assumption that stress-induced leakage current due to the oxide degradation is the total of the trap-assisted tunnel current of all the electron traps. Soft breakdown will occur if a critical number of electron traps is reached locally, to form a conductive path between the poly-Si gate and silicon substrate. Total destruction of the capacitor structure takes place when the energy that is dissipated in this localized area exceeds the amount needed for thermal breakdown of the oxide. The oxide breakdown is determined by the large voltage drop in the applied voltage that suddenly occurs. The experimental results for a 0.18-μm nMOS device show that its dc parameters such as threshold voltage V_{th}, transconductance g_m and drain saturation current I_{dsat} are not drastically changed during both soft and hard breakdown [59]. Also, the device transition frequency f_T is only reduced from 45 to 42 GHz at gate bias voltage $V_g = 0.9$ V and drain supply voltage $V_{dd} = 1.1$ V. However, large deterioration of the minimum noise figure F_{min} can be observed under hard breakdown, increasing from 1 dB to almost 4 dB at 12 GHz. Such a drastic increase of the device high-frequency noise after gate oxide breakdown can be attributed to the significant increase in the contribution of gate shot noise due to increased gate leakage current.

Integration of digital baseband and RF front-end in a single chip makes the RF circuits vulnerable to coupling of the digital switching noise through the common substrate. Measurements up to 10 GHz show that a simple resistive network is not sufficient to represent the substrate. Therefore, a more complicated model, although simple enough, based on the complex π-network with parallel resistance and capacitance in each branch, can be used [60]. Depending on the test structure, the equivalent circuit for substrate coupling modelling can also be represented in the form of the T-network equivalent [61]. Comparison of the phase noise of the test cross-coupled differential tail-biased 3.5-GHz VCO with a symmetric inductor using 0.18-μm CMOS process technology as a figure-of-merit to evaluate different protection schemes such as p^+ ground bars, deep n-wells and their combination, demonstrates that deep n-wells can be considered as the most effective isolators for the substrate coupling noise. The noise was injected into the substrate by the digital switching noise emulator using a p^+ substrate trap. When the digital noise emulator is switched on without any protection, the VCO phase noise degrades by 12 dB.

7.7 PRACTICAL SCHEMATICS OF CMOS VCOs

Since the oscillator phase noise depends primarily on the quality factor of the tank circuit, it can be improved by using bondwire inductors. Figure 7.28 shows the schematic of the differential

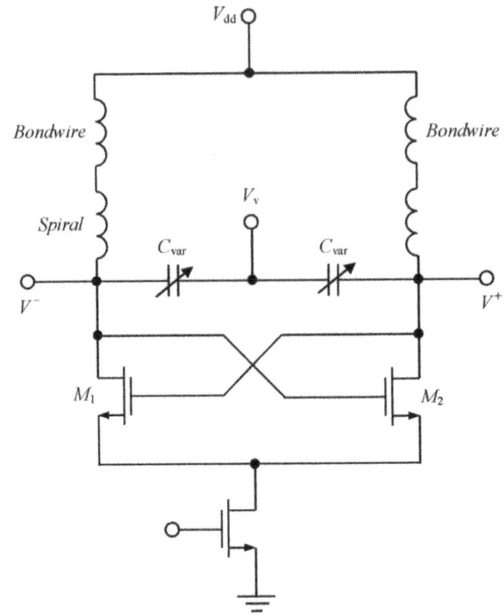

Figure 7.28 Schematic of a differential *LC* CMOS VCO

tail-biased CMOS VCO, where the resonant circuit inductors are realized by series connections of the spiral on-chip inductors and bondwires connecting the die pads to the package leads [62]. Two accumulation-mode MOS varactors are used in a differential configuration, while the cross-coupled *n*MOS differential pair provides positive feedback, feeding the tank to compensate for the losses. The main concern in the use of bondwires as tank inductors is that their value is affected by a large spread. Therefore, to be able to use bondwire inductors, a tuning element with a larger tuning range is required. However, this causes a significant VCO gain variation. Hence, despite the very low phase noise performance of such a VCO, it can hardly be used within a frequency synthesizer due to potentially large loop gain variation, which can result in severe stability issues. However, a compromise solution is possible when the bondwire inductor represents only a part of the overall tank inductor. As a result, using the combination of a spiral inductor of 3.7 nH with Q-factor of 3 and a bondwire inductor of 3 nH with Q-factor of 35 results in a phase noise of −119 dBc/Hz at 600 kHz from carrier of 1.3 GHz with oscillator current consumption of 6 mA and supply voltage of 2 V. The oscillation frequency can be varied from 1.1 to 1.45 GHz. If a solution that uses only bondwire inductors were acceptable, the quality factor of 17 can be achieved for the tank circuit that gives almost 10 dB phase noise improvement for a given current consumption using 0.35-μm digital CMOS technology.

The quality factor of the spiral inductor can be improved by optimization of its geometry for a given technology to minimize both its series metal resistance and extra losses caused by the substrate spread resistance, skin effect and eddy currents. The low resistance of 3.7 Ω was achieved for a three-terminal balanced octagonal inductor with simulated inductance of 2.857 nH in a 0.25-μm standard CMOS technology with only two metal layers, 0.6 and 1 μm thick, respectively [63]. To minimize the slope variation of the frequency tuning curve, *p*MOS

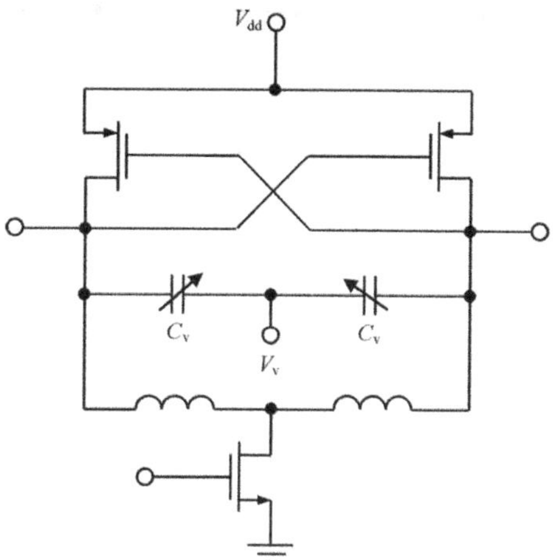

Figure 7.29 Differential *p*MOS VCO circuit topology

devices in conjunction with two p^+/n diode varactors were used, as shown in Figure 7.29. As a result, by using a differential tail-biased *p*MOS topology, the phase noise was as low as −127.5 dBc/Hz at 600 kHz and −142 dBc/Hz at 3 MHz offsets from the carrier of 1.8 GHz. A 28% wide tuning range was achieved with a supply voltage of 1.8 V and current consumption of 18 mA.

Using a 0.13-μm partially depleted silicon-on-insulator (SOI) technology provides reduction of the fixed transistor capacitance by 30% compared with bulk CMOS and reduced parasitic substrate capacitance with wider capacitance tuning ratio of the accumulation-mode MOS varactors [64]. Improved phase noise performance by 2 dB can be realized using high-resistivity substrate (∼ 300 Ω cm) compared with using low resistivity substrate (10 Ω cm). Finally, the further noise improvement can be achieved by using a single-turn inductor topology eliminating interwinding capacitance. The schematics of a complementary cross-coupled balanced VCO is shown in Figure 7.30, with *n*MOS devices having $f_T = 148$ GHz and $f_{max} = 162$ GHz and *p*MOS devices having $f_T = 68$ GHz and $f_{max} = 102$ GHz, respectively. The resonator consists of a single-turn copper inductor, with quality factor greater than 50 and estimated inductance of 60 pH at 40 GHz, and the accumulation-mode MOS varactors, which have a nominal capacitance of 0.4 pF. The output signals are buffered using source followers based on *n*MOS transistors. The measured phase noise was −109 dBc/Hz at 4 MHz offset from carrier of 40 GHz with output power of −8 dBm, power dissipation of 11.25 mW in the core at supply voltage of 1.5 V. The VCO achieves a tuning range of 15% with a control voltage of $\pm 0.5V_{dd}$. By making the metal interconnection between the two cross-coupled transistors shorter, which lowers the losses and parasitic capacitance, and optimizing the device finger width, the VCO operating frequencies can be increased, and 99 GHz VCO can demonstrate a tuning range of 2.5 GHz with the phase noise varying from −99.5 to −102.7 dBc/Hz at 10 MHz offset [65].

Using a complementary Colpitts oscillator topology allows one to achieve lower phase noise performance of the oscillator by providing reduced hot-carrier effect and higher transconductance of the constitutent transistors. Figure 7.31a shows a complementary Colpitts oscillator schematic, which is composed of a complementary M_1 (*n*MOS) and M_2 (*p*MOS) transistor pair, a tank inductor and an accumulation-mode varactor [66]. The transistors M_3 (*n*MOS)

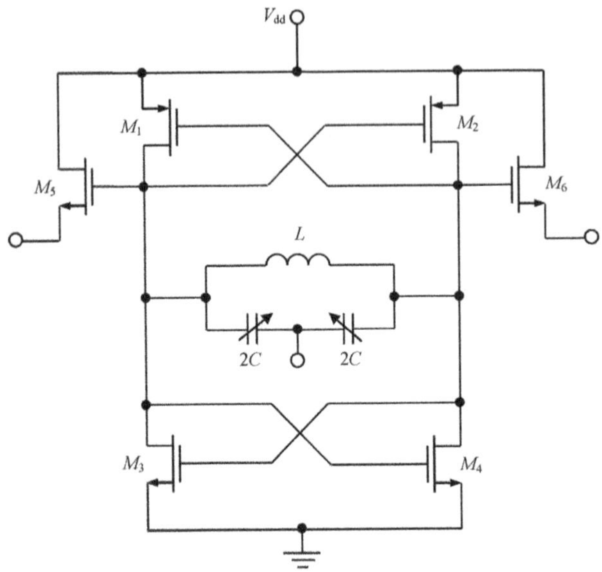

Figure 7.30 Schematic of a 40-GHz complementary VCO

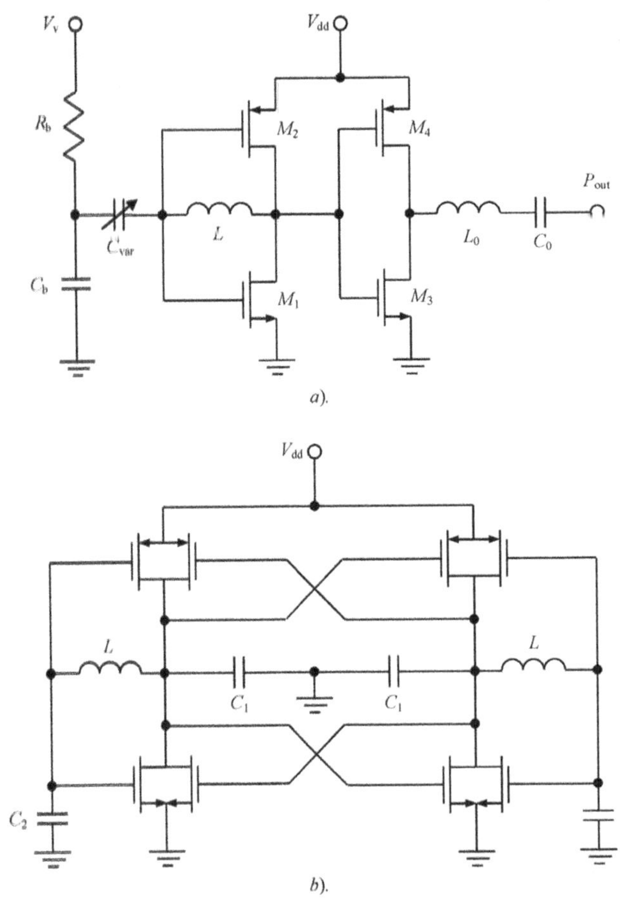

Figure 7.31 Complementary Colpitts oscillator topologies

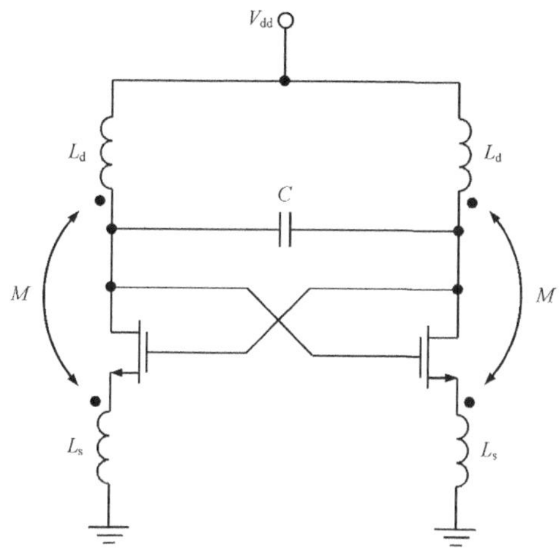

Figure 7.32 Transformer-feedback VCO schematic

and M_4 (pMOS) represent the inverter buffer stage. The varactor is connected between ac ground through the bypass capacitor C_B and transistor gates to obtain a wider tuning range. The series circuit consisting of an inductor L_0 and a capacitor C_0 is added for impedance matching and dc blocking. The feedback elements of such an oscillator are the overall gate–source capacitance C_{gs} and substrate capacitance C_{sub}. In this case, the oscillation conditions with negative conductance can be established only at frequencies above $\omega_{min} = 1/\sqrt{LC_{gs}}$. The measurement results for an oscillator using 0.35-μm CMOS technology show the phase noise of −118.1 dBc/Hz at 1 MHz offset from carrier of 6 GHz and dc current of 4.6 mA from a supply voltage of 2 V. Figure 7.31b shows the balanced cross-coupled Colpitts oscillator topology combining two identical single-ended Colpitts oscillators [67]. By optimizing the oscillator feedback factor defined by the ratio between the capacitances C_1 and C_2, the phase noise performance can be improved by up to 10 dB.

To improve the VCO performance in terms of low supply voltage, low power and low phase noise, a CMOS VCO with transformer feedback can be used. As shown in Figure 7.32, such an oscillator employs an integrated transformer instead of a conventional spiral inductor [68]. The transformer primary coils with self-inductance L_d are connected to the transistor drains, each forming an LC tank circuit. The secondary coils with self-inductance L_s are connected to the transistor sources each. Both the primary and secondary coils are magnetically coupled to each other with a mutual coupling factor M. With the transformer feedback, the drain voltage can swing above the supply voltage and the source voltage can swing below the ground potential, and they are in phase. Hence, for the same voltage amplitude and phase noise, the supply voltage can be reduced, or better phase noise can be achieved for the same supply voltage. The optimum self-inductance ratio of $k = L_d/L_s$ is chosen to minimize the phase noise level. The oscillator topology with transformer feedback representing two identical Hartley oscillators cross-coupled connected is an alternative to the cross-coupled Colpitts oscillator with capacitive feedback. Using a single differential transformer to maximize quality factor and minimize chip area, the phase noise was measured at the level of −119 dBc/Hz at 1 MHz offset from the carrier of 3.8 GHz at a supply voltage of 0.5 V with power consumption of 0.57 mW. Differential coils were used in both primary and secondary coils, which have a ratio of 5 to 2. The frequency

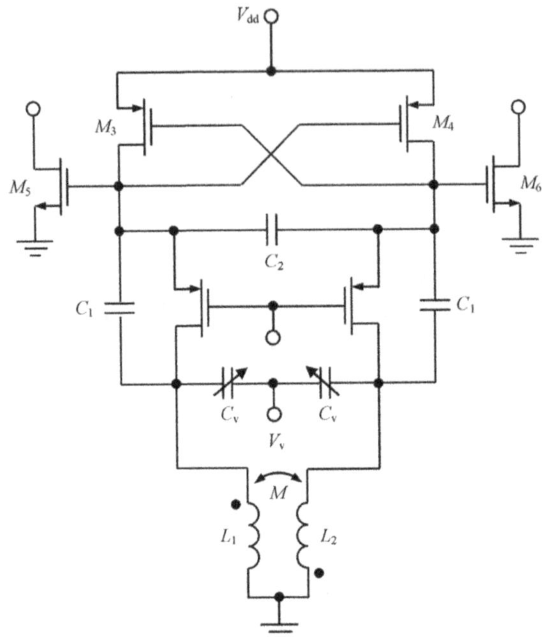

Figure 7.33 Schematic of differential Colpitts *p*MOS VCO

tuning range from 3.65 to 3.76 GHz was achieved using two accumulation-mode varactors instead of the fixed capacitance C.

Since *p*MOS transistors provide low-frequency flicker noise that is lower by an order of magnitude compared with their *n*MOS counterparts and the hot-carrier effect in *p*MOS devices is typically smaller, which is especially important for the submicrometre CMOS process, there is a possibility of improving the phase noise performance by designing a fully *p*MOS-based differential VCO. Figure 7.33 shows such a differential Colpitts VCO, which is realized by coupling of two common gate identical single-ended Colpitts VCOs [69]. The common node of each identical VCO in a differential configuration represents virtual ground at odd mode and open at even mode. The feedback capacitances C_1 and C_2 provide optimum feedback coefficient to realize the maximum negative small-signal resistance. Being implemented in a 0.18-μm CMOS process, the VCO had a tuning capability from 4.6 to 5 GHz with measured phase noise better than −120.42 dBc/Hz at 1 MHz offset over the entire tuning range. At a supply voltage of 1.5 V, the power consumption of the VCO core was only 3 mW.

The full integration of RF transceivers into a single CMOS chip implies the use of low intermediate frequency (IF) or zero-IF architecture that require quadrature local oscillator signals for image rejection and demodulation. The conventional technique based on using two VCOs with coupling transistors suffers from a trade-off between quadrature accuracy and phase noise. Moreover, the coupling transistors increase the power consumption. Figure 7.34 shows the schematic of a quadrature CMOS VCO based on second-order harmonic coupling to enforce the quadrature relation between two differential cross-coupled oscillators [70]. Since the quadrature coupling in this case is established by means of coupled inductors rather than by transistors, the coupling elements do not introduce significant extra sources of noise. The

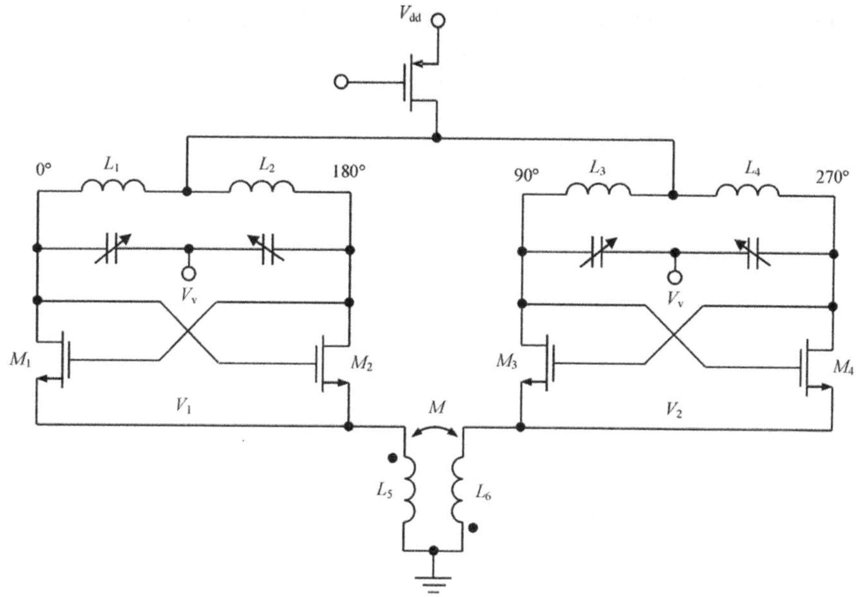

Figure 7.34 Schematic of a superharmonic-coupled quadrature VCO

quadrature superharmonic-coupled VCO consists of two separate identical differential oscillators whose common-mode even-order harmonics are coupled the pair of inductors L_5 and L_6. Transistor M_5 supplies the bias current from the top side of the circuit, such that a rail-to-rail oscillation waveform is possible without exceeding its peak value at the breakdown voltage limits of transistors from M_1 to M_4. The inductors L_1 and L_2 (as well as L_3 and L_4) are deliberately not combined into a single symmetric inductor, since a symmetric inductor exhibits a larger quality factor when driven differentially [71]. This would result in a larger suppression of the common-mode second harmonic, which is instead beneficial for maximizing the quadrature coupling in this topology. Such a superharmonic-coupled quadrature VCO implemented in a 0.25-μm CMOS technology demonstrates a measured phase noise of −124 dBc/Hz at 1 MHz offset over the frequency tuning range from 4.57 to 5.21 GHz at 22 mW power consumption with the worst-case image rejection of 33 dB.

For implementation in highly integrated transceivers with reduced size and cost intended for wireless handset applications, it is important to analyse the performance comparison of CMOS versus bipolar VCO in SiGe BiCMOS technology. BiCMOS is an excellent technology to use for such a comparison, since it offers a full suite of bipolar and CMOS transistors as well as high-performance passive components. For a comparative analysis, a CMOS VCO operating at a fundamental frequency of 2.3 GHz and a bipolar VCO operating at twice the fundamental frequency followed by a frequency divider by a factor of two were chosen, both designed in IBM 0.25-μm SiGe BiCMOS technology [72]. Both VCO core circuits represent fully differential cross-coupled topologies having comparable $f_T = 48$ GHz for bipolar and $f_T = 40$ GHz for CMOS transistors. Frequency tuning is provided by varactors representing by the bipolar collector–base junctions capacitances, while CMOS transistors are used as digital varactors to provide the digital band switching to extend the total frequency coverage. Although low-frequency flicker noise is much larger in CMOS devices than in silicon or SiGe bipolar devices, its effect on VCO phase noise can be minimized by properly sizing the CMOS devices as well as the ratio between *n*MOS and *p*MOS transistors used. As a result, the phase noise of the bipolar VCO at smaller offsets is better than that of the CMOS VCO, but it becomes worse at higher frequency offsets. For example, the phase noise level becomes equal at approximately 10 kHz offset, while CMOS VCO achieves better phase noise by 8 dB at 100 kHz offset. Excluding the bond pads, CMOS VCO has a slightly smaller chip size, mostly because of the frequency divider circuit required for a bipolar VCO. The frequency sensitivity to supply voltage, or pushing factor, is similar for both CMOS and bipolar VCO. However, the phase noise sensitivity to temperature and supply voltage is different, with significant degradation of the CMOS VCO phase noise when the supply voltage drops from its initial value of 3 V to 2.5 V. This sensitivity mainly originates from the biasing scheme used in CMOS VCO where current is self-biased by *n*MOS and *p*MOS transistors. This also affects the phase noise degradation versus process and temperature variation, which is much stronger compared with its bipolar VCO counterpart. However, the simplicity of CMOS VCO biasing eliminates the noise contribution from a separate biasing circuit required for a bipolar VCO.

REFERENCES

1. Y. P. Tsividis, *Operation and Modeling of the MOS Transistor*, New York: McGraw-Hill (1987).
2. P. Andreani and S. Mattisson, 'On the Use of MOS Varactors in RF VCO's', *IEEE J. Solid-State Circuits*, **SC-35**, 905–910 (2000).
3. F. Svelto, P. Erratico and R. Castello, 'A Metal–Oxide–Semiconductor Varactor', *IEEE Electron Device Lett.*, **20**, 164–166 (1999).
4. A.-S. Porret, T. Melly, C. C. Enz and E. A. Vittoz, 'Design of High-*Q* Varactors for Low-Power Wireless Applications Using a Standard CMOS Process', *IEEE J. Solid-State Circuits*, **SC-35**, 337–345 (2000).
5. S.-S. Song and H. Shin, 'An RF Model of the Accumulation-Mode MOS Varactor Valid in Both Accumulation and Depletion Regions', *IEEE Trans. Electron Devices*, **ED-50**, 1997–1999 (2003).
6. A. V. Grebennikov and F. Lin, 'An Efficient CAD-Oriented Large-Signal MOSFET Model', *IEEE Trans. Microwave Theory Tech.*, **MTT-48**, 1732–1742 (2000).
7. C. Enz, 'An MOS Transistor Model for RF IC Design Valid in All Regions of Operation', *IEEE Trans. Microwave Theory Tech.*, **MTT-50**, 342–359 (2002).
8. B. Senapati, K. Ehwald, W. Winkler and F. Furnhammer, 'High performance MOS Varactor SPICE Model', *Electronics Lett.*, **38**, 1416–1417 (2002).
9. K. Molnar, G. Rappitsch, Z. Huszka and E. Seebacher, 'MOS Varactor Modeling With a Subcircuit Utilizing the BSIM3v3 Model', *IEEE Trans. Electron Devices*, **ED-49**, 1206–1211 (2002).
10. J. Maget, M. Tiebout and R. Kraus, 'Influence of Novel MOS Varactors on the Performance of a Fully Integrated UMTS VCO in Standard 0.25-μm CMOS Technology', *IEEE J. Solid-State Circuits*, **SC-37**, 953–958 (2002).
11. V. Vashchenko, P. Francis and P. J. Hopper, Varactor and Method of Forming a Varactor With an Increased Linear Tuning Range, U.S. Patent 6,653,716 (Nov. 2003).
12. F. Svelto, S. Manzini and R. Castello, 'A Three Terminal Varactor for RF IC's in Standard CMOS Technology', *IEEE Trans. Electron Devices*, **ED-47**, 893–895 (2000).
13. J. Maget, M. Tiebout and R. Kraus, 'MOS Varactors with *n*- and *p*-Type Gates and Their Influence on an *LC*-VCO in Digital CMOS', *IEEE J. Solid-State Circuits*, **SC-38**, 1139–1147 (2003).
14. F. Herzel, M. Pierschel, P. Weger and M. Tiebout, 'Phase Noise in a Differential CMOS Voltage-Controlled Oscillator for RF Applications', *IEEE Trans. Circuits and Systems — II: Analog and Digital Signal Process.*, **CAS-II-47**, 11–15 (2000).

15. B. Razavi, 'A Study of Phase Noise in CMOS Oscillators', *IEEE J. Solid-State Circuits*, **SC-31**, 331–343 (1996).
16. J. J. Rael and A. A. Abidi, 'Physical Processes of Phase Noise in Differential LC Oscillators', *Proc. 2000 IEEE Custom Integrated Circuits Conf.*, 569–572 (2000).
17. S. K. Magierowski and S. Zukotynski, 'CMOS *LC*-Oscillator Phase-Noise Analysis Using Nonlinear Models', *IEEE Trans. Circuits and Systems — I: Regular Papers*, **CAS-I-51**, 664–677 (2004).
18. A. Jerng and C. G. Sodini, 'The Impact of Device Type and Sizing on Phase Noise Mechanisms', *IEEE J. Solid-State Circuits*, **SC-40**, 360–369 (2005).
19. Y. Nemirovsky, I. Brouk and C. G. Jakobson, '$1/f$ Noise in CMOS Transistors for Analog Applications', *IEEE Trans. Electron Devices*, **ED-48**, 921–927 (2001).
20. K. K. O, N. Park and D.-J. Yang, '$1/f$ Noise of NMOS and PMOS Transistors and Their Implications to Design of Voltage Controlled Oscillators', 2002 *IEEE RFIC Symp. Dig.*, 59–62 (2002).
21. J. Chang, A. A. Abidi and C. R. Viswanathan, 'Flicker Noise in CMOS Transistors from Subthreshold to Strong Inversion at Various Temperature', *IEEE Trans. Electron Devices*, **ED-41**, 1965–1971 (1994).
22. S. Levantino, C. Samori, A. Bonfani, S. L. J. Gierkink, A. L. Lacaita and V. Bocuzzi, 'Frequency Dependence on Bias Current in 5-GHz CMOS VCOs: Impact on Tuning Range and Flicker Noise Upconversion', *IEEE J. Solid-State Circuits*, **SC-37**, 1003–1011 (2002).
23. J. M. Lopez-Villegas, J. Samitier, C. Cane, P. Losantos and J. Bausells, 'Improvement of the Quality Factor of RF Integrated Inductors by Layout Optimization', *IEEE Trans. Microwave Theory Tech.*, **MTT-48**, 76–83 (2000).
24. J. R. Long and M. A. Copeland, 'The Modeling, Characterization, and Design of Monolithic Inductors for Silicon RF IC's', *IEEE J. Solid-State Circuits*, **SC-32**, 357–369 (1997).
25. N. A. Talwalkar, C. P. Yue and S. S. Wong, 'Analysis and Synthesis of On-Chip Spiral Inductors', *IEEE Trans. Electron Devices*, **ED-52**, 176–182 (2005).
26. N. M. Nguyen and R. G. Meyer, 'Si IC-Compatible Inductors and *LC* Passive Filters', *IEEE J. Solid-State Circuits*, **SC-25**, 1028–1031 (1990).
27. J. Craninckx and M. S. J. Steyaert, 'A 1.8-GHz Low-Phase-Noise CMOS VCO Using Optimized Hollow Spiral Inductors', *IEEE J. Solid-State Circuits*, **SC-32**, 736–744 (1997).
28. J. Y.-C. Chang, A. A. Abidi and M. Gaitan, 'Large Suspended Inductors on Silicon and Their Use in a 2-μm CMOS RF Amplifier', *IEEE Electron Devices Lett.*, **14**, 246–248 (1993).
29. Y.-H. Kao, M.-T. Hsu, M.-C. Hsu and P.-A. Wu, 'A Systematic Approach for Low Phase Noise CMOS VCO Design', *IEICE Trans. Electron.*, **E86-C**, 1427–1432 (2003).
30. Y.-H. Kao and M.-T. Hsu, 'Theoretical Analysis of Low Phase Noise Design of CMOS VCO', *IEEE Microwave and Wireless Components Lett.*, **15**, 33–35 (2005).
31. T. S. D. Cheung and J. R. Long, 'Shielded Passive Devices for Silicon-Based Monolithic Microwave and Millimeter-Wave Integrated Circuits', *IEEE J. Solid-State Circuits*, **SC-41**, 1183–1200 (2006).
32. C. P. Yue and S. S. Wong, 'On-Chip Spiral Inductors with Patterned Ground Shields for Si-Based RF IC's', *IEEE J. Solid-State Circuits*, **SC-33**, 743–752 (1998).
33. S.-M. Yim, T. Chen and K. K. O, 'The Effects of a Ground Shield on the Characteristics and Performance of Spiral Inductors', *IEEE J. Solid-State Circuits*, **SC-37**, 237–244 (2002).
34. E. Hernandez, R. Berenguer, J. Aguilera, J. Ibarguren and I Gutierrez, 'A Fully Integrated CMOS VCO for DCS-1800 Direct Conversion Receivers', *Microwave J.*, **46**, 182–192 (2003).
35. N. H. W. Fong, J.-O. Plouchart, N. Zamdmer, D. Liu, L. F. Wagner, C. Plett and N. G. Tarr, 'Design of Wide-Band CMOS VCO for Multiband Wireless LAN Applications', *IEEE J. Solid-State Circuits*, **SC-38**, 1333–1342 (2003).
36. R. S. Muller and T. I. Kamins, *Device Electronics for Integrated Circuits*, New York: John Wiley & Sons (1977).
37. A. Grebennikov, *RF and Microwave Power Amplifier Design*, New York: McGraw-Hill (2004).
38. M. Bucher, C. Lallement and C. C. Enz, 'An Efficient Parameter Extraction Methodology for the EKV MOST Model', *Proc. 1996 IEEE Int. Conf. Microelectronic Test Structures*, 145–150 (1996).

39. M. C. Ho, K. Green, R. Culbertson, J. Y. Yang, D. Ladwig and P. Ehnis, 'A Physical Large Signal Si MOSFET Model for RF Circuit Design', *1997 IEEE MTT-S Int. Microwave Symp. Dig.*, 391–394 (1997).
40. D. Ham and A. Hajimiri, 'Concepts and Methods in Optimization of Integrated *LC* VCOs', *IEEE J. Solid-State Circuits*, **SC-36**, 896–909 (2001).
41. A. Hajimiri and T. H. Lee, 'Design Issues in CMOS *LC* Oscillators', *IEEE J. Solid-State Circuits*, **SC-34**, 717–724 (1999).
42. A. J. Cote, 'Matrix Analysis of Oscillators and Transistor Applications', *IRE Trans. Circuit Theory*, **CT-5**, 181–188 (1958).
43. J. K. Clapp, 'Frequency Stable *LC* Oscillators', *Proc. IRE*, **42**, 1295–1300 (1954).
44. N. Itoh, 'Low Voltage Low Phase Noise CMOS VCO and Its Flicker Noise Influence', *IEICE Trans. Electron.*, **E86-C**, 1062–1068 (2003).
45. N. H. W. Fong, J.-O. Plouchart, N. Zamdmer, D. Liu, L. F. Wagner, C. Plett and G. Tarr, 'A Low-Voltage Multi-GHz VCO with 58% Tuning Range in SOI CMOS', *Proc. 2002 IEEE Custom Integrated Circuits Conf.*, 423–426 (2002).
46. N. H. W. Fong, J.-O. Plouchart, N. Zamdmer, D. Liu, L. F. Wagner, C. Plett and N. G. Tarr, 'A 1-V 3.8-5.7-GHz Wide-Band VCO with Differentially Tuned Accumulation MOS Varactors for Common-Mode Noise Rejection in CMOS SOI Technology', *IEEE Trans. Microwave Theory Tech.*, **MTT-51**, 1952–1959 (2003).
47. J. Mira, T. Divel, S. Ramet, J.-B. Begueret and Y, Deval, 'Distributed MOS Varactor Biasing for VCO Gain Equalization in 0.13 μm CMOS Technology', 2004 *IEEE RFIC Symp. Dig.*, 131–134 (2004).
48. A. D. Berny, A. M. Niknejad and R. G. Meyer, 'A 1.8-GHz *LC* VCO with 1.3-GHz Tuning Range and Digital Amplitude Calibration', *IEEE J. Solid-State Circuits*, **SC-40**, 909–917 (2005).
49. F. Herzel, H. Erzgraeber and N. Ilkov, 'A New Approach to Fully Integrated CMOS LC-Oscillator with a Very Large Tuning Range', *Proc. 2000 IEEE Custom Integrated Circuits Conf.*, 573–576 (2000).
50. Z. Li and K. O, 'A 900-MHz 1.5-V CMOS Voltage-Controlled Oscillator Using Switched Resonators with a Wide Tuning Range', *IEEE Microwave and Wireless Components Lett.*, **13**, 137–139 (2003).
51. Z. Li and K. K. O, 'A Low-Phase-Noise and Low-Power Multiband CMOS Voltage-Controlled Oscillator', *IEEE J. Solid-State Circuits*, **SC-40**, 1296–1302 (2005).
52. A. Rofougaran, J. Rael, M. Rofougaran and A. A. Abidi, 'A 900 MHz CMOS *LC*-Oscillator with Quadrature Outputs', *Proc. 1996 IEEE Int. Solid-State Circuits Conf.*, 392–393 (1990).
53. P. Andreani, A. Bonfanti, L. Romano and C. Samori, 'Analysis and Design of a 1.8-GHz CMOS *LC* Quadrature VCO', *IEEE J. Solid-State Circuits*, **SC-37**, 1737–1747 (2002).
54. P. Andreani, 'A 2 GHz, 17% Tuning Range Quadrature CMOS VCO with High Figure-of-Merit and 0.6° Phase Error', *Proc. 2002 Europ. Solid-State Circuits Conf.*, 815–818 (2002).
55. M. Tiebout, 'Low-Power Low-Phase-Noise Differentially Tuned Quadrature VCO Design in Standard CMOS', *IEEE J. Solid-State Circuits*, **SC-36**, 1018–1024 (2001).
56. S. Naseh, M. J. Deen and O. Marinov, 'Effects of Hot-Carrier on the Performance of the LC-Tank CMOS Oscillators', *IEEE Trans. Electron Devices*, **ED-50**, 1334–1339 (2003).
57. R. Brederlow, W. Weber, D. Schmitt-Landsiedel and R. Thewes, 'Hot-Carrier Degradation of the Low-Frequency Noise in MOS Transistors Under Analog and RF Operating Conditions', *IEEE Trans. Electron Devices*, **ED-49**, 1588–1596 (2002).
58. M. Depas, T. Nigam and M. M. Heyns, 'Soft Breakdown of Ultra-Thin Gate Oxide Layers', *IEEE Trans. Electron Devices*, **ED-43**, 1499–1504 (1996).
59. R. Zeng and H. Wang, 'Effect of FN Stress and Gate-Oxide Breakdown on High-Frequency Noise Characteristics in Deep-Submicrometer *n*MOSFETs', *IEEE Electron Device Lett.*, **26**, 390–393 (2005).
60. H. Liao, S. Rustagi, J. Shi and Y. Z. Xiong, 'Characterization and Modeling of the Substrate Noise and Its Impact on the Phase Noise of VCO', 2003 *IEEE MTT-S Int. Microwave Symp. Dig.*, 247–250 (2003).

61. M. Pfost and H.-M. Rein, 'Modeling and Measurements of Substrate Coupling in Si-Bipolar IC's up to 40 GHz', *IEEE J. Solid-State Circuits*, **SC-33**, 582–591 (1998).
62. F. Svelto, S. Deantoni and R. Castello, 'A 1.3 GHz Low-Phase Noise Fully Tunable CMOS LC VCO', *IEEE J. Solid-State Circuits*, **SC-35**, 356–361 (2000).
63. B. De Muer, N. Itoh, M. Borremans and M. Steyaert, 'A 1.8 GHz Highly Tunable Low-Phase-Noise CMOS VCO', *Proc. 2000 IEEE Custom Integrated Circuits Conf.*, 585–588 (2000).
64. N. Fong, J. Kim, J.-O. Plouchart, N. Zamdmer, D. Liu, L. Wagner, C. Plett and G. Tarr, 'A Low-Voltage 40-GHz Complementary VCO with 15% Frequency Tuning Range in SOI CMOS Technology', *IEEE J. Solid-State Circuits*, **SC-39**, 841–846 (2004).
65. C. Cao and K. K. O, 'Millimeter-Wave Voltage-Controlled Oscillators in 0.13-μm CMOS Technology', *IEEE J. Solid-State Circuits*, **SC-41**, 1297–1304 (2006).
66. C.-Y. Cha and S.-G. Lee, 'A Complementary Colpitts Oscillator in CMOS Technology', *IEEE Trans. Microwave Theory Tech.*, **MTT-53**, 881–887 (2005).
67. C.-Y. Cha and S.-G. Lee, 'Overcome the Noise Optimization Limit of Differential LC Oscillator with Asymmetric Capacitance Tank Structure', *2004 IEEE RFIC Symp. Dig.*, 583–586 (2004).
68. K. Kwok and H. C. Luong, 'Ultra-Low-Voltage High-Performance CMOS VCOs Using Transformer Feedback', *IEEE J. Solid-State Circuits*, **SC-40**, 652–660 (2005).
69. M.-D. Tsai, Y.-H. Cho and H. Wang, 'A 5-GHz Low Phase Noise Differential Colpitts CMOS VCO', *IEEE Microwave and Wireless Components Lett.*, **15**, 327–329 (2005).
70. S. L. J. Gierkink, S. Levantino, R. C. Frye and V. Bocuzzi, 'A Low-Phase-Noise 5-GHz Quadrature VCO Using Superharmonic Coupling', *IEEE J. Solid-State Circuits*, **SC-38**, 1148–1154 (2003).
71. M. Danesh and J. R. Long, 'Differentially Driven Symmetric Microstrip Inductors', *IEEE Trans. Microwave Theory Tech.*, **MTT-50**, 332–340 (2002).
72. D. Wang and X. Wang, 'The Performance Comparison of CMOS vs Bipolar VCO in SiGe BiCMOS Technology', *2003 IEEE MTT-S Int. Microwave Symp. Dig.*, 101–104 (2003).

8

Wideband voltage-controlled oscillators

Wideband voltage-controlled oscillators (VCOs) are used in a variety of RF and microwave systems, including broadband measurement equipment, wireless and TV applications and military electronic countermeasures (ECM) systems. In modern ECM systems, they serve as the frequency-agile local oscillators in receiver subsystems and fast-modulation noise sources in active jamming subsystems. Among wideband tunable signal sources such as YIG-tuned oscillators, wideband VCOs are preferable because of their small size, low weight, high settling time speed and capability of fully monolithic integration. Therefore, modern radar and communication applications demand VCOs that are capable of being swept across a wide range of potential threat frequencies with a speed and settling time far beyond those of the YIG-tuned oscillators. This chapter discusses the basic concepts of wideband VCO circuit design and gives specific circuit solutions using lumped elements and transmission lines to improve their frequency tuning characteristics. Various examples of RF and microwave VCO circuit configurations using bipolar, MOSFET and MESFET devices are analysed, their circuit parameters are calculated or optimized to provide maximum tuning bandwidth or minimum tuning linearity. Also included are numerous practical examples of wideband VCOs for RF and microwave application in radar or telecommunication systems.

8.1 MAIN REQUIREMENTS

In many applications, particularly ECM, system performance significantly improves when VCOs are capable of wide tuning bandwidths with linear varactor tuning, as well as high-speed switching. High-speed frequency agility, while affording small size and more than octave tuning bandwidths, can be achieved only by considering the most important VCO parameters, such as post tuning drift and settling time, frequency coverage and linearity, frequency pulling and pushing, spectral purity and temperature stability.

How do the basic circuit parameters affect VCO tuning properties? Consider a series negative resistance oscillator resonant circuit with fully connected varactor, as shown in Figure 8.1. The oscillation frequency f_0 as a function of the VCO parameters in a steady-state condition when $R_{in} + R_L = 0$ can be expressed as

$$f_0 = \frac{1}{2\pi}\sqrt{\frac{C_{in} + C_v}{LC_{in}C_v}} = \frac{1}{2\pi}\sqrt{\frac{1}{LC_v}\frac{P_L Q_L}{P_v Q_v}} \tag{8.1}$$

RF and Microwave Transistor Oscillator Design A. Grebennikov

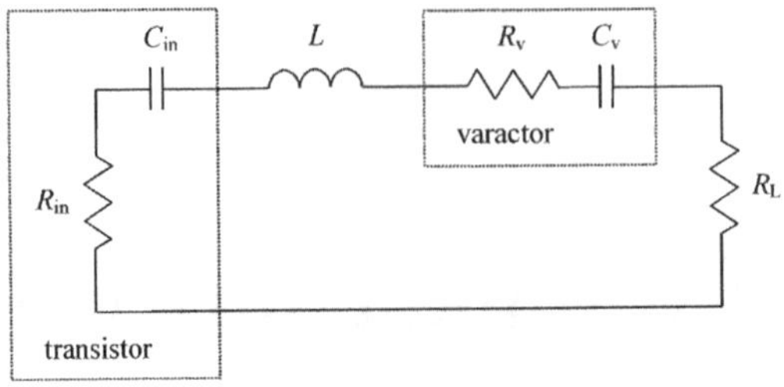

Figure 8.1 Series negative resistance oscillator resonant circuit

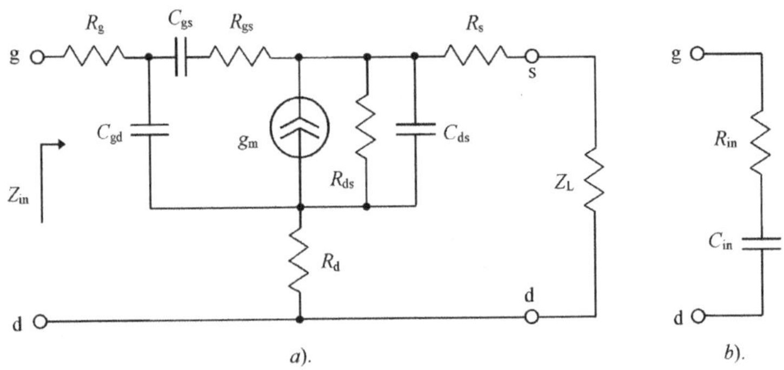

Figure 8.2 Small-signal loaded common-drain FET equivalent circuits

where

$$Q_v = \frac{1}{\omega_0 R_v C_v}$$

is the varactor quality factor, $\omega_0 = 2\pi f_0$ is the angular resonant frequency, R_v is the series parasitic varactor capacitance, C_v is the varactor junction capacitance,

$$Q_L = \frac{C_{in} + C_v}{\omega_0 R_L C_v C_{in}}$$

is the quality factor associated with the load resistance R_L, C_{in} is the input equivalent capacitance of the active device, P_v is the power dissipated in the varactor, P_L is the power delivered to the load.

In this case, VCO is considered as a negative resistance oscillator based on the equivalent one-port active device configuration (the transistor, to the input terminal of which a load is connected, is replaced by an impedance with zero active and negative reactive parts). Figure 8.2 shows (a) the small-signal equivalent transistor circuit using the common drain FET device that provides (b) the one-port negative resistance model. The equivalent input impedance of a loaded common drain FET device can be written, using the common source Y-parameters, as

$$Z_{in} = R_g + \frac{1 + R_d Y_{22} + (R_s + Z_L)(Y_{11} + Y_{12} + Y_{21} + Y_{22} + R_d \Delta Y)}{Y_{11} + \Delta Y\,(R_d + R_s + Z_L)} \tag{8.2}$$

where $Z_L = 1/Y_L$, $Y_L = G_L + jB_L$.

For sufficiently small resistances R_g, R_s and R_d, the input reactance $\mathrm{Im}Z_{in}$ is simplified to

$$\mathrm{Im}Z_{in} = \frac{B_{11} + B_{12} + B_{21} + B_{22} + B_L}{G_{11}(G_{22} + G_L) - G_{12}G_{21} + B_{12}B_{21} - B_{11}(B_{22} + B_L)} \tag{8.3}$$

Hence, the input equivalent capacitance $C_{in} = -1/\omega \mathrm{Im}Z_{in}$ can be written as

$$C_{in} = C_{gs}\frac{C'_{ds} + C_{gd}\left(1 - \dfrac{C_{gd}}{C_{gs}} - g_m R_{gs}\right) - R_{gs}C_{gs}\left(\dfrac{1}{R_{ds}} + \dfrac{1}{R_L}\right)}{C_{gs}\left(1 - g_m R_{gs}\right) + C'_{ds}\left[1 + \left(\omega C_{gs} R_{gs}\right)^2\right]} \tag{8.4}$$

where $C'_{ds} = C_{ds} + C_L$, $C_L = B_L/\omega$. The numerical results show that, for constant gate and drain bias voltages, the value of the capacitance C_{in} varies within 20–30% in a frequency range up to 0.5–0.6 f_T. In this case, taking into account that $C_{gd} << C_{gs}$ and that the effect of the gate–source resistance R_{gs} is negligible at lower frequencies, the equivalent input capacitance C_{in} can be approximately calculated by $C_{in} \cong C_{gs}C'_{ds}/(C'_{ds} + C_{gs})$.

The tuning sensitivity of the VCO resonant circuit to the variation of the varactor capacitance is written as

$$\frac{df}{dC_v} = -\frac{f_0}{2C_v}\frac{P_v Q_v}{P_L Q_L} \tag{8.5}$$

which indicates that, for the given values of the resonant frequency f_0 and ratio of P_v/P_L, a wider frequency bandwidth will be realized using a varactor with lower capacitance value and higher quality factor. Maximum tuning range can be provided with $C_v/C_{in} \ll 1$ when

$$\frac{P_v Q_v}{P_L Q_L} \to 1$$

As a result, the relative tuning sensitivity becomes maximally equal to

$$\frac{df}{f_0}\bigg/\frac{dC_v}{C_v} = -\frac{1}{2} \tag{8.6}$$

Generally, the VCO tuning characteristics are essentially nonlinear. Using the abrupt varactors for a wideband frequency tuning up to 45% results in the slopes of the frequency tuning characteristics within the limits of 10:1. Therefore, to linearize varactor tuning characteristics, analogue or digital linearizers are used, resulting in an increase of settling time and post-tuning drift of the whole VCO subsystem. However, the analogue linearizer enables one to improve intrinsic VCO linearity by only an order of magnitude [1]. Prospects to improve tuning linearity can be considered, either with the use of varactors with a special doping profile to give a required voltage–charge characteristic or with the use of hyperabrupt varactors [2, 3]. In both cases, technologically, the problem is not easily solvable and the required C–V dependence occurs only in a narrow range of varactor bias. As a result, a proper choice of the structure and type of the VCO resonant circuit to provide linear frequency tuning is an extremely difficult problem. It was found that the use of abrupt varactors in the oscillation system with two coupled resonant circuits allows linearizing the modulation curve within limits of 1% for VCOs with varactor tuning bandwidths of 1–3% [4]. Use of hyperabrupt varactors can help substantially to minimize the frequency tuning nonlinearity in a wide frequency range. For an ideal lossless oscillation system with optimum values of circuit parameters, it is possible to realize more than one octave varactor tuning band with tuning slope ratio of about 1.5.

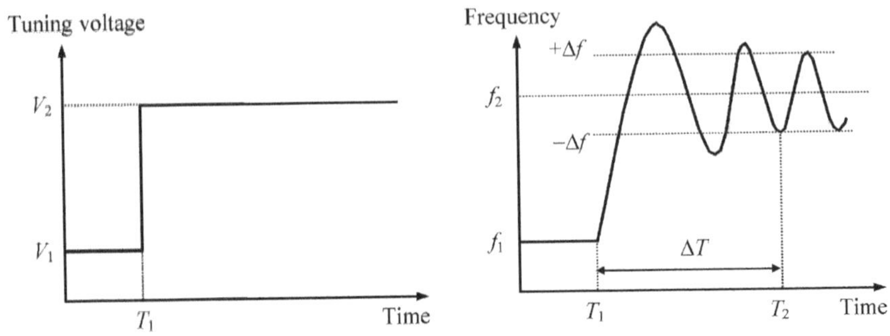

Figure 8.3 Transient response of oscillator resonant frequency

For microwave wideband VCOs, it is convenient to use oscillation systems with two coupled resonant circuits using coupled microstrip lines or T-transformers [5]. The advantage is that it yields dc isolation without using an additional blocking capacitor. This circuit is very suitable for monolithic integration. The parameter optimization of the MESFET VCO oscillation system with a T-transformer between the gate and the varactor contributes to the realization of the tuning nonlinearity of ±30 MHz or ±0.4% for a varactor frequency tuning bandwidth of 7–10 GHz [6]. Improvement of the VCO frequency tuning linearity can be achieved by providing high values of the coupling coefficient between the two coupled resonant circuits near the hysteresis region.

Figure 8.3 shows the transient response of the oscillator resonant frequency following a step change in tuning voltage, where ΔT is the settling time and Δf is the post-tuning drift [1, 7]. Settling time ΔT is defined as the time required to reach a specific error bandwidth around the final value of operating frequency, whereas post-tuning drift Δf is the frequency deviation over a prescribed time interval. In order to be effective, a microwave VCO must be able to change operating frequencies in a time interval lasting from tenths of a nanosecond to several tenths of a microsecond with a set-on accuracy of a few megahertz. Increased tuning speed and reduced output power variations require a larger tuning capacitance range. A varactor bias circuit should be designed to have a negligible effect on the fundamental oscillator performance and minimize the frequency transition time simultaneously. For example, for a varactor bias circuit with series resistance of 100 Ω and series inductance of 50 nH consisting of a low-pass filter together with the varactor capacitance, the frequency rise and fall times to step in a tuning voltage are about 2 ns with a slew rate of 600 MHz/ns to tune the oscillation frequency from 2.2 to 3.4 GHz [8]. Generally, when designing wideband VCOs with minimum post-tuning drift, it is necessary to take into account simultaneously several factors, such as active device characterization, circuit topology, thermal considerations, and load pulling effect.

The strongest technical requirements are associated with military microwave VCO systems. For instance, these systems are required to operate over a temperature range from −54°C to +95°C [1]. In this case, temperature stability can be obtained either by active compensation on the tuning or bias control network or by stabilization of the base plate temperature. The latter approach utilizes an appropriate heating element applied directly to the VCO circuit and isolated with some insulating material, with thermal characteristics that allow the overall temperature excursion of the VCO base plate to be limited to a few degrees. A critical factor in minimizing post-tuning drift is the active device attachment method. For example, the use of a eutectic device die attachment is preferable over epoxy attachment because of the epoxy's higher thermal resistance. For Au–Si eutectic attachment instead of silver epoxy, it is possible to reduce the device junction temperature significantly.

To minimize the load pulling effect, it is necessary to provide sufficient isolation between the VCO and reactive components such as switches, filters or input matching circuits of multipliers and power amplifiers. Typically, about 40–50 dB of load isolation is required to minimize frequency pulling. This can be achieved by using a broadband lossy match single-ended amplifier or balanced power amplifier as a buffer amplifier stage. For maximum isolation level, it is necessary to use an additional wideband attenuator with 10 dB or higher dissipation.

8.2 SINGLE-RESONANT CIRCUITS WITH LUMPED ELEMENTS

8.2.1 Series resonant circuit

Figure 8.4a shows the VCO series resonant circuit representing a simple series connection of an inductance L and a varactor to the transistor, where C_{in} is the equivalent transistor input capacitance and C_v is the varactor junction capacitance [9]. The oscillation frequency ω_0 that corresponds to the intersection point of two curves shown in Figure 8.4b, which are the frequency dependencies of varactor and device impedances, can be found from

$$\frac{1}{\omega C_v} - X_{in}(\omega) = 0 \tag{8.7}$$

where

$$X_{in} = \omega L - \frac{1}{\omega C_{in}} = L\frac{\omega^2 - \omega_1^2}{\omega}$$

is the reactance seen by the one-port network and expressed in terms of zeros and poles of the

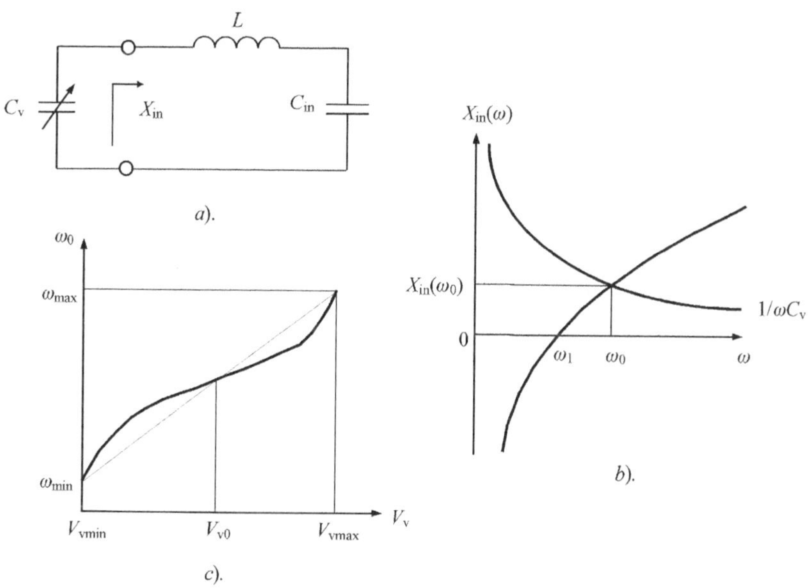

Figure 8.4 Equivalent VCO series resonant circuit and its frequency performance

reactance function for the second-order reactive circuit, where $\omega_1 = 1/\sqrt{LC_{\text{in}}}$ and $\omega = 0$ are the zero and pole of the reactance X_{in}, respectively.

Solving Equation (8.7) and using the varactor voltage–capacitance dependence defined by Equation (6.1) in Chapter 6 give the following equation for the oscillation frequency of the series resonant circuit:

$$\omega_0 = \omega_1 \sqrt{1 + \frac{C_{\text{in}}}{C_{\text{v0}}}\left(1 + \frac{V_{\text{v}}}{\varphi}\right)^{\gamma}} \tag{8.8}$$

The frequency tuning bandwidth is calculated in general form through the varactor capacitance ratio at maximum and minimum bias conditions. Equation (8.8) for maximum and minimum varactor bias conditions can be rewritten as

$$K_{\text{f}} = \sqrt{K_{\text{c}} \frac{1+q}{K_{\text{c}}+q}} \tag{8.9}$$

where $q = C_{\text{in}}/C_{\text{vmin}}$, $K_{\text{f}} = \omega_{\max}/\omega_{\min}$ is the frequency tuning ratio, $\omega_{\max}$ is the high tuning bandwidth frequency, $\omega_{\min}$ is the low tuning bandwidth frequency, $K_{\text{c}} = C_{\text{vmax}}/C_{\text{vmin}}$ is the varactor capacitance ratio, C_{vmax} is the maximum varactor capacitance, C_{vmin} is the minimum varactor capacitance.

From Equation (8.8) it follows that close to linear frequency tuning can be realized using varactors with sensitivity of $\gamma = 2$. However, even for hyperabrupt varactors, sensitivities typically range from 1.1 to 1.6. Therefore, it is very important to define a circuit solution minimizing the tuning nonlinearity. To approximate $\omega_0(V_{\text{v}})$ to the linear function within a voltage range of $V_{\text{vmin}} \le V_{\text{v}} \le V_{\text{vmax}}$, such a function ideally should be continuous and must have equal derivatives at each point within this range in accordance with

$$S(V) = \frac{d\omega_0(V_{\text{v}})}{dV_{\text{v}}} = \frac{\omega_{\text{v0}}^2}{2\omega_0}\frac{\gamma}{\varphi}\left(1 + \frac{V_{\text{v}}}{\varphi}\right)^{\gamma-1} = \text{const} \tag{8.10}$$

where $\omega_{\text{v0}} = 1/\sqrt{LC_{\text{v0}}}$. The tuning slope characteristic $S(V_{\text{v}})$ will be constant if

$$\frac{dS(V_{\text{v}})}{dV_{\text{v}}} = 0 \tag{8.11}$$

at each point in the voltage range $V_{\text{vmin}} \le V_{\text{v}} \le V_{\text{vmax}}$.

Applying the condition stated in Equation (8.11) to Equation (8.10) results in

$$\frac{C_{\text{in}}}{C_{\text{v0}}}\left(1 + \frac{V_{\text{v}}}{\varphi}\right)^{\gamma}(\gamma - 2) + 2(\gamma - 1) = 0 \tag{8.12}$$

which has a singular solution only. Since $(C_{\text{in}}, C_{\text{v0}}) > 0$ and $(1 + V_{\text{v}}/\varphi)^{\gamma} > 0$, there is a bend point of the dependence $\omega_0(V_{\text{v}})$, shown in Figure 8.4c for varactor sensitivities within the range

$$1 < \gamma < 2 \tag{8.13}$$

Consequently, using varactors with sensitivity γ defined by Equation (8.13) gives the capability of reducing the tuning slope variation and improving VCO linearity. The tuning slope variation can be defined as the ratio of maximum slope value of the varactor tuning characteristic

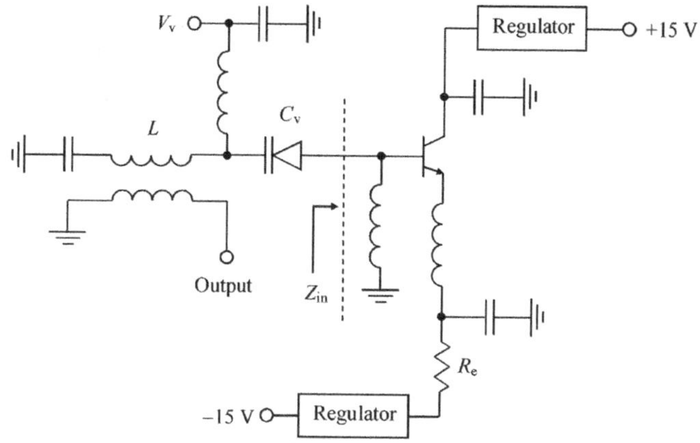

Figure 8.5 Series resonant circuit VCO with linear varactor tuning range [10]

to its minimum value in accordance with

$$\alpha = \left(\frac{d\omega_0}{dV_v}\right)_{max} \Big/ \left(\frac{d\omega_0}{dV_v}\right)_{min} \tag{8.14}$$

Since $|dC_v/dV_v|$ decreases with the increase of V_v, for a theoretical evaluation of the frequency tuning nonlinearity for the series resonant circuit when $\gamma \leq 1$, it is convenient to use the expression obtained from Equation (8.14) in the form

$$\alpha = K_f \left(\frac{\varphi + V_{vmax}}{\varphi + V_{vmin}}\right)^{1-\gamma} \tag{8.15}$$

For a given value of K_f, the tuning slope ratio α decreases with the increase of γ and becomes minimally equal to $\alpha = K_f$ when $\gamma = 1$. As a result, the series resonant circuit with properly chosen varactor enables one to improve VCO tuning linearity ideally in an octave frequency tuning bandwidth, obtaining a minimum value of $\alpha = 2$.

Figure 8.5 shows the circuit schematic of a wideband VCO with linear varactor tuning using a series resonant circuit configuration [10]. For such a VCO, varactor voltage tuning from 0.9 to 8.8 V provides a tuning frequency bandwidth from 2.6 to 3.9 GHz. This VCO circuit also includes two voltage regulators to vary the transistor input capacitance as its value depends on the dc bias conditions. Consequently, the total circuit capacitance $C_T = C_v C_{in}/(C_v + C_{in})$ is a function of both the varactor bias and dc supply voltages. Such an approach enables one to maximize the effective sensitivity γ, making its value close to 2. As a result, a linear tuning characteristic over a frequency bandwidth of more than 40% with a tuning slope ratio of less than 1.25 or tuning linearity relative to a best-fit straight line of less than ±0.5% was achieved.

8.2.2 Parallel resonant circuit

Figure 8.6a shows the parallel equivalent VCO resonant circuit, representing a parallel connection of a circuit inductance L, a varactor junction capacitance C_v and equivalent input transistor capacitance C_{in} [9]. The oscillation frequency ω_0 that corresponds to the intersection point of two curves, i.e., the frequency dependencies of varactor and device reactances shown in

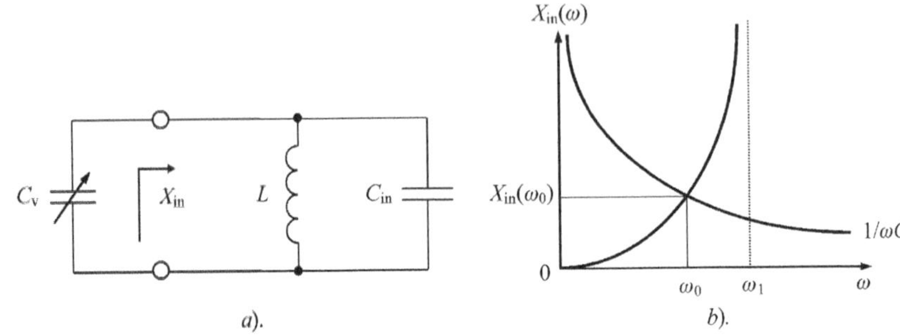

Figure 8.6 Equivalent VCO parallel resonant circuit and its frequency performance

Figure 8.6b, can be defined using Equation (8.7). In this case, the reactance X_{in} for the parallel resonant circuit is expressed as

$$X_{in} = \frac{\omega L}{1 - \omega^2 L C_{in}} = \frac{1}{C_{in}} \frac{\omega}{\omega_1^2 - \omega^2} \tag{8.16}$$

where $\omega = 0$ and $\omega = \omega_1$ are the zero and pole of the reactance X_{in}, respectively.

Solving Equation (8.7) for a parallel resonant circuit yields the following equation for the resonant frequency as a function of the circuit parameters:

$$\omega_0 = \omega_1 \Bigg/ \sqrt{1 + \frac{C_{v0}}{C_{in}} \left(1 + \frac{V_v}{\varphi}\right)^{-\gamma}} \tag{8.17}$$

By using Equation (8.17), the frequency tuning ratio K_f as a function of the varactor capacitance ratio K_c can be written as

$$K_f = \sqrt{\frac{K_c + q}{1 + q}} \tag{8.18}$$

Following the derivation procedure which was applied to evaluate the linearity properties of the series resonant circuit, we can obtain the following equations required to define the parallel resonant circuit linearity conditions:

$$S(V_v) = \frac{\omega_0^3}{2\omega_{v0}^2} \frac{\gamma}{\varphi} \left(1 + \frac{V_v}{\varphi}\right)^{-\gamma-1} \tag{8.19}$$

$$\frac{C_{v0}}{C_{in}}(\gamma - 2) - 2(\gamma + 1)\left(1 + \frac{V_v}{\varphi}\right)^{\gamma} = 0 \tag{8.20}$$

From Equation (8.20) it follows that the bend point of a function $\omega_0(V_v)$ exists only for

$$\gamma > 2 \tag{8.21}$$

The tuning nonlinearity of parallel resonant circuit in the range $\gamma \leq 2$ can be evaluated as

$$\alpha = \frac{1}{K_f^3} \left(\frac{\varphi + V_{vmax}}{\varphi + V_{vmin}}\right)^{\gamma+1} \tag{8.22}$$

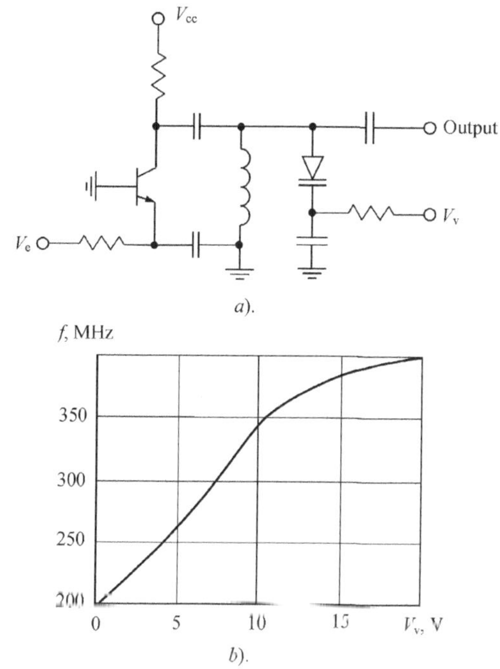

Figure 8.7 Parallel resonant circuit VCO with octave varactor tuning range

A comparison of Equations (8.15) and (8.22) shows that using the series resonant circuit for the VCO design enables one to reduce significantly the frequency tuning nonlinearity in a wideband frequency range compared with the parallel resonant circuit. If an abrupt varactor with $\gamma = 0.5$ is used and an octave frequency bandwidth for maximum varactor bias range is realized, a tuning slope ratio for the series resonant circuit can be reduced by a factor of 3–4 t. Using hyperabrupt varactors with sensitivity $1 \leq \gamma \leq 2$, the tuning slope ratio can be reduced by an order of magnitude.

Figure 8.7a shows an octave-band VCO common base schematic using a parallel resonant circuit configuration. In this case, we should take into account that, in order to achieve full-octave coverage, the equivalent transistor input capacitance should be relatively low. At the same time, the varactor voltage tuning range must be high. However, it is necessary to be very careful when VCO is tuned below a varactor bias voltage of 1 V where the varactor quality factor becomes very low. An example of the frequency tuning characteristic is given in Figure 8.7b [11]. It is clearly seen that the frequency range 200–400 MHz is achieved at the expense of the significant variation of the tuning slope ratio. Thus, when using a parallel circuit configuration to realize a stable linear VCO frequency tuning, the tuning voltage at the varactor must not be too low or too high.

To improve VCO tuning performance, it is advisable to connect an additional varactor between the base and the emitter of the transistor, resulting in a T-type connection of the varactors. Figure 8.8a shows a common collector VCO schematic with a parallel inductor, a T-type varactor connection and an inductively coupled load [12]. In this case, by varying the varactor capacitance from 15 to 5 pF, a frequency tuning bandwidth from 250 to 400 MHz was

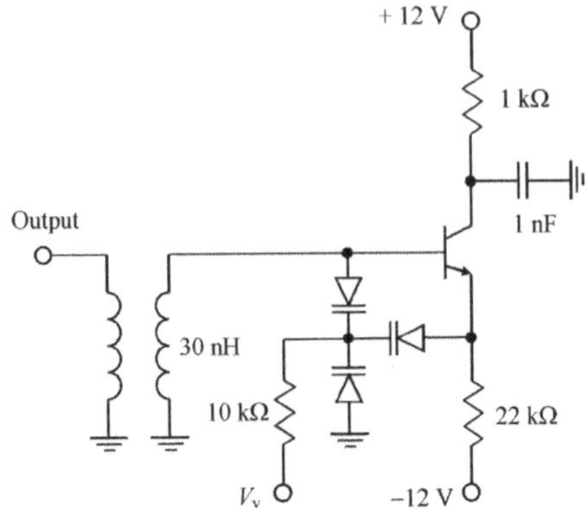

Figure 8.8 VCO schematic with T-type varactor connection [12]

achieved with stable oscillations, high purity of the output frequency spectrum and flat output power over the entire frequency tuning bandwidth.

To minimize the VCO phase noise level, it is very important to correctly choose elements in the varactor bias circuits. For instance, it is better to replace the large resistor or choke inductor between the varactor and its dc bias by a series connection of the inductor, whose inductance value should be high in the tuning range and low below the tuning range, and the resistor, whose resistance value is large enough to prevent parasitic oscillations that can be caused by a realized double-resonant circuit configuration. In this case, inductance works as an effective short-circuit for low-frequency noise through the varactor bias circuit. As an example, the choke inductor in the VCO schematic shown in Figure 8.5, and large-value resistors in VCO schematics, shown in Figure 8.8, should be replaced by series RL circuits. By using the series-connected 50-Ω resistor and 80-nH inductor between varactor and its dc bias, noise reduction from 8 to 12 dB was achieved at 10 kHz offsets from the carrier over the frequency tuning range 750–1250 MHz [13].

8.3 DOUBLE-RESONANT CIRCUIT WITH LUMPED ELEMENTS

Figure 8.9a shows two coupled resonant circuits (double-resonant circuit), where the equivalent transistor input capacitance and the varactor junction capacitance are connected through the inductive T-transformer [9]. To define the linearity conditions for such an oscillation system in a wide frequency range, it is necessary to solve Equation (8.7) with the reactance $X_{\rm in}$ defined by

$$X_{\rm in} = \omega L_{\rm v} + \frac{\omega L(1 - \omega^2 L_{\rm in} C_{\rm in})}{1 - \omega^2 C_{\rm in}(L_{\rm in} + L)} = \omega(L + L_{\rm v})(1 - k^2)\frac{\omega^2 - \omega_2^2}{\omega^2 - \omega_1^2} \tag{8.23}$$

where $\omega = 0$ and $\omega = \omega_2$ are zeros and $\omega = \omega_1$ is a pole of the reactance $X_{\rm in}$,

$$k = L\Big/\sqrt{(L + L_{\rm in})\,(L + L_{\rm v})}$$

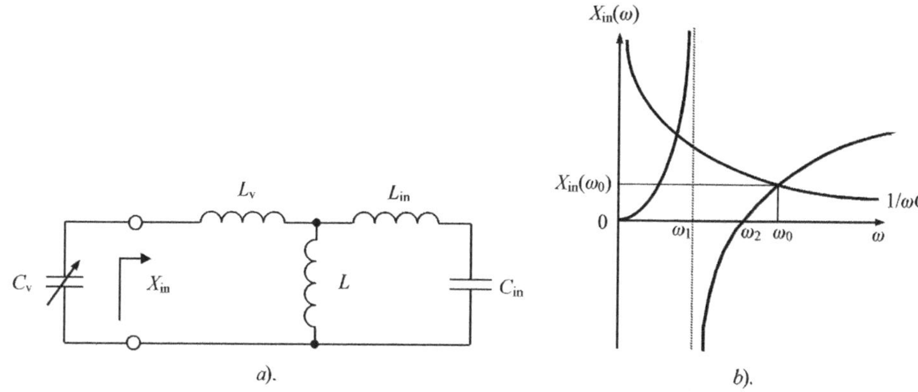

Figure 8.9 Equivalent VCO double-resonant circuit and its frequency performance

is the inductive coupling coefficient,

$$\omega_2 = \omega_1 / \sqrt{1 - k^2}$$

$$\omega_1 = 1 / \sqrt{C_{in}(L + L_{in})}$$

In this case, the values of the oscillation frequency ω_0 are located above the second zero ω_2, as shown in Figure 8.9b.

The resonant frequencies of a double-resonant circuit with lumped elements can be obtained as

$$\omega_0 = \omega_1 \sqrt{1 + \left(\frac{\omega_{v0}}{\omega_1}\right)^2 \left(1 + \frac{V_v}{\varphi}\right)^\gamma + \left(\frac{\omega_k}{\omega_1}\right)^2} \Big/ \sqrt{2(1 - k^2)} \tag{8.24}$$

where

$$\left(\frac{\omega_k}{\omega_1}\right)^2 = \sqrt{\left[\left(\frac{\omega_{v0}}{\omega_1}\right)^2 \left(1 + \frac{V_v}{\varphi}\right)^\gamma - 1\right]^2 + \left(2k\frac{\omega_{v0}}{\omega_1}\right)^2 \left(1 + \frac{V_v}{\varphi}\right)^\gamma}$$

$$\omega_{v0} = 1/\sqrt{C_{v0}(L + L_v)}$$

Applying a condition given by Equation (8.11) to Equation (8.24) yields the following equation to evaluate linearity properties of double-resonant circuit:

$$\frac{2k^2}{\left(\frac{\omega_0}{\omega_1}\right)^2 - 1} + (\gamma - 1)\left(1 + \frac{V_v}{\varphi}\right)^{-\gamma} \left(\frac{\omega_k}{\omega_1}\frac{\omega_{v0}}{\omega_1}\right)^2 = \left(\frac{\omega_k}{\omega_1}\right)^2 \frac{\left(\frac{\omega_0}{\omega_1}\right)^2 - 1}{2\left(\frac{\omega_0}{\omega_1}\right)^2} \tag{8.25}$$

which shows a bend point of the frequency tuning characteristic for

$$\gamma < 1. \tag{8.26}$$

Unlike the single series resonant circuit, such a result can be achieved by an appropriate choice of the inductive coupling coefficient k. To calculate the frequency tuning characteristic,

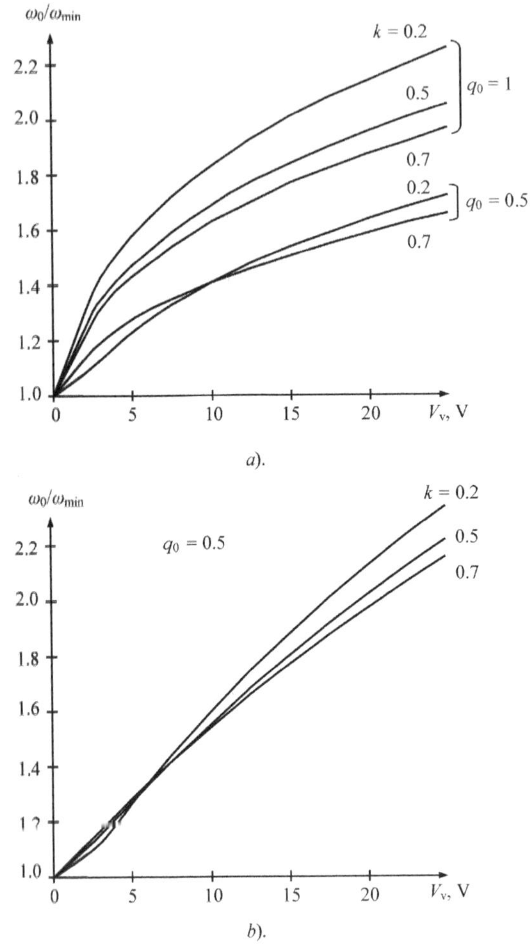

Figure 8.10 Frequency tuning bandwidth versus varactor bias voltage

it is best to rewrite Equation (8.24) in normalized form as

$$\frac{\omega_0}{\omega_{\min}} = \sqrt{\frac{1 + \left(\dfrac{\omega_{v0}}{\omega_1}\right)^2 \left(1 + \dfrac{V_v}{\varphi}\right)^\gamma + \left[\dfrac{\omega_0(V_v)}{\omega_1}\right]^2}{1 + \left(\dfrac{\omega_{v0}}{\omega_1}\right)^2 \left(1 + \dfrac{V_{v\min}}{\varphi}\right)^\gamma + \left[\dfrac{\omega_0\left(V_{v\min}\right)}{\omega_1}\right]^2}} \tag{8.27}$$

where $(\omega_{v0}/\omega_1)^2 = q_0(1 + L_{in}/L)^2 k^2$, $q_0 = C_{in}/C_{v0}$.

Figure 8.10 shows the frequency tuning curves of a double-resonant circuit for $V_{v\min} = 0$, $L_{in} = L_v$, various values of q_0 and k using (a) an abrupt varactor with the values of $\gamma = 0.5$ and $\varphi = 0.6$ V and (b) a hyperabrupt varactor with the approximated values of $\gamma = 1$ and $\varphi = 2.5$ V. From comparison of curves shown in Figure 8.10a for $q_0 = 0.5$ it follows that a decrease of k from 0.7 to 0.2 results in a decrease of the tuning slope ratio by a factor of almost two, and reaches its minimum value of $\alpha = 3.2$ in a frequency range of $K_f = 1.7$. For $q_0 = 1$, at which the octave tuning bandwidth is realized, α does not practically depend on coupling

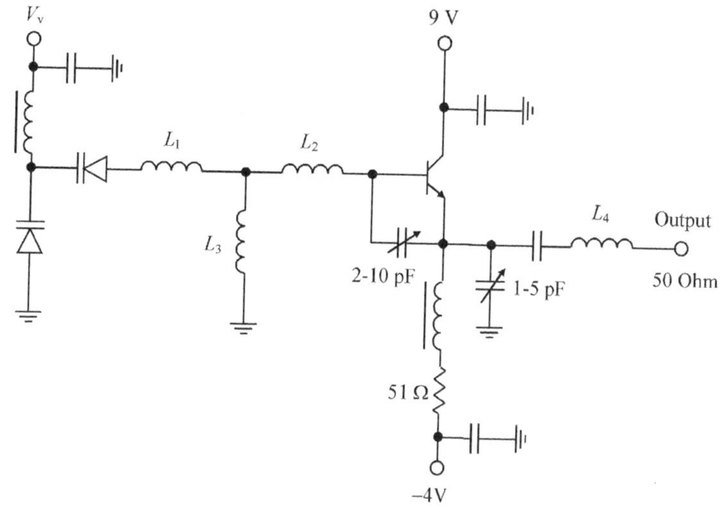

Figure 8.11 L-band bipolar VCO circuit schematic

coefficient k having a value of $\alpha > 6$. Using the oscillation system with $q_0 < 0.5$ results in a significantly smaller frequency tuning bandwidth with abrupt varactors. However, it follows from Figure 8.10b that using hyperabrupt varactors enables one to improve tuning linearity, and there is an optimum value of $k = 0.5$ for $q_0 = 0.5$ when $\alpha \leq 1.4$ in a frequency range of $K_f = 2.2$. The main advantage of such an oscillation system with two coupled resonant circuits is a wide linear varactor frequency tuning for different values of q_0.

Figure 8.11 shows the electrical schematic of the L-band bipolar VCO circuit. The optimum q_0 was realized using two back-to-back connected silicon hyperabrupt varactors with $C_{v0} =$ 20 pF. The active device represents a microwave bipolar transistor characterized by transition frequency $f_T = 2.3$ GHz and collector capacitance $C_c = 5$ pF. In this case, when the collector terminal of the bipolar transistor is RF grounded, the collector capacitance can be considered as the first approximation for the equivalent input device capacitance C_{in}. The calculated values of lumped inductances are identical and equal to $L_1 = L_2 = L_3 = 9$ nH. To provide the frequency stable and monotonic varactor tuning, the load circuit includes the variable feedback capacitance in the range 1–5 pF and series inductance L_4 of 3–10 nH.

Figure 8.12 shows the comparative tuning bandwidth characteristics and modulation sensitivities of L-band VCOs with a double-resonant circuit (curves 1 and 3) and a single-resonant circuit (curves 2 and 4). The single-resonant circuit is represented by a series inductance $L = L_1 + L_2$ and an ideal choke instead of inductance L_3. Comparison of the curves shows that using the oscillation system with two coupled resonant circuits widens the frequency tuning bandwidth (curve 1) and reduces the tuning nonlinearity. The tuning slope ratio $\alpha = 9$ (curve 4) for the single-resonant circuit decreases to $\alpha = 2.3$ (curve 3) for the double-resonant circuit. This means that the tuning linearity relative to a best-fit straight line improves from $\pm 3.5\%$ to $\pm 0.7\%$ for a tuning bandwidth of 45%.

By minimizing the varactor lead inductances and optimizing the inductances L_2 and L_3, it is possible to realize maximum tuning bandwidth using a common base device configuration. Figure 8.13 shows the electrical schematic of the microwave X-band bipolar VCO designed for a very wide frequency range monotonically tunable from 3.0 to 9.4 GHz [14]. The possible negative resistance bandwidth was calculated and measured in a frequency range 2.5–11 GHz

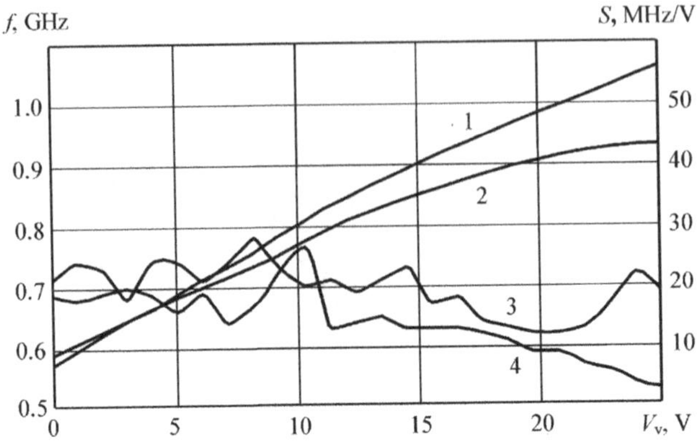

Figure 8.12 Varactor tuning bandwidth characteristics

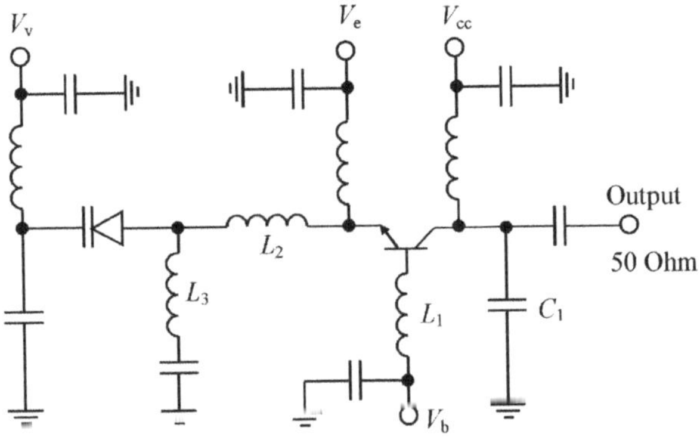

Figure 8.13 Microwave *X*-band bipolar VCO circuit schematic

for a NEC567 bipolar transistor using a feedback inductor L_1 and a capacitor C_1. Such a wideband frequency tuning was provided by a GaAs hyperabrupt varactor with $C_{v0} = 1.2$ pF and $K_c = 11$.

8.4 TRANSMISSION LINE CIRCUIT REALIZATION

At microwave frequencies, the VCO oscillation systems very often represent structures with distributed parameters using microstrip transmission lines. It allows one to achieve, for both hybrid and monolithic VCO realizations, higher quality factors of the oscillation systems with optimum temperature stability and noise characteristics in comparison with *LC*-oscillators [15]. One of the basic problems for microstrip VCO design is the realization of a wide tuning bandwidth with a high tuning linearity.

8.4.1 Oscillation system with uniform transmission line

Among VCO oscillation systems with distributed elements the structures with uniform transmission line are the most popular in view of simplicity of their practical microstrip implementation. For the first time an opportunity arose to use a uniform transmission line as a two-port

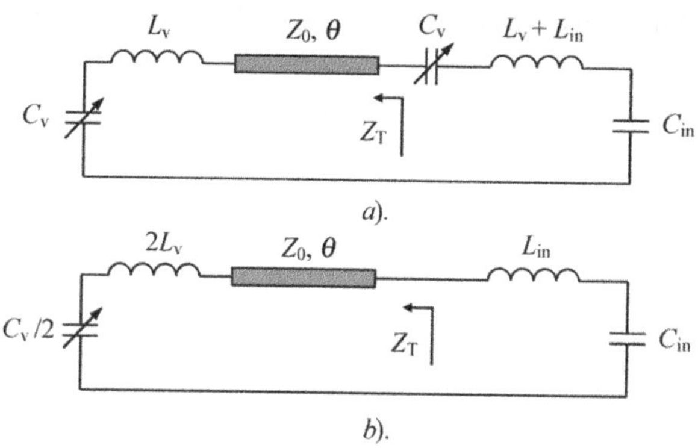

Figure 8.14 Equivalent circuits of oscillation systems with uniform transmission line

transforming network between the transistor and varactor in order to increase the oscillation system sensitivity to a varactor capacitance tuning [16]. In some cases, the sensitivity of such an oscillation system to varactor capacitance tuning becomes higher than that of the oscillation system without a transmission line, taking into account the varactor and transistor lead inductances [17]. In practice, the two back to back varactors are often used, being connected either (a) at both sides of a transmission line or (b) at its opposite side, as shown in Figure 8.14. It is crucial to determine the relationship between the varactor capacitance, equivalent input device capacitance and the parameters of the transmission line to achieve maximum tuning bandwidth for minimum voltage range of the varactor tuning. Note that an increase in the number of varactors is undesirable, because it leads to a growth of losses in the oscillation system and their complexity.

For the circuit shown in Figure 8.14a, the oscillation frequency can be obtained as

$$\frac{1}{j\omega C_{\text{in}}} + j\omega\left(L_{\text{v}} + L_{\text{in}}\right) + \frac{1}{j\omega C_{\text{v}}} + Z_{\text{T}}\left(\omega\right) = 0 \tag{8.28}$$

where

$$Z_{\text{T}} = Z_0 \frac{Z_{\text{v}} + jZ_0 \tan\theta}{Z_0 + jZ_{\text{v}} \tan\theta}$$

is the input impedance of a loaded lossless transmission line, θ is the electrical length of a transmission line and $Z_{\text{v}} = j\omega L_{\text{v}} + (1/j\omega C_{\text{v}})$ is the load impedance.

Solving Equation (8.28) for $\tan\theta$ allows us to determine the oscillation frequencies of the resonant circuit with two varactors connected between a transmission line as

$$\tan\theta = \omega C_{\text{v}} Z_0 \frac{A + (1 - \omega^2 L_{\text{v}} C_{\text{v}}) C_{\text{in}}/C_{\text{v}}}{\omega^2 C_{\text{in}} C_{\text{v}} Z_0^2 - A(1 - \omega^2 L_{\text{v}} C_{\text{v}})} \tag{8.29}$$

where $A = 1 - \omega^2 C_{\text{in}}\left(L_{\text{in}} + L_{\text{v}}\right) + C_{\text{in}}/C_{\text{v}}$.

Similarly, the same equation for the resonant circuit with two varactors connected to the opposite side of the transmission line, shown in Figure 8.14b can be written as

$$\tan\theta = \omega C_{\text{v}} Z_0 \frac{A + (1 - \omega^2 L_{\text{v}} C_{\text{v}}) C_{\text{in}}/C_{\text{v}}}{\omega^2 C_{\text{in}} C_{\text{v}} Z_0^2 - 2(1 - \omega^2 L_{\text{in}} C_{\text{in}})(1 - \omega^2 L_{\text{v}} C_{\text{v}})} \tag{8.30}$$

A comparison of Equations (8.29) and (8.30) shows that both resonant circuits provide the same tuning properties if

$$\omega^2 C_{\text{in}} C_{\text{v}} Z_0^2 \gg A(1 - \omega^2 L_{\text{v}} C_{\text{v}})$$

when the oscillation frequency can be determined from

$$\omega = \frac{2C_{\text{in}} + C_{\text{v}}}{C_{\text{in}} C_{\text{v}}} \frac{1}{\omega(L_{\text{in}} + 2L_{\text{v}}) + Z_0 \tan\theta} \tag{8.31}$$

For a short transmission line with electrical length of $\theta < \pi/4$, the frequency tuning coefficient K_{f} depends on varactor capacitance tuning coefficient K_{c} as

$$K_{\text{f}} = \sqrt{K_{\text{c}} \frac{1 + 2q}{K_{\text{c}} + q}} \tag{8.32}$$

The required value of K_{c} for a given value of K_{f} can be obtained as

$$A_2 K_{\text{c}}^2 + A_1 K_{\text{c}} + A_0 = 0 \tag{8.33}$$

where

$$\begin{aligned}
A_2 &= \omega_{\max} C_{\text{vmin}} Z_0 K_{\text{f}} \left[K_{\text{f}}^2 - \omega_{\max}^2 C_{\text{in}} (L_{\text{in}} + 2L_{\text{v}})\right] \\
&\quad + \left[\omega_{\max}^4 C_{\text{in}} L_{\text{v}} (L_{\text{in}} + L_{\text{v}}) - K_{\text{f}}^2 \omega_{\max}^2 \left(Z_0^2 C_{\text{in}} + L_{\text{v}}\right)\right] C_{\text{vmin}} \tan\theta \\
A_1 &= K_{\text{f}}^2 \left\{2K_{\text{f}} \omega_{\max} C_{\text{in}} Z_0 + \left[K_{\text{f}}^2 - \omega_{\max}^2 C_{\text{in}} (L_{\text{in}} + 2L_{\text{v}})\right] \tan\theta\right\} \\
A_0 &= K_{\text{f}}^4 \frac{C_{\text{in}}}{C_{\text{vmin}}} \tan\theta
\end{aligned}$$

where

$$\theta = \frac{1}{K_{\text{f}}} \tan^{-1} \frac{1 - \omega_{\max}^2 C_{\text{vmin}} L_{\text{v}}}{\omega_{\max} C_{\text{vmin}} Z_0} + \frac{1}{K_{\text{f}}} \tan^{-1} \frac{1 - \omega_{\max}^2 C_{\text{in}} (L_{\text{in}} + L_{\text{v}}) + \dfrac{C_{\text{in}}}{C_{\text{vmin}}}}{\omega_{\max} C_{\text{in}} Z_0}$$

However, it is advisable to analyse resonant circuits with transmission lines in a general form to obtain the common regularities and particular features. In this case, the coefficients A_2, A_1 and A_0 in quadratic Equation (8.33) can be rewritten in a general form [18]

$$\begin{aligned}
A_2 &= mK_{\text{f}} \left[K_{\text{f}}^2 - n^2 (1 + 2q)\right] + \left\{p^2 \left[(1 + 2q) n^2 - qp^2\right] - K_{\text{f}}^2 (qm^2 + p^2)\right\} \tan\theta \\
A_1 &= K_{\text{f}}^2 \left\{2K_{\text{f}} qm + \left[K_{\text{f}}^2 - n^2 (1 + 2q)\right] \tan\theta\right\} \\
A_0 &= K_{\text{f}}^4 q \tan\theta
\end{aligned}$$

where

$$\theta = \frac{1}{K_{\text{f}}} \tan^{-1} \frac{1 - p^2}{m} + \frac{1}{K_{\text{f}}} \tan^{-1} \frac{1 + (1 + p^2) q - (1 + 2q) n^2}{mq}$$

using the following independent normalized parameters:

$$m = \omega_{\max} C_{\text{vmin}} Z_0 \qquad n = \omega_{\max}/\omega_{\max 0}$$

$$p = \omega_{\max}/\omega_{\text{vmax}} \qquad q = C_{\text{in}}/C_{\text{vmin}}$$

where

$$\omega_{\text{vmax}} = 1/\sqrt{L_{\text{v}}C_{\text{vmin}}}$$

$$\omega_{\text{max0}} = 1 \Bigg/ \sqrt{\frac{C_{\text{in}}C_{\text{vmin}}}{2C_{\text{in}} + C_{\text{vmin}}}(L_{\text{in}} + 2L_{\text{v}})}$$

The electrical length of a transmission line at high bandwidth frequency can be determined from

$$\theta_{\max} = \tan^{-1}\left[m\frac{A + (1-p^2)q}{qm^2 - A(1-p^2)}\right] \tag{8.34}$$

where $A = 1 + q(1+p^2) - (1+2q)n^2$.

Figure 8.15a shows the dependencies $K_{\text{f}}(K_{\text{c}})$ for the resonant circuit with a transmission line between two varactors with various values of normalized parameters: $0.25 \leq m \leq 2$, $n = 0.8$, $p = 0.7$ and $1 \leq q \leq 10$. The dotted lines indicate the dependencies $K_{\text{f}}(K_{\text{c}})$ for the

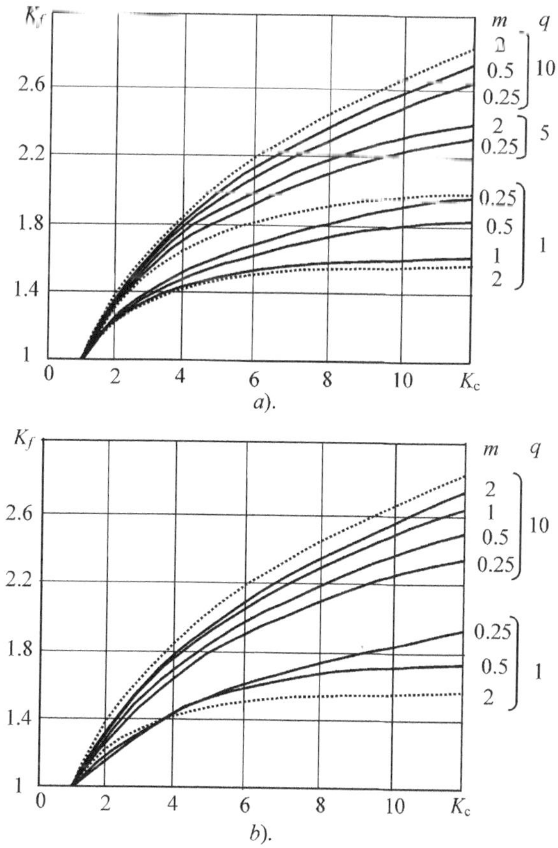

Figure 8.15 Dependencies $K_{\text{f}}(K_{\text{c}})$ of oscillation systems with uniform transmission line

oscillation system without a transmission line merged with the dependencies $K_f(K_c)$ for the oscillation system with a transmission line with $m = 2$ ($q = 1, 10$). The same dependencies for the resonant circuit when a transmission line is connected between the active device and two back-to-back varactors are shown in Figure 8.15b. A comparison of curves shows that in both cases their behaviour is practically similar. For high q, it is preferable to use an oscillation circuit without a transmission line or to choose a transmission line with high characteristic impedance. However, at low q and m, it is possible to widen the frequency tuning bandwidth using a transmission line. This means that, in this case, it is necessary to choose a transmission line with the lowest possible characteristic impedance. For some intermediate value of $q(q \cong 2.5$ for the first case and $q \cong 1.5$ for the second), the band properties of both oscillation systems are the same and do not depend on m.

Nevertheless, the oscillation system with a transmission line between two varactors has some advantages in band properties compared with the oscillation system with a transmission line between the active device and the varactors. A comparison of the curves $K_c(m)$ shown in Figure 8.16 for $p = 0.7, n = 0.8$ shows that, at low m, the same octave tuning bandwidth for the oscillation system (a) with a transmission line between two varactors is realized with lower K_c than (b) with a transmission line between the active device and two varactors.

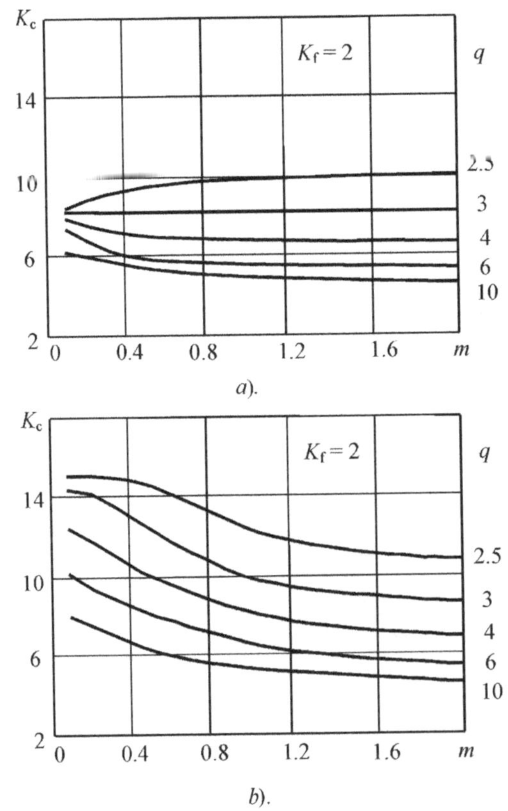

Figure 8.16 Dependencies $K_f(m)$ of oscillation systems with uniform transmission line

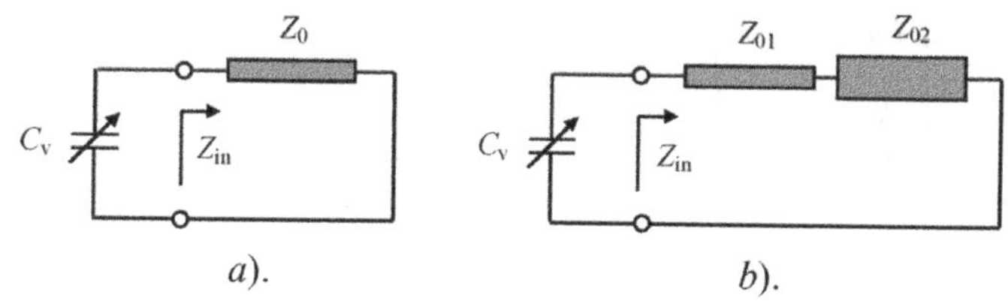

Figure 8.17 Oscillation systems with (a) uniform and (b) two-section transmission lines

8.4.2 Oscillation system with multi-section transmission line

The band properties of a resonant circuit containing a short-circuited transmission line shown in Figure 8.17a can be improved if, instead of a uniform transmission line, a nonuniform transmission line is used, the spectrum of natural frequencies of which is nonequidistant. Figure 8.17b shows an example of the short-circuited two-section transmission line. Among the oscillation systems with nonuniform transmission lines, the oscillation systems with a multi-section line with alternating high and low characteristic impedance of its uniform sections are ideal from the standpoint of maximum frequency tuning bandwidth. A sensitivity variation of such a multi-section line is determined by a spectrum rarefaction near the resonant frequency for a tuning band widening and spectrum narrowing in the case of frequency stabilization.

The main difference in frequency properties between the oscillation system with (a) a short-circuited uniform transmission line with an equidistant spectrum and (b) a multi-section line with a nonequidistant spectrum is shown in Figure 8.18. The graphical results show that it is preferable to use a multi-section line for wideband varactor tuning on a fundamental resonant frequency when a smaller change in K_c causes a larger change in K_f. The section adjacent to the varactor should have high characteristic impedance [17].

A comparative analysis of various structures of oscillator resonant circuits with multi-section lines shows that the simplest in practical implementation, ensuring at the same time close to the maximum frequency tuning bandwidth for a given value of K_c, is the oscillator resonant circuit with a two-section line (see Figure 8.17b). Figure 8.19 shows the resonant circuit with a two-section transmission line, where M is the characteristic impedance ratio, θ_1 and θ_2 are the electrical lengths of the high-impedance and low-impedance sections, respectively.

The tuning frequency bandwidth can be found from the phase balance condition given by

$$2\left(j\omega L_v + \frac{1}{j\omega C_v}\right) + Z_{in}(\omega) = 0 \tag{8.35}$$

which can be represented in general form

$$K_c = \frac{2K_f^2}{2p^2 + \dfrac{mK_f}{Z_0}Z_{in}(K_f)} \tag{8.36}$$

where

$$Z_{in} = Z_0 \tan\left(\tan^{-1}\left(\frac{1}{M}\tan\left(\tan^{-1}(MX_L) + \frac{2-k}{K_f}\theta_{max}\right)\right) + \frac{k\theta_{max}}{K_f}\right)$$

$$X_L = \left[(1+2q)n^2 - 2p^2q - K_f^2\right]/mqK_f$$

$k = 2\theta_1/(\theta_1+\theta_2)$, $\theta_{max} = (\theta_{max1}+\theta_{max2})/2$, θ_{max1} and θ_{max2} are the electrical lengths of the high-impedance and low-impedance sections at a high bandwidth frequency, respectively [17].

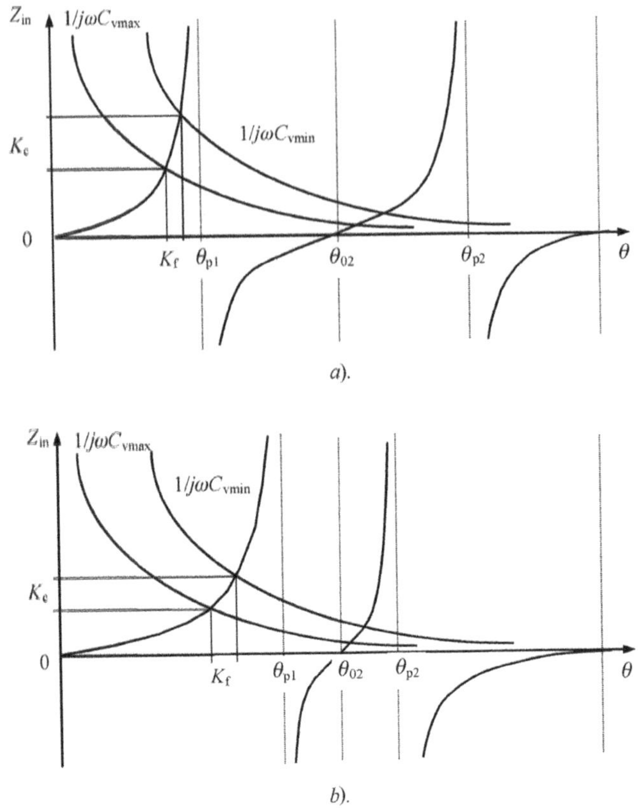

Figure 8.18 Frequency behaviour of oscillation system with (a) short-circuited uniform and (b) stepped transmission line

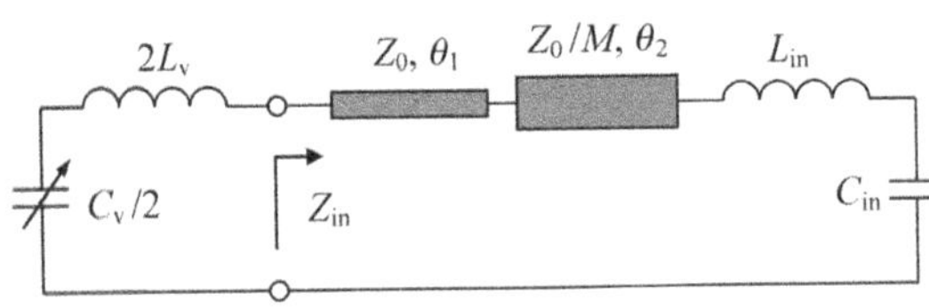

Figure 8.19 Equivalent VCO resonant circuit with two-section line

The maximum electrical length θ_{max} is determined as a result of numerical solution of Equation (8.36) with a single unknown parameter at fixed values of normalized parameters m, n, p, q and k when $K_f = K_c = 1$.

The numerical calculations have shown that, for various values of a normalized parameter m, there is an optimum value k, at which K_c is minimal. For $n = 0.8$, $p = 0.7$, $M = 20$, $q = 1$ and $m = 1$, an octave bandwidth is realized for $K_c = 4.7$ with $\theta_2/\theta_1 = 2.3$. For equal electrical lengths when $\theta_2 = \theta_1$, the octave bandwidth is provided with $K_c = 5$. The dependencies $K_f(K_c)$ for various values of (a) $0.25 \leq m \leq 2$, (b) $1 \leq M \leq 30$ and (c) $0 \leq n \leq 0.8$ at fixed values of $p = 0.7$ and $k = 1$ are shown in Figure 8.20. Comparison of the curves shows that the

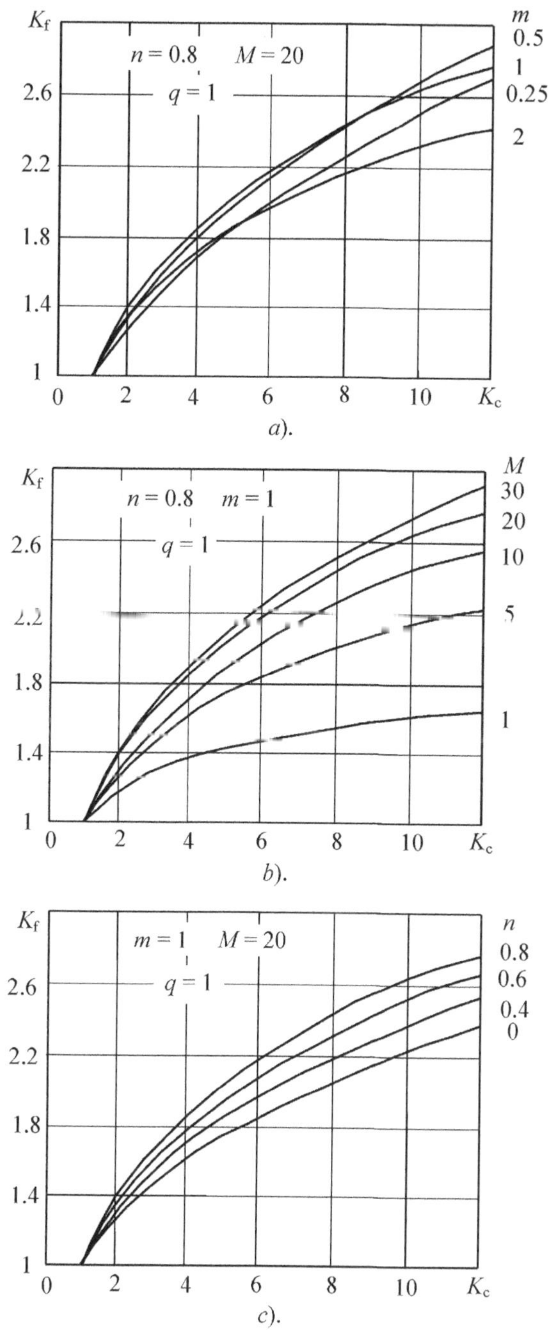

Figure 8.20 Frequency dependencies of oscillation system with two-section line

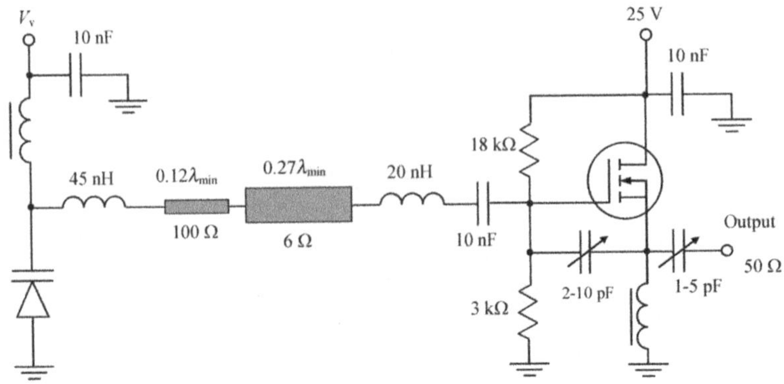

Figure 8.21 Circuit schematic of common drain MOSFET VCO with two-section line

oscillation system with a two-section line provides wideband tuning for both low and high q. Maximum sensitivity to varactor capacitance tuning is realized in the range $m = 0.5$–1, as follows from Figure 8.20a. An increase in the characteristic impedance ratio M leads to a substantial widening of the tuning bandwidth, shown in Figure 8.20b. Besides, a smaller value of the parameter n results in a narrower frequency tuning bandwidth, as shown in Figure 8.20c. This means that the lead lumped inductances contribute to an additional widening of the tuning bandwidth.

Figure 8.21 shows the electrical circuit of the common drain MOSFET VCO with an optimal two-section line. The active device is characterized by the transition frequency $f_T = 1$ GHz and the equivalent input capacitance $C_{in} = 2$ pF. The varactor minimum capacitance C_{vmin} is equal to 2 pF with tuning coefficient $K_c = 4.5$–6.5 in a bias range from 3 to 25 V. A high value of the characteristic impedance ratio $M > 16$ was provided using a high-impedance coaxial transmission line and low-impedance symmetric stripline. The measured frequency tuning characteristics of the MOSFET VCO with a two-section line (curve 1) and with a lumped inductance (curve 2) are shown in Figure 8.22. In the latter case, a two-section line is replaced by a series lumped inductance to realize a lumped VCO with the series resonant LC-circuit.

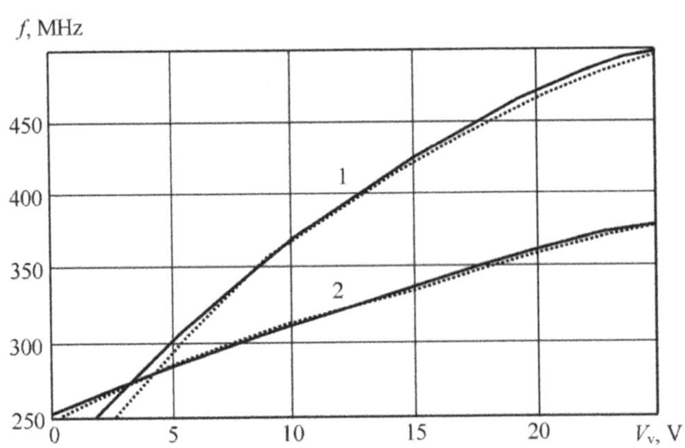

Figure 8.22 Frequency tuning characteristics of MOSFET VCO with two-section line

The calculated frequency tuning characteristics are shown by dotted lines for both cases. Using a two-section line allows one to widen significantly the frequency tuning bandwidth. A full-octave frequency bandwidth from 250 to 500 MHz was achieved within a varactor bias range from 2 to 25 V. The tuning slope ratio $\alpha < 2$ can be provided for a tuning bandwidth of 46% with $K_f = 1.6$.

8.5 VCO CIRCUIT DESIGN ASPECTS

8.5.1 Common gate MOSFET and MESFET VCOs

To analyse the frequency tuning possibility of the MOSFET VCO resonant circuit, it is sufficient to consider the simple device equivalent circuit, especially when the operating frequencies are not too high. Figure 8.23 shows the common gate MOSFET VCO equivalent circuit where $Z_1 = 1/j\omega C$ is the capacitive source reactance, $Z_2 = j\omega L$ is the inductive gate feedback reactance and $Z_L = R_L + j\omega L_L$ is the load impedance [19]. Since the device feedback gate–drain capacitance C_{gd} is usually much smaller than the device gate–source capacitance C_{gs}, its influence on the VCO impedance characteristic can be neglected, which simplifies the analytical calculations significantly.

The start-up conditions for such an oscillator can be written as

$$\mathrm{Re} Z_{in} < 0 \tag{8.37}$$

$$\mathrm{Im}\,(Z_{in} + Z_1) = 0 \tag{8.38}$$

where the input impedance Z_{in} is determined through the device small-signal common source Y-parameters by

$$Z_{in} = \frac{1 + Y_{11} Z_2 + (Y_{22} + Z_2 \Delta Y)\, Z_L}{Y_{11} + Y_{12} + Y_{21} + Y_{22} + \Delta Y\,(Z_2 + Z_L)} \tag{8.39}$$

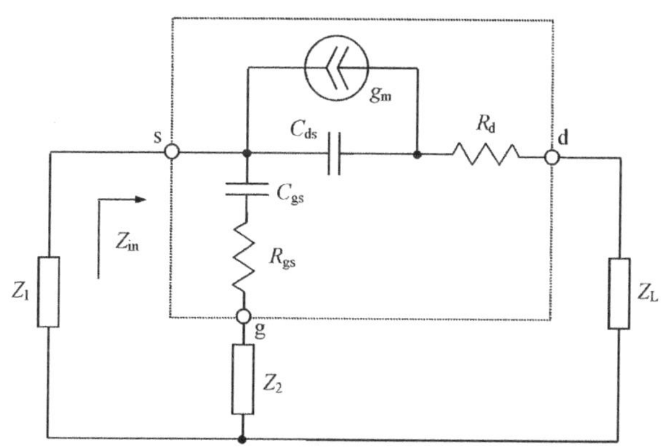

Figure 8.23 Common gate MOSFET VCO equivalent circuit

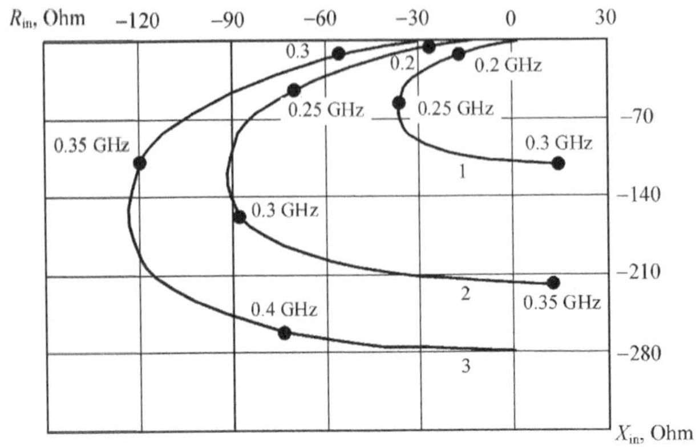

Figure 8.24 Frequency dependencies of impedance curves for different feedback parameters

Substituting the small-signal Y-parameters expressed through the parameters of the MOSFET equivalent circuit into Equation (8.39) yields

$$Z_{in} = \frac{(1 - \omega^2 C_{ds} L_L)(1 - \omega^2 C_{gs} L) - \omega^2 C_{ds} C_{gs}(R_L + R_d) R_{gs} +}{g_m - \omega^2 C_{ds} C_{gs}(R_L + R_d + R_{gs}) +} \rightarrow$$

$$\rightarrow \frac{j\omega[(1 - \omega^2 C_{gs} L) C_{ds}(R_L + R_d) + (1 - \omega^2 C_{ds} L_L) C_{gs} R_{gs}]}{j\omega[C_{gs} + C_{ds} - \omega^2 C_{gs} C_{ds}(L + L_L)]} \tag{8.40}$$

Figure 8.24 shows the theoretical impedance (resistance and reactance) curves versus frequency for different values of L and L_L and $R_L = 50\ \Omega$ (curve 1: $L = 200$ nH, $L_L = 50$ nH; curve 2: $L = 200$ nH, $L_L = 0$ nH, curve 3: $L = 100$ nH, $L_L = 0$ nH). The MOSFET device has a 5-μm gate length and its small-signal equivalent circuit parameters are of $g_m = 27$ mA/V, $C_{gs} = 5$ pF, $R_{gs} = 25\ \Omega$, $C_{ds} = 3.6$ pF, and $R_d = 70\ \Omega$. These dependencies show that, for different values of the inductances L and L_L, the input resistance R_{in} becomes negative in a frequency range from 200 to 400 MHz. With decrease of the inductance L_L, a frequency range with negative resistance extends to higher frequencies. The same influence is challenged by the feedback inductance L, but at the same time this inductance has practically no effect on the width of the frequency range with the negative resistance. The input reactance is inductive at low frequencies, whereas it becomes capacitive at higher frequencies.

To analyse the possibility of the maximum frequency tuning bandwidth, it is best to consider the required variation of the feedback inductance L when $\text{Re}Z_{in} = 0$ in Equation (8.40). As a result, the feedback inductance can be calculated from

$$L = \frac{1}{\omega^2 C_{gs}} \left(1 + \frac{p}{2} - \sqrt{\frac{p^2}{4} - q} \right) \tag{8.41}$$

where

$$p = \frac{g_m(1 - \omega^2 C_{ds} L_L)}{\omega^2 C_{ds}^2 (R_L + R_d)}$$

$$q = \left(\frac{C_{gs}}{C_{ds}} \right)^2 \frac{R_{gs}(1 - \omega^2 C_{ds} L_L)^2}{R_L + R_d} + R_{gs} \left[\omega^2 C_{gs} C_{ds}(R_L + R_d + R_{gs}) - g_m \right] \frac{C_{gs}}{C_{ds}}$$

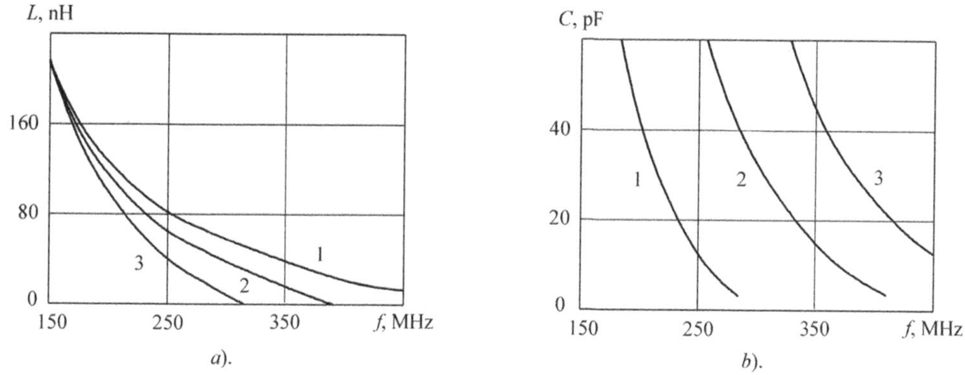

Figure 8.25 Frequency dependencies of feedback (a) inductance and (b) capacitance

Figure 8.25a shows the frequency dependencies of the feedback inductance L for different values of the load inductance L_L (curve 1: $L_L = 0$. curve 2: $L_L = 50$ nH, curve 3: $L_L = 75$ nH). An improved tuning sensitivity to the variations of the feedback inductance is realized with minimum load inductance. To provide such a behaviour of the feedback inductance, resulting in a wideband tuning, it is necessary to connect a varactor in series to the gate inductor L. At high bias voltages, the variations of the varactor capacitance are not so significant. However, at low bias voltages, the greater variations of the varactor capacitance result in a significant increase of the equivalent inductance.

By using Equation (8.38), the value of the capacitance C expressed through the device equivalent circuit parameters can be obtained for a given value of the feedback inductance L from

$$C = \frac{C_{gs}(1 - \omega^2 C_{ds} L_L) + C_{ds}(1 - \omega^2 C_{gs} L)}{\omega^2 C_{gs} C_{ds}(R_L + R_d)R_{gs} - (1 - \omega^2 C_{ds} L_L)(1 - \omega^2 C_{gs} L)} \tag{8.42}$$

Figure 8.25b shows the frequency dependencies of the source capacitance C for different values of the device transconductance g_m (curve 1: $g_m = 10$ mA/V. curve 2: $g_m = 20$ mA/V, curve 3: $g_m = 30$ mA/V). It is clearly seen that, at fixed frequency, smaller values of the device transconductance g_m result in a significantly smaller value of the capacitance C. As a result, by setting the minimum capacitance value at high bandwidth frequency with maximum transconductance, it is possible to achieve more than octave-band frequency tuning using a varactor connected in series to the feedback inductor. In this case, the transconductance reduces at lower frequencies, contributing to the minimum variations of the output power. Analysis of Equation (8.41) shows that the variations of the device transconductance g_m have negligible effect on the value of gate feedback inductance L.

For clear qualitative demonstration of the appearance of the negative resistance and its dependence on the main parameters of the device equivalent circuit, it is enough to set to zero the values of C_{ds}, R_d and L_L. Then, Equation (8.40) can be simplified to

$$Z_{in} = \frac{g_m(1 - \omega^2 C_{gs} L) + \omega^2 C_{gs}^2 R_{gs}}{g_m^2 + \omega^2 C_{gs}^2} + \frac{j\omega C_{gs}(g_m R_{gs} - 1 + \omega^2 C_{gs} L)}{g_m^2 + \omega^2 C_{gs}^2} \tag{8.43}$$

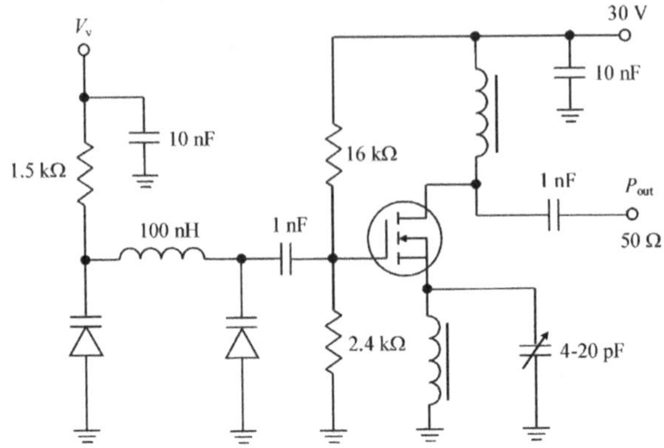

Figure 8.26 Circuit schematic of common gate MOSFET VCO

From Equation (8.43) it follows that the negative resistance is a result of the presence of the feedback gate inductance L, and it occurs at frequencies

$$\omega > \frac{1}{\sqrt{C_{gs}\left(L - \dfrac{C_{gs}R_{gs}}{g_m}\right)}} \tag{8.44}$$

Hence, the low tuning bandwidth frequency can be chosen by using a proper value of the gate feedback inductance L for a particular active device with measured or modelled values of the gate time constant $\tau_g = C_{gs}R_{gs}$ and small-signal transconductance g_m. Depending on values of the gate–source resistance R_{gs}, the input reactance can change its values from inductive to capacitive.

Figure 8.26 shows the practical realization of such a common gate MOSFET VCO using two hyperabrupt varactors with $K_c = 10$ and $C_{vmin} = 2$ pF. This allows one to achieve output power of 17 ± 5 dBm in the frequency tuning bandwidth from 170 to 390 MHz with varactor bias range 0.4–30 V. The output power reduces at low bandwidth frequencies for low varactor bias voltages when the oscillator loaded quality factor degrades due to decrease in the varactor quality factor. The variable capacitance connected to the device source terminal is used to fulfil the phase balance condition over the entire tuning bandwidth. Despite some limitations in applying the small-signal analysis and using the simplified device equivalent circuit, the experimental results are in a very good agreement with the theoretical estimate of the frequency tuning bandwidth.

Similar small-signal analysis can be applied to the common gate MESFET oscillator. By using the NE32684 device with $g_m = 38$ mA/V, $C_{gs} = 0.24$ pF and $C_{ds} = 0.108$ pF (under bias conditions of $V_{dd} = 2$ V and $I_d = 20$ mA) and gate feedback inductance of 7 nH, the negative resistance occurs above 3 GHz and disappears at around 7 GHz [20]. The major part of input reactance is capacitive and needs an inductive part to provide the phase compensation. Figure 8.27 shows the practical circuit schematic of the common gate MESFET VCO. By using a tank resonator with parallel inductor of 1 nH and a varactor with capacitance tuning range from 0.2 to 1.4 pF, the frequency tuning bandwidth from 2.6 to 5.2 GHz with output power of 3 ± 5 dBm at 50-Ω load (connected through blocking capacitance to the drain terminal) were achieved. The difference between practical schematics of the MOSFET VCO shown in

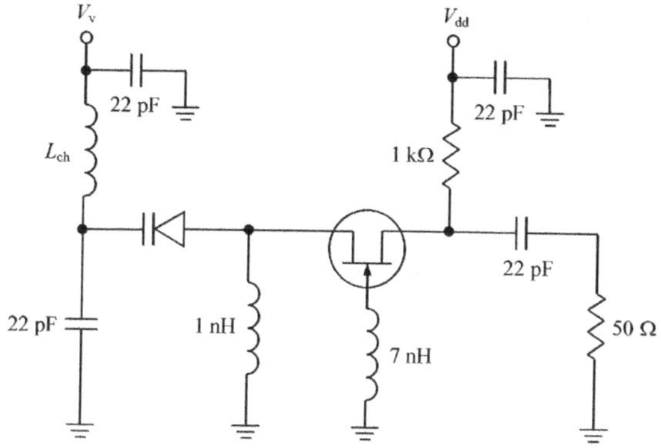

Figure 8.27 Circuit schematic of common gate MESFET VCO (permission by IEICE)

Figure 8.26 and the MESFET VCO shown in Figure 8.27 is in the location of the varactors. The frequency tuning of the MOSFET VCO is realized by varactors located in the gate circuit, whereas a varactor located in the source circuit provides frequency tuning for the MESFET VCO.

8.5.2 Common collector bipolar VCO

The tuning possibility of a bipolar VCO resonant circuit can be evaluated by using the simple device equivalent circuit shown in Figure 8.28 [19]. Here, the collector terminal is common and usually RF grounded in the practical realization with a bypass capacitor. The simplified equivalent circuit includes the collector capacitance C_c, the base–emitter capacitance C_π (including diffusion and junction capacitances) and a current source described by the small-signal transconductance g_m. To provide wideband frequency tuning, the two varactors, C_v^b and C_v^e, are used in the base and emitter circuits, respectively.

For a common collector bipolar VCO, the equation for resonant frequencies in a steady-state operation mode can be given by

$$\omega^2 L C_v^b \left(C_c + \frac{C_\pi C_v^e}{C_\pi + C_v^e} \right) = C_v^b + C_c + \frac{C_\pi C_v^e}{C_\pi + C_v^e} \tag{8.45}$$

To characterize the VCO band properties, it is convenient to use the generalized dependencies $K_f(K_{c1}, K_{c2})$, where $K_{c1} = C_{vmax}^b / C_{vmin}^b$, $K_{c2} = C_{vmax}^e / C_{vmin}^e$. To obtain the results regardless of the particular values of the circuit parameters, it is best to introduce normalized

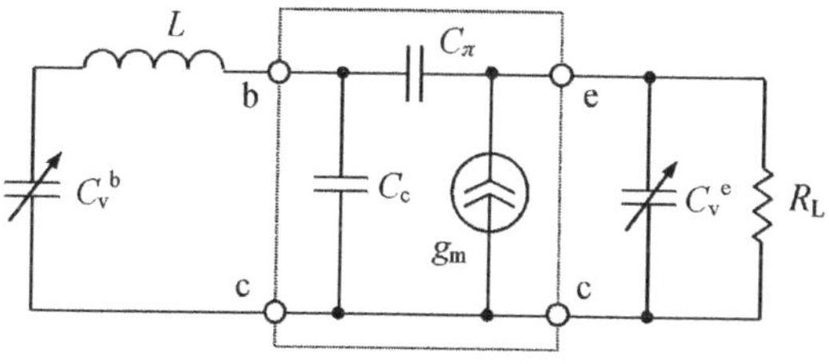

Figure 8.28 Equivalent circuit of common collector bipolar VCO

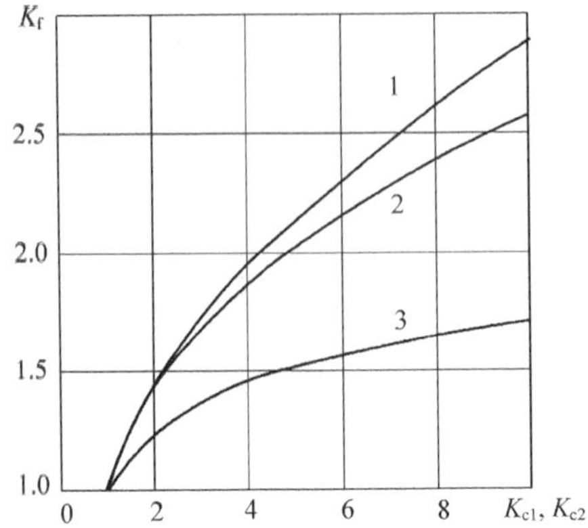

Figure 8.29 Frequency tuning performance of common collector bipolar VCO

parameters. By using the normalized parameters $m_0 = \omega_T C_c/g_m$, $q_1 = C_c/C^b_{vmin}$ and $q_2 = C_c/C^e_{vmin}$, Equation (8.45) can be rewritten in general form as

$$K_f = \sqrt{K_{c1}\frac{(1+q_1)(m_0+q_2)+q_1}{(q_1+K_{c1})(m_0K_{c2}+q_2)+q_1K_{c2}}\frac{K_{c2}(1+m_0)+q_2}{1+m_0+q_2}} \tag{8.46}$$

Figure 8.29 shows the different dependencies $K_f(K_{c1}, K_{c2})$ for various values of the normalized parameters q_1, q_2 and $m_0 = 0.012$. Here, curve 1 is plotted for $q_1 = 1$ and $q_2 = 0.5$ with varactor tuning in the base and emitter circuits simultaneously. Curve 2 is characterized by $q_1 = 1$ and $q_2 = 0.05$, with varactor tuning only in the base circuit when $K_{c2} = 1$. Curve 3 is calculated for $q_1 = 0.1$, $q_2 = 0.5$ with varactor tuning only in the emitter circuit when $K_{c1} = 1$. A comparison of the curves shows that, for varactor tuning in the base and emitter circuits simultaneously, maximum tuning bandwidth is achieved (curve 1). Using varactors only in the base circuit (curve 2) gives larger tuning bandwidth than in the case of the varactor tuning only in the emitter circuit (curve 3). In this case, decreasing q_2 and increasing q_1 can increase the tuning bandwidth. To increase the tuning bandwidth by emitter varactor tuning only, it is necessary to reduce the parameter q_1 significantly, provided $q_2 = 1$.

Figure 8.30a shows a typical common collector lumped VCO circuit where the two back-to-back varactors provide wideband tuning [21]. In this circuit, the load is conductively connected to the resonant circuit inductor. The choke inductors L_{ch} and bypass capacitors C_b form low-pass filters with high impedance at the fundamental frequency to isolate voltage supplies and low impedance for the modulation frequencies. By using abrupt varactors with $K_c = 3$ in a bias voltage range 0–5 V and minimum capacitance $C_{vmin} = 0.6$ pF, it is possible to provide wideband tuning in a frequency range with $K_f = 1.6$. Similar tuning bandwidth is predicted by curve 2 in Figure 8.29. Taking into account that the equivalent device input capacitance is equal to 1.5 pF, the tuning bandwidth from 5.5 to 8.0 GHz was realized with a tank inductor of $L = 1.9$ nH. To maintain the output voltage flat with frequency, a third varactor can be connected between collector and emitter terminals, as shown in Figure 8.30b [22]. As the tuning frequency increases, the capacitance of the varactor decreases, thus minimizing the variations of the output impedance across the collector of the transistor. Without this varactor, the collector voltage amplitude decreases with increasing frequency.

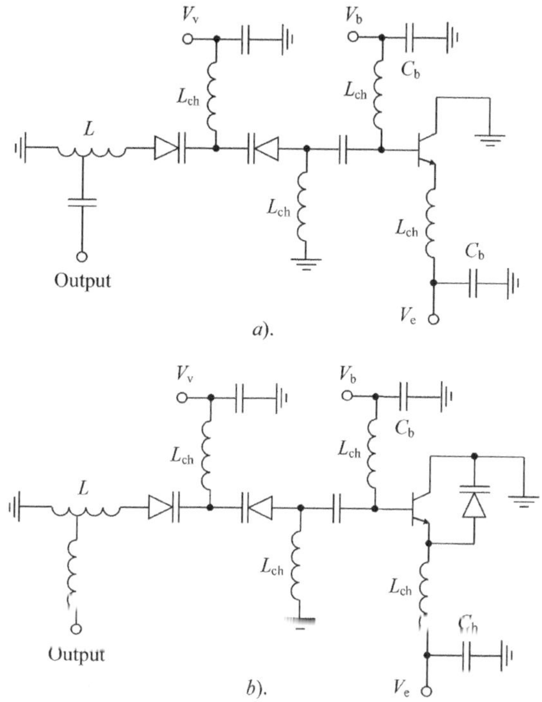

Figure 8.30 Typical common collector lumped VCO circuits

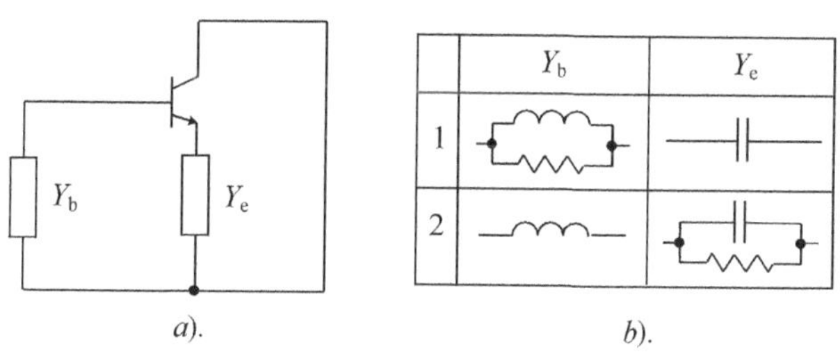

Figure 8.31 Simplified VCO schematic with two combinations of base and emitter circuits

However, for a common collector VCO, the conductive load connection is not the only way to realize a maximum level of output power. In addition, it is very important to provide its minimum flatness over the entire tuning bandwidth. The load can also be connected to the emitter terminal, thus decreasing the influence of the load impedance on the resonant circuit, which enables one to realize its higher quality factor. Figure 8.31 shows (a) the simplified common collector VCO schematic and (b) two possible combinations of the admittances in the base and emitter circuits. In the first case, the load is connected to the resonant circuit conductively or inductively, provided the impedance in the emitter circuit is capacitive. The second combination requires inductive impedance in the base circuit when the load is connected in parallel to the device emitter and collector terminals.

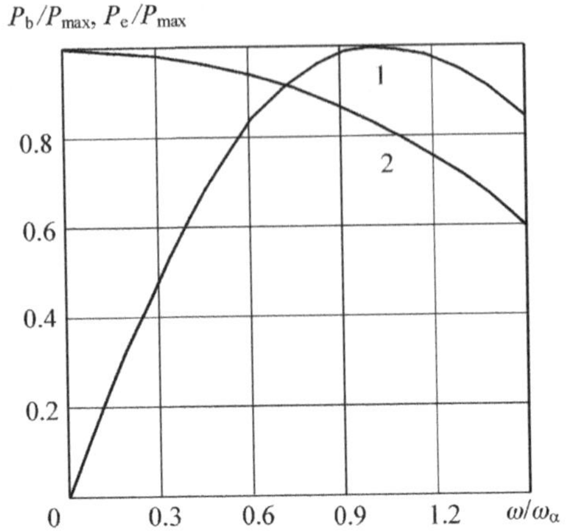

Figure 8.32 Output power versus oscillation frequency

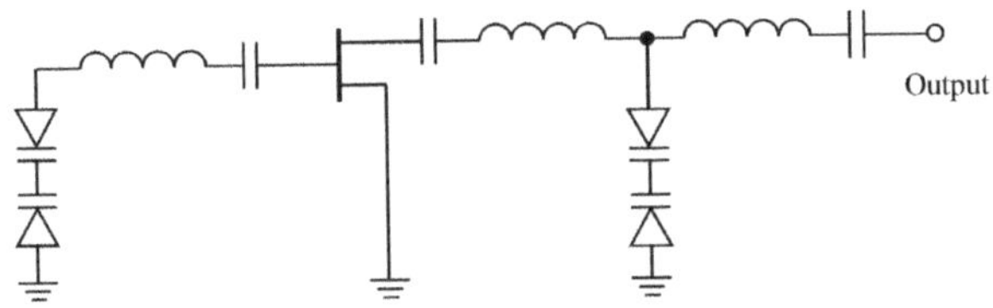

Figure 8.33 Circuit schematic of common drain MESFET VCO

Figure 8.32 shows the calculated dependencies of the normalized output power versus normalized frequency for both above-mentioned cases [23]. For case 1, the load is connected to the base circuit and the output power has maximum when $\omega = \omega_\alpha$. For case 2, the output power comes from the device emitter and its level changes negligibly up to $\omega = 0.5\omega_\alpha$. As a result, for the latter case, VCO can be tuned easily in a very wide frequency range by a simple tuning of the value of the series inductance in the base circuit using a varactor diode in reverse-bias operation. The series or parallel RC-circuit with constant capacitance and load resistance can realize the capacitive impedance in the emitter circuit.

To match device output impedance to the standard load and make the output power flat over a very wide frequency range, the T-type matching circuit consisting of a shunt pair of the two back-to-back varactors and two series inductors can be used. Figure 8.33 shows the simplified schematic (without bias circuits) of the monolithic MESFET oscillator where hyperabrupt varactors with a capacitance ratio of 10:1 provide wideband frequency tuning from approximately 7 to 18 GHz [24].

8.5.3 Common base bipolar VCO

Figure 8.34 shows the schematic of the common base VCO with varactors in both the base and emitter circuits [19, 25]. The resonant frequencies can be found from

$$\omega^2 LC_v^b\left[1+\frac{C_v^e}{C_c}+\frac{\omega_T C_v^e}{g_m}(1+g_m R_L)\right] = \left(1+\frac{\omega_T C_v^e}{g_m}\right)\left(1+\frac{C_v^b}{C_c}\right)+\frac{C_v^e}{C_c}+\frac{C_v^e+C_v^b}{C_c}g_m R_L \tag{8.47}$$

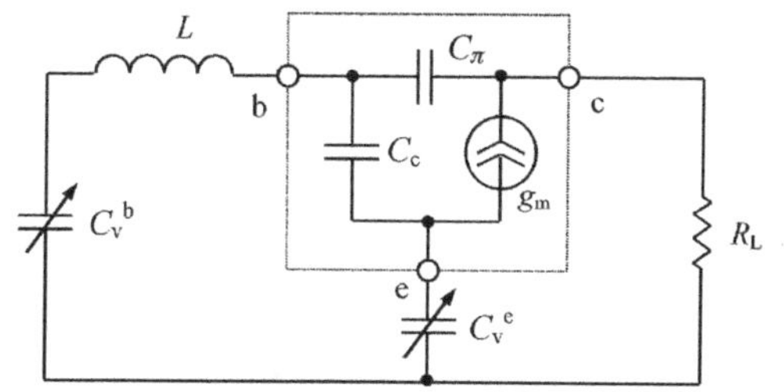

Figure 8.34 Circuit schematic of common base bipolar VCO

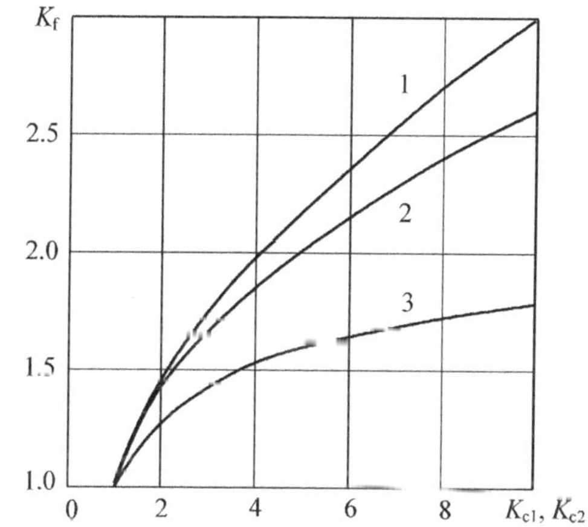

Figure 8.35 Frequency dependencies of common base bipolar VCO

In a general form, Equation (8.47) can be rewritten as

$$K_f = \sqrt{K_{c1} \frac{q_2(q_1 + n_0) + q_1(n_0 + m_0) + m_0}{q_2(q_1 + n_0 K_{c1}) + q_1 K_{c2}(n_0 + m_0) + m_0 K_{c1} K_{c2}} \frac{q_2 + K_{c2}(n_0 + m_0)}{q_2 + n_0 + m_0}} \quad (8.48)$$

where $n_0 = 1 + \omega_T R_L C_c$.

Figure 8.35 shows the dependencies $K_f(K_{c1}, K_{c2})$ for various values of the normalized parameters q_1, q_2 and $m_0 = 0.012$, $n_0 = 1.6$. A comparison of the curves shows that, for the simultaneous varactor tuning in the base and emitter circuits, three-fold frequency overlapping (curve 1: $q_1 = 1$, $q_2 = 0.5$) can be achieved. At the same time, the bipolar VCO with varactors only in the base circuit (curve 2: $q_1 = 1$, $q_2 = 0.05$, $K_{c2} = 1$) is characterized by a larger sensitivity to the varactor capacitance tuning than the emitter–varactor VCO configuration (curve 3: $q_1 = 0.1$, $q_2 = 0.5$, $K_{c1} = 1$). For the purpose of frequency bandwidth widening, it is best to reduce the parameter q_2 and increase the parameter q_1. The influence of the parameter n_0 is negligible and need not be taken into consideration.

Figure 8.36 shows the practical L-band bipolar VCO manufactured using hybrid technology. To achieve wideband tuning, silicon hyperabrupt varactors with a minimum capacitance value of $C_{vmin} = 1.2$ pF and capacitance ratio of $K_c > 10$ within a varactor bias range 0–25 V were used. Using the lead inductances of the bipolar transistor and varactors provides the required value of the base inductance L. Experimental results are shown in Figure 8.37. The minimum

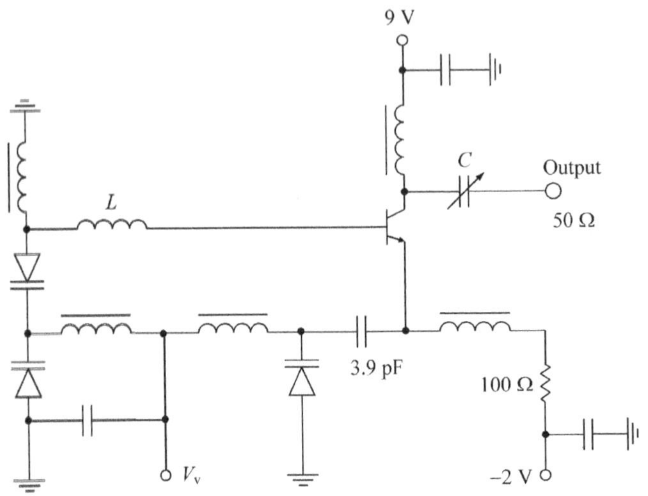

Figure 8.36 Practical L-band bipolar VCO schematic

flatness of the VCO output power can be achieved by proper tuning of the variable capacitor C within its capacitance range from 0.5 to 2 pF. To provide a frequency tuning range from 0.65 to 1.76 GHz, it is necessary to tune the varactor bias voltage V_v from 0 to 25 V. The resulting tuning slope ratio α is less than 6 over the entire frequency tuning range.

8.6 WIDEBAND NONLINEAR DESIGN

There are many different computer-based approaches to design and optimize VCOs according to their parameter specifications. For example, a simple quasilinear method can be applied

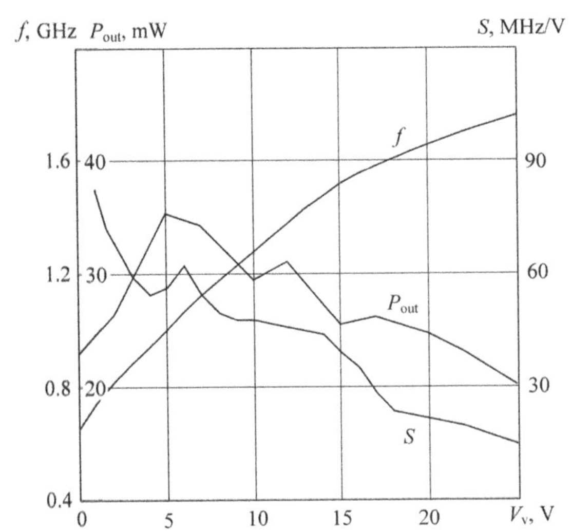

Figure 8.37 Experimental frequency tuning characteristics of L-band bipolar VCO

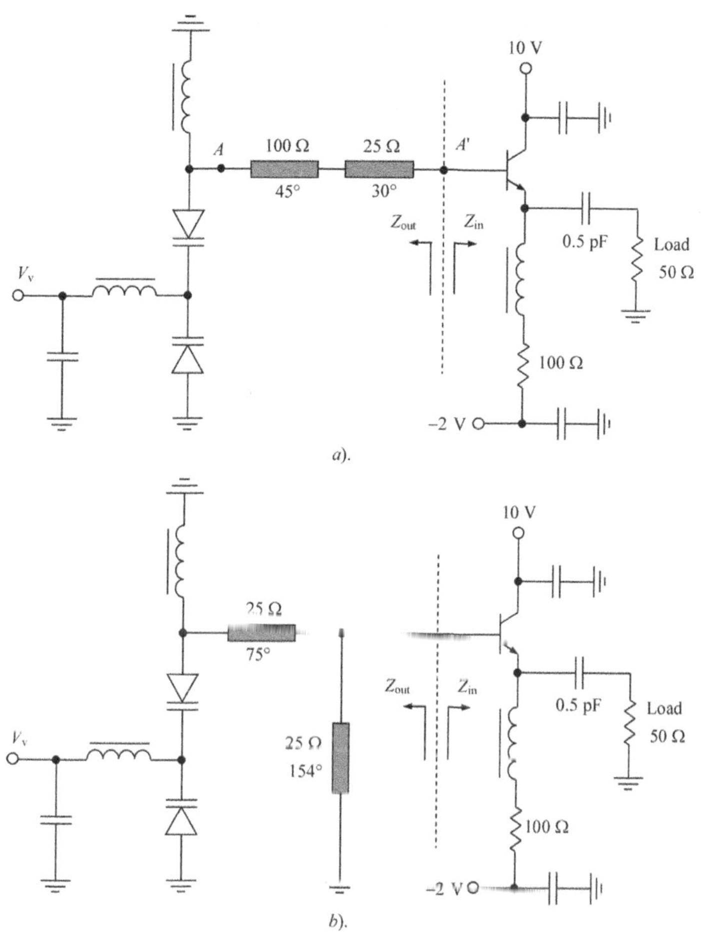

Figure 8.38 Electrical bipolar VCO circuits with (a) two-section line and (b) L-transformer

to provide the nonlinear design of wideband linear bipolar VCO [17]. Figure 8.38 shows the common collector bipolar VCO circuits with (a) two-section line and (b) L-transformer. The two back-to-back hyperabrupt varactors are capable of realizing a wide tuning bandwidth in a full-octave frequency range from 4 to 8 GHz. The nonlinear properties of the bipolar transistors can be described with the help of the simplified equivalent Gummel–Poon model of the bipolar transistor including the device lead inductances. A frequency-tuning bandwidth is determined iteratively for each value of varactor bias V_v using the common-collector bipolar transistor Z-parameters under the numerical solution of two nonlinear equations with two unknown parameters in the form

$$\mathrm{Re}[Z_{\mathrm{in}}(V_{\mathrm{be}}, \omega)] + \mathrm{Re}[Z_{\mathrm{out}}(\omega)] = 0 \tag{8.49}$$

$$\mathrm{Im}[Z_{\mathrm{in}}(V_{\mathrm{be}}, \omega)] + \mathrm{Im}[Z_{\mathrm{out}}(\omega)] = 0 \tag{8.50}$$

where V_{be} is a fundamental voltage amplitude across the emitter–base junction.

As an example, let the bipolar transistor be characterized by the following parameters: transition frequency $f_T = 6$ GHz, collector capacitance $C_c = 0.5$ pF, small-signal transconductance $g_m = 1.6$ A/V, base resistance $r_b = 4\,\Omega$, collector resistance $r_c = 1.75\,\Omega$, emitter resistance $r_e = 0.3\,\Omega$, lead inductances $L_b = 0.3$ nH and $L_e = L_c = 0.5$ nH. To provide wideband varactor tuning, the hyperabrupt varactors were chosen with capacitance $C_{v0}(V_v = 0) = 3$ pF, sensitivity $\gamma = 0.9$, contact potential $\varphi = 1.5$ V and lead inductance $L_v = 0.5$ nH. It is assumed that the values of the choke inductances and bypass capacitances are too large to have any significant influence on VCO band properties, and the losses in the resonant circuit are negligible. The required value of the series emitter capacitance was chosen to keep the oscillator amplitude balance condition given by Equation (8.49) in an entire tuning frequency bandwidth. To satisfy the phase balance condition expressed by Equation (8.50), it is necessary to choose the appropriate value of the lumped inductance or electrical lengths of transmission lines at high bandwidth frequency.

The numerical design approach, including the circuit analysis and its optimization based on an iterative computation procedure should include the following main steps:

- choosing the device equivalent circuit and determination its linear and nonlinear elements;
- choosing the varactor with maximum tuning capacitance factor to provide maximum tuning bandwidth;
- calculating the dc and fundamental current components using a Fourier series expansion on the basis of the exponential bipolar current model;
- calculating the average nonlinear device elements, for the simplest case, limiting them to the large-signal transconductance and collector capacitance;
- for a given maximum varactor bias voltage and maximum bandwidth frequency, the elements of the resonant circuit are calculated by solving a system of Equations (8.49) and (8.50);
- by changing the varactor bias voltage, the next bandwidth frequency is calculated with the appropriate calculation of the tuning slope ratio; this process continues as long as the varactor bias voltage becomes minimal required;
- to provide a minimum tuning slope ratio, an optimization procedure is undertaken when a system of two equations is solved for different values of the circuit parameters, including the characteristic impedances and electrical lengths of the transmission lines.

The frequency dependencies calculated numerically are shown in Figure 8.39. The oscillation system (curve 3) with a uniform transmission line ($Z_0 = 50\,\Omega$, $\theta = 54°$) is less sensitive to a varactor capacitance tuning compared with the oscillation system (curve 5) with a lumped inductance ($L = 2.4$ nH) connected between the points A–A', as shown in Figure 8.38a. To increase the frequency tuning bandwidth, the transistor base lead inductance is minimized and the characteristic impedance Z_0 is increased. The oscillation system with an L-transformer shown in Figure 8.38b is characterized by approximately the same band properties (curve 2) with $k = 2 \times 75°/(75° + 154°) = 0.65$. However, in this case, it is possible to improve tuning nonlinearity by changing the ratio between the electrical lengths of the transmission lines. As a result, an increase of the value k from 0.65 to 1.2 leads to linear frequency tuning with $\alpha \leq 1.5$ in the frequency bandwidth 10–15% (curve 1). The oscillation systems with the L-transformer or two coupled transmission lines instead of a single transmission line have a higher phase slope and, therefore, a higher quality factor. The resonator consisting of open-circuited coupled

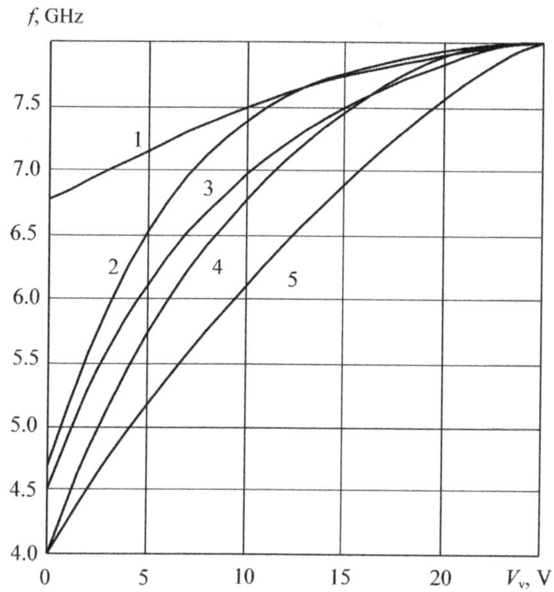

Figure 8.39 Theoretical frequency dependencies for various oscillation systems

microstrip lines exhibits a quality factor twice as high as the quality factor of a single microstrip line resonator in the frequency range 10–20 GHz and contributes to the phase noise reduction of about 8 dB at a varactor bias of −4 V [5]. The tuning bandwidth can be greater by 25% for the oscillation system with two short-circuited coupled transmission lines [26]. The oscillation system with a two-section line has the greatest sensitivity to varactor capacitance tuning. Already at a value of the characteristic impedance ratio $M = 4$ (curve 4), it provides a full-octave frequency tuning similar to the oscillation system with lumped inductance. In contrast to the latter case, the oscillation system with a two-section line enables one to realize the frequency tuning bandwidth up to 45% with $\alpha \leq 3$ in a smaller varactor bias range from 3 to 16 V.

8.7 DUAL MODE VARACTOR TUNING

In the oscillation system with two coupled transmission lines, a dual mode control can result in a very wide frequency bandwidth. In this case, the frequency tuning is realized by means of separate adjustment of even and odd mode fields describing the propagation of an electromagnetic wave in a system of two coupled transmission lines in a transverse electromagnetic (TEM) approximation. For example, the mechanical tuning of the microwave bipolar oscillator with two coupled coaxial transmission lines could provide an octave frequency tuning from 1.5 to 3.0 GHz [27].

The circuit schematic of the oscillation system using a system of two coupled transmission lines is shown in Figure 8.40, where C_v is the varactor capacitance. Generally, it is required to define the *ABCD*-parameters of the two-port network, which characterizes the VCO oscillation system and is derived from a four-port network with the appropriate boundary conditions. Then, it is necessary to express the parameters of the equivalent two-port network through these

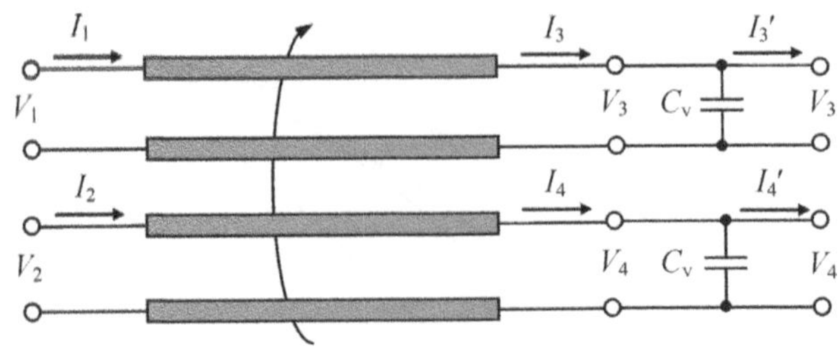

Figure 8.40 Equivalent oscillation system with two-coupled transmission lines

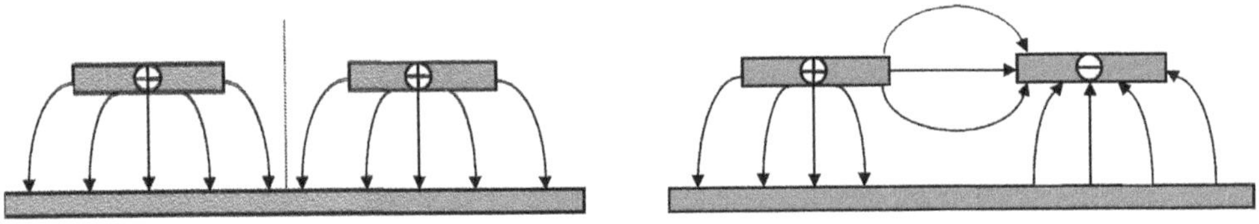

a). *b*).

Figure 8.41 Electric field distribution for (a) even and (b) odd TEM modes

ABCD-parameters and to determine the VCO characteristics under the steady-state oscillation conditions in view of the nonlinear active device equivalent circuit.

It is known, that two mode fields exist in a system of two coupled transmission lines: even (symmetric) and odd (antisymmetric) excitation modes. The TEM (a) even and (b) odd mode electric field distributions in a homogeneous dielectric medium are shown in Figure 8.41. Figure 8.41 demonstrates that the stronger coupling between the transmission lines takes place at an odd excitation mode. For each excitation mode, the characteristic impedance Z_{0e} corresponds to an even mode, while the characteristic impedance Z_{0o} corresponds to an odd mode.

In this case,

$$Z_{0e} > Z_0 > Z_{0o} \tag{8.51}$$

where $Z_0 = \sqrt{Z_{0e}Z_{0o}}$ is the characteristic impedance of a single transmission line. To derive the four-port *ABCD*-matrix for lossless transmission lines, symmetric and antisymmetric port excitations are assumed. The overall result is defined by a superposition of the results at both types of excitation.

Then, the four-port chain *ABCD*-matrix of a system of two identical lossless transmission lines in an inhomogeneous dielectric medium can be written as

$$\begin{bmatrix} V_1 \\ V_2 \\ I_1 \\ I_2 \end{bmatrix} = \begin{bmatrix} a_{11} & a_{12} & b_{11} & b_{12} \\ a_{21} & a_{22} & b_{21} & b_{22} \\ c_{11} & c_{12} & d_{11} & d_{12} \\ c_{21} & c_{22} & d_{21} & d_{22} \end{bmatrix} \begin{bmatrix} V_3 \\ V_4 \\ I_3 \\ I_4 \end{bmatrix} \tag{8.52}$$

where

$$a_{11} = a_{22} = d_{11} = d_{22} = \frac{1}{2}(\cos\theta_e + \cos\theta_o)$$

$$a_{12} = a_{21} = d_{12} = d_{21} = \frac{1}{2}(\cos\theta_e - \cos\theta_o)$$

$$b_{11} = b_{22} = \frac{j}{2}(Z_{0e}\sin\theta_e + Z_{0o}\sin\theta_o)$$

$$b_{12} = b_{21} = \frac{j}{2}(Z_{0e}\sin\theta_e - Z_{0o}\sin\theta_o)$$

$$c_{11} = c_{22} = \frac{j}{2}(Y_{0e}\sin\theta_e + Y_{0o}\sin\theta_o)$$

$$c_{12} = c_{21} = \frac{j}{2}(Y_{0e}\sin\theta_e - Y_{0o}\sin\theta_o)$$

where $Y_{0e} = 1/Z_{0e}$, $Y_{0o} = 1/Z_{0o}$, θ_e is the even mode electrical length, and θ_o is the odd mode electrical length of the transmission lines, respectively [28]. In a homogeneous dielectric medium, the electrical lengths for even and odd modes are equal, i.e., $\theta = \theta_e = \theta_o$. In this case, the coupling coefficient between transmission lines can be written as

$$C = \frac{Z_{0e} - Z_{0o}}{Z_{0e} + Z_{0o}} \tag{8.53}$$

where $C = 0$ for zero coupling and $C = 1$ for completely superposed transmission lines [29].

The four-port *ABCD*-matrix of the circuit, consisting of two parallel varactors, is written by

$$[ABCD] = \begin{bmatrix} 1 & 0 & 0 & 0 \\ 0 & 1 & 0 & 0 \\ j\omega C_v & 0 & 1 & 0 \\ 0 & j\omega C_v & 0 & 1 \end{bmatrix} \tag{8.54}$$

Multiplying the two *ABCD*-matrices from Equations (8.52) and (8.54) and imposing the boundary conditions $I_2 = -I_2$, and $I_3' = I_4' = 0$ yield the resulting *ABCD*-matrix of the remaining two-port network in the form

$$[ABCD]_T = \begin{bmatrix} \dfrac{Y_{0o}\tan\theta_{0o} + Y_{0e}\tan\theta_{0e}}{Y_{0o}\tan\theta_{0o} - Y_{0e}\tan\theta_{0e}} & \dfrac{2j}{Y_{0e}\tan\theta_{0e} - Y_{0o}\tan\theta_{0o}} \\ \dfrac{2j}{Z_{0e}\cot\theta_{0e} - Z_{0o}\cot\theta_{0o}} & \dfrac{Y_{0o}\tan\theta_{0o} + Y_{0e}\tan\theta_{0e}}{Y_{0o}\tan\theta_{0o} - Y_{0e}\tan\theta_{0e}} \end{bmatrix} \tag{8.55}$$

where $\theta_{0e} = \theta_e + \tan^{-1}(Z_{0e}\omega C_v)$, $\theta_{0o} = \theta_o + \tan^{-1}(Z_{0o}\omega C_v)$.

The admittance elements of the two-port π-network shown in Figure 8.42 which are expressed through the *ABCD*-matrix elements as

$$Y_1 = \frac{D-1}{B} \qquad Y_2 = \frac{A-1}{B} \qquad Y_3 = \frac{1}{B} \tag{8.56}$$

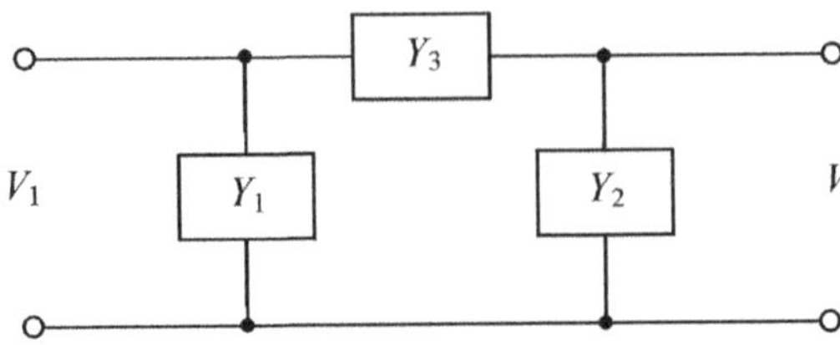

Figure 8.42 Equivalent π-network of dual mode oscillation system

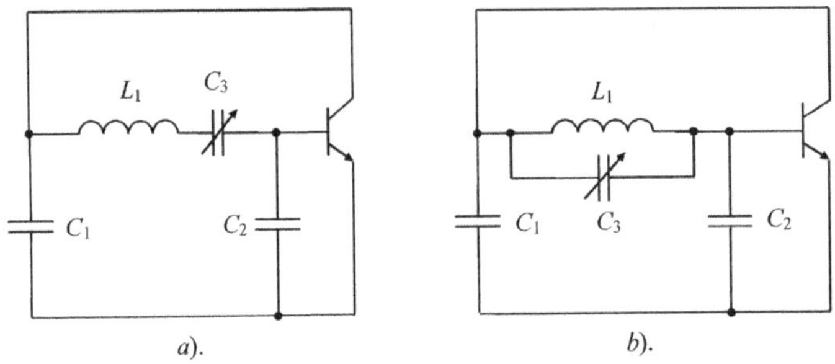

Figure 8.43 Equivalent circuits of (a) Clapp and (b) Seiler oscillators

can be rewritten as

$$Y_1 = Y_2 = jY_{0e} \tan\theta_{0e} \tag{8.57}$$

$$Y_3 = -j\frac{Y_{0e}}{2}\tan\theta_{0e} + j\frac{Y_{0o}}{2}\tan\theta_{0o} \tag{8.58}$$

From Equations (8.57) and (8.58) it follows that the dual mode VCO oscillation system represents a three-point capacitive resonant circuit. When tuning is provided by two varactors connected to the ends of resonators, a simultaneous change of the inductive and capacitive reactances occurs. In contrast to well-known three-point capacitive oscillation systems with (a) series (Clapp oscillator) and (b) parallel (Seiler oscillator) tuning circuits, which are shown in Figure 8.43, the dual mode VCO provides the minimum amplitude variations in a wide frequency tuning range. This can be explained by the fact that all feedback admittances Y_1, Y_2, and Y_3 change their values simultaneously, resulting in substantially smaller variations of the amplitude and phase feedback factors. To improve frequency stability and to decrease a phase noise at a high power output level the bipolar transistor in the VCO oscillation circuit can be configured with a common emitter.

Numerical calculations were performed for the common emitter bipolar oscillator with emitter bias $V_e = -2$ V and self-bias resistor $R_e = 100\ \Omega$. The transistor equivalent circuit parameters and detailed circuit design procedure using an analytical optimum approach for a dual mode bipolar oscillator are given elsewhere [30].

Figure 8.44 shows the circuit schematic of such a microwave dual mode wideband bipolar VCO. The values of the oscillation system parameters are chosen at a maximum oscillation frequency $f = f_{max}$ that corresponds to a minimum varactor capacitance $C_v = C_{vmin}$ at a maximum bias varactor voltage $V_v = V_{vmax}$. The wide frequency tuning bandwidth is realized using the two hyperabrupt varactors with the following approximation parameters: $C_{v0} = 3$ pF, $\gamma = 0.9$ and $\varphi = 1.5$ V. Such approximation parameters are easily provided with commercially

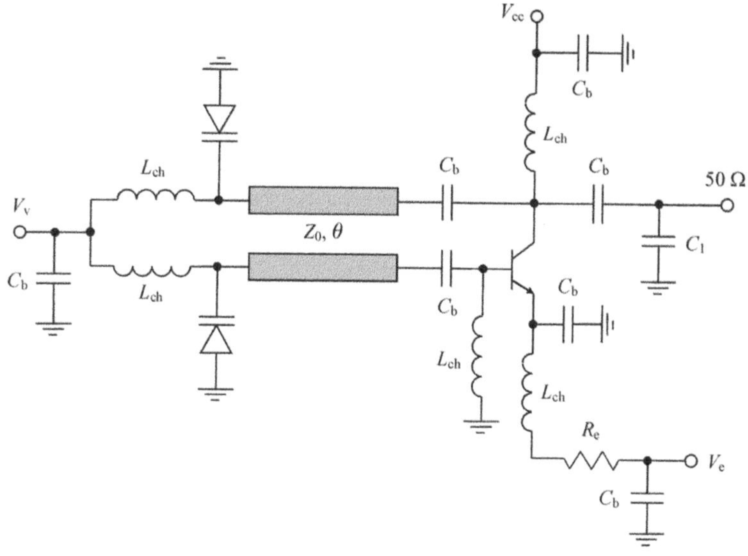

Figure 8.44 Circuit schematic of microwave dual mode wideband bipolar VCO

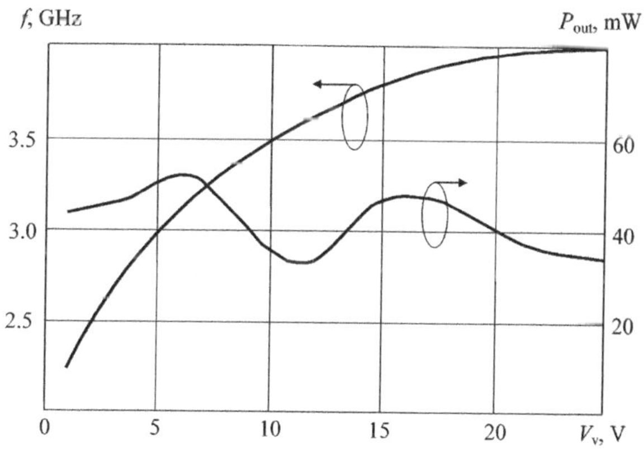

Figure 8.45 Output power and frequency versus varactor bias voltage

available silicon or GaAs hyperabrupt varactors having capacitance ratio more than 10:1. According to the numerical calculations, the value of a lead varactor inductance L_v must be equal to 2.5 nH at the high tuning bandwidth frequency of 4 GHz. In that case, the required value of electrical lengths of the two coupled transmission lines with the characteristic impedances $Z_{0e} = 75\ \Omega$ and $Z_{0o} = 25\ \Omega$ in a homogeneous dielectric medium is equal to $\theta = \theta_e = \theta_o = 102.5°$. The value of the additional capacitor C_1 in accordance with a balance phase condition must be equal to 1.2 pF, which enables connection of the VCO output directly to the standard load of 50 Ω without impedance matching.

Figure 8.45 shows the calculated dual mode VCO tuning oscillation frequency f and fundamental output power P_{out} versus varactor bias voltage V_v. The values of choke inductances L_{ch} and bypass or blocking capacitances C_b are assumed to be quite large, having negligible influence on the VCO band properties. As a result, the frequency tuning bandwidth of

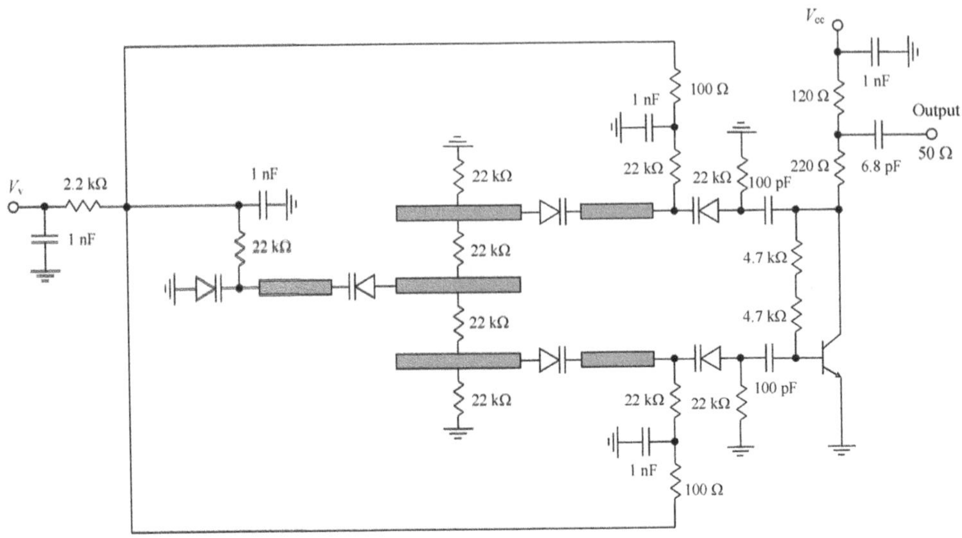

Figure 8.46 VCO schematic with interdigital bandpass filter [31]

2.25–4.00 GHz with the output power from 30 to 50 mW can be realized when the varactor bias voltage V_v changes from 1 to 25 V.

To maximize the frequency tuning bandwidth, the feedback circuit can contain a bandpass filter as a frequency-selective element. For example, for a microstrip oscillator shown in Figure 8.46, the feedback network includes a varactor–tuned interdigital bandpass filter [31]. The phase velocities corresponding to the odd and even excitation modes in the coupled microstrip line array are controlled by three silicon hyperabrupt varactors BB833 with minimum capacitance of 0.75 pF each. Another three varactors of the same type are connected in series, each to separate microstrip lines to adjust both the centre bandwidth frequency and overall phase shift of the feedback circuit. Due to the low quality factor of the varactors used, the insertion loss of the feedback network is rather high. Therefore, the bipolar transistor BFP420 with $f_T = 25$ GHz is required as the active device. The two resistors of 4.7 kΩ, connected in series to simplify the printed-circuit-board layout, provide the dc biasing. By changing the varactor bias voltage from 0 to 30 V, a frequency tuning bandwidth from 2.05 to 4.05 GHz was achieved. The tuning curve is quite nonlinear, having tuning slope of more than 100 MHz/V at a approximately 7 V and below 10 MHz/V at a tuning voltage of approximately 30 V. The varactor bias resistors of 22 kΩ cause an increased phase noise. However, lower values of these resistances result in additional losses and output power reduction.

Figure 8.47 shows the schematic of a microstrip oscillator where the feedback network includes the varactor-tuned combline bandpass filter [32]. The advantage of using a combline filter is its compact size, wide tuning range and sizeable spacing between resonator lines to provide adequate coupling. In this case, the combline bandpass filter with only one resonator (2) with series connected varactor and two short-circuited microstrip lines (1 and 3) are used. In such an oscillator architecture based on combline bandpass filter, the single varactor can tune the oscillation frequency within the tuning range of 1:3.5. To maintain the total phase shift of 360° in the oscillator loop, so that oscillation conditions could be satisfied over the entire frequency tuning range, two phase shifters in the form of low-pass filters using the same varactors are connected. By using the bipolar transistor BFP520 with $f_T = 45$ GHz and silicon

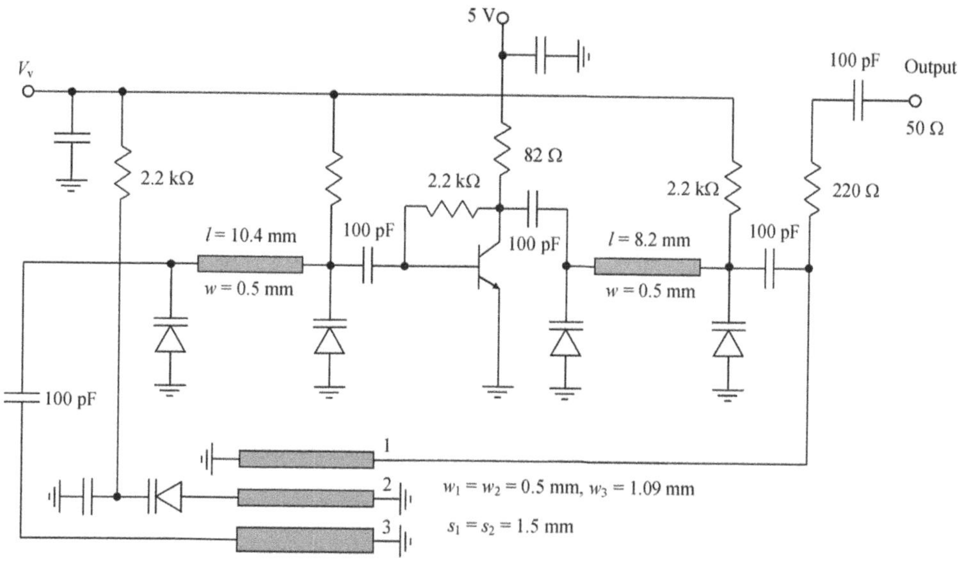

Figure 8.47 VCO schematic with variable combline bandpass filter [32]

hyperabrupt varactors BB857 with a capacitance tuning range from 0.55 pF at 28 V to 6.5 pF at 1 V, more than three-fold frequency tuning range from 580 to 2200 MHz was achieved. The output power varied from −10 dBm at 600 MHz to 5 dBm at 2200 MHz. The 1-mm-thick FR-4 board was used for oscillator circuit hybrid implementation.

Wideband frequency tuning and low phase noise are assumed to be opposing design parameters since an increase in a tuning range normally results in an increase in VCO phase noise. However, using an improved complex design approach based on either a single-device or a push–push topology, a coupled resonator with optimized loaded quality factor and noise filtering in the emitter of the bipolar device has resulted in wideband VCOs capable of delivering stable, low-noise output signals with tuning range as wide as 4.2 GHz [33]. For example, by changing the varactor bias voltage from 0 to 25 V, a frequency tuning bandwidth from 350 to 1050 MHz with typical tuning sensitivity of 20 to 48 MHz/V and phase noise of −112 dBc/Hz at 10 kHz offset from the carrier can be achieved. The tuning bandwidth from 1.5 to 3.2 GHz with tuning sensitivity of 100 to 200 MHz/V and phase noise of −92 dBc/Hz at 10 kHz offset is provided via varactor tuning voltages of 0.5 to 20 V.

8.8 PRACTICAL RF AND MICROWAVE WIDEBAND VCOs

8.8.1 Wireless and satellite TV applications

A typical receiver for wireless handset applications requires converting frequency bandwidths of 850/900 MHz or 1800/1900 MHz to the intermediate frequencies covering a range of 90–400 MHz. Satellite TV broadcasting in different countries usually covers frequency bandwidths of 620–790 MHz, 2.5–2.7 GHz or 11.7–12.5 GHz. For a frequency bandwidth of 11.7–12.5 GHz, the typical TV set-top dual down-converter after the first conversion of complete frequency-modulated TV signal requires a wideband VCO usually tuned from approximately 1.0 to 2.0 GHz to provide the second conversion to the intermediate frequency in a range of

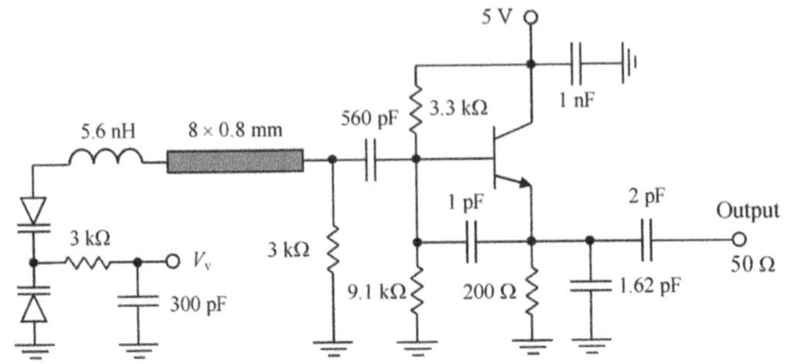

Figure 8.48 Circuit schematic of common collector wideband VCO [34]

480–612 MHz. In this case, a control voltage for VCO usually varies from 1 to 20 V, with high linearity of the varactor tuning characteristic. Consequently, the wider frequency tuning bandwidth, the better linearity for a given range of the control bias voltage can be provided.

Figure 8.48 shows the circuit schematic of a simple wideband VCO structure with grounded collector where a microstrip line is used between the transistor and two back-to-back varactors [34]. The lumped inductor of 5.6 nH and the geometry of the microstrip line were optimized to fit the desired frequency bandwidth, whereas the feedback capacitances of 1 and 1.62 pF are chosen to provide minimum flatness of the output power. As a result, by using the bipolar transistor NE68519 and silicon hyperabrupt varactors with capacitances changed from 22.5 pF at zero bias voltage to 0.8 pF at $V_v = 20$ V, a large frequency coverage of 980–2120 MHz with the output power of 4±2 dBm was realized. The circuit is very sensitive to the choice of the transistor type due to the very wide bandwidth requirements.

The very wide frequency tuning bandwidth can also be achieved with a push–pull VCO configuration, shown in Figure 8.49 [34]. In the push–pull operation mode, the current circulates between two transistor bases so as to drive them with a 180° shift. By using the same two

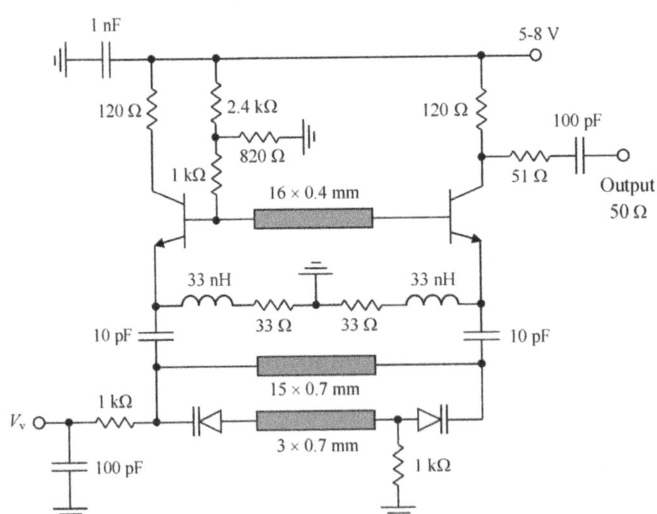

Figure 8.49 Push–pull VCO configuration with emitter varactors [34]

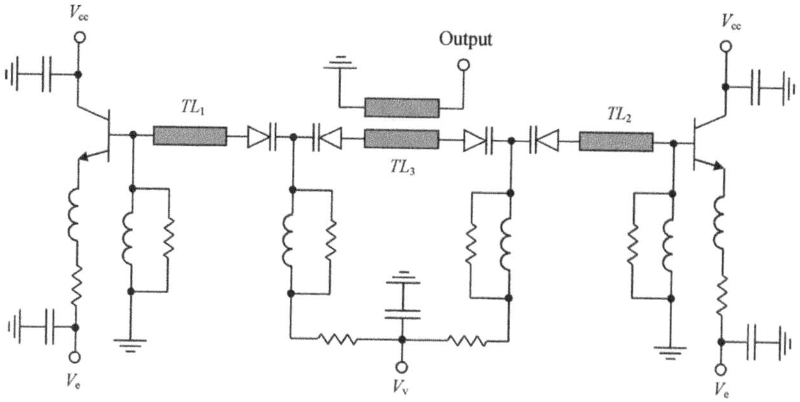

Figure 8.50 Push–pull VCO configuration with base varactors

back-to-back varactors in emitter circuits and the bipolar devices NE68119, a frequency bandwidth from 820 to 2120 MHz can be achieved. In this case, the output power was 4.5±2 dBm for the supply voltage of 7 V. The choke inductances of 33 nH and series resistances of 33 Ω provide the device self-biasing conditions. Unlike the common collector VCO with one active device, the push–pull VCO circuit very often does not require the feedback capacitances to provide the necessary phase shift. The series circuit capacitances of 10 pF were optimized to provide smooth tuning over the design frequency bandwidth. The necessary circuit response at high bandwidth frequencies is achieved by using a microstrip line with geometry of 15 × 0.7 mm, which has only a small effect on the lower bandwidth frequencies due to its dominantly capacitive reactance. The function of the microstrip line connected directly between the transistors is both feedback and phase alignments to minimize output power response. All microstrip lines were fabricated on a 1-mm-thick FR4 substrate. A relatively large thickness was selected to minimize the shunting effects of the mounting pads.

An alternative push–pull common collector VCO configuration, where two pairs of back-to-back varactors are located in base circuits incorporated into the transmission line structure, is shown in Figure 8.50 [35]. Due to the symmetric disposition of transmission lines TL_1 and TL_2 having equal electrical lengths and similar varactors resulting in a virtual ground at the centre point of the transmission line TL_3, the self-excitation of the fundamental frequency in a push–pull mode is established. In each bias circuit connected to the device base or back-to-back varactor pair, a damping resistor is connected in parallel with the choke coil to provide sufficient isolation between voltage supply and resonant circuit, also inhibiting spurious modes of oscillation. A negative bias voltage V_e is applied to each emitter terminal through the series connected choke coil and resistor, which value regulates the dc collector current. Applying wide-range varactor tuning voltage could provide a wideband frequency tuning from 940 to 2060 MHz. Wider frequency tuning bandwidth can be achieved by using hyperabrupt varactors with smaller minimum capacitance and greater capacitance ratio.

Figure 8.51 shows the electrical circuit of the monolithic common drain MESFET VCO developed for phase-locked loop application in the telecommunication system at 14 GHz [36]. A planar Schottky-barrier diode with 0.5 × 280 μm stripe suitable for monolithic integration with the MESFET device of the same gate geometry is used as a varactor. The varactor junction capacitance is tuned from 1.0 to 0.5 pF by applying a reverse-bias voltage from 0 to 5 V. To

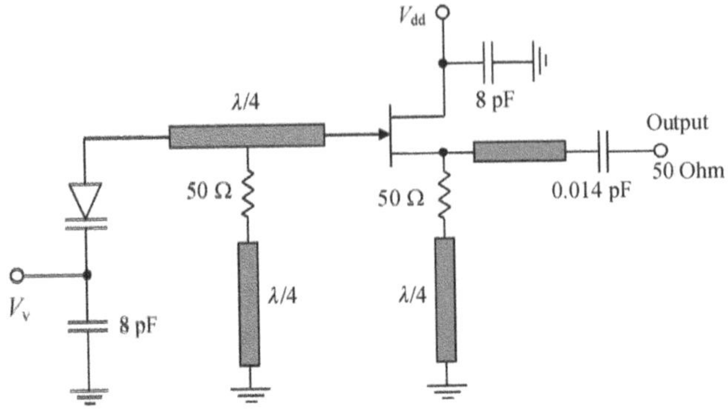

Figure 8.51 Electrical circuit of monolithic common-drain MESFET VCO (© 1987 IEEE)

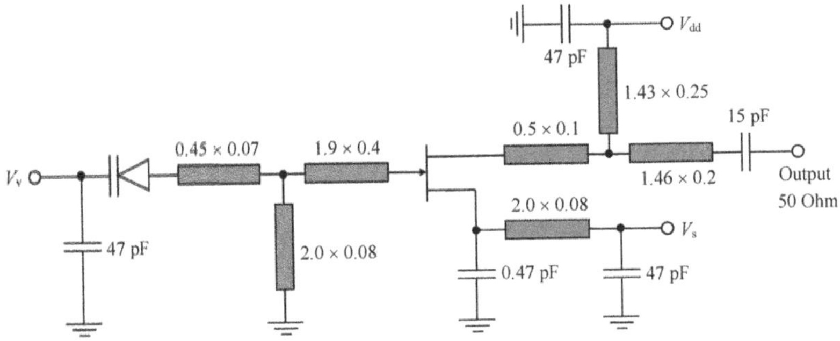

Figure 8.52 X-band common-gate GaAs MESFET VCO schematic (© 1986 IEEE)

provide inductive reactance in the gate circuit, the microstrip line is used as a quarter-wave transformer between the varactor diode and the MESFET device. A 50-Ω resistor was inserted into the source circuit for self-biasing, resulting in a drain current stabilization. Another 50-Ω resistor was connected to the microstrip impedance transformer to provide a gate zero bias point. These resistors are necessary to prevent low-frequency oscillations when they contribute to a significant decrease of the loaded quality factor for frequencies much lower or higher than resonant frequency. A output series capacitance is small enough to reduce the pulling effect, as well as the output power. A frequency tuning bandwidth from 11.3 to 14.3 GHz was realized by varactor tuning in a bias range from 0 to 7 V with output power of −4 dBm.

Using a T-type two-port network between the varactor and active device helps to improve linearity of the VCO frequency tuning characteristics. Figure 8.52 shows the X-band common gate GaAs MESFET VCO schematic where the T-type two-port network is represented by microstrip lines [6]. The length and width of microstrip lines realized on 0.25-mm-thick alumina substrate are given in millimetres. The GaAs MESFET device with a geometry of 0.8 × 200 μm is determined by its equivalent circuit small-signal parameters: $g_m = 35$ mA/V, $C_{gs} = 0.23$ pF, $C_{gd} = 0.017$ pF and $R_g + R_{gs} + R_s = 6.6\ \Omega$. To provide a wideband frequency tuning, a hyperabrupt GaAs varactor was chosen with a maximum capacitance $C_{vmax} = 2.5$ pF, a minimum capacitance $C_{vmin} = 0.15$ pF and a series resistance $R_s = 6\ \Omega$. For a maximum varactor bias range from 0 to 30 V, the frequency tuning bandwidth from 7.055 to 11.93 GHz with a frequency tuning coefficient $K_f = 1.67$ was realized. If the same two back-to-back

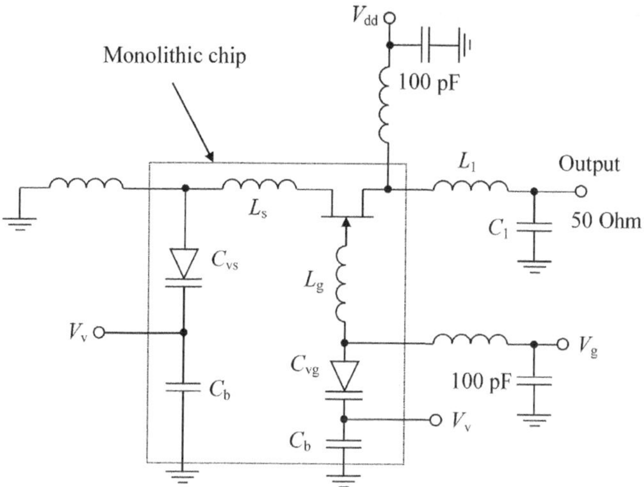

Figure 8.53 Microwave monolithic wideband VCO circuit schematic

varactors are used in the source circuit (instead of the 0.47 pF capacitance), it is possible to improve the linearity in a sufficiently wide frequency bandwidth. In the frequency range from 7 to 10 GHz, a frequency deviation from linear tuning law of ±30 MHz was obtained, which yields a reduced tuning slope ratio of about $\alpha = 1.5$ compared with $\alpha = 2.1$ for the VCO without additional varactors.

8.8.2 Microwave monolithic VCO design

Figure 8.53 shows the wideband VCO circuit that was used to design and fabricate four GaAs monolithic VCO chips that cover 2–4, 4–7, 7–12 and 12–18 GHz frequency ranges, respectively [37]. Each monolithic chip includes a MESFET device, two varactors, gate inductance L_g, source inductance L_s and bypass capacitances C_b. The varactor diode represents a single implanted structure into semi-insulating material and is formed with the same active region as for the MESFET device. Its capacitance ratio is typically of 8:1 or greater. The common gate VCO configuration using the gate inductor as the regenerative feedback element exhibits negative impedance at the source terminal across the entire tuning bandwidth for the equivalent load resistance connected to the drain terminal of approximately 15 Ω. To provide output matching with standard load of 50 Ω, the L-type matching circuit with the series inductance L_1 and shunt capacitance C_1 is used. Due to the required additional area, the output matching circuit was not included on the chip. A summary of the electrical performances and the circuit parameters of the four designed VCOs is given in Table. 8.1.

A similar design philosophy and circuit schematic was applied to the design of a GaAs MMIC VCO with varactor tuning bandwidth of 2.5–6.0 GHz [38]. For wideband frequency tuning, the GaAs hyperabrupt varactor with high capacitance ratios from 0.2 to 3 pF in the gate and the silicon hyperabrupt varactor from 0.5 to 3 pF in the source were used. The values of the varactor capacitance are chosen so that they provide simultaneous tuning from one biasing voltage. The frequency tuning bandwidth can be easily shifted down, for example, to 1.5–4.0 GHz by increasing the length of the gate inductance. Using a silicon varactor in the

Table 8.1 Electrical performances and circuit parameters of wideband VCOs [37]

Tuning bandwidth (GHz)	2.04–4.46	3.56–6.74	6.97–12.2	11.6–20.0
Output power (dBm)	12.6–19.4	13.3–17.1	15.2–20.1	5.5–15.8
MESFET gate width (μm)	2400	1200	600	300
$C_{vs}(V_v = 0)$ (pF)	2.9	2.3	1.5	1.1
$C_{vg}(V_v = 0)$ (pF)	5.1	3.6	2.2	1.4
L_s(nH)	1.3	1.4	0.17	0.06
L_g(nH)	13.5	5.9	2.1	0.6
C_b(pF)	45	45	40	21

gate circuit leads to the slight decrease of the frequency bandwidth from 85 to 78%. However, the post-tuning drift when using a GaAs varactor is about 50 times slower in reaching the final frequency within 2 MHz.

To improve a loaded quality factor of the VCO resonant circuit and minimize the effects of frequency pulling, it is best to incorporate an additional buffer amplifier stage. This also gives the possibility of connecting its output terminal to the standard load of 50 Ω. A very wideband tuning from 6.9 to 15 GHz with minimum output power of 9 dBm via varactor bias range from 0.4 to 25 V was achieved for the VCO with buffer amplifier stage using both HBT devices [39]. The circuit functional schematic is shown in Figure 8.54, where the HBT device of VCO circuit is designed with a common base configuration. The buffer amplifier stage has a common emitter configuration and operates at saturation to minimize output power variation with temperature. The circuit, which is realized on the chip of size 0.8 × 1 mm, requires only two external varactors to form a complete wideband VCO.

To minimize the excessive losses in the resonant circuit under wideband varactor tuning, it is necessary to choose the varactors with maximum quality factors and provide the required

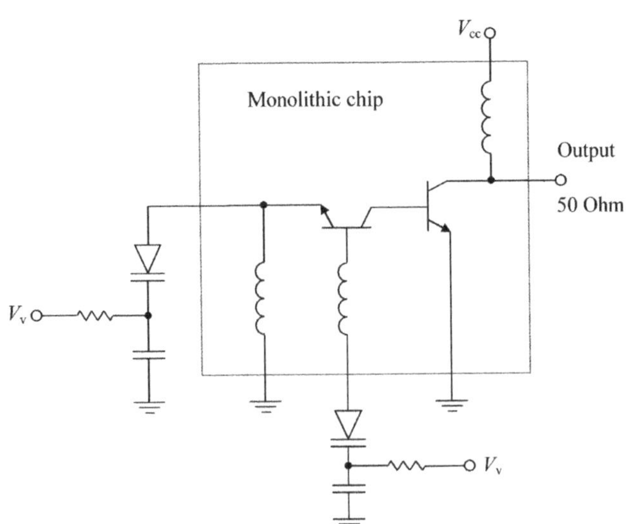

Figure 8.54 Circuit schematic of wideband VCO with buffer amplifier

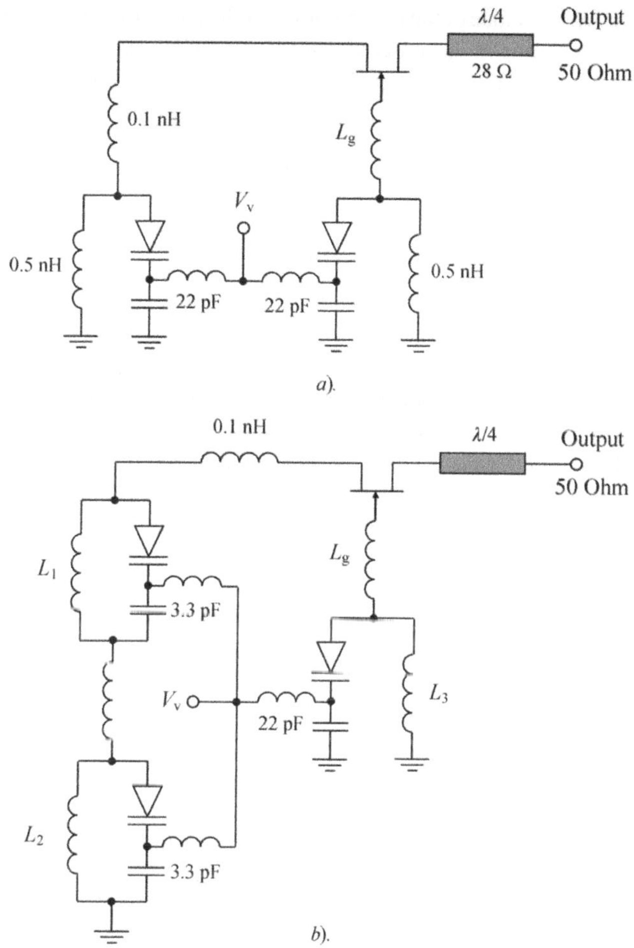

Figure 8.55 Microwave monolithic very wideband VCO circuits

broadband matching. For example, the matching of the device output impedance with the 50-Ω load can be provided by a microstrip transformer of a quarter-wavelength at highest bandwidth frequency, as shown in Figure 8.55a. Using the commercially available hyperabrupt GaAs varactor diodes with a quality factor of 3000 at a bias voltage of 4 V provides a frequency tuning bandwidth from 7.25 to 14.65 GHz with a tuning linearity of ±5% and harmonic level of less than –20 dBc [40]. Such an octave frequency tuning requires the varactor voltage bias range from 2 to 22 V because it was impossible to provide the steady-state oscillation process below bias voltage of 2 V. A three-stage buffer amplifier connected to the MESFET VCO output contributed to the increased level of the output power of 16 to 19 dBm. Using the lumped inductances in parallel to both varactor diodes provides an additional improvement of the circuit sensitivity to the varactor tuning. Improvement occurs when the VCO oscillation frequency is close to the natural resonant frequency of the parallel circuit consisting of the varactor capacitance and parallel inductance. However, to avoid the unstable operation mode accompanied by jumping effects, it is necessary to provide the capacitive reactance of the source circuit and the inductive reactance of the gate circuit over the entire frequency tuning bandwidth. Greater than full-octave tuning range at frequencies above 8 GHz can be achieved using a triple varactor design, as shown in Figure 8.55b. In this case, the ultrawideband frequency tuning from 6.5 to 16.1 GHz with tuning linearity within ±1% up to 14 GHz for a varactor bias range of 0 to 25 V was realized [41].

8.8.3 Push–push oscillators and oscipliers

Push–push VCOs are used in many applications due to their wide tuning bandwidth, low noise, low frequency pulling and possibility of increasing tuning frequencies by a factor of two. Such an oscillator represents a balanced circuit, in which two transistors provide opposite-phase oscillations at the fundamental and odd frequencies and in-phase oscillations at even harmonics. As a result, provided there is a sufficient suppression of fourth and higher harmonic components, the second harmonic signal is delivered to the load, which is connected appropriately to the circuit inductive element, lumped inductance or microstrip line. Therefore, to realize maximum output power, the generating current waveform should contain a high level of second harmonic component. A push–push oscillator combines both: a balanced oscillator and a balanced doubler. Therefore, it is called *osciplier* [42].

The main advantage of a push–push oscillator is a high quality factor of the resonant circuit because the fundamental resonator is decoupled from the second harmonic load that improves both load pulling and noise performance. To use a common collector configuration, a series tank circuit connected between the device bases can be realized. The overall resonant circuit also includes transistor collector–base capacitances, varactor capacitances and bondwire inductances. In this case, the load can be connected conductively to the centre point of the circuit inductance when the cancellation of the even harmonic can be provided, as shown in Figure 8.56a. For symmetric parts in the odd mode of the operation, this centre point becomes a virtual ground. Such a push–push VCO enables one to provide very wideband varactor tuning over a frequency range from 6.94 to 19.22 GHz with the fundamental harmonic suppression of about 10 dB when additional varactors are used instead of collector–emitter feedback capacitances [43, 44]. Such a wideband frequency tuning can be realized using double pairs of GaAs hyperabrupt varactors with a sensitivity of $\gamma = 1.1$ in a bias range of 0–20 V.

An inductive connection of the resonant circuit to the load shown in Figure 8.56b makes it possible to further increase the output operating frequency. By properly unbalancing both transistor circuits, the fundamental can be mixed with the second harmonic to enhance the third harmonic [45]. In this case, the resonant circuits of the two transistors are different since the resonant frequency of one is chosen to be at the fundamental frequency and the other at the doubled frequency. The osciplier designed on the basis of such a technique uses high-impedanced bias lines of a quarter-wavelength at the fundamental on the top half of the circuit and a quarter-wavelength at the second harmonic on the bottom half. The varactors have different capacitance values at zero bias voltage. The tuning voltage is applied to the varactors through a low-pass filter to minimize the *RC* time constant that limits frequency tuning speed. To optimize VCO performance, it is best to use the feedback capacitances C_f connected between the collector and emitter terminals. The output matching circuit tuned for the third harmonic provides matching to a 50-Ω load

Using a microstrip line between the varactors enables to provide simultaneous fundamental and second harmonic tuning when the fundamental components from both transistors are

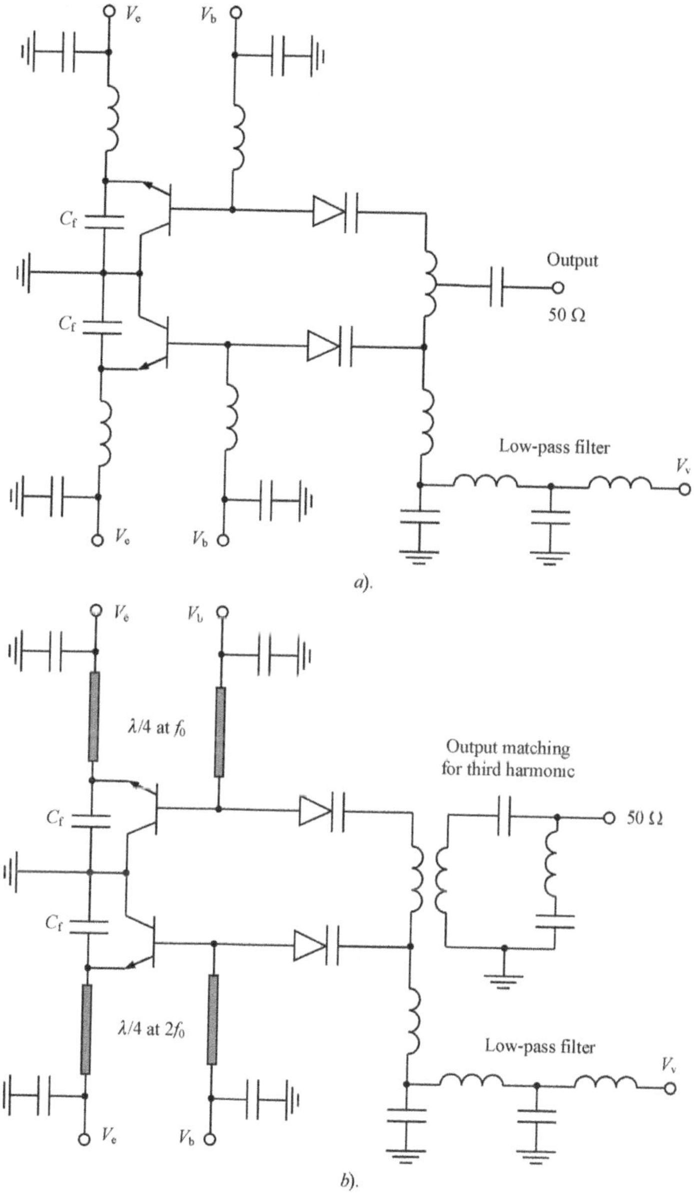

Figure 8.56 Circuit schematics of push–push wideband VCOs

delivered to the load using a coupled microstrip line, whereas the second harmonic components are delivered to the other load conductively connected to the centre point of the resonant circuit microstrip line. Such a balanced microstrip VCO using a high-voltage MOSFET device with the gate geometry of 5 μm × 1 mm, small-signal transconductance $g_m = 29$ mA/V and $f_T = 1$ GHz, which provides both push–pull and push–push operation modes, is shown in Figure 8.57 [18]. By using the hyperabrupt varactors with minimum capacitance of 2 pF and

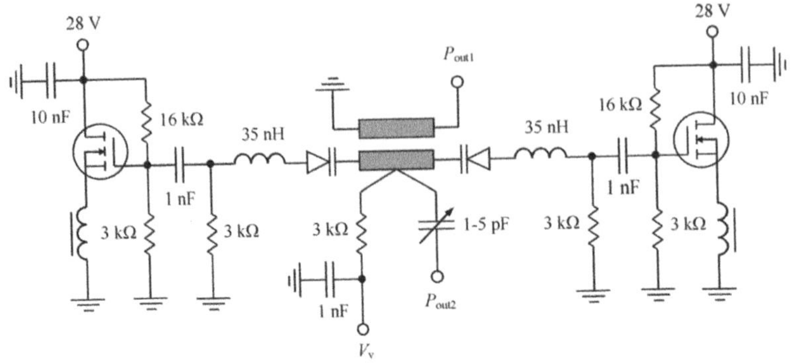

Figure 8.57 Balanced MOSFET VCO with gate varactors

capacitance ratio of about 10, the microstrip lines with a characteristic impedance of 70 Ω, an electrical length of about 45° and a coupling coefficient $C = 0.45$, stable operation conditions were realized when the variable capacitor is properly tuned within its tuning range of 1–5 pF. As a result, the fundamental frequency can be tuned from 325 to 600 MHz with output power of 24±1.5 dBm and second harmonic suppression of better than 25 dB. At the same time, the second harmonic signal is delivered to the port P_{out2} tuned from 650 to 1200 MHz with output power of 21±2 dBm and suppression of the fundamental-frequency component better than −15 dBc.

REFERENCES

1. D. A. Boyd, 'VCO Techniques for EW Systems', *Microwave J.*, **31**, 129–136 (1988).
2. D. F. Peterson, 'Varactor Properties for Wide-Band Linear-Tuning Microwave VCO's', *IEEE Trans. Microwave Theory Tech.*, **MTT-28**, 110–118 (1980).
3. A. E. Moysenko and C. A. Barrat, 'Computer Aided Design and Manufacture of GaAs Hyperabrupt Varactors', *Microwave J.*, **25**, 99–103 (1982).
4. D. Kaifez and E. J. Hwan, 'Linearity Limits of the Varactor-Controlled Osc-Mod Circuits', *IEEE Trans. Microwave Theory Tech.*, **MTT-33**, 620–625 (1985).
5. V. Gungerich, F. Zinkler, W. Anzill and P. Russer, 'Noise Calculations and Experimental Results of Varactor Tunable Oscillators with Significantly Reduced Phase Noise', *IEEE Trans. Microwave Theory Tech.*, **MTT-43**, 278–285 (1995).
6. W. El-Kamali, J.-P. Grimm, R. Meierer and C. Tsironis, 'New Design Approach for Wide-Band FET Voltage-Controlled Oscillators', *IEEE Trans. Microwave Theory Tech.*, **MTT-34**, 1059–1063 (1986).
7. D. A. Boyd, 'Design Considerations for Post-Tuning Drift (PTD) Reduction in Modern VCO Subsystems', *Microwave J.*, **29**, 121–130 (1986).
8. D. F. Peterson, 'Tuning Speed Limitations in Wide-Band Varactor-Tuned Oscillators', *1978 IEEE MTT-S Int. Microwave Symp. Dig.*, 297–299 (1978).
9. A. V. Grebennikov, 'Wideband linear varactor tuning of *LC* resonant circuits', *Int. J. Electrical Engineering Education*, **36**, 37–45 (1999).
10. R. Beach, 'High Speed Linear Oscillators', *Microwave J.*, **21**, 59–65 (1978).
11. J. Grosjean, 'Hyperabrupt Diodes Make Simple VCO's', *Microwaves & RF*, **22**, 145–150 (1983).
12. R. Charbonnier, Oscillateur a Trios Diodes a Capacite Variable, France Patent 73 09015 (1974).

13. R. L. Michels, Method for Reducing Phase Noise in Oscillators, U.S. Patent 5,138,285 (August 1992).
14. J. Obregon, P. G. Marechal, Y. Le Trou and R. Funck, 'Ultrabroadband Electronically Tunable Oscillators', *Proc. 11th Europ. Microwave Conf.*, 475–479 (1981).
15. V. Mannassewitsch, *Frequency Synthesizers: Theory and Design*, London: John Wiley & Sons (1987).
16. C. N. Herbert and J. Chernega, 'Broadband Varactor Tuning of Transistor Oscillators', *Microwaves*, **6**, 28–32 (1967).
17. A. V. Grebennikov, 'The Frequency Tuning Characteristics of Transistor Voltage-Controlled Oscillators with Distributed Parameters', *Int. J. Circuit Theory Appl.*, **27**, 393–414 (1999).
18. A. V. Grebennikov and V. V. Nikiforov, 'Octave-Band UHF MOSFET Oscillators (in Russian)', *Poluprovodnikovaya Elektronika v Tekhnike Svyazi*, **26**, 188–194 (1986).
19. A. V. Grebennikov, 'Wideband VCO Designs Are Independent of Circuit Parameters', *Microwaves & RF*, **40**, 147–155 (2001).
20. Y.-H. Kao and T.-H. Wu, 'Behaviors of Negative Resistances and Its Influences on VCO Design', *IEICE Trans. Electron.*, **E84-C**, 243–247 (2001).
21. A. Shipow, 'Linearity in Solid State Microwave Voltage Tuned Oscillators', *Microwave J.*, **26**, 130–137 (1983).
22. R. F. Pasos, Voltage Controlled Oscillator in Which Capacitive Diodes Become Resistive During Portions of Each Cycle, U.S. Patent 3,582,823 (June 1971).
23. H.-N. Toussaint and P. Olfs, 'Transistor Power Oscillator, Electronically Tunable from 250 to 500 MHz', *Proc. IEEE*, **56**, 226–227 (1968).
24. H. C. Huang and D. D. Mawhinney, Monolithic Voltage Controlled Oscillator, U.S. Patent 4,458,215 (July 1984).
25. A. V. Grebennikov, 'An Analytic Method of Microwave Bipolar Single-Frequency and Voltage-Controlled Oscillator Design', *Int. J. RF and Microwave Computer-Aided Eng.*, **9**, 403–414 (1999).
26. J. L. Merenda, Varactor Tuned Strip Line Resonator and VCO Using Same, U.S. Patent 5,942,950 (August 1999).
27. R. G. Rogers, 'A Dual Mode Tuning Circuit for Microwave Transistor Oscillators', *IEEE Trans. Microwave Theory Tech.*, **25**, 120–127 (1977).
28. G. I. Zysman and A. K. Johnson, 'Coupled Transmission Line Networks in an Inhomogeneous Dielectric Medium', *IEEE Trans. Microwave Theory Tech.*, **17**, 753–759 (1969).
29. H. Howe, *Stripline Circuit Design*, Dedham: Artech House (1974).
30. A. V. Grebennikov, 'A Dual Mode Wideband Bipolar Microwave Transistor Voltage-Controlled Oscillator Design', *Int. J. RF and Microwave Computer-Aided Eng.*, **9**, 424–434 (1999).
31. M. Vidmar, 'A Wideband Varactor-Tuned Microstrip VCO', *Microwave J.*, **42**, 80–86 (1999).
32. G. A. Hofbauer, 'A Low Noise Wideband Microwave Oscillator Using a Tunable Microstrip Combline Filter', *Microwave J.*, **46**, 82–97 (2003).
33. U. L. Rohde, A. K. Poddar and G. Bock, *The Design of Modern Microwave Oscillators for Wireless Applications: Theory and Optimization*, Hoboken, New Jersey: John Wiley & Sons (2005).
34. P. Shveshkeyev, 'A Wideband VCO for Set-Top Applications', *Microwave J.*, **42**, 74–88 (1999).
35. F. W. Kruse and D. L. Baskins, Oscillator with Separate Voltage Controls for Narrow and Wide Band Tuning, U.S. Patent 3,397,365 (May 1968).
36. T. Ohira, T. Hiraoka and H. Kato, 'MMIC 14-GHz VCO and Miller Frequency Divider for Low-Noise Local Oscillators', *IEEE Trans. Microwave Theory Tech.*, **MTT-35**, 657–661 (1987).
37. B. N. Scott, M. Wurtele and B. B. Creiger, 'A Family of Four Monolithic VCO MIC's Covering 2-18 GHz', *1984 IEEE Microwave and Millimeter-Wave Monolithic Circuits Symp. Dig.*, 58–61 (1984).
38. J. E. Andrews, T. J. Holden, K. W. Lee and A. F. Podell, '2.5–6.0 GHz Broadband GaAs MMIC VCO', *1988 IEEE MTT-S Int. Microwave Symp. Dig.*, 491–494 (1998).
39. A. Adar and R. Ramachandran, 'An HBT Wideband VCO', *1991 IEEE MTT-S Int. Microwave Symp. Dig.*, 247–250 (1991).
40. J. Kitchen, 'Octave Bandwidth Varactor-Tuned Oscillators', *Microwave J.*, **30**, 347–353 (1987).

41. K. Adams and C. Oxley, 'Broadband FET VCO Design', *Electronic Eng.*, **61**, S25–S30 (1989).
42. Z. Nativ and Y. Shur, 'Push–Push VCO Design with CAD Tools', *Microwave J.*, **32**, 127–132 (1989).
43. R. G. Winch, 'Very Broadband Bipolar VCO', *Electronics Lett.*, **17**, 871–873 (1981).
44. R. G. Winch, 'Wide-Band Varactor-Tuned Oscillators', *IEEE J. Solid-State Circuits*, **SC-17**, 1214–1219 (1982).
45. R. G. Winch, 'Oscipliers: *K*-Band VCOs You Build with Bipolars', *Microwaves*, **16**, 62–67 (1977).

9

Noise reduction techniques

This chapter discusses phase noise reduction techniques and gives specific resonant circuit solutions using lumped and distributed parameters for frequency stabilization and phase noise reduction. Phase noise improvement can also be achieved by appropriate low-frequency loading and feedback circuitry optimization. The feedback system incorporated into the oscillator bias circuit can provide significant phase noise reduction over a wide frequency range from high frequencies to microwaves. Particular discrete implementations of a bipolar oscillator with collector and emitter noise feedback circuits are described. Also a filtering technique based on a passive *LC*-filter to lower the phase noise in the differential oscillator is presented. Several topologies of fully integrated CMOS voltage-controlled oscillators using filtering technique are shown and discussed. A novel noise-shifting differential VCO based on a single-ended classical three-point circuit configuration with common base can improve the phase noise performance by appropriate circuit realization. An optimal design technique using an active element based on a tandem connection of a common source FET device and a common base bipolar transistor with optimum coupling of the active element to the resonant circuit is presented. The phase noise in microwave oscillators can also be reduced using negative resistance compensation, increasing the loaded quality factor of the oscillator resonant circuit. Finally, a new approach utilizing a nonlinear feedback loop for phase noise suppression in microwave oscillators is discussed.

9.1 RESONANT CIRCUIT DESIGN TECHNIQUE

Microwave transistor oscillators have received much attention in recent years due to significant progress in device technology which enables the development of local oscillator and VCO components with reduced size, power consumption and improved overall electrical performance. For example, by using HBT devices, excellent low-noise characteristics are realized in monolithic oscillator applications up to about 20 GHz. Despite the high level of low-frequency noise, the MESFET and HEMT devices provide a widening of the oscillator operating frequency range to more than 100 GHz. JFET devices are a good choice to achieve low-noise oscillator performance at UHF band. One way to improve oscillator noise characteristics is to use the optimal bias condition, whereas a second way lies in the increase of the loaded quality factor of the oscillator resonant circuit by appropriate choice of the circuit topology.

RF and Microwave Transistor Oscillator Design A. Grebennikov

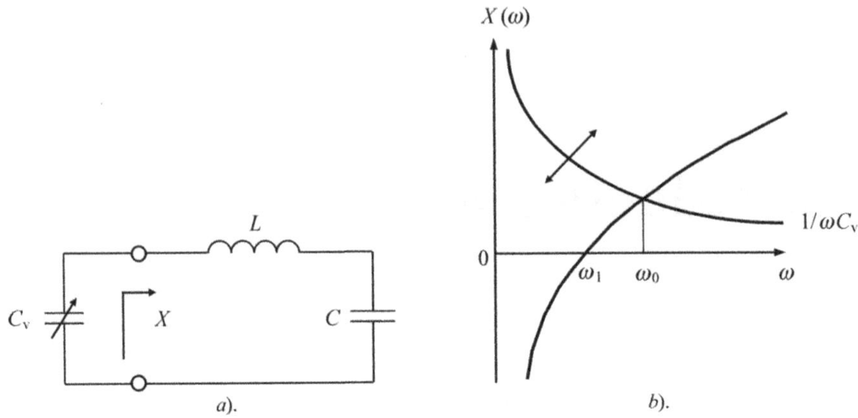

Figure 9.1 Oscillator series resonant circuit with variable capacitance and its frequency performance

An increase in the quality factor implies a decrease in the frequency sensitivity of the resonant circuit to the variation of its parameters.

9.1.1 Oscillation systems with lumped elements

Figure 9.1a shows the oscillation system with a series resonant circuit representing a simple series connection of the variable capacitance C_v, circuit inductance L and capacitance C, where C_v is considered the equivalent active device nonlinear capacitance varying over temperature, signal amplitude or supply voltage. The frequency behaviour of such a resonant circuit is given by Figure 9.1b, where $\omega = 0$ and $\omega = \omega_1 = 1/\sqrt{LC}$ are the pole and zero of the circuit reactance X, respectively. The oscillation frequency ω_0 of such a series resonant circuit can be found from

$$\omega_0 = \sqrt{\frac{C + C_v}{LCC_v}} \tag{9.1}$$

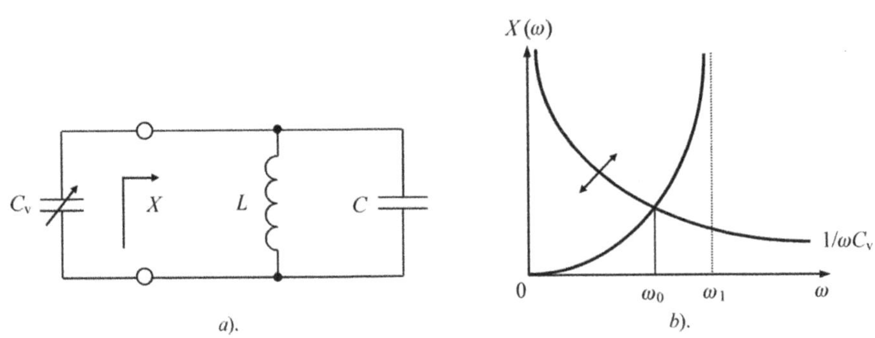

Figure 9.2 Oscillator parallel resonant circuit with variable capacitance and its frequency performance

The relative sensitivity of the series resonant circuit to the variation of the variable capacitance C_v can be obtained from Equation (9.1) in the form

$$\frac{d\omega}{\omega_0}\Big/\frac{dC_v}{C_v} = -\frac{1}{2}\frac{C}{C + C_v} \tag{9.2}$$

From Equation (9.2) it follows that maximum sensitivity is realized for the series resonant circuit representing the only inductance and variable capacitance when $C \to \infty$, being equal to

$$\frac{d\omega}{\omega_0}\Big/\frac{dC_v}{C_v} = -\frac{1}{2} \tag{9.3}$$

At the same time, minimum sensitivity of the series resonant circuit to the variation of capacitance C_v is realized under the extreme condition when $C \to 0$.

Figure 9.2a shows the oscillation system with parallel resonant circuit representing a simple parallel connection of the variable capacitance C_v and circuit inductance L and capacitance C. The frequency behaviour of such a resonant circuit is given by Figure 9.2b, where $\omega = 0$ and $\omega = \omega_1 = 1/\sqrt{LC}$ are the zero and pole of the reactance X, respectively. The oscillation frequency ω_0 of such a parallel resonant circuit can be found from

$$\omega_0 = \frac{1}{\sqrt{L(C + C_v)}} \tag{9.4}$$

The relative frequency sensitivity of the parallel resonant circuit to the variation of the variable capacitance C_v can be obtained from Equation (9.4) in the form

$$\frac{d\omega}{\omega_0}\Big/\frac{dC_v}{C_v} = -\frac{1}{2}\frac{C_v}{C + C_v} \tag{9.5}$$

From Figure 9.2b it follows that the tuning sensitivity of the parallel circuit can be easily reduced by moving pole of its reactance $X(\omega)$ to the left, i.e., increasing the value of the resonant circuit capacitance C. Further reduction of the circuit sensitivity can be achieved by moving the zero of its reactance X to the right, towards its pole. Such a possibility can be realized for the series-parallel resonant circuit with a variable capacitance C_v shown in Figure 9.3a.

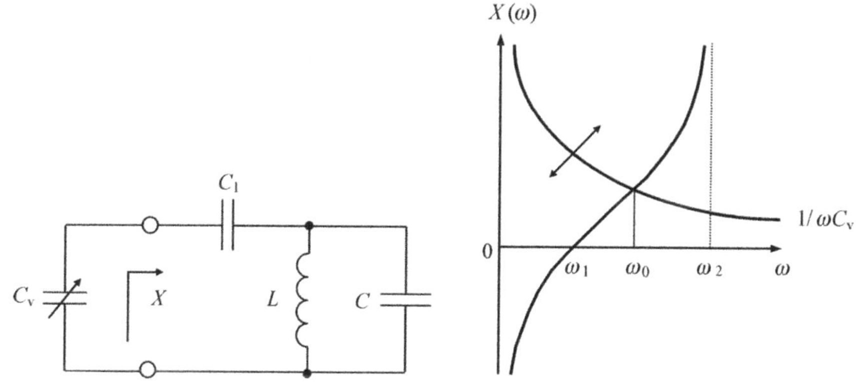

a). b).

Figure 9.3 Oscillator series-parallel resonant circuit with variable capacitance and its frequency performance

The frequency behaviour of such a series-parallel resonant circuit is illustrated by Figure 9.3b, where $\omega = \omega_1 = 1/\sqrt{L(C + C_1)}$ and $\omega = \omega_2 = 1/\sqrt{LC}$ are the zero and pole of the reactance X, respectively. The oscillation frequency ω_0 of the series-parallel resonant circuit can be found from

$$\omega_0 = \frac{1}{\sqrt{L\left(C + \dfrac{C_v C_1}{C_1 + C_v}\right)}} \tag{9.6}$$

The relative frequency sensitivity of the series-parallel resonant circuit to the variation of the variable capacitance C_v can be obtained from Equation (9.6) in the form

$$\frac{d\omega}{\omega_0} \bigg/ \frac{dC_v}{C_v} = -\frac{1}{2}\frac{C_1}{C_1 + C_v}\frac{C_v C_1}{C(C_v + C_1) + C_v C_1} \tag{9.7}$$

From Figure 9.3b it follows that the tuning frequency sensitivity of the series-parallel oscillator resonant circuit can be significantly reduced by the mutual approaching of the zero and pole of its reactance X. For the same resonant frequency ω_0, reducing the series capacitance C_1 and increasing the shunt capacitance C can achieve this. This contributes to the oscillator frequency stabilization by minimizing the effect of the variation of the equivalent active device nonlinear capacitance on the oscillator resonant frequency.

9.1.2 Oscillation systems with transmission lines

It is known that the frequency sensitivity of the resonant circuit containing a short-circuited transmission line can be reduced, if instead of the uniform transmission line a nonuniform one is used, the spectrum of natural frequencies of which is significantly nonequidistant. Figure 9.4 shows the resonant circuits with (a) uniform and (b) two-section transmission lines, where θ_1 is the electrical length of the low-impedanced section with the characteristic impedance Z_1 and θ_2 is the electrical length of the high-impedanced section with the characteristic impedance Z_2. The main difference in the frequency properties between the oscillation system with (a) short-circuited uniform transmission line with an equidistant spectrum and (b) short-circuited two-section transmission line with a nonequidistant spectrum in the case of $Z_2 \gg Z_1$ is illustrated by Figure 9.5. The graphic results show that it is advisable to use a two-section line for frequency stabilization at odd resonant frequencies, including fundamental in the case of a low-impedanced section adjacent to the variable capacitance C_v.

The sensitivity of the oscillation system with a two-section line to capacitance variation can be qualitatively evaluated according to two-by-two rapprochement or remoteness of zeros and poles of function $Z_{in}(\omega)$ along a frequency axis. The input impedance of short-circuited

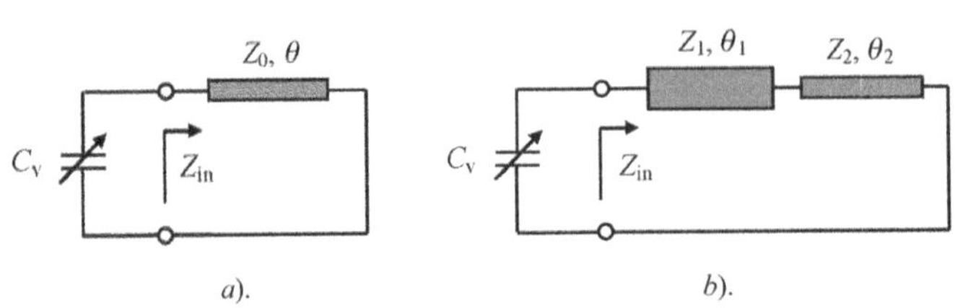

Figure 9.4 Oscillation systems with (a) uniform and (b) two-section transmission line

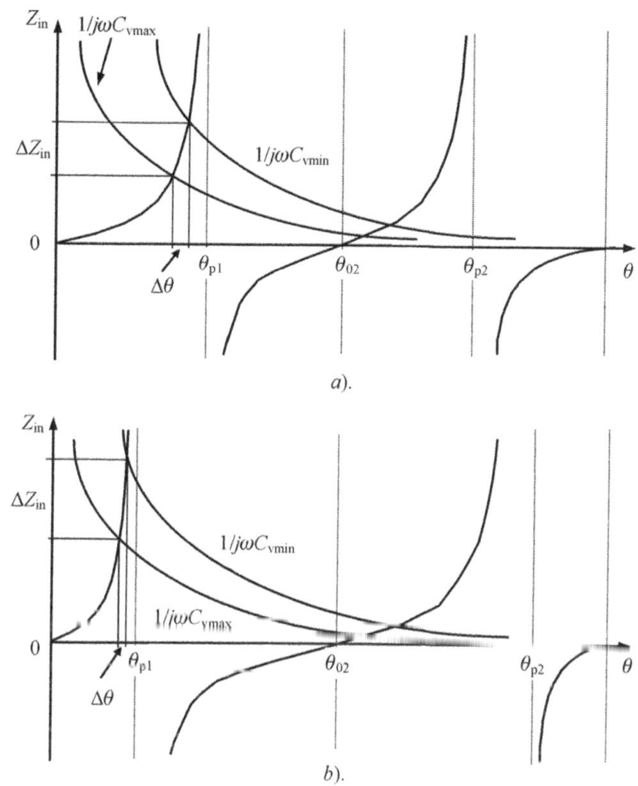

Figure 9.5 Frequency dependencies of oscillation system with two-section transmission line

two-section transmission line, provided $\theta_1 = \theta_2 = \theta/2$, can be written as

$$Z_{in} = jZ_1 \frac{1+M}{1 - M\tan^2\dfrac{\theta}{2}} \tan\frac{\theta}{2} \tag{9.7}$$

where $M = Z_2/Z_1$ is the characteristic impedance ratio. From Equation (9.7) it follows that the two-section transmission line has its zeros at $\theta = k\pi$ and poles at $\theta = k\pi \pm 2\tan^{-1}(1/\sqrt{M})$, $k = 0, \pm 1, \pm 2, \ldots$. The first pole of the function $Z_{in}(\theta)$ for the fundamental resonant frequency operation at $M = 20$ is displaced substantially to the left and becomes equal to $\theta_{p1} = 25.2°$. In this case, the same frequency deviation corresponds to significantly larger capacitance variation in comparison with the resonant circuit with uniform transmission line. This effect contributes to the increase of the sensitivity of the oscillator phase characteristic and loaded quality factor, resulting in reduction of the oscillator phase noise.

It follows from an analysis of the parallel feedback or negative resistance oscillator noise spectrum that the oscillator phase noise is inversely proportional to the square of the loaded quality factor Q_L of the oscillator resonant circuit. Consequently, for a given load conductance G_L and output power P_{out}, the value of Q_L has a significant effect upon the oscillator phase noise level. Generally, the loaded quality factor Q_L of the oscillator parallel resonant circuit can be defined by

$$Q_L = \frac{\omega_0 S_0}{2G_L} \tag{9.8}$$

where

$$S_0 = \left.\frac{\partial(B_{out} + B_L)}{\partial\omega}\right|_{\omega=\omega_0}$$

is the susceptance slope parameter or susceptance sensitivity of the oscillator parallel resonant circuit at the resonant frequency ω_0, B_{out} and B_L are the equivalent output active device and load susceptances, respectively [1]. In Equation (9.8) it is assumed that the real part of the equivalent output active device and load admittances are frequency independent in the vicinity of the steady-state oscillation conditions. Such a simplification has usually been accepted as a basis for oscillation analysis. A similar expression can be written in terms of the impedance parameters for the series resonant circuit. Hence, in order to improve the oscillator loaded quality factor, it is necessary to increase the susceptance sensitivity S_0.

In order to compare the oscillators with lumped and distributed parameters for the purpose of the best phase noise characteristics, it is sufficient to determine the susceptance sensitivity S_0 for each oscillator circuit with equal capacitances. For a simple parallel lumped resonant circuit shown in Figure 9.6a,

$$S_0 = \frac{\partial}{\partial\omega}\left(\omega C - \frac{1}{\omega L}\right) = C\left(1 + \frac{1}{\omega^2 LC}\right)\bigg|_{\omega=\omega_0} = 2C \tag{9.9}$$

The oscillator parallel resonant circuit with uniform transmission line is shown in Figure 9.6b. Since the ratio between the transmission line electrical length θ and the frequency

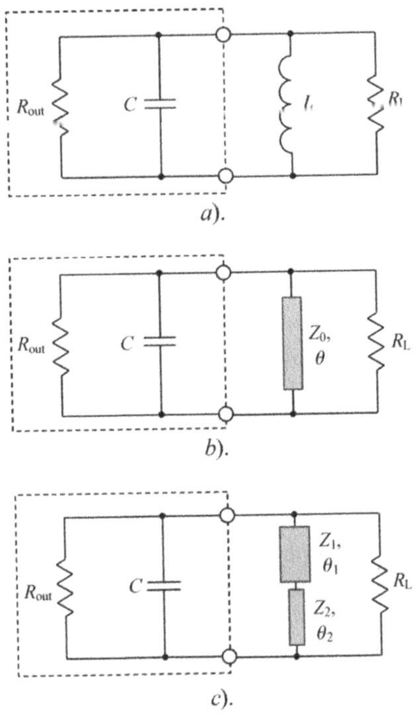

Figure 9.6 Equivalent parallel oscillator resonant circuits with (a) lumped parameters and (b, c) distributed parameters

ω is defined as

$$\theta = \omega l \sqrt{\varepsilon_r}/c \tag{9.10}$$

where l is the length of the transmission line, ε_r is the dielectric permittivity, c is the free-space velocity of light, the following condition is met:

$$\frac{\partial \theta}{\partial \omega} = \frac{\theta}{\omega} \tag{9.11}$$

Then, by taking into account that $C = 1/\omega_0 Z_0 \tan\theta$ at resonant frequency, the sensitivity S_0 for the parallel resonant circuit with a lossless uniform transmission line is obtained as

$$S_0 = \frac{\partial}{\partial \omega}\left(\omega C - \frac{1}{Z_0 \tan\theta}\right) = C\left(1 + \frac{\theta}{\omega C Z_0 \sin^2\theta}\right)\Bigg|_{\omega=\omega_0} = C\left(1 + \frac{2\theta}{\sin 2\theta}\right) \tag{9.12}$$

Similarly, the susceptance sensitivity S_0 for the parallel resonant circuit with lossless two-section transmission line of equal section lengths $\theta_1 = \theta_2 = \theta/2$ shown in Figure 9.6c can be written as

$$S_0 = \frac{\partial}{\partial \omega}\left(\omega C - \frac{1}{Z_1 \tan\dfrac{\theta}{2}}\,\frac{1 - M\tan^2\dfrac{\theta}{2}}{1+M}\right)\Bigg|_{\omega=\omega_0} = C\left(1 + \frac{\theta}{\sin\theta}\,\frac{1 + M\tan^2\dfrac{\theta}{2}}{1 - M\tan^2\dfrac{\theta}{2}}\right) \tag{9.13}$$

Figure 9.7 shows the frequency dependencies of the normalized sensitivity S_0/C for different types of oscillator circuits. For a lumped resonant circuit, $S_0/C = 2$, which means that it is independent of frequency. For the resonant circuits with transmission lines, the value of the sensitivity S_0 for a given capacitance C can be increased significantly by an appropriate increase of θ. If $\theta = 40°$, $S_0/C = 2.42$ for the resonant circuit with uniform transmission line,

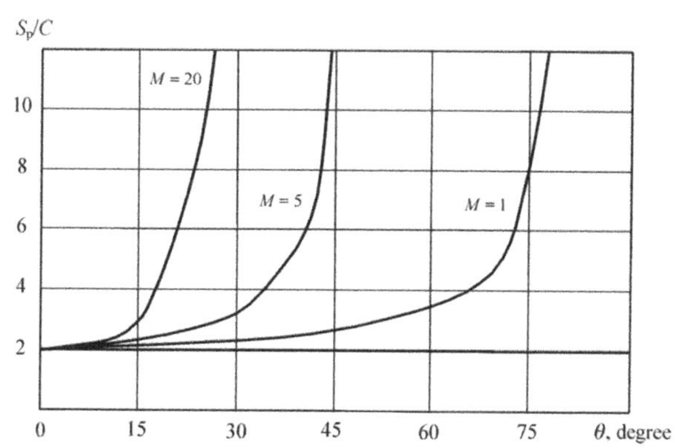

Figure 9.7 Frequency dependencies of the normalized susceptance sensitivity S_0/C for different characteristic impedance ratio M

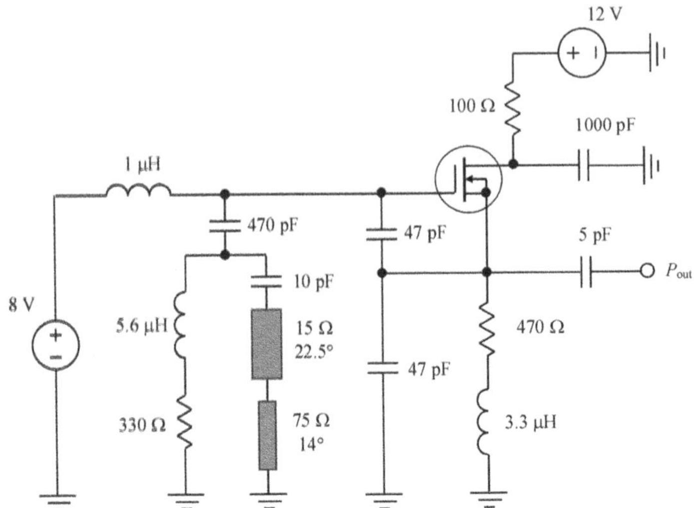

Figure 9.8 Equivalent circuit of the MOSFET oscillator with two-section line

whereas the use of the two-section transmission line with $M = 5$ results in $S_0/C = 6.35$. In the case of $\theta = 24°$, by using of the two-section transmission line with $M = 20$, it is possible to increase the value of S_0/C by a factor of more than ten. For the same values of S_0/C, the use of the two-section transmission line with a high value of M enables one to reduce the total electrical length of the transmission line by a factor of three to four.

Verification of the theoretical assumptions regarding the improvement of the oscillator noise characteristics in the case of the resonant circuit with transmission lines was made on the basis of the MOSFET microwave oscillator, for example, in the circuit simulator Serenade 7.5 [2]. The equivalent circuit of the MOSFET oscillator with a two-section transmission line connected instead of lumped inductance into the gate circuit is shown in Figure 9.8. The oscillation frequency is 400 MHz, and the output power is 11 dBm.

Figure 9.9 shows the results of computer simulation of the oscillator phase noise characteristics, where the phase noise level of the oscillator with ideal lumped inductance $L = 22$ nH is the same as that of the oscillator using an ideal uniform transmission line with characteristic

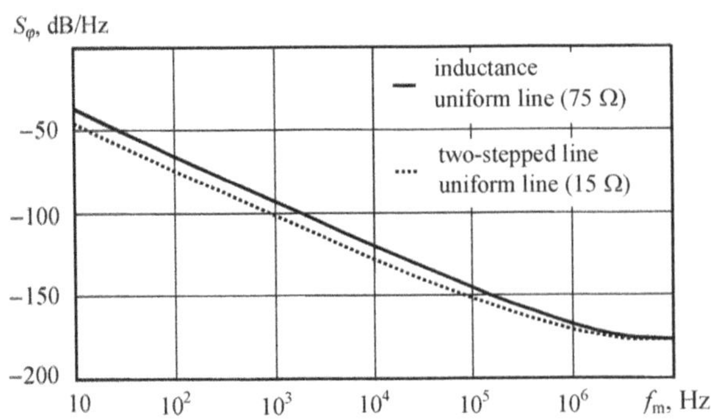

Figure 9.9 MOSFET oscillator phase noise characteristics

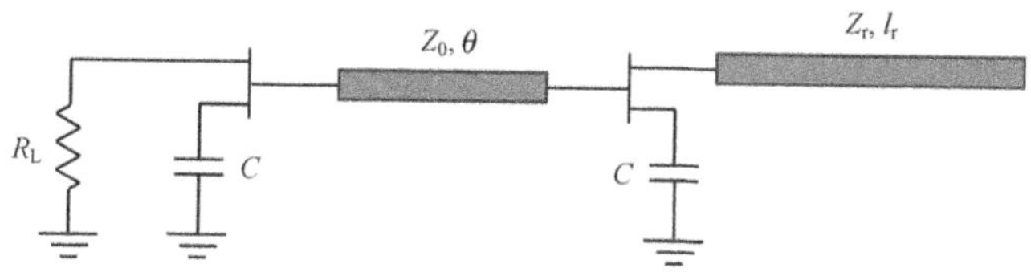

Figure 9.10 Block diagram of microwave oscillator with open-circuit transmission line

impedance $Z_0 = 75\ \Omega$ and electrical length $\theta = 37°$ (solid line). The phase noise characteristics can be improved by decrease of the characteristic impedance of the transmission line according to theoretical predictions. In this case, the phase noise level is reduced by approximately 7–8 dB up to the frequency offset $f_m = 10$ kHz for the value of $Z_0 = 15\ \Omega$ (dotted line). For such a low characteristic impedance of the uniform transmission line, to maintain the same value of the input impedance of its short-circuited configuration, the value of its electrical length must be increased to $\theta = 74°$. However, the same improved phase noise level can be obtained by using a two-section transmission line with half the total electrical length: $Z_1 = 15\ \Omega, \theta_1 = 23°$ and $Z_2 = 75\ \Omega, \theta_2 = 14°$. In practice, it is difficult to realize an inductor with a high value of its quality factor, especially for microwaves, which leads to a substantial deterioration of the oscillator noise characteristics. Therefore, it is better to use transmission lines with as short as possible electrical length for monolithic microwave integrated circuit oscillators.

From Equation (9.10–9.12) it follows that, with the significant increase in the length of the transmission line, the sensitivity S_0 and, consequently, loaded quality factor Q_L can also be increased significantly. Figure 9.10 shows a schematic diagram of a microwave push–pull oscillator containing an open-circuited transmission line with characteristic impedance Z_r and geometrical length l_r as a resonator [3]. The microwave oscillator consists of two identical FET devices whose gates are interconnected by a transmission line with characteristic impedance Z_0 and electrical length θ. The length of this transmission line is chosen such that the two FET devices resonate with each other [4]. A negative conductance at the desired oscillation frequency is generated by a capacitive source positive feedback. The oscillator outputs are taken from the drains of two FETs for the load resistance R_L and transmission-line resonator, respectively, providing the required output port isolation. Stable oscillations at around 20 GHz with output power of 0 dBm were observed at gate and drain bias voltages of 0 and 3 V, respectively.

Three coaxial cables of different lengths were used as the open-circuited resonators. Spurious oscillations at unwanted signal frequencies were not observed in any of these cases. Table 9.1 shows a significant 10 dB reduction in phase noise with the connection of a

Table 9.1 Measured parameters of the oscillator with open-circuit cable resonators [3]

Cable length (cm)	Number of wavelength (N)	Cable insertion loss (dB)	Phase noise improvement (dB)
10	23	0.35	10
50	89	1.7	18
100	171	3.2	21

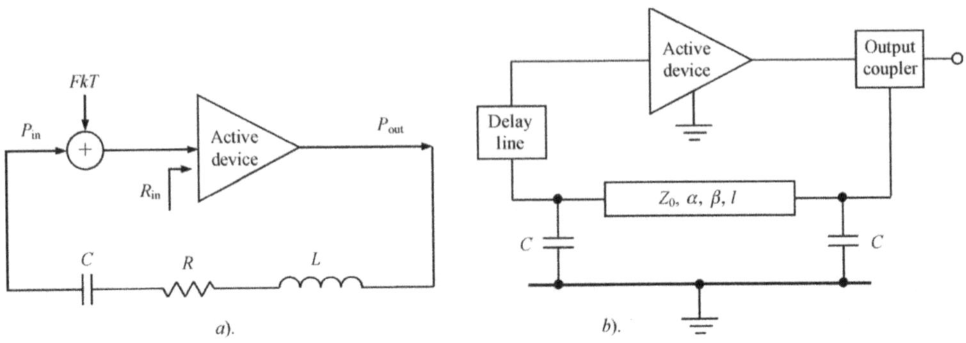

Figure 9.11 Equivalent circuits of *LC* and transmission line oscillators

transmission-line resonator of length 10 cm compared with the open-circuit output port. Further improvement in phase noise can be achieved by increasing the resonator length. Similar results were observed when short-circuited transmission lines were used as the resonators instead of open-circuited transmission lines [3].

To minimize the sideband noise in the oscillators, it is necessary to provide an optimum ratio of the oscillator loaded quality factor Q_L to its unloaded quality factor Q_0 [5, 6]. Figure 9.11 shows a general equivalent oscillator model. It is assumed that the thermal noise can be modelled as a single noise source *FkT* at the input of the active device (operating as a current source in a common case), where *F* is the noise figure (generally dependent on the input impedance of the active device). The feedback resonator represents a series *LCR* resonant circuit with an equivalent loss resistance *R*. The active device has zero output admittance, which can be achieved by using a switched-mode operation mode, and known input impedance. It should be noted that such an approach can be applied only when the thermal (additive) noise is the major noise source. Therefore, the obtained results cannot be valid for small frequency offsets where low-frequency flicker noise dominates.

The single sideband noise-to-carrier ratio of the oscillator can be defined as

$$L(f_m) = \frac{FkT}{8Q_0^2\left(\frac{Q_L}{Q_0}\right)^2\left(1-\frac{Q_L}{Q_0}\right)^2 P_{out}}\left(\frac{f_0}{f_m}\right)^2 \tag{9.14}$$

where $Q_0 = 2\pi f_0 L/R$ is the quality factor of the *LCR* resonant circuit, $Q_L = 2\pi f_0 L/(R + R_{in})$ is the loaded quality factor where R_{in} is the active device input resistance.

Equation (9.14) has its minimum when differentiated with respect to Q_L/Q_0 for constant Q_0 and F when

$$\frac{\partial L(f_m)}{\partial (Q_L/Q_0)} = 0 \tag{9.15}$$

As a result, a minimum single sideband phase noise power occurs when $Q_L/Q_0 = 1/2$ [5]. Therefore, Equation (9.14) simplifies to

$$L(f_m) = \frac{2FkT}{Q_0^2 P_{out}}\left(\frac{f_0}{f_m}\right)^2 \tag{9.16}$$

If the oscillator operates in a switched-mode high-efficiency operation with zero low output admittance (the device output is represented by a lossless switch), a minimum single sideband phase noise power occurs when $Q_L/Q_0 = 2/3$[6]. Consequently, Equation (9.14) can be rewritten as

$$L(f_m) = \frac{27FkT}{32Q_0^2 P_{out}} \left(\frac{f_0}{f_m}\right)^2 \tag{9.17}$$

The same results obtained by Equations (9.16) and (9.17) for minimum phase noise level are achieved for the transmission line oscillator shown in Figure 9.11b, where Z_0 is the characteristic impedance, α is the attenuation coefficient, β is the phase constant and l is the length of the transmission line, respectively [7, 8]. It is assumed that the active device and transmission-line resonator have both inputs and outputs matched to Z_0. The delay line is necessary to provide the phase compensation. In this case, for the small values of $\alpha l < 0.05$ and $f_m/f_0 \ll 1$, the loaded quality factor is defined by

$$Q_L = \frac{\pi}{1 + 2\alpha l \, (\pi f_0 C)^2} (\pi f_0 C)^2 \tag{9.18}$$

From Equation (9.18) it can be seen that, as the value of the shunt capacitance C is increased, the loaded quality factor Q_L increases to a limiting factor of

$$Q_0 = \frac{\pi}{2\alpha l} \tag{9.19}$$

which is defined as the unloaded quality factor.

A microstrip oscillator with oscillation frequency of 1.49 GHz based on a bipolar transistor NE68135 was fabricated using a RT Duroid substrate with dielectric permittivity of $\varepsilon_r = 10.2$. To deliver power to the external load, a 3 dB Wilkinson divider was used. Phase compensation is accomplished by means of a short length of transmission line and is finely tuned using a trimming capacitor. For a transmission line with $\alpha l = 0.019$, resulting in the unloaded quality factor $Q_0 = 83$ and noise figure $F = 3$ dB, Equation (9.16) gives the theoretical limit in terms of sideband noise power of approximately −102.6 dBm/Hz at 10 kHz offset, which is only 1.7 dB greater than the measured phase noise power level. An increase in the noise level was observed if the optimum condition was not met.

To improve the oscillator phase noise characteristic, a parallel circuit resonator based on open-circuited transmission lines can be added to the oscillator resonant circuit at its output, as shown in Figure 9.12 [9, 10]. In this case, the resonator is simply a combination of two

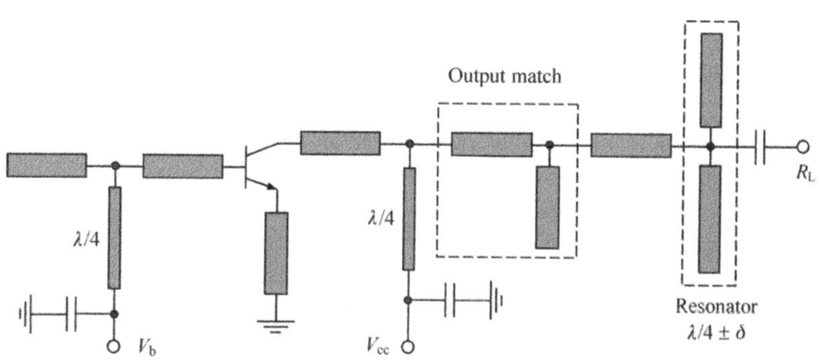

Figure 9.12 Circuit schematic of bipolar oscillator with parallel resonator (© 2000 IEEE)

open-ended stubs having lengths of $l_\pm = \lambda/4(1 + \Delta\omega/\omega_0) \pm \delta$, where λ is the wavelength corresponding to the oscillator resonant frequency ω_0. The input impedance of the lossless resonator with one end terminated by the load resistance R_L can be written as

$$Z_r = R_L \frac{1}{1 + j(\tan\theta_- + \tan\theta_+)} \quad (9.20)$$

where $\theta_\pm$ are the electrical lengths of the open-ended stubs corresponding to their lengths $l_\pm$, respectively. The two tangent functions in Equation (9.20) cancel each other at the resonant frequency ω_0 when $\Delta\omega = 0$ due to their odd properties. However, when $\Delta\omega \neq 0$, the electrical lengths $\theta_\pm$ are not symmetrical relatively 90°. As a result, when $\Delta\omega$ increases in a forward positive direction, both tangent functions demonstrate a sharp increase simultaneously. According to Equation (9.20), the susceptance sensitivity of such a parallel resonant circuit will increase rapidly when $\Delta\omega$ is sufficiently small, thus resulting in a significant increase of the resonator loaded quality factor. The unloaded quality factor of the resonant circuit depends on the attenuation coefficient of the stubs α in accordance with Equation (9.19), where $l = l_+ + l_-$, and can reach a value of 205 using a 10 mil alumina substrate or a value of 77 using a 80 μm GaAs substrate.

The hybrid 18 GHz oscillator using an AlGaAs/InGaAs HBT active device with a resonator fabricated on alumina substrate with $\delta = \lambda/30$ and a loaded quality factor of 45 exhibited an output power of 10.3 dBm and an efficiency of 19.3% at a supply voltage of 2 V with a phase noise level of −120 dBc/Hz at 1 MHz offset [9]. The effect of the resonator resulted in about 5 dB phase noise reduction. In addition, the oscillator pulling factor defined as $(f_{max} - f_{min})/f_0$, where f_{max} and f_{min} are the maximum and minimum oscillation frequencies respectively, is improved by factor of 6 when the load phase angle is changed over the range of 2π. The monolithic 38 GHz oscillator using the similar HBT technology with a resonator fabricated on GaAs substrate with $\delta = \lambda/36$ and a loaded quality factor of 34 achieved an output power of 11.9 dBm and an efficiency of 10% at a supply voltage of 3.2 V with a phase noise level of −114 dBc/Hz at 1 MHz offset [10]. The effect of the resonator was about 5 dB phase noise reduction.

9.2 LOW-FREQUENCY LOADING AND FEEDBACK OPTIMIZATION

Optimization of low-frequency loading and feedback circuit can contribute to minimize the $1/f$ noise in the transistor oscillators. In a first approximation, a gate voltage noise generator connected in series with a noise-free nonlinear two-port circuit can model the low-frequency $1/f$ noise in FET devices. So, if the low-frequency voltages applied to gate–source and drain–source terminals are reduced, the resulting sideband components around the oscillation resonance frequency due to nonlinear mixing of these voltages will be reduced as well. To realize such an approach, it is necessary to provide short-circuiting of the drain port and open-circuiting of the gate port at low frequencies [11]. The phase noise improvement under these conditions was verified experimentally by connecting external variable resistances R_g and R_d to the gate and drain ports, as shown in Figure 9.13a, where C_b are the blocking capacitances. The minimum phase noise conditions are realized for the ideal case when $R_g = \infty$ and $R_d = 0$. This means that in practice a high value of external resistance R_g at low frequencies is needed when it is bypassed by the choke inductance L_{ch}, and no converted noise variations will be observed.

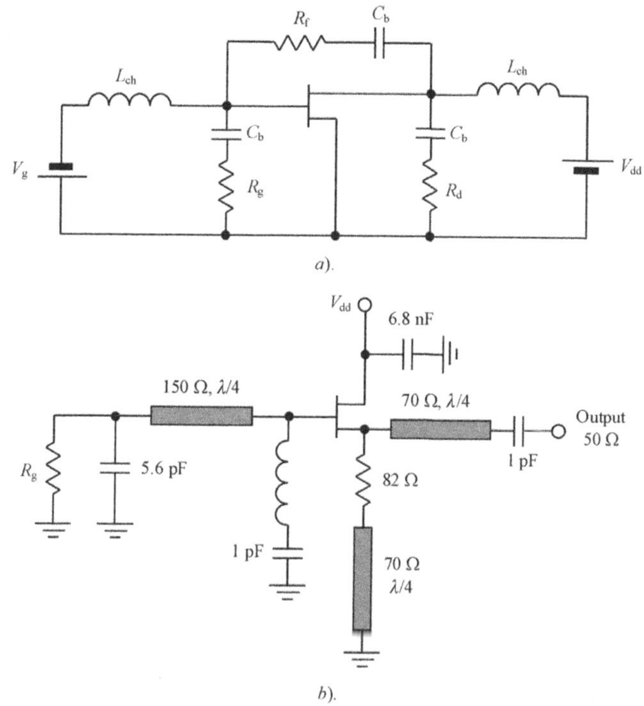

Figure 9.13 Diagram of low frequency circuit reducing phase noise

It was found that $1/f$ noise could be approximated by a quasistationary phenomenon having a quasiconstant autocorrelation function [12]. Consequently, it is possible to reduce phase noise by applying a feedback voltage in series with the $1/f$ noise generator using a parallel resistor R_f shown in Figure 9.13a. The higher voltage gain from input to output, the stronger feedback may be applied and the lower variations of the gate–source voltage can be observed. The low-frequency feedback circuit is also capable of cancelling thermal noise at low frequencies. The amount of cancellation is determined by the circuit delay and the frequency. Due to gate and drain loading, an 8 dB phase noise improvement is obtained at 1 kHz offset from the carrier for 10 GHz microwave oscillator. Optimizing the values of the gate, drain and feedback resistances results in an overall 11 dB phase noise improvement with optimum values of $R_f = 0$, $R_g > 100\,\Omega$ and $R_d < 100\,\Omega$ [11].

Figure 9.13b shows the circuit schematic of a common drain microwave MESFET oscillator with a direct current resistor R_g between the gate and the ground [13]. This low-frequency resistor, bypassed at high frequencies, also affects the noise up-conversion by allowing the gate rectification current to produce an optimum reverse bias on the gate. The lowest phase noise was obtained when a gate resistor $R_g = 1$ kΩ was used in the 5.6 GHz oscillator and a gate resistor $R_g = 200$ kΩ was used in the 7.4 GHz oscillator, providing a 14 dB noise reduction. Thus, there is an optimum value of the bias voltage minimizing the oscillator phase noise. In order to obtain low phase noise at low offset frequencies with dominating $1/f$ noise, the bias point should be chosen so as to avoid the increasing output conductance at the transition between the active and saturation regions, and operation in the pinch-off region should be

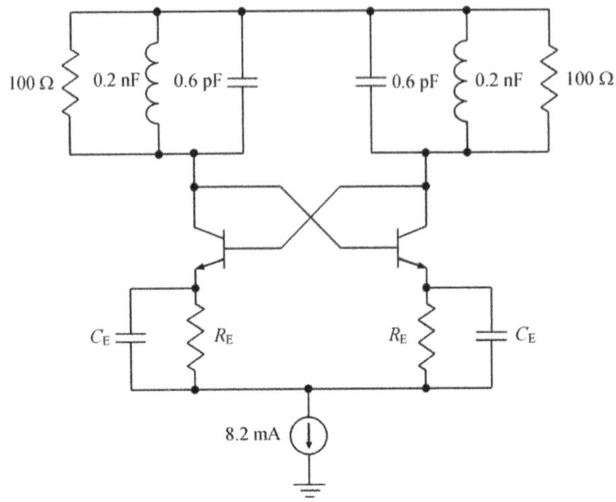

Figure 9.14 VCO topology incorporating emitter degeneration

minimized, corresponding to a Class AB mode with high quiescent current [14]. Within these limitations, it is necessary to maximize the oscillator output power. To keep the MESFET device operating in the active region, the technique utilizing a pair of limiting diodes can be used [15]. Two limiting diodes are placed between the source of the MESFET device and resonator to clip the oscillation amplitude before driving the transistor into nonlinear operation. Although adding a pair of diodes introduces new nonlinearities in the circuit, overall $1/f$ noise will be reduced since a MESFET is operated in a linear active region and these nonlinearities are small compared with those of the MESFET device.

Figure 9.14 shows a typical *LC* oscillator circuit based on a cross-coupled transistor pair incorporating emitter degeneration [16]. The proper choice of the parameters of the oscillation circuit provides the tuning range from 12 to 15 GHz using bipolar devices with transition frequency $f_T = 45$ GHz from SiGe 6HP process. The oscillator phase noise at 2 MHz offset is illustrated in Figure 9.15. If degeneration capacitance C_E is too small, the noise generated by

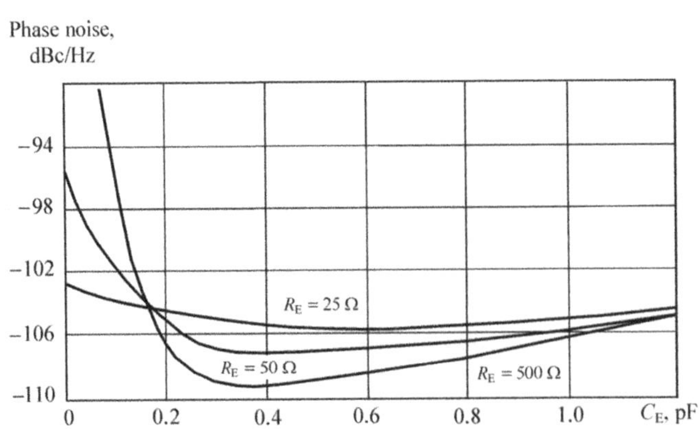

Figure 9.15 Simulated phase noise of *LC* oscillator with emitter degeneration

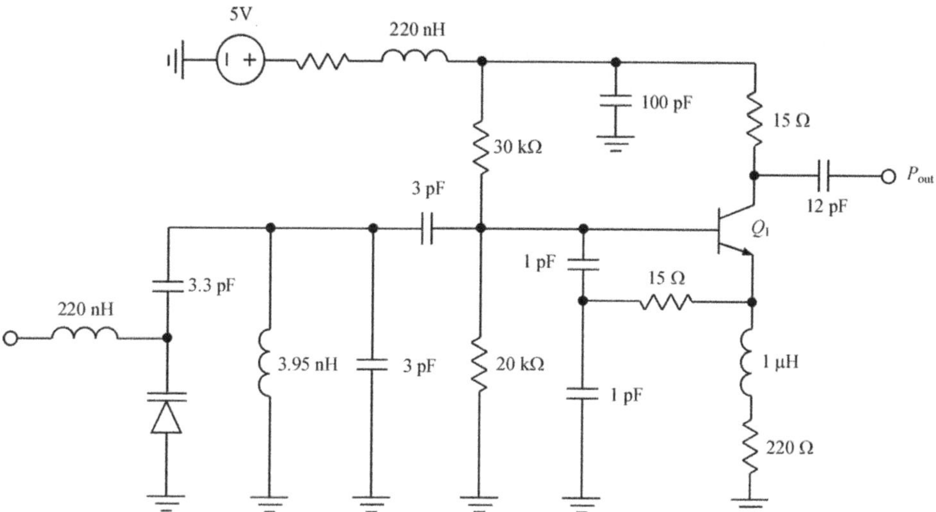

Figure 9.16 Bipolar oscillator schematic with passive feedback (© 1994 IEEE)

the feedback resistance R_E is not adequately filtered, resulting in noise degradation. For a degenerated VCO with $R_E = 300\ \Omega$, varying capacitance C_E from 0.3 to 0.8 pF provides a tuning range of the oscillation frequency from 14.6 to 12.6 GHz. The choice of optimum combination of the values of the degeneration resistance R_E and capacitance C_E enables one to improve the phase noise performance by approximately 7 dB compared with the nondegenerated design.

The experimental results show that, when using a negative feedback from unbypassed emitter resistance in a bipolar oscillator, the effect of flicker noise up-conversion can be significantly reduced [17]. Theoretically, this can be explained by the linearization of the device transfer function when dc and fundamental components of the collector current become linear functions of the base voltage, beginning from very small values, as follows from Figure 4.13 and Equations (4.76–4.78) given in Chapter 4. This leads to a significant reduction of the low-frequency noise-to-carrier modulation or flicker noise up-conversion. However, a high value of the unbypassed emitter resistor also reduces the active device transconductance, resulting in lower output power and regeneration factor. Therefore, it is preferable to use the emitter resistance together with noise feedback circuits.

Sufficient noise suppression can be achieved by using a passive feedback RF circuit composed of the series resistor connected to the device emitter and a capacitive voltage divider between the base and feedback resistor and the feedback resistor and ground [18, 19]. For a bipolar 800 MHz VCO, the circuit schematic of which is shown in Figure 9.16, the value of the feedback resistor is equal to 15 Ω and each capacitance is equal to 1 pF loosely coupled with the tuned resonant circuit. By optimizing the feedback elements and dc operating point, the close-in phase noise can be improved by 20–30 dB, achieving more than −110 dBc/Hz at 10 kHz offset.

Figure 9.17 shows a bipolar oscillator schematic with active emitter noise feedback circuit [19–21]. Due to the feedback loop, the noise across the emitter resistor is sampled by a sampling transistor Q_2, inverted by 180°, amplified and fed back into the base of the RF transistor Q_1. The simulation results shown in Figure 9.18 demonstrate of about 40 dB phase noise improvement

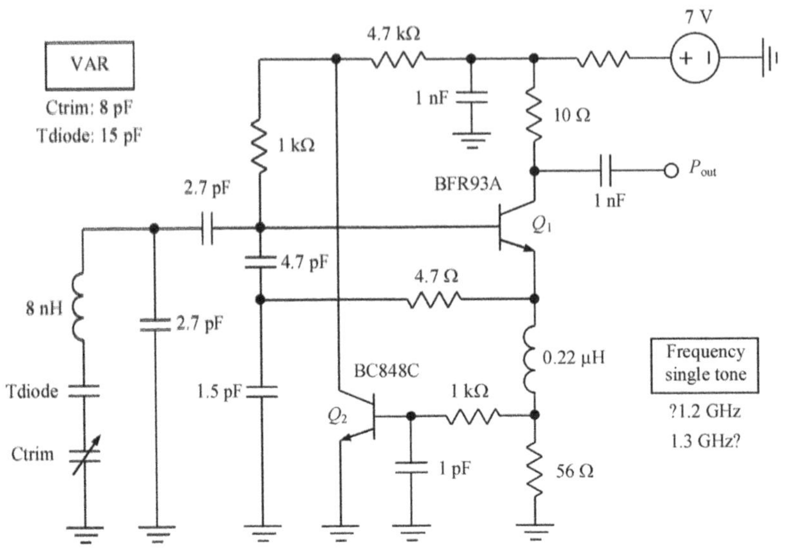

Figure 9.17 Bipolar oscillator schematic with active emitter noise feedback (© 2000 IEEE)

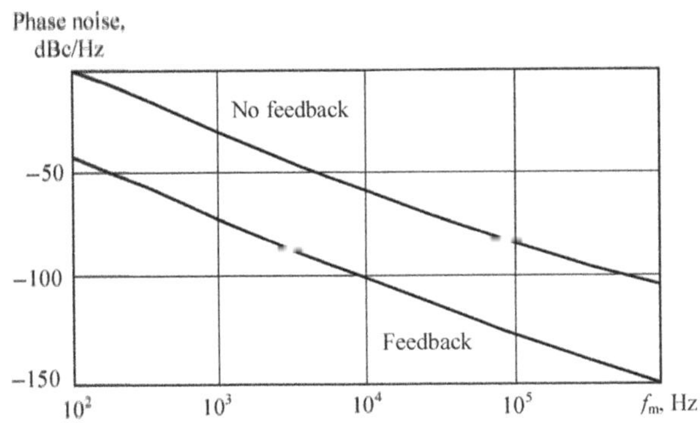

Figure 9.18 Phase noise of bipolar oscillator with and without feedback

for a perfect phase shift of 180° at the oscillation frequency of 1.2 GHz. The same result can be achieved using noise sampling from the collector resistance shown in Figure 9.19. Such a voltage-controlled oscillator provides a frequency tuning range from 612 to 1124 MHz within the varactor bias voltage range from +1 to +22 V. Since the bandwidth for a single transistor can only be about 1 MHz with guaranteed phase shift close to 180°, practical circuits can show a noise improvement between 15 and 20 dB [19].

A fully integrated RF CMOS VCO with an on-chip low-frequency feedback loop for flicker noise suppression is shown in Figure 9.20 [22]. Here, the emitter resistance R_6 first samples the low phase noise from the dc current. Then, a low-pass filter represented by a resistor R_5 and a capacitor C_3 is used to suppress the high-frequency components appearing across the emitter resistor R_6. The sampling nMOS transistor M_3 provides 180° inversion, amplification of the low phase noise signal and its final delivery to the gate of the nMOS transistor M_1 through the resistor R_1. The capacitances C_1 and C_2 form a positive feedback loop to provide

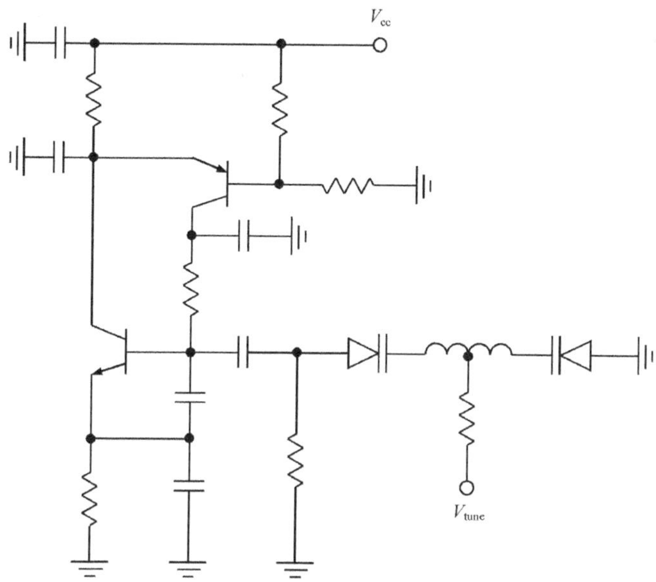

Figure 9.19 Bipolar oscillator schematic with collector noise feedback

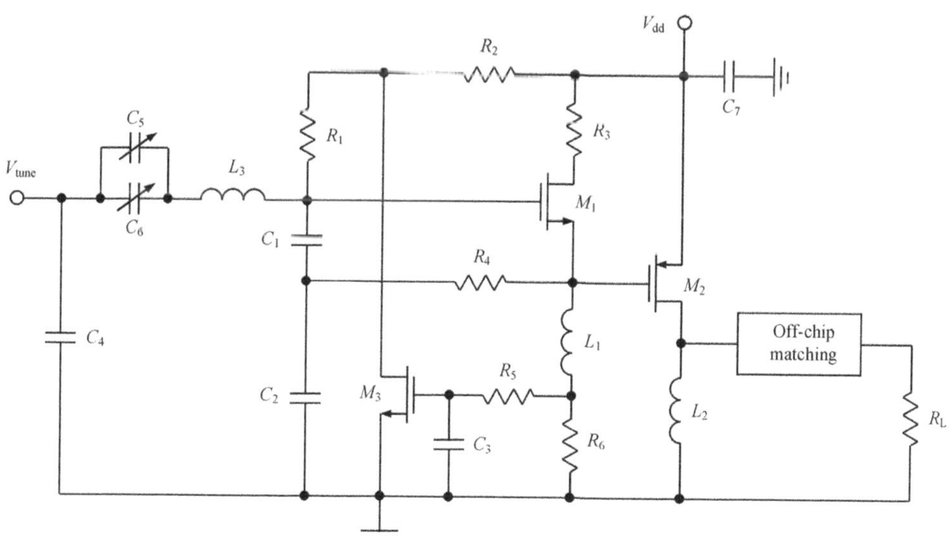

Figure 9.20 CMOS VCO with low-frequency feedback

the negative resistance, resulting in the start-up and steady-state oscillation conditions. The *LC* tank consists of the inductor L_3 with a quality factor around 10 at 5 GHz and the varactors C_5 and C_6 with a tuning range from 0.65 to 1.4 pF within 2 V voltage range. The resistors R_1 and R_3 provide the biasing of the MOS transistors. The inductor L_1 serves as RF choke. The *p*MOS transistor M_2 and inductor L_2 are the components of the buffer amplifier. Being implemented in 0.18 μm CMOS technology, such a VCO with low noise feedback loop provides the output power of 15 mW at supply voltage of 1.8 V, the frequency tuning range of 6.1% around the centre frequency of 5.74 GHz and the phase noise of −69 dBc/Hz at 10 kHz offset and −98 dBc/Hz at 100 kHz offset. The size of the chip area is 1.6 × 1.5 mm. It should be noted that using such a low noise feedback loop enabled the phase noise reduction of 4–6 dB.

9.3 FILTERING TECHNIQUE

The current source in the differential *LC* oscillators is required to set the bias current and provides high impedance in series with the switching active devices of the differential pair. In a perfectly balanced circuit, odd harmonics circulate in a differential path with no current flowing through the current source (out-of-phase operation). At the same time, even harmonics flow in a common-mode path through the active devices, resonator circuit and current source (in-phase operation). Because of the mixing effect provided by the nonlinearities (nonlinear transconductance and intrinsic capacitances) of the oscillator transistors, the low-frequency noise of the current source is initially upconverted to high-frequency noise around even harmonics and then down-converted to the phase noise around the fundamental. Since the level of the third- and higher-order harmonics in the resonant *LC* oscillator is negligibly small, the effect of the second harmonic can be taken into account. Thus, to prevent the effect of current source low noise modulation of the second harmonic, it is necessary to provide low impedance for the second harmonic. In other words, it is necessary to create a condition of current source bypassing for the second harmonic. Such an approach to the phase noise improvement is called a filtering technique. Several examples of the filtering technique applied to the differential *LC* oscillators are given elsewhere [23].

The simplest circuit solution which can be applied to the differential tail-biased VCO shown in Figure 9.21 is to place the shunt capacitance C_s with a large value (resulting in a small reactance at the second harmonic) in parallel to the current source M_1. However, one needs to be careful with the large value of the shunt capacitance in order to eliminate the self-modulation

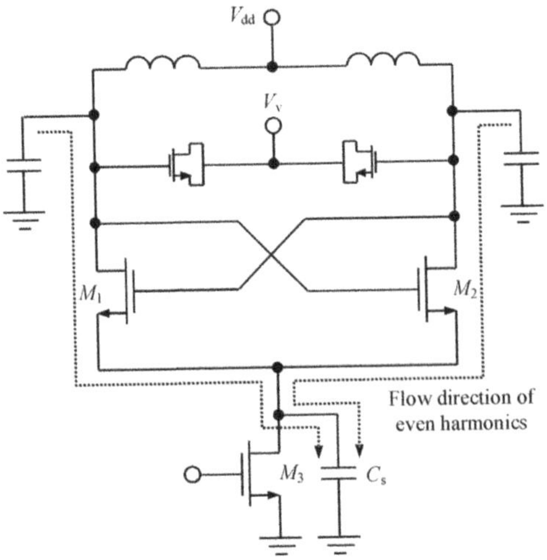

Figure 9.21 Differential tail-biased VCO with capacitive noise filter

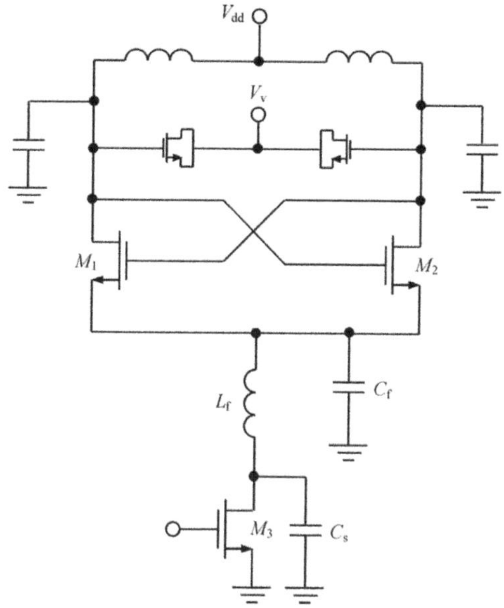

Figure 9.22 Differential tail biased VCO with complete noise filter

phenomenon resulting in AM–FM conversion (see Chapter 2). As an alternative, inserting the series inductance between the current source and the tail creates high impedance for the second harmonic, minimizing its contribution to the signal spectrum. Figure 9.22 shows the circuit schematic with a parallel filter based on the series inductance L_f connected in parallel to the capacitance C_f in addition to the shunt capacitance C_s. This parallel filter resonates on the second harmonic. Its impedance is limited only by the quality factor of inductance. As a result, the inserted inductance and two capacitances comprise a complete noise filter for a tail-biased differential *LC* oscillator.

Figure 9.23 shows the differential top-biased *LC* oscillator schematic where the current source is connected between the positive voltage supply and centre tap of the differential inductor. From the principle of dc operation, both tail-biased and top-biased schematics are identical and the position of the current source can be exchanged. However, in practical implementation, their RF performances are different. For instance, the top-biased VCO is more immune to substrate noise because the current source is placed in an *n*-well, rather than in the substrate [23]. However, from an analysis of the flow directions of even harmonics shown in Figure 9.21, it can be seen that the top-biased oscillator up-converts less flicker noise into phase noise around the fundamental frequency. This means that the level of the second harmonic flowing through the current source for the top-biased differential VCO is less than for the tail-biased oscillator. To minimize the phase noise, the complete noise filter for the top-biased VCO represents the large shunt capacitance connected in parallel to the current source and the second-harmonic filter comprising the parallel inductance L_f and capacitance C_f having high reactance at the second harmonic.

In the extreme operation when it is necessary to increase the VCO output power, a very large gate voltage is applied to the current-source transistor, resulting in an almost short-circuit to ground for the *LC* resonant circuit. This reverts to the voltage-biased differential VCO

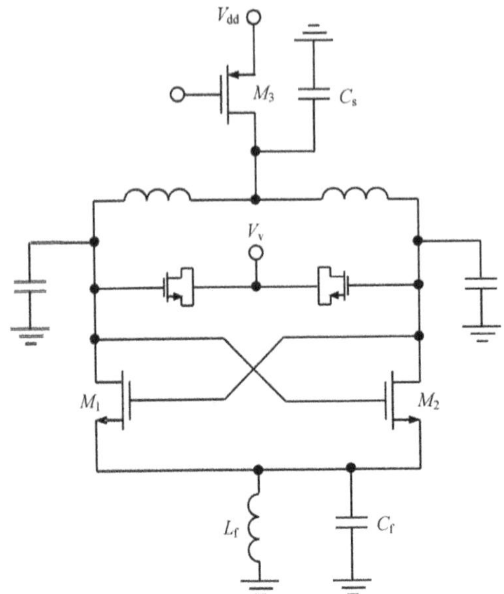

Figure 9.23 Differential top-biased VCO with complete noise filter

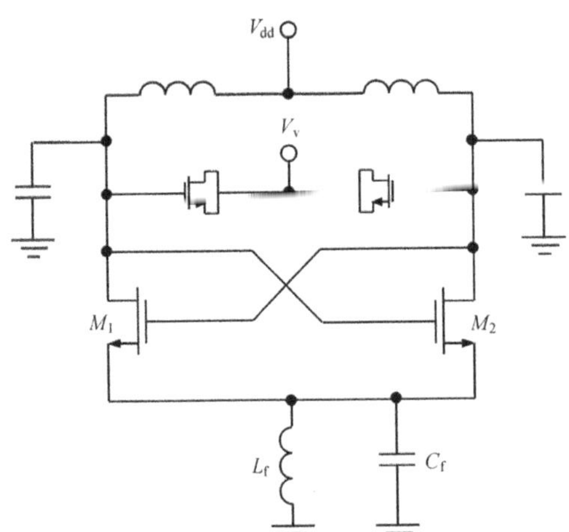

Figure 9.24 Differential voltage-biased VCO with noise filter

shown in Figure 9.24 where a noise filter in the tail is tuned to the second harmonic. In a steady-state operation, this oscillator provides the highest voltage amplitude because there is no voltage drop across the current-source transistor. At the same time, there is no low phase noise up-conversion due to the effect of the second harmonic. As a result, the phase noise of the voltage-biased VCO is least with the largest amplitude of the oscillation signal.

A tail-biased differential oscillator implemented in 0.35 μm CMOS technology can provide phase noise of −153 dBc/Hz at 3 MHz offset from the oscillation frequency of 1.2 GHz and consumes 3.7 mA from a supply voltage of 2.5 V. The phase noise of the identical reference

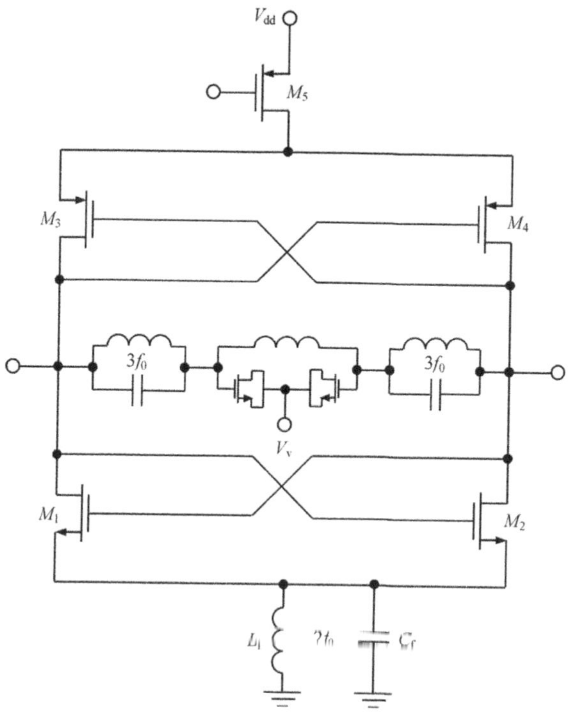

Figure 9.25 Differential top-biased VCO with harmonic tuned *LC* tank

oscillator without noise filter is 7 dB less at the same offset. A top-biased oscillator using the same technology with tuning oscillation frequency from 1.0 to 1.2 GHz provides the phase noise of −152 dBc/Hz at 3 MHz offset, an 8 dB improvement over its reference oscillator. The noise filter uses a 10 nH on-chip spiral inductance of square form and a 40 pF MIM capacitor [23].

Since both the sideband noise-to-carrier ratio in general and the device noise figure in particular can be reduced by increasing the voltage amplitude for the same tank resistance and oscillator quality factor, the harmonic tuning concept based on using additional resonant circuits tuned to the one of the harmonics of the fundamental frequency can be useful [24]. Using the third harmonic tuning resonators connected to the device drains contributes to the Class F approximation with the flattened drain waveform. In this case, the voltage fundamental-frequency and third harmonic components are out-of-phase, thus resulting in an increased fundamental-frequency voltage amplitude. As a result, the differential top-biased CMOS VCO with third harmonic *LC*-resonators and second harmonic noise filtering technique, the circuit schematic of which is shown in Figure 9.25, can provide the phase noise improvement of more than 10 dB compared with the standard VCO, achieving −140.8 dBc/Hz at 3 MHz offset from a carrier of 2 GHz in 0.35 μm CMOS technology.

Figure 9.26a shows the equivalent circuit of the tail-biased differential VCO where, instead of FET current source, a polysilicon resistor R_s, which is substantially free from $1/f$ noise, defines a tail current [25]. The resistor value is small enough to lower the loaded quality factor of the resonant circuit. Therefore, to provide RF isolation of the resistor from the resonant circuit,

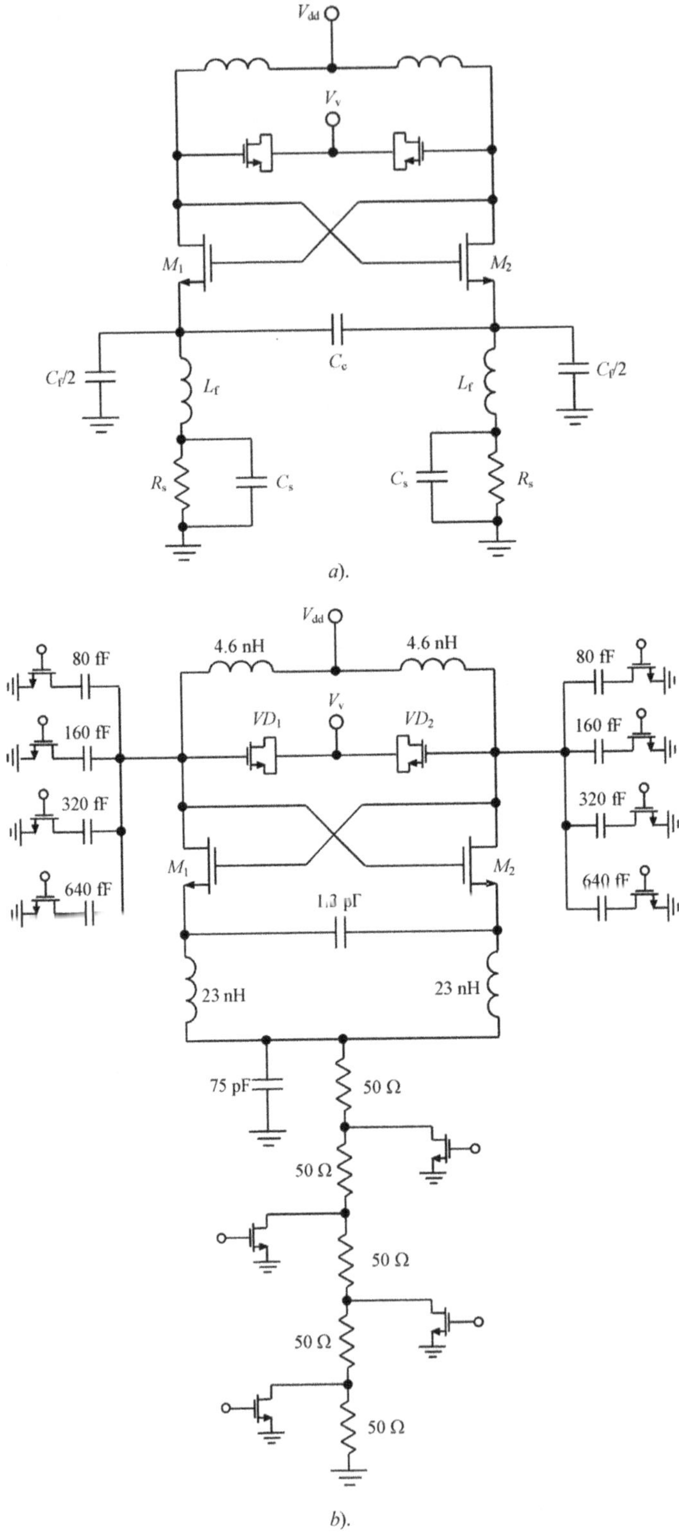

Figure 9.26 Differential VCO with noiseless current sources (© 2003 IEEE)

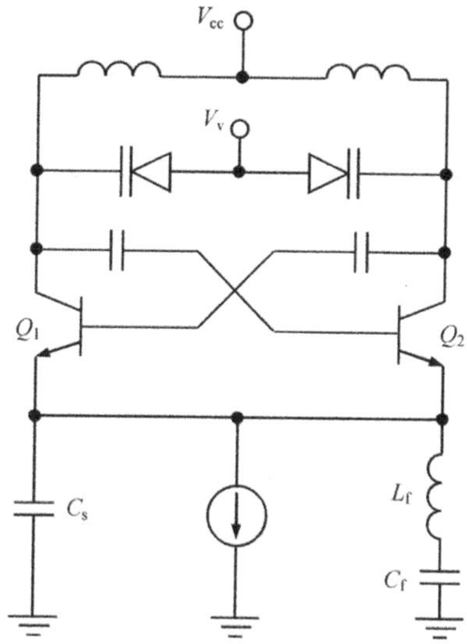

Figure 9.27 Differential tail-biased VCO with second harmonic trap

a high value of inductance L_f is used in each part. Together with a parallel capacitance $C_f/2$, it can compose the parallel resonant filter tuned to the second harmonic. A large capacitor across each resistor C_s shunts the wideband resistor noise, preventing its up-conversion. To improve the balanced operation conditions of two *n*MOS devices, M_1 and M_2, when the fundamental frequency voltages at their sources should be exactly out of phase, the decoupling capacitor C_c is used. When this capacitance is too large, the circuit behaves like a conventional differential pair. However, for zero value of the capacitance C_c, the oscillator represents the push–pull schematic when, to provide the soft start-up conditions, it is necessary to use large values of the filter capacitances $C_f/2$, minimizing their reactances at the fundamental frequency.

In the practical differential VCO using 0.35 μm CMOS technology, the dc current is regulated by segmenting a fixed resistor and selectively shorting the segments with digital bits, as shown in Figure 9.26b [25]. The circuit can be tuned from 1.43 to 1.64 GHz with two *n*MOS varactors, VD_1 and VD_2, connected in parallel with a 4-bit binary-switched capacitor array. The oscillator operates at 2.7 V with drain current of 6 mA. The measured phase noise is lower by 20 dB in the flicker noise-dominated area (less than 10 kHz offset) and by 15 dB at 50 kHz offset compared with a conventional differential VCO. At 1 MHz offset, the phase noise becomes the same for both configurations. Across the entire discrete tuning range, the phase noise changes by only about 2 dB. The degree of suppression is limited by the capacitor C_f, which is determined by interconnects in a practical configuration.

Figure 9.27 shows the equivalent circuit of the bipolar cross-coupled tail-biased differential VCO where the second harmonic signal generated at the transistor emitters is flowing through the series L_fC_f filter tuned to the second harmonic rather than the current source [26]. The shunt capacitor C_s connected in parallel to the current source suppresses the level of higher-order harmonics, which may appear in the current source. Such a differential VCO configuration is capable of improving the phase noise of −140 dBc/Hz at 3 MHz offset from the oscillation frequency of 900 MHz, corresponding to a conventional bipolar differential VCO without second harmonic filter by 1.9 dB.

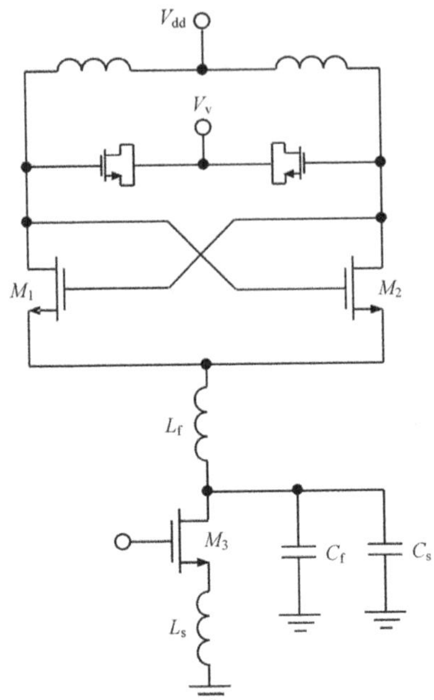

Figure 9.28 Differential VCO with tail current noise suppression

Figure 9.28 shows the tail-biased differential VCO using a low noise suppression technique, which includes inductive degeneration and low-pass filtering [27]. An off-chip degeneration inductor L_s is placed between the source of the tail transistor M_3 and ground. In this case, the noise current power of the current source transistor is reduced by a factor of $|1 + jg_m\omega L_s|^2$, where g_m is the transconductance of the transistor M_3. For the values of $g_m = 50$ mS and $L_s = 30$ μm, the noise reduction begins at about 100 kHz. The cutoff frequency of the low-pass filter with the series inductance L_f and shunt capacitance C_f can be significantly reduced by using an additional off-chip large-value shunt capacitance C_s. The overall phase noise reduction in simulation is 3 dB at the 3 MHz offset from the oscillation frequency of 2 GHz, of which 2 dB is due to the inductive degeneration and 1 dB is due to the on-chip low-pass filter. A supply voltage of 1.4 V and dc current of 9 mA were used, with the off-chip inductance $L_s = 100$ μm and the following values of the on-chip elements: tank inductor of 2.3 nH with a quality factor of 9 at 2.2 GHz, inductance $L_f = 3$ nH and capacitance $C_f = 10$ pF. It should be noted that the advantages yielded by the inductive degeneration increase at lower offset frequencies. The practical VCO was fabricated using standard 0.35 μm CMOS technology with frequency tuning from 1.96 to 2.36 GHz for the same supply voltage and dc current. As can be easily predicted, the best results are obtained at the highest frequencies when the tuning capability of the varactors becomes the smallest with minimal nonlinearity of their voltage–capacitance characteristic. Using either shunt capacitance $C_s = 30$ nF or off-chip inductance $L_s = 100$ μm results in 5 and 6.5 dB phase noise improvement at 3 MHz offset, respectively, compared with the phase noise of a conventional tail-biased differential VCO of -131 dBc/Hz. At 100 kHz offset, the major contributor to the phase noise reduction is the degeneration inductance L_s providing an improvement from -94 to -103.5 dBc/Hz.

It should be noted that the transistor equivalent circuit parameters such as base–emitter capacitance, device transconductance or collector capacitance are nonlinear, having different nonlinear behaviour versus bias voltage. Therefore, there is a possibility to further improve the noise performance by optimizing tail current and resonant circuit elements when the oscillation frequency sensitivity to the tail current is minimized. For example, there is an optimum bias point for MOS transistors where the intermodulation distortions can be minimized due to quadratic dependence of their transfer characteristic in the region close to the device threshold voltage [28]. For bipolar devices, there is a region with medium values of the tail current where the phase noise level of the bipolar VCO can be minimized. Without an emitter degeneration resistor and a low-pass filter, by only optimizing the resonant circuit elements and choosing the optimum tail current of 6.5 mA and power consumption of 14 mW for the 2.6 GHz differential VCO using Si-bipolar process with $f_T = 20$ GHz, the phase noise level was improved from -92 to -104 dBc/Hz at 100 kHz offset [29].

9.4 NOISE-SHIFTING TECHNIQUE

In modern wireless communication systems, cross-coupled oscillators have been preferred over other topologies for monolithic integrated circuit implementation due to their simplicity of realization using CMOS technology and differential operation. However, because the current-source transistor is top or tail located, to improve the phase noise performance, it is necessary to use the special filtering techniques for second harmonic suppression. Therefore, the new oscillator topologies based on classical types of the oscillator can overcome this problem when one of the active device ports is grounded [30]. In addition, these topologies provide larger oscillation amplitude for a given bias current because there is no voltage drop of the dc current across the current-source transistor, similar to the differential voltage-biased VCO shown in Figure 9.24.

Figure 9.29a shows a single-ended common base bipolar Colpitts oscillator configuration. The required regeneration factor for the start-up oscillation conditions can be chosen using a proper ratio of the feedback capacitances C_1 and C_2. To maximize the loaded quality factor of the resonant circuit, the choke inductance can be connected between the bias resistor R_{bias} and

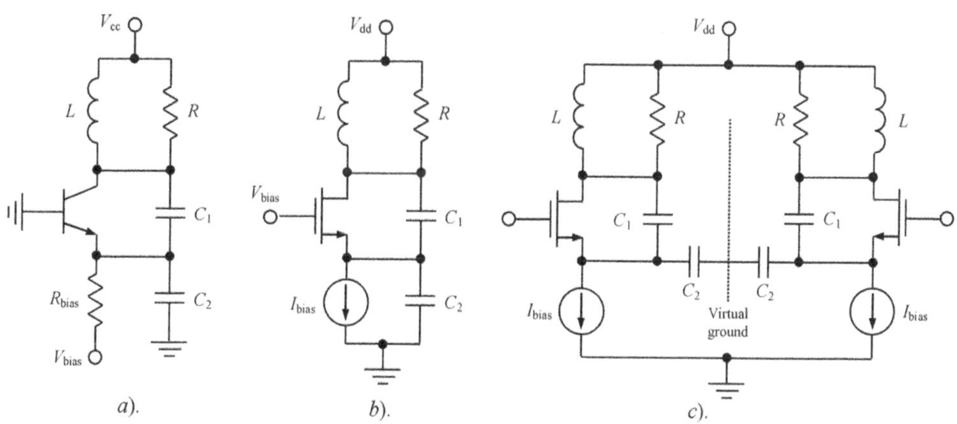

Figure 9.29 Single-ended and differential oscillator topologies

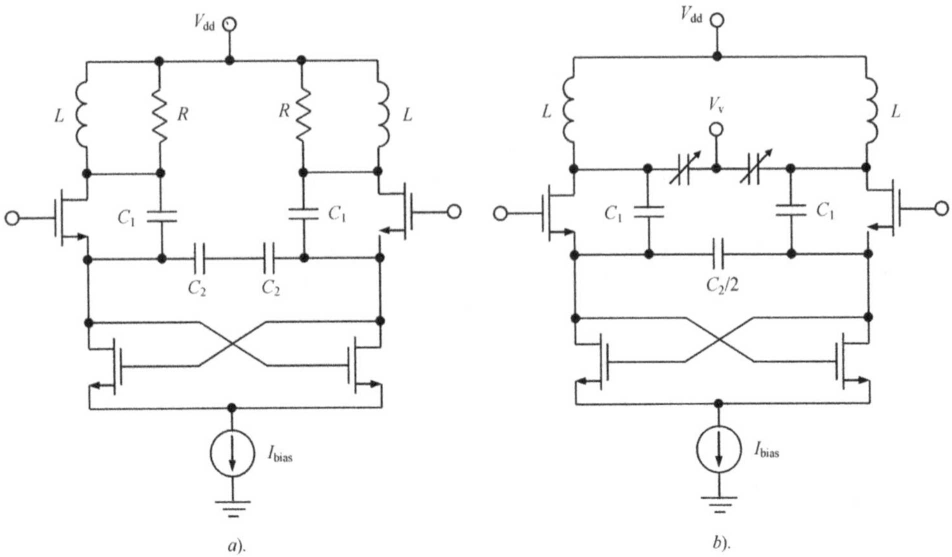

Figure 9.30 Noise-shifting differential common gate oscillator topologies

emitter. In CMOS implementation, the single-ended bipolar common base oscillator can be replaced by the schematic using an nMOS device shown in Figure 9.29b with RF grounded gate and current source instead of bias resistor. Connecting two identical single-ended oscillators, as shown in Figure 9.29c, can provide the differential output. In a perfectly balanced circuit, identical sides carry 180° out-of-phase signals of equal magnitude. Consequently, there is a midpoint (the centre node between C_2 capacitances) where the signal magnitude is zero. For absolutely identical circuits in each balanced side, the difference between signal magnitudes becomes equal to zero in each midpoint (dotted line) referred to as a virtual ground.

Because of a push–pull operation, the dc current flows through each current-source transistor for only half a period. Therefore, these current sources can be replaced by a pair of cross-coupled nMOS transistors providing a synchronized current switching from one side to another, as shown in Figure 9.30a. To add frequency-tuning capability to the new oscillator topology, it is possible to include two varactors connected in parallel with the tank inductor. Finally, the two shared C_2 capacitors connected in series can be replaced by an equivalent capacitor with half of their individual value. Final differential noise-shifting common gate oscillator schematic is shown in Figure 9.30b [30]. It should be noted that C_2 capacitances with virtual ground node serve as shunt capacitances used in a filtering technique. These capacitances have small reactances at higher-order harmonics, minimizing their flow through the current-source transistors. In addition, a loaded quality factor of such an oscillator can be increased by connecting series inductances with high reactance at fundamental between the main and cross-coupled transistors.

The differential noise-shifting topology was implemented in a common gate VCO fabricated using 0.35 μm BiCMOS process technology to operate from 1.8 to 2.45 GHz [30]. The tank inductors have quality factors of 6. To maximize the oscillation voltage amplitude and regeneration factor, the capacitance C_2 was chosen to be four times capacitance C_1. The oscillator shows a phase noise of −139 dBc/Hz at 3 MHz offset from the carrier of 1.8 GHz drawing

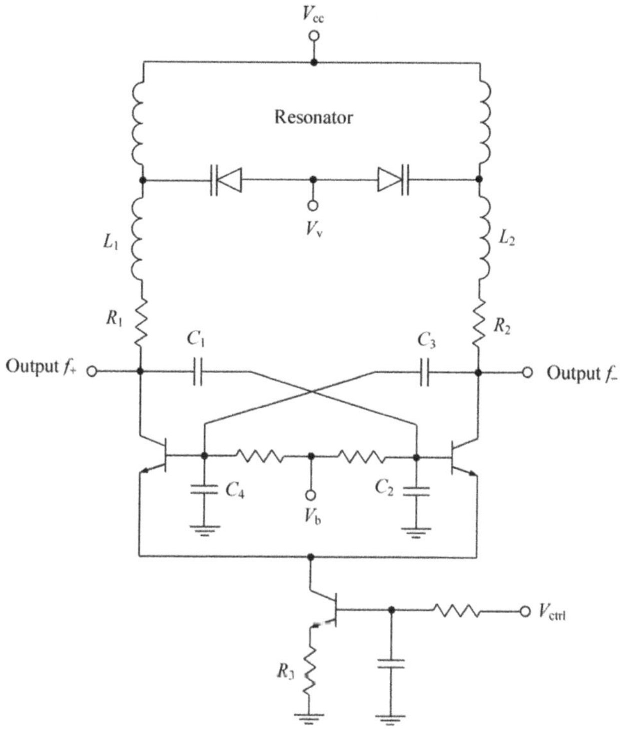

Figure 9.31 Bipolar cross-coupled differential VCO schematic

dc current of 4 mA from supply voltage of 2.5 V. The application of *LC* filtering technique to this oscillator when the added *LC* network was designed to resonate at second harmonic shows a negligible effect on the phase noise performance.

To minimize the phase noise of the cross-coupled differential VCO, it is important to optimize the capacitive feedback ratio as well. Figure 9.31 shows the circuit schematic of the cross-coupled differential bipolar VCO designed for wireless application, where L_1 and L_2 represent the parasitic bondwire inductances, resistors R_1 and R_2 are used to damp the spurious oscillations, capacitors C_1, C_2 and C_3, C_4 provide positive feedback [31]. The optimum feedback ratio for this design is $C_2/C_1 = C_4/C_3 = 3.5$. For a maximum varactor frequency tuning bandwidth, the values of feedback capacitances should be as low as possible. The optimization of the feedback ratio is important to set operation conditions with optimum regeneration factor and symmetric waveforms of the collector current of each half-circuit as a dominant noise contributor. For bipolar devices with $f_T = 25$ GHz, its contribution at 100 kHz offset from the carrier of 800 MHz was of 68% of total noise, while the contribution of the base resistance thermal noise was simulated to be less than 5%. Due to the loaded Q_L of the resonant circuit of about 16, the noise contribution from the resonator is of 21%. Although the noise contribution from the tail current is sufficiently small, about 7%, it becomes the major noise source at offset frequencies less than 3 kHz. The sum of all noise contributors results in the noise-to-signal ratio of −106.2 dBc/Hz. The VCO consumes dc current of 1.6 mA from a 2.7 V power supply.

Figure 9.32 shows a balanced common gate Colpitts VCO fabricated using InGaP/GaAs HBT process with $f_T = 60$ GHz and $f_{max} = 110$ GHz [32]. By varying base bias voltage V_b

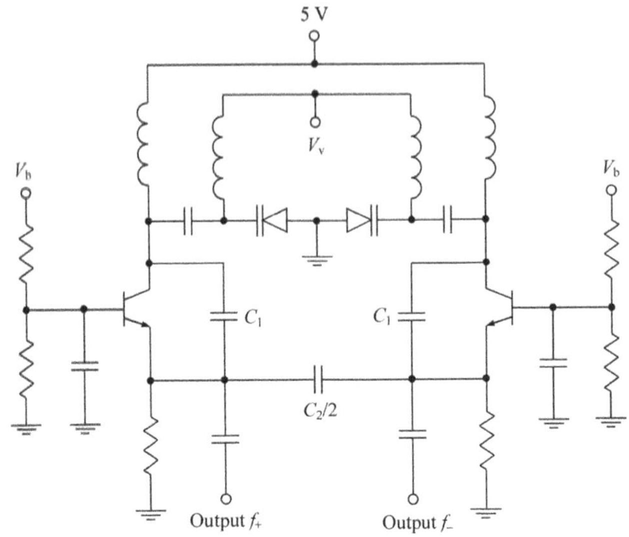

Figure 9.32 Bipolar balanced Colpitts VCO schematic

from 2 to 3.5 V, it was found that there is an optimum value of approximately 3 V, resulting in a minimum level of the phase noise and close to a maximum tank voltage amplitude just before the onset of the transistor saturation. Such a critical operation mode, corresponding to the border between the active and saturation regions, provides compromise conditions when the output power is high enough to minimize the phase noise level, but the collector–base junction has not yet been forward-biased to resistively shunt the oscillator resonant circuit, making its loaded quality factor significantly worse. By varying varactor bias voltage V_v from 0 to 4 V, the frequency tuning of 150 MHz with output power of about 6 dBm with small variations within 1 dB was achieved. Minimum phase noise was less than −112 dBc/Hz at 100 kHz offset from the carrier of 6.4 GHz with the slope of 20 dB/decade at higher offsets. This shows that the balanced Colpitts VCO schematic is a favourable topology for low phase noise monolithic design.

9.5 IMPEDANCE NOISE MATCHING

It is known that to achieve maximum signal-to-noise ratio in an oscillator, the optimum relationships between the active device, resonant circuit and feedback parameters should be established. For example, the minimum noise figure is obtained when the load conductance, including the real part of the device output admittance, is transformed via the feedback circuit into the optimum source admittance at the input of the active device. Therefore, generally for low-noise oscillator design, it is necessary to consider the following aspects [33]:

- proper choice of the active device with
  - small noise figure combined with a small correlation coefficient
  - higher output power
  - small output conductance
  - reasonably high input impedance;

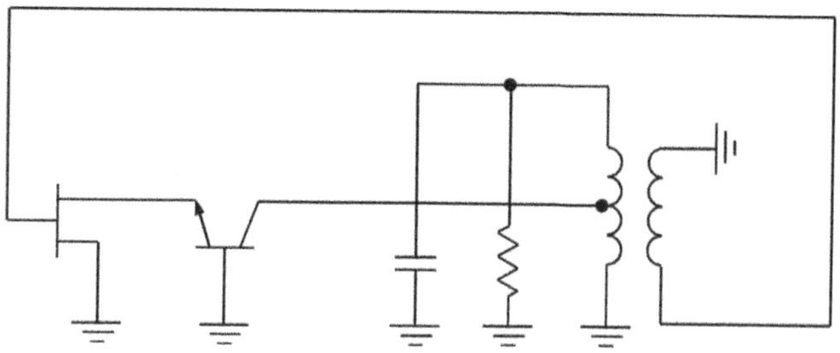

Figure 9.33 Equivalent circuit of optimally designed transformer-coupled oscillator

- meeting an impedance condition at the input of the active device which can be achieved by optimization of the feedback factor and which leads to optimum impedance noise matching;
- high quality factor of the resonant circuit;
- optimum coupling coefficient of the feedback transformer.

For example, choosing the optimum value of the voltage transformation ratio of the transformer in a transformer-coupled oscillator, it is possible to improve the signal-to-noise ratio by 30 dB [33]. In a common case, it is difficult to satisfy all requirements simultaneously. In this case, the best signal-to-noise ratio can be achieved by optimization of the interdependent parameters.

Considering the criteria for the choice of an optimum active element, sometimes it is much more promising to combine several active devices rather than using a single one, because one requires a low noise input stage with sufficiently high input impedance and a final stage with high output power capability and high output impedance. Figure 9.33 shows the simplified equivalent circuit of the transformer-coupled oscillator with dual-stage active element based on a tandem connection of the common source MESFET device with high input impedance and low noise figure and common base bipolar transistor with high output power capability. Compared with other transistor configurations, the common base connection shows the highest output impedance for the same output power. The minimum noise figure of the MESFET device is provided by optimum choice of the biasing conditions: drain supply voltage and dc current. To minimize the up-conversion of the $1/f$ noise, it is necessary to provide a nearly linear operation of the MESFET device. In addition, by appropriate partial coupling of the active element to the resonant circuit, the signal-to-noise ratio can be significantly improved. As an example, for a 150 MHz oscillator with the output power of -8 dBm, supply voltage of 35 V, dc current of 60 mA, coupling coefficient of 0.1 and quality factor of the resonant circuit of 300, the noise-to-carrier ratio (including both amplitude and phase noises) of -166 dBc measured in 1 Hz bandwidth at 10 kHz offset can be obtained at 50 Ω load [33].

The cascode configuration of two bipolar devices, one with a common emitter and the other with a common base, makes it possible to increase the output resistance of the active element, resulting in higher loaded quality factor of the oscillator resonant circuit without degradation of the noise performance provided by the common emitter transistor. Figure 9.34 shows the circuit schematic of the cascode bipolar oscillator where the negative power supply is connected to the emitter of the first transistor, while the collector of the second transistor is dc grounded [34]. By using a low-power bipolar device with $f_T = 3$ GHz and a resonant circuit with quality factor $Q_0 = 200$, the output power of 10 dBm at the oscillation frequency of 230 MHz with dc current

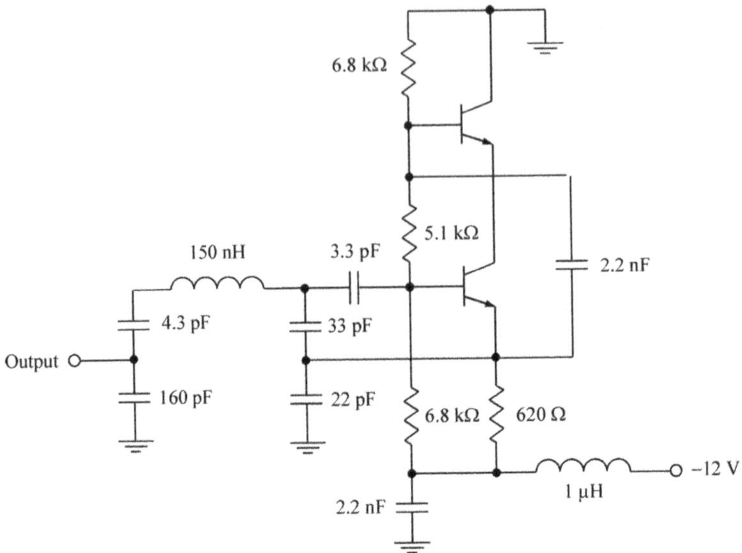

Figure 9.34 Bipolar cascode low-noise oscillator [34]

of 6 mA and phase noise of less than −120 dBc/Hz at 10 kHz offset was obtained. In order to achieve a low level of the phase noise, it is very important to properly isolate the oscillator circuit from the buffer amplifier using electric shielding. In addition, to minimize the effect of board parasitics, the grounding of the circuit elements must as close to each other as possible.

Figure 9.35 shows the cascode bipolar VCO where the first transistor is connected in a common collector configuration whereas the second transistor is configured for a common

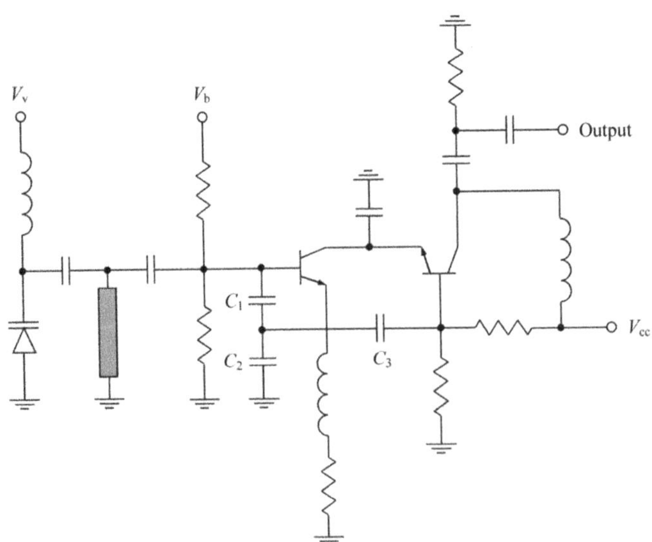

Figure 9.35 Bipolar cascode VCO with optimum phase noise

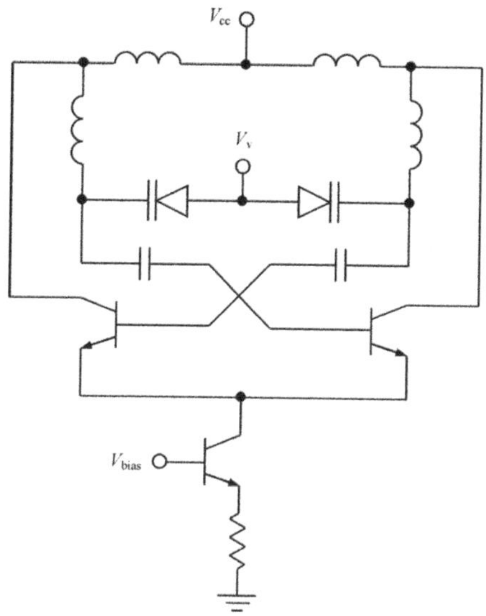

Figure 9.36 Differential tail-biased VCO with tapped inductor

emitter operation [35]. The common collector device is used as a gain element and biased to operate in the active region by maximizing its quiescent current, preventing it from going into saturation. If it were to saturate, it would load the tank circuit connected to its base terminal, thus affecting the VCO phase noise performance. The common emitter transistor is used as a limiting element, providing the maximum VCO output power. Its partial connection to the tank circuit through the feedback capacitances C_1, C_2 and C_3 is necessary to minimize the VCO phase noise. By optimizing the dc bias and RF feedback conditions, an improvement in 10 dB phase noise can be achieved without degrading the frequency tuning bandwidth.

In the differential VCO, due to capacitive cross-coupling between the transistors, the resonant circuit is fully loaded on the equivalent input device impedance. However, if the gain capability of the active device is sufficiently high at the operating frequency, an optimum partial connection of the device into the resonant circuit can improve the noise performance and also makes it possible to achieve higher signal amplitude while avoiding breakdown and keeping approximately the same tuning range. Figure 9.36 shows the circuit schematic of the differential VCO with tapped resonant circuit inductor. As expected, the phase noise will be reduced by 6 dB using a taping ratio 1:1. Being fabricated in a SiGe process, such a differential VCO with a tapped inductor and the varactors represented by collector–base junctions with a hyperabrupt collector doping profile provides a frequency tuning range of 23% around 2.4 GHz at a supply voltage of 2.5 V with current consumption of 16.5 mA and phase noise of −128 dBc/Hz at 1 MHz offset [36].

An additional active device can be used to compensate for the loss in the oscillation system. For example, by using the inverted common collector transistor configuration, a stable negative resistance is generated, extracting the additional energy into the resonant circuit compensating for the loss due to the load resistance. This results in a significant increase in the loaded quality factor of the resonant circuit. Figure 9.37 shows the schematic diagrams of the oscillators (a) without compensation circuit and (b) with compensation circuit [37]. The soft start-up and

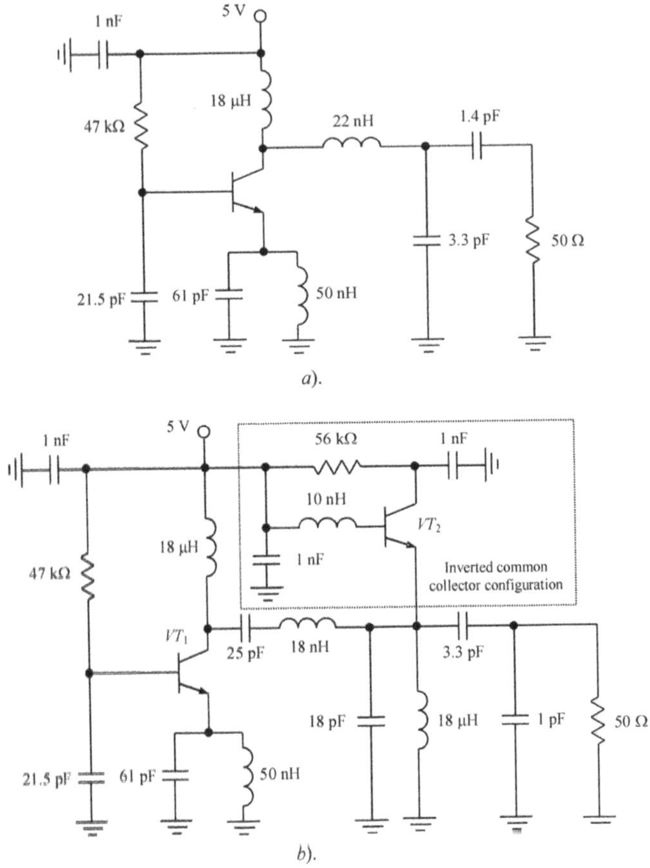

Figure 9.37 Schematic diagrams of oscillator (a) without and (b) with compensation circuit (permission by IET)

steady-state oscillation conditions are realized by using the capacitive feedback reactance in the base and inductive reactances in the emitter and collector circuits. Both circuits have the same terminal impedances at the emitter and base ports, but different in the collector port. After incorporating the negative resistance circuit based on a common collector transistor VT_2, the collector of the main transistor VT_1 sees the inductive impedance with the resistive part approaching zero. At oscillation frequency of 800 MHz with output power of -6.8 dBm, the oscillator with compensation negative resistance circuit provides 6 dB phase noise improvement at 10 kHz offset.

9.6 NONLINEAR FEEDBACK LOOP NOISE SUPPRESSION*

There is a real possibility of improving the oscillator phase noise significantly by using a special nonlinear feedback loop which can create the proper phase conditions for output noise

* This section is based on the extended version presented by courtesy of T. Banky and T. Berceli from Budapest University of Technology and Economics [38].

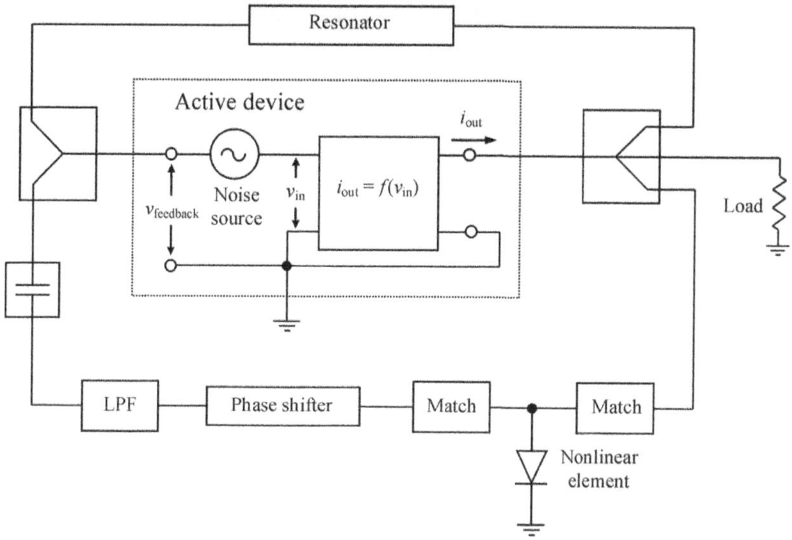

Figure 9.38 Block diagram of microwave oscillator with nonlinear feedback loop

components, the mixing effect of which contributes to the subtraction of the noise components around the fundamental. Figure 9.38 shows a block diagram of the microwave oscillator with an additional nonlinear feedback. The resonator can be represented by a filter or any configuration of the lumped or distributed *LC* elements that can provide the proper conditions for the soft start-up and steady-state oscillation conditions. An active device is shown as a noise-free nonlinear two-port network with a low-noise voltage source connected to its input. A nonlinear feedback path includes a diode as a nonlinear element, two matching circuits to match its input and output impedances, phase shifter, low-pass filter to suppress the fundamental, second and higher-order harmonics and dc blocking capacitor.

For the sake of the simplicity of the analytical representation, let us consider only one low-noise component. Then, the entire signal entering the nonlinear two-port network can be written as

$$v_{in} = V_0 + V_{in} \cos \omega t + V_{1/f} \cos(\Omega t + \Phi) \tag{9.21}$$

where V_0 is the dc bias voltage, V_{in} and ω are the voltage amplitude and frequency of the self-oscillations, $V_{1/f}$ and Ω are the voltage amplitude and frequency of the low-noise component with initial phase Φ, respectively.

The transfer characteristic of the nonlinear two-port network can be represented by the power series

$$i_{out} = a_0 + a_1 v_{in} + a_2 v_{in}^2 + \ldots + a_n v_{in}^n \tag{9.22}$$

where $i_{out}(t)$ is the output collector or drain current and $v_{in}(t)$ is the voltage at the active device input. For a polynomial representation of the transfer function, it is convenient to apply a power-series analysis, which is relatively easy to use and which gives a good intuitive sense of the nonlinear behaviour of the active device.

Substituting Equation (9.21) into Equation (9.22) and using trigonometric identities, for the dc (constant), first (linear), second (quadratic) and third (cubic) voltage terms, we can write

$$\begin{aligned}
i_{\text{out}} &= a_0 + a_1 V_0 + a_2 V_0^2 + a_3 V_0^3 + \frac{1}{2} a_2 \left(V_{\text{in}}^2 + V_{1/f}^2 \right) \\
&+ \left[a_1 + \frac{3}{4} a_3 \left(V_{\text{in}}^2 + 2V_{1/f}^2 \right) \right] V_{\text{in}} \cos \omega t + \left[a_1 + \frac{3}{4} a_3 \left(V_{1/f}^2 + 2V_{\text{in}}^2 \right) \right] V_{1/f} \cos \Psi \\
&+ \frac{1}{2} a_2 V_{\text{in}}^2 \cos 2\omega t + \frac{1}{2} a_2 V_{1/f}^2 \cos 2\Psi + \frac{1}{4} a_3 V_{\text{in}}^3 \cos 3\omega t + \frac{1}{4} a_3 V_{1/f}^3 \cos 3\Psi \\
&+ a_2 V_{\text{in}} V_{1/f} [\cos(\omega t + \Psi) + \cos(\omega t - \Psi)] \\
&+ \frac{3}{4} a_3 V_{\text{in}}^2 V_{1/f} [\cos(2\omega t + \Psi) + \cos(2\omega t - \Psi)] \\
&+ \frac{3}{4} a_3 V_{\text{in}} V_{1/f}^2 [\cos(2\Psi + \omega t) + \cos(2\Psi - \omega t)]
\end{aligned} \tag{9.23}$$

where $\Psi = \Omega t + \Phi$. For simplicity of numerical calculation, the values of a_0 and V_0 are considered zero, whereas the second and third orders of low-noise voltage components, $V_{1/f}^2$ and $V_{1/f}^3$, are assumed to be negligible compared with others. It can be seen from Equation (9.23) that the coefficients a_1 and a_3 have a significant effect on the amplitude of the fundamental component of the output current whereas the amplitudes of sidebands near fundamental ($\omega + \Omega$ and $\omega - \Omega$) are functions of coefficient a_2.

By converting all components, including the noise sidebands down to the baseband again through a properly set nonlinearity, we can obtain a signal identical or very close to identical to the signal that was produced by the original noise source. To properly design the feedback loop, it is necessary to provide the following experimental verification:

- nonlinear measurements for coefficients a_n;
- choosing required elements for the power splitter and feedback *LC*-filter for a particular oscillation frequency;
- choosing a type of nonlinearity (proper active element in a nonlinear feedback loop) in accordance with values of the coefficients a_n;
- phase measurements to choose the proper phase shifter needed to achieve an inverted noise signal to add to the signal at the active device input.

It is very important to make a correct choice of the nonlinear element in a nonlinear feedback loop for effective compensation of the noise components. The feedback signal from the nonlinear element can be written as

$$v_{\text{feedback}} = b_0 + b_1 i_{\text{out}} + b_2 i_{\text{out}}^2 + \ldots + b_k i_{\text{out}}^k \tag{9.24}$$

where $i_{\text{out}}(t)$ is the output collector or drain current defined by Equation (9.23).

According to Equation (9.23), the new frequency components are produced as the mixing products of the terms appearing in Equation (9.23). Consequently, it is necessary to investigate these products to be able to determine the required optimum values for coefficients b_n. In this case, it is enough to take those products that contain less than second power of the voltage amplitude $V_{1/f}$ representing low frequencies (since this signal is passed through a low-pass filter before entering the active device again).

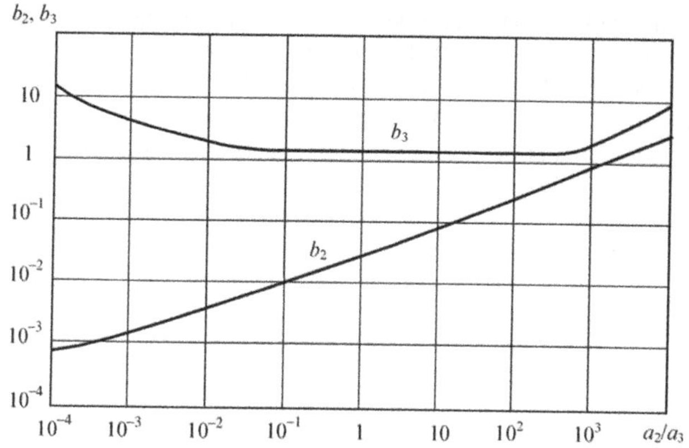

Figure 9.39 Feedback nonlinear coefficients b_2 and b_3 versus active device nonlinearity

Figure 9.39 shows a diagram defining the feedback nonlinear coefficients b_2 and b_3 through the ratio of the active device transfer function coefficients a_2/a_3. Here, we can see that, for an active device with strong third-order nonlinearity when $a_2 \ll a_3$, it is necessary to use the nonlinear element in a feedback loop with strong third-order nonlinearity when $b_2 \ll b_3$ as well. A limiting antiparallel diode pair can easily realize such a nonlinearity, for example. For the other side of diagram where $a_2 \gg a_3$ (when sidebands have sufficiently high amplitudes which is not typical), it is necessary to provide an active element with strong second- and third-order nonlinearities at the same time, which can not be effectively realized.

The experimental verification was performed using a 5-GHz microstrip oscillator based on ATF36077 pHEMT device. For this type of a transistor, the gate-source capacitance and drain current source are the major contributors to the low-frequency flicker noise, whereas the main nonlinearity is provided by a nonlinear transconductance. From the measured active device nonlinearity, a value of 0.095 was found for the ratio a_2/a_3. As a result, to provide a third-order nonlinearity in the feedback loop, the two diodes HSMS-8002 were connected in antiparallel positions. The measured phase noise shows 5 dB improvement compared with 7-dB improvement obtained by simulation. This difference can be caused by the fact that only the noise source at the active device input was taken into account, and the expected conversion loss in the feedback loop is higher.

The major feature of this technique can be clearly seen from the simulation and measurement results: the noise suppression effect is working over the entire noise bandwidth of interest, since the conversion and phase shift have constant values for all noise components throughout the frequency offset region (from kHz to MHz). Also, it is convenient in circuit implementation, as utilization (or later addition) of the low frequency feedback loop has a negligible effect on the amplitude and frequency of the self-oscillations. In addition, for a transistor configuration with common emitter, there is no need to use a phase shifter in the feedback loop since such a configuration provides the required phase inversion of 180° for a low-frequency signal.

REFERENCES

1. G. L. Matthaei, L. Young and E. M. T. Jones, *Microwave Filters, Impedance-Matching Networks, and Coupling Structures*. New York: McGraw-Hill (1980).
2. Serenade Design Suite, *Examples Volume, Version 7.5*, Ansoft Corporation, New Jersy (1998).

3. K. S. Ang, M. J. Underhill and I. D. Robertson, 'Transmission-Line Stabilized Monolithic Oscillator', *IEEE Trans. Microwave Theory Tech.*, **MTT-49**, 395–398 (2001).
4. K. S. Ang, M. J. Underhill and I. D. Robertson, 'Balanced Monolithic Oscillators at *K*- and *Ka*-Band', *IEEE Trans. Microwave Theory Tech.*, **MTT-48**, 187–193 (2000).
5. J. K. A. Everard, 'A Review of Low Noise Oscillator: Theory and Design', *Proc. IEEE Int. Frequency Control Symp.*, 909–918 (1997).
6. J. K. A. Everard, 'Minimum Sideband Noise in Oscillators', *Proc. 40th Annual Frequency Control Symp.*, 336–339 (1986).
7. J. K. A. Everard and K. K. M. Cheng, 'Low Noise Transmission Line Oscillators', *Proc. 2nd Int. Frequency Control and Synthesis Conf.*, 58–60 (1989).
8. J. K. A. Everard and K. K. M. Cheng, 'Novel Low-Noise 'Fabry-Perot' Transmission Line Oscillator', *Electronics Lett.*, **25**, 1106–1108 (1989).
9. K. Hosoya, S. Tanaka, Y. Amamiya, T. Niva and H. Shimawaki, 'A Low Phase-Noise 18-GHz HBT Oscillator Utilizing a ($\lambda/4 \pm \delta$) Open Stubs Resonator', *Proc. 1999 Asia-Pacific Microwave Conf.*, 64–67 (1999).
10. K. Hosoya, S. Tanaka, Y. Amamiya, T. Niva and H. Shimawaki, 'A Low Phase-Noise 38-GHz HBT Oscillator Utilizing a Novel Transmission Line Resonator', (2000) *IEEE MTT-S Int. Microwave Symp. Dig.*, 47–50 (2000).
11. M. Prigent and J. Obregon, 'Phase Noise Reduction in FET Oscillators by Low-Frequency Loading and Feedback Circuitry Optimization', *IEEE Trans. Microwave Theory Tech.*, **MTT-35**, 349–352 (1987).
12. M. S. Keshner, '$1/f$ Noise', *Proc. IEEE*, **70**, 212–218 (1982).
13. A. N. Riddle and R. J. Trew, 'A New Method of Reducing Phase Noise in GaAs FET Oscillators', (1984) *IEEE MTT-S Int. Microwave Symp. Dig.*, 274–276 (1984).
14. V. Gunderich, J. Deuringer, W. Anzill and P. Russer, 'Phase Noise Reduction of Microwave Oscillators by Optimization of the Dynamic Behaviour', (1994) *IEEE MTT-S Int. Microwave Symp. Dig.*, 953–956 (1994).
15. A. M. Darwish, A. K. Ezzeddine, H.-L. A. Hung and F. R. Phelleps, 'A New Phase Noise Reduction Technique for MMIC Oscillators', (1992) *IEEE MTT-S Int. Microwave Symp. Dig.*, 463–466 (1992).
16. J.-H. C. Zhan, K. Maurice and J. Duster and K. T. Kornegay, 'Analysis and Design of Negative Impedance *LC* Oscillators Using Bipolar Transistors', *IEEE Trans. Circuits and Systems — I: Fundamental Theory App.*, **CAS-I-50**, 1461–1464 (2003).
17. D. J. Healy III, 'Flicker of Frequency and Phase and White Frequency and Phase Fluctuations in Frequency Sources', *Proc. 26th Annual Frequency Control Symp.*, 43–49 (1972).
18. U. L. Rohde, C.-R. Change and J. Gerber, 'Design and Optimization of Low-Noise Oscillators Using Nonlinear CAD Tools', *Proc. 1994 IEEE Int. Frequency Control Symp.*, 548–554 (1994).
19. U. L. Rohde and G. Klage, 'Analyze VCOs and Fractional-N Synthesizers', *Microwaves & RF*, **39**, 57–78 (2000).
20. U. L. Rohde and F. Hagemeyer, 'Feedback Technique Improves Oscillator Phase Noise', *Microwaves & RF*, **37**, 61–70 (1998).
21. U. L. Rohde, 'A Novel RFIC for UHF Oscillators', *2000 IEEE RFIC Symp. Dig.*, 53–56 (2000).
22. Y. Lin, K. H. To, J. S. Hamel and W. M. Huang, 'Fully Integrated 5 GHz CMOS VCOs with On Chip Low Frequency Feedback Circuit for $1/f$ Induced Phase Noise Suppression', *Proc. 2002 Europ. Solid-State Circuits Conf.*, 551–554 (2002).
23. E. Hegazi, H. Sjoland and A. A. Abidi, 'A Filtering Technique to Lower *LC* Oscillator Phase Noise', *IEEE J. Solid-State Circuits*, **SC-36**, 1921–1930 (2001).
24. H. Kim, S. Ryu, Y. Chung, J. Choi and B. Kim, 'A Low Phase-Noise CMOS VCO with Harmonic Tuned *LC* Tank', *IEEE Trans. Microwave Theory Tech.*, **MTT-54**, 2917–2924 (2006).
25. A. Ismail and A. A. Abidi, 'CMOS Differential *LC* Oscillator with Suppressed Up-Converted Flicker Noise', *2003 IEEE Int. Solid-State Circuits Sym. Dig.*, 98–99 (2003).

26. Xu. Wang and Xi. Wang, Cross-Coupled Voltage Controlled Oscillator with Improved Phase Noise Performance, U.S. Patent 6,680,657 (January 2004).
27. P. Andreani and H. Sjoland, 'Tail Current Noise Suppression in RF CMOS VCOs', *IEEE J. Solid-State Circuits*, **SC-37**, 342–348 (2002).
28. A. Grebennikov, *RF and Microwave Power Amplifier Design*, New York: McGraw-Hill (2004).
29. C. Samori, A. Zanchi, S. Levantino and A. L. Lacaita, 'A Fully-Integrated Low-Power Low-Noise 2.6-GHz Bipolar VCO for Wireless Applications', *IEEE Microwave and Wireless Components Lett.*, **11**, 199–201 (2001).
30. R. Aparicio and A. Hajimiri, 'A Noise-Shifting Differential Colpitts VCO', *IEEE J. Solid-State Circuits*, **SC-37**, 1728–1736 (2002).
31. M. A. Margarit, J. L. Tham, R. G. Meyer and M. J. Deen, 'A Low-Noise, Low-Power VCO with Automatic Amplitude Control for Wireless Applications', *IEEE J. Solid-State Circuits*, **SC-34**, 761–771 (1999).
32. H. Zirath, R. Kozhuharov and M. Ferndahl, 'Balanced Colpitts Oscillator MMICs Designed for Ultra-Low Phase Noise', *IEEE J. Solid-State Circuits*, **SC-40**, 2077–2086 (2005).
33. G. Braun and H. Lindenmeier, 'Transistor Oscillators with Impedance Noise Matching', *IEEE Trans. Microwave Theory Tech.*, **MTT-39**, 1602–1610 (1991).
34. A. S. Luchinin, 'Low-Noise Bipolar Oscillator Design (in Russian)', *Izvestiya Vuzov USSR, Ser. Radioelektronika*, **30**, 3–8 (1987).
35. J. K. McKinney, P. J. Yeh and B. Avanic, Cascode Oscillator Having Optimum Phase Noise and Bandwidth Performance, U.S. Patent 5,621,362 (April 1997).
36. P. W. Lai, L. Dobos and S. Long, 'A 2.4 GHz SiGe Low Phase-Noise VCO Using On Chip Tapped Inductor', *Proc. 2003 Europ. Solid-State Circuits Conf.*, 505–508 (2003).
37. V. M. T. Lam and P. C. L. Yip, 'Microwave Oscillator Phase Noise Reduction Using Negative Resistance Compensation', *Electronics Lett.*, **29**, 379–381 (1993).
38. T. Banky and T. Berceli, 'Investigations on Noise Suppression Effects of Nonlinear Feedback Loops in Microwave Oscillators', (2004) *IEEE MTT-S Int. Microwave Symp. Dig.*, 2015–2018 (2004).

Index

Printed and bound by CPI Group (UK) Ltd, Croydon, CR0 4YY

16/09/2026

14963387-0004